Lecture Notes in Physics

Springer-Verlag Berlin Heidelberg GmbH

Physics and Astronomy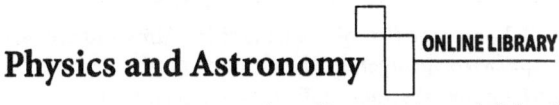

http://www.springer.de/phys/

Editorial Policy

The series *Lecture Notes in Physics* (LNP), founded in 1969, reports new developments in physics research and teaching -- quickly, informally but with a high quality. Manuscripts to be considered for publication are topical volumes consisting of a limited number of contributions, carefully edited and closely related to each other. Each contribution should contain at least partly original and previously unpublished material, be written in a clear, pedagogical style and aimed at a broader readership, especially graduate students and nonspecialist researchers wishing to familiarize themselves with the topic concerned. For this reason, traditional proceedings cannot be considered for this series though volumes to appear in this series are often based on material presented at conferences, workshops and schools (in exceptional cases the original papers and/or those not included in the printed book may be added on an accompanying CD ROM, together with the abstracts of posters and other material suitable for publication, e.g. large tables, colour pictures, program codes, etc.).

Acceptance

A project can only be accepted tentatively for publication, by both the editorial board and the publisher, following thorough examination of the material submitted. The book proposal sent to the publisher should consist at least of a preliminary table of contents outlining the structure of the book together with abstracts of all contributions to be included.

Final acceptance is issued by the series editor in charge, in consultation with the publisher, only after receiving the complete manuscript. Final acceptance, possibly requiring minor corrections, usually follows the tentative acceptance unless the final manuscript differs significantly from expectations (project outline). In particular, the series editors are entitled to reject individual contributions if they do not meet the high quality standards of this series. The final manuscript must be camera-ready, and should include both an informative introduction and a sufficiently detailed subject index.

Contractual Aspects

Publication in LNP is free of charge. There is no formal contract, no royalties are paid, and no bulk orders are required, although special discounts are offered in this case. The volume editors receive jointly 30 free copies for their personal use and are entitled, as are the contributing authors, to purchase Springer books at a reduced rate. The publisher secures the copyright for each volume. As a rule, no reprints of individual contributions can be supplied.

Manuscript Submission

The manuscript in its final and approved version must be submitted in camera-ready form. The corresponding electronic source files are also required for the production process, in particular the online version. Technical assistance in compiling the final manuscript can be provided by the publisher's production editor(s), especially with regard to the publisher's own Latex macro package which has been specially designed for this series.

Online Version/ LNP Homepage

LNP homepage (list of available titles, aims and scope, editorial contacts etc.):
http://www.springer.de/phys/books/lnpp/
LNP online (abstracts, full-texts, subscriptions etc.):
http://link.springer.de/series/lnpp/

H. M. J. Boffin   D. Steeghs   J. Cuypers   (Eds.)

Astrotomography

Indirect Imaging Methods
in Observational Astronomy

Springer

Editors

Henri M.J. Boffin
Jan Cuypers
Royal Observatory of Belgium
Avenue Circulaire, 3
1180 Brussels, Belgium

Danny Steeghs
Department of Physics and Astronomy
University of Southampton
Highfield
SO17 1BJ Southampton, United Kingdom

Cover Picture: Spiral arms in the accretion disc of a close binary system as obtained from numerical simulations (see H.M.J. Boffin, p. 69) and the emission-line profiles such a disc would produce when observed at different orbital phases of the binary system.

Library of Congress Cataloging-in-Publication Data.

Die Deutsche Bibliothek - CIP-Einheitsaufnahme

Astrotomography : indirect imaging methods in observational astronomy / H. Boffin ... (ed.).

(Lecture notes in physics ; 573)
(Physics and astronomy online library)
ISBN 978-3-662-14317-9   ISBN 978-3-540-45339-0 (eBook)
DOI 10.1007/978-3-540-45339-0

ISSN 0075-8450

Typesetting: Camera-ready by the authors/editors
Camera-data conversion by Steingraeber Satztechnik GmbH Heidelberg
Cover design: *design & production*, Heidelberg

Printed on acid-free paper
SPIN: 10834370   54/3141/du - 5 4 3 2 1 0

Preface

The idea for this book originated in the La Silla observatory where two of the editors were doing phase-resolved spectroscopy of some cataclysmic variable stars using the NTT. We realized that although indirect imaging techniques such as eclipse mapping and Doppler tomography had been around for more than a decade and had provided some of the most interesting discoveries, no book existed which covered these techniques. Moreover, no colloquium had ever been organized specifically on these topics. The implementation of tomographic methods in astrophysics, in order to probe structures on angular scales of micro-arcseconds, started about 15 years ago with the development of the eclipse mapping method. This method is able to reconstruct light distributions in eclipsing binaries by exploiting the regular obscuration of the light source by one of the binary components. A similar approach to regularised data fitting lead to a variety of related methods in order to resolve light distributions of the accretion flows in binaries, the surface structures of stars and the inner regions of active galaxies. The scientific output of these methods is considerable and they are increasingly becoming versatile tools for a wide community of researchers.

A specialised workshop seemed highly desirable, so we decided to organise the first international workshop on astrotomography. The idea of the meeting, which took place in Brussels in early July 2000, was to bring together researchers sharing an interest in applying indirect imaging methods in astronomy, and to compare the methods used in different fields. During the meeting, a large amount of time was devoted to extensive reviews of the various reconstruction techniques. In conjunction with the reviews, short contributed talks highlighted recent results and developments. Due to the small number of participants, 60, there was plenty of opportunity for discussion and interaction. Moreover, we wanted that the proceedings of this meeting could be used as a handbook on these methods. The reviewers were therefore asked to provide extensive accounts of their field. The proceedings thus consist of 13 reviews of about 25 pages each as well as 15 contributed talks of 6~8 pages. A wide range of topics are discussed, mostly on the properties of accretion flows in semi-detached binary systems containing a compact stellar remnant. Other topics include the surface and magnetic field structure of single stars, the shock waves of Mira stars, the accretion flows around black holes in binaries and active galactic nuclei and the structure of Algol systems. The large variety of subjects covered is a clear illustration of the importance that indirect imaging techniques have gained in astrophysics. A new generation of optical telescopes and spectrographs is coming on-line which will push the possibilities of indirect imaging even further. In conjunction with that, specialised instruments and projects on existing telescopes will deliver data sets with high time and wavelength resolutions tailored for accurate mapping experiments. We hope that these proceedings will provide a helpful overview for any researcher interested in such techniques. With the same spirit of producing more than just proceedings, we also include a list of some useful resources on the Internet. We also hope that the web page of the workshop will be kept alive and become a useful reference on astrotomography.

We would like to thank all the participants for making this workshop a success, and in particular all the contributing review authors for having generously agreed to come to the meeting at their own expense, and for their efforts in providing a balanced set of review papers. Many thanks to all the members of the local organising committee for the hard work before, during and after the workshop. The Brussels Planetarium is thanked for providing us with a meeting venue and excellent support. We also wish to thank the Director of the Royal Observatory of Belgium, Prof. Paul Pâquet, for his efforts. Rob Hynes provided us with a superb 'scientific impression' of an interacting binary that featured on the workshop poster and various other locations. Finally, we are grateful for financial support from project G.0265.97 of the Research Programme of the Fund for Scientific Research – Flanders (F.W.O. – Vlaanderen).

Brussels, Southampton,   *Henri Boffin,*
November 2000   *Danny Steeghs,*
                                                          Jan Cuypers

Workshop webpage: http://www.astro.oma.be/DopplerWorkshop/

Contents

Contents

List of Contributors

Timothy M.C. Abbott
Nordic Optical Telescope
Apartado 474
38700 Santa Cruz de La Palma
Canary Islands, Spain
tabbot@not.iac.es

Rodrigo Alvarez
Institut d'Astronomie et
d'Astrophysique
Université Libre de Bruxelles
C.P. 226, Boulevard du Triomphe
1050 Brussels, Belgium
Rodrigo.Alvarez@oma.be

Raymundo Baptista
Departemento de Física, UFSC
Campus Trinidade
88040-9000 Florianópolis, Brazil
bap@fsc.ufsc.br

John R. Barnes
Centro de Astrofisica
da Universidade do Porto
Rua das Estrelas
4150 Porto, Portugal
jrb@astro.up.pt

Henri M.J. Boffin
Royal Observatory of Belgium
3 av. Circulaire
1180 Brussels, Belgium
Henri.Boffin@oma.be

David A.H. Buckley
South African Astronomical
Observatory
PO Box 9, Observatory 7935
Cape Town, South Africa
dibnob@saao.ac.za

Elsa Bertino
South African Astronomical
Observatory
PO Box 9, Observatory 7935
Cape Town, South Africa
eb@saao.ac.za

Kim B. Bruce
Mullard Space Science Laboratory
University College London
Holmbury St Mary, Dorking, UK
kbb@mssl.ucl.ac.uk

Andrew Collier Cameron
School of Physics and Astronomy
University of St Andrews
Scotland KY16 9SS, UK
acc4@st-andrews.ac.uk

Jorge Casares
Instituto de Astrofisica de Canarias
38200 La Laguna
Tenerife, Spain
jcv@ll.iac.es

Craig Chambers
Mullard Space Science Laboratory
University College London
Holmbury St Mary, Dorking, UK
cch@mssl.ucl.ac.uk

Philip A. Charles
Department of Physics and Astronomy
University of Southampton
Southampton, SO17 1BJ, UK
pac@astro.soton.ac.uk

Mark Cropper
Mullard Space Science Laboratory
University College London
Holmbury St Mary, Dorking, UK
msc@mssl.ucl.ac.uk

Vik S. Dhillon
Department of Physics and Astronomy
University of Sheffield
Sheffield S3 7RH, UK
vik.dhillon@sheffield.ac.uk

Jean-François Donati
Observatoire Midi-Pyrénées
14 Avenue E. Belin
31400 Toulouse, France
jean-francois.donati@obs-mip.fr

André Fokin
Institute for Astronomy of the Russia
Academy of Sciences
48 Pjatnitskaja
109017 Moscow, Russia
fokin@inasan.rssi.ru

Hidekazu Fujiwara
Department of Earth and Planetary
Sciences, Kobe University
Nada-ku
Kobe 657-8501, Japan
fujiwara@jet.planet.sci.kobe-u.ac.jp

Denis Gillet
Observatoire de Haute-Provence
04870 Saint-Michel l'Observatoire,
France
gillet@obs-hp.fr

Petr Hadrava
Astronomical Institute of the
Academy of Sciences of the Czech
Republic, 251 65 Ondřejov,
Czech Republic
had@sunstel.asu.cas.cz

Pasi Hakala
Observatory and Astrophysics
Laboratory
FIN-00014, University of Helsinki
Finland
Pasi.Hakala@astro.utu.fi

Reinhard Hanuschik
ESO
Karl-Schwarzschild-Str. 2
85748 Garching, Germany
rhanusch@eso.org

Emilios T. Harlaftis
Institute of Astronomy and Astro-
physics
National Observatory of Athens
P.O. Box 20048, Thession
Athens - 11810, Greece
ehh@astro.noa.gr

Carole A. Haswell
Department of Physics and Astronomy
The Open University
Walton Hall, Milton Keynes
MK7 6AA, UK
C.A.Haswell@open.ac.uk

Herman Hensberge
Royal Observatory of Belgium
Ringlaan 3
1180 Brussel, Belgium
Herman.Hensberge@oma.be

Keith Horne
University of St Andrews
Scotland KY16 9SS, UK
kdh1@st-andrews.ac.uk

Robert I. Hynes
Department of Physics and Astronomy
University of Southampton
Southampton, SO17 1BJ, UK
rih@astro.soton.ac.uk

S. Ilijïc
Faculty of Geodesy
University of Zagreb
Kačiiceva 26
10000 Zagreb, Croatia
silijic@geodet.geof.hr

Alain Jorissen
Institut d'Astronomie et
d'Astrophysique, Université
Libre de Bruxelles,
C.P. 226, Boulevard du Triomphe,
1050 Brussels, Belgium
ajorisse@astro.ulb.ac.be

Oleg Kochukhov
Uppsala Astronomical Observatory
Box 515No.
75120 Uppsala, Sweden
Oleg.Kochukhov@astro.uu.se

U. Kolb
Department of Physics and
Astronomy, The Open University
Walton Hall, Milton Keynes
MK7 6AA, UK
U.C.Kolb@open.ac.uk

Jens Kube
Universitäts-Sternwarte Göttingen
Geismar Landstrasse 11
37073 Göttingen, Germany
jkube@uni-goettingen.de

J.D. Landstreet
Physics and Astronomy Department
The University of Western Ontario
London, Ontario
Canada N6A 3K7
jlandstr@uwo.ca

Makoto Makita
Department of Astronomy, Kyoto
University
Sakyo-ku
Kyoto 606-8502, Japan
makita@jet.planet.sci.kobe-u.ac.jp

Tom R. Marsh
Dpt of Physics and Astronomy
Southampton University
Highfield, Southampton SO17 1BJ
trm@astro.soton.ac.uk

Takuya Matsuda
Department of Earth and Planetary
Sciences, Kobe University
Nada-ku
Kobe 657-8501, Japan
matsuda@jet.planet.sci.kobe-u.ac.jp

Luisa Morales-Rueda
Dpt of Physics and Astronomy
Southampton University
Highfield, Southampton SO17 1BJ
lmr@astro.soton.ac.uk

Chris K. J. Moran
Dpt of Physics and Astronomy
Southampton University
Highfield, Southampton SO17 1BJ
ckjm@astro.soton.ac.uk

Rachel C. North
Dpt of Physics and Astronomy
Southampton University
Highfield, Southampton SO17 1BJ
rcn@astro.soton.ac.uk

Kieran O'Brien
University of St Andrews
Scotland KY16 9SS, UK
kso@st-and.ac.uk

J.M. Oliveira
ESA Space Science Department
SCI-SO/ESTEC, PB 299
2200 AG Noordwijk, The Netherlands
joliveir@estec.esa.nl

K. Pavlovski
Department of Physics
University of Zagreb
Bijenička 32
10000 Zagreb, Croatia
kpavlovski@geodet.geof.hr

Pascal Petit
Laboratoire d'Astrophysique
Observatoire Midi-Pyrénées
14 avenue Edouard Belin
31400 Toulouse, France
petit@ast.obs-mip.fr

Nikolai Piskunov
Uppsala Astronomical Observatory
Box 515No.
75120 Uppsala, Sweden
piskunov@astro.uu.se

Bertrand Plez
GRAAL,Université Montpellier II,
cc072
34095 Montpellier cedex 05, France
plez@graal.univ-montp2.fr

Stephen Potter
South African Astronomical
Observatory
P.O. Box 9, Observatory 7935
Cape Town, South Africa
sbp@sirius.saao.ac.za

Mercedes T. Richards
Department of Astronomy
University of Virginia
P.O. Box 3818, Charlottesville
VA 22903-0818, USA
mrichards@virginia.edu

Daniel J. Rolfe
Dpt of Physics and Astronomy
The Open University
Walton Hall
Milton Keynes, MK7 6AA
d.j.rolfe@open.ac.uk

Encarni Romero-Colmenero
South African Astronomical
Observatory
P.O. Box 9, Observatory 7935
Cape Town, South Africa
erc@saao.ac.za

Axel Schwope
Astrophysikalisches Institut Potsdam
An der Sternwarte 16
Potsdam 14482, Germany
aschwope@aip.de

S.L.S. Shorlin
Physics and Astronomy Department
The University of Western Ontario
London, Ontario
Canada N6A 3K7
sshorlin@astro.uwo.ca

T.A.A. Sigut
Physics and Astronomy Department
The University of Western Ontario
London, Ontario
Canada N6A 3K7
asigut@astro.uwo.ca

R.C. Smith
University of Sussex
Astronomy Centre
Brighton, BN1 9QJ, UK
rcs@star.cpes.susx.ac.uk

Danny Steeghs
Dpt of Physics and Astronomy
Southampton University
Southampton, SO17 1BJ, UK
ds@astro.soton.ac.uk

Rudi Stehle
University of Leicester
Astronomy Group
University Road, Leicester, LE1 7RH
rst@star.le.ac.uk

Claus Tappert
Dipartimento di Astronomia
Vicolo dell Ósservatorio 5
I-35122 Padova, Italy
claus1@sole.pd.astro.it

Sonja Vrielmann
Dept. of Astronomy
University of Cape Town
Rondebosch 7700, South Africa
sonja@pinguin.ast.uct.ac.za

G.A. Wade
Département de Physique
Université de Montréal

C.P.6128, Succ. Centre Ville
Montréal H3C 3J7, Canada
wade@astro.umontreal.ca

Christopher A. Watson
Dpt of Physics and Astronomy
University of Sheffield
Sheffield S3 7RH, UK
c.watson@sheffield.ac.uk

Graham Wynn
Dpt of Physics and Astronomy
University of Leicester
Leicester LE1 7RH, UK
gwy@star.le.ac.uk

Cristina Zurita
Instituto de Astrofisica de Canarias
38200 La Laguna
Tenerife, Spain
czurita@ll.iac.es

List of Participants

- **Raymundo Baptista** UFSC Trindade, Brazil
- **John Barnes** Universidade de Porto, Portugal
- **Henri Boffin** Royal Observatory of Belgium
- **Jan Cuypers** Royal Observatory of Belgium
- **Andrew Collier Cameron** University of St Andrews, UK
- **Jean-Pierre De Cuyper** Royal Observatory of Belgium
- **Vik Dhillon** University of Sheffield, UK
- **Jean-Francois Donati** Observatoire Midi Pyrenees, France
- **Lars Freyhammer** Royal Observatory of Belgium
- **Michael Goad** University of Leicester, UK
- **Paul Groot** Harvard Smithsonian CfA, USA
- **Petr Hadrava** Academy of Sciences of the Czech Republic
- **Pasi Hakala** Tuorla Observatory, Finland
- **Emilios Harlaftis** National Observatory of Athens, Greece
- **Herman Hensberge** Royal Observatory of Belgium
- **Frederic V. Hessman** Universitaets-Sternwarte Goettingen, Germany
- **Donald W. Hoard** Cerro Tololo Inter-American Observatory, Chile
- **Keith Horne** University of St Andrews, UK
- **Gaitee Hussain** University of St Andrews, UK
- **Robert Hynes** University of Southampton, UK
- **Sasa Ilijic** Zagreb University, Croatia
- **Alain Jorissen** Universite Libre de Bruxelles, Belgium
- **Pavel Koubsky** Ondrejov Observatory, Czech Repulic
- **Akiko Koyama** Kobe University, Japan
- **Jens Kube** Universitaets-Sternwarte Goettingen, Germany
- **Markus Kuster** Tuebingen, Germany
- **Patricia Lampens** Royal Observatory of Belgium
- **Stuart Littlefair** University of Sheffield, UK
- **Makoto Makita** Kobe University, Japan
- **Tom Marsh** Southampton University, UK
- **Takuya Matsuda** Kobe University, Japan
- **Ronald Mennickent** Universidad de Concepcion, Chile
- **Luisa Morales-Rueda** University of Southampton, UK
- **Vitaly Neustroev** Udmurt State University, Russia
- **Rachel North** University of Southampton, UK
- **Kieran O'Brien** University of St Andrews, UK
- **Manuel A. Perez-Torres** University College Cork, Ireland
- **Pascal Petit** Observatoire Midi-Pyrenees, France
- **Nikolai Piskunov** Uppsala Astronomical Observatory, Sweden
- **Stephen Potter** South African Astronomical Observatory
- **Gavin Ramsay** Mullard Space Science Lab, UK
- **Mercedes Richards** University of Virginia, USA
- **Pablo Rodriguez-Gil** IAC Tenerife, Spain
- **Daniel Rolfe** The Open University, UK

- **Robert Schwarz** Astrophysical Institute Potsdam, Germany
- **Axel Schwope** Astrophysical Institute Potsdam, Germany
- **Warren Skidmore** University of St Andrews, UK
- **Vallery Stanichev** Bulgarian Academy of Sciences
- **Danny Steeghs** University of Southampton, UK
- **Claus Tappert** Universita di Padova, Italy
- **Gaghik Tovmassian** IA UNAM, USA
- **Eduardo Unda** University of Southampton, UK
- **Sonja Vrielmann** University of Cape Town, SA
- **Christopher Watson** University of Sheffield, UK
- **Graham Wynn** University of Leicester, UK
- **Cristina Zurita** IAC, Spain

Some Useful Resources on the Internet

- http://www.astro.oma.be/DopplerWorkshop
  The web page of the workshop in which updated information will be available as well as useful links to astrotomography resources.

- http://www.astro.soton.ac.uk/~trm/software.html
  Software from Tom Marsh, including *doppler*, for doppler imaging of accretion discs, *molly* for 1D spectrum analysis, and *pamela*, for reduction from 2D to 1D astronomical spectra.

- http://ibm-2.MPA-Garching.MPG.DE/~henk/
  Henk Spruit preliminary web page, containing his fast Doppler mapping program.

- http://star-www.st-and.ac.uk/~kdh1/
  The minimalist web page of Keith Horne.
  In http://star-www.st-and.ac.uk/schedar/kdh1/doptom/doptom.html, a paper about Doppler Tomography can be found as well as the source code.

- http://sunkl.asu.cas.cz/~had/korel.html
  KOREL is a code for spectra disentangling using Fourier transforms, available from P. Hadrava.

- http://www.astro.soton.ac.uk/~trm/doppler_table.html
  Up-to-date list of publications using Doppler Tomography, maintained by Tom Marsh.

- http://www.astro.univie.ac.at/~kgs/research.html
  Home page of the stellar activity working group of the Institute for Astronomy at the University of Vienna. Includes an impressive collection of Doppler images of stars.

- http://www.shef.ac.uk/~phys/people/vdhillon/
  Home page of Vik Dhillon with some online presentations, including the one he gave in Brussels.

- http://www.astro.virginia.edu/people/faculty/mtr8r/index.html
  The web page of Mercedes T. Richards with information about doppler tomography of Algols and hydrodynamic simulations of mass transfer.

- http://star-www.st-and.ac.uk/~acc4/coolpages/imaging.html
  Mapping starspots of A. Collier Cameron with the slides of his presentation in Brussels and some eclipsing binaries star mapping movies.

- http://webast.ast.obs-mip.fr/people/donati/
  The animated homepage of J.-F. Donati.

- http://www.shef.ac.uk/~phys/people/vdhillon/ultracam/
  ULTRACAM is an ultra-fast, triple-beam CCD camera which has been designed to study one of the few remaining unexplored regions of observational parameter space – high temporal resolution. The camera, which has recently been funded in full (292 k) by PPARC, will see first light during 2001 and will be used on 2-m, 4-m and 8-m class telescopes in Australia, the Canary Islands, Chile, Greece, South Africa and Spain to study astrophysics on the fastest timescales. ULTRACAM is a project of V. Dhillon and T. Marsh.

- http://astro.esa.int/SA-general/Research/Detectors_and_optics/
  detectors_scam.html
  S-Cam is the prototype of a cryogenic camera for ground-based astronomy, based around a 6x6 array of Ta–Al superconducting tunnel junction (STJ) devices, photon-counting array detectors with intrinsic energy resolution. The detector presently provides individual photon arrival time accuracy to about 5 μs, and a wavelength resolution of about 60 nm at 500 nm, with each array element capable of counting up to $\sim$5000 photons s^{-1}.

- http://www.astro.soton.ac.uk/~rih/binsim.html
  Rob Hynes's binary star visualisation.

Doppler Tomography

T.R. Marsh

Department of Physics and Astronomy, Southampton University,
Highfield, Southampton SO17 1BJ

Abstract. I review the method of Doppler tomography which translates binary-star line profiles taken at a series of orbital phases into a distribution of emission over the binary. I begin with a discussion of the basic principles behind Doppler tomography, including a comparison of the relative merits of maximum entropy regularisation versus filtered back-projection for implementing the inversion. Following this I discuss the issue of noise in Doppler images and possible methods for coping with it. Then I move on to look at the results of Doppler Tomography applied to cataclysmic variable stars. Outstanding successes to date are the discovery of two-arm spiral shocks in cataclysmic variable accretion discs and the probing of the stream/magnetospheric interaction in magnetic cataclysmic variable stars. Doppler tomography has also told us much about the stream/disc interaction in non-magnetic systems and the irradiation of the secondary star in all systems. The latter indirectly reveals such effects as shadowing by the accretion disc or stream. I discuss all of these and finish with some musings on possible future directions for the method. At the end I include a tabulation of Doppler maps published in refereed journals.

1   Introduction

Many rapidly rotating single and binary stars change little during the course of a single rotation or orbit. The spots on single stars can persist for many days, while cataclysmic variable stars may stay in outburst for over 100 orbits and in quiescence for ten times longer still. However, for the observer, orbital rotation can cause considerable variability both in flux and spectra. This arises from a combination of changes in aspect angle and visibility, caused by geometrical effects, and the rotation of all velocity vectors with the binary orbit. These effects are a blessing and a curse: without them we would know considerably less than we do about such stars, however the complex variability can be hard to interpret.

The method of Doppler tomography was developed to unravel the emission line variations of cataclysmic variable stars (CVs) [48]. CVs are short period binary stars, with orbital periods typically between 1.5 and 10 hours, which are beautifully set up to allow us to study accretion. The stellar components of the binary, a white dwarf and a low-mass main-sequence star, are faint, and their semi-detached configuration means that the geometry is entirely specified by the mass ratio and orbital inclination alone. Unfortunately, CVs are far too small to be resolved directly – they typically subtend $< 10^{-4}$ seconds of arc at Earth – and we can learn nothing of their structure from direct imaging. Instead we must turn to more indirect methods. Two key methods are "eclipse mapping", introduced by Horne [39] and reviewed in this volume by Baptista, and "Doppler tomography", the subject of this chapter. Eclipse mapping relies on the geometrical information contained in eclipse light curves; Doppler Tomography uses the velocity information contained in Doppler-shifted light curves.

In this paper I detail the principles behind Doppler tomography and hope to give the reader a full picture of what is now a widely-applied tool in the field. I follow the fundamentals with a short section on accounting for stochastic noise, which has usually been ignored to date, before finally moving onto a survey of results. Although I will attempt to be as self-contained as possible, the subject has become too large to cover every application of Doppler tomography and so I instead focus upon cataclysmic variable stars. I start this endevour with a potted history of the development of Doppler tomography.

2 History

Although not restricted in application to CVs, since Doppler tomography was developed for CVs and has so far mostly been applied to them (with several honourable exceptions covered elsewhere in this volume), for completeness I start by describing our standard picture of CVs. Fig. 1 shows a schematic representation of our model of non-magnetic CVs with a white dwarf surrounded by a flat disc, orbiting a tidally-distorted main-sequence star. In addition a stream of matter flows from the main sequence-star and hits the disc in a spot. All of these components, as well as others, have been seen in Doppler images. In eclipsing CVs, as seen in Fig 1, the main-sequence star occults the accreting regions allowing us to locate the chief sources of emission. It was precisely this that led to the

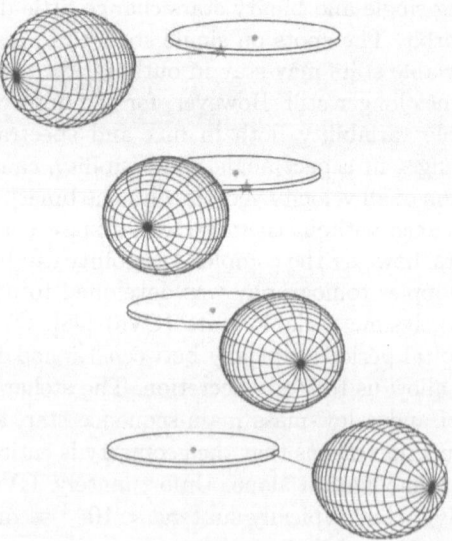

Fig. 1. A schematic illustration of a cataclysmic variable viewed at four orbital phases.

development of the standard model in the early 1970s [89,72]. In systems with compact emission sources, the sharp ingress and egress features during eclipse allow the source location to be pin-pointed, and thus, for instance, it is possible to determine the binary mass ratio from the observed position of the gas stream impact [73,94]. What about more diffuse emission – the disc for instance? An obvious possibility is to fit a parameterised model of the disc emission to light-curves. However this is plagued by our lack of good *a priori* models. For instance, a symmetric distribution following a power-law in radius is a simple and obvious choice for modelling discs, but are discs really symmetric and what if they do not follow power-laws? These problems were solved by Keith Horne who during his thesis work introduced the powerful idea of regularised fitting to the study of CVs [39]. Horne modelled the accretion regions with a grid of many independent pixels, effectively giving a model of great flexibility. To escape the degeneracy engendered by such an approach, Horne selected the image of "maximum entropy" where the entropy S is given (in the simplest case) by

$$S = -\sum_{i=1}^{M} p_i \ln p_i. \tag{1}$$

Here p_i is given by

$$p_i = \frac{I_i}{\sum_{j=1}^{M} I_j}, \tag{2}$$

where I_i is the image value assigned to pixel i. This is the method of eclipse mapping.

In eclipse mapping a one-dimensional light curve leads to two-dimensional map. Emission lines on the other hand, as I will explain, contain the extra dimension of velocity and so can give better constrained images (albeit with some disadvantages). Horne realised that the formation of the line profile from a disc was analogous to the formation of medical X-rays used in computed tomography to image the human brain. This led to the development of Doppler tomography [48], implemented in the first instance in a way analogous to eclipse mapping, using maximum entropy regularisation.

Since that time, and following articles focussing upon the more accessible filtered back-projection inversion [40,62], the use of Doppler tomography has exploded, with over 100 refereed publications making use of it or containing theoretical simulations of Doppler images (Fig. 2). Doppler maps have been published for some 16 dwarf novae, 13 AM Her stars, 11 DQ Her stars, 16 nova-like variables and 5 black-hole systems, along with other types such as Algols and Super-Soft sources. Doppler tomography is approaching the status of a standard tool; I now begin discussion of its underlying principles.

3   The Principles of Doppler Tomography

Doppler tomography arose from a desire to interpret the emission-line profile variations of accretion discs. Past papers on tomography have started by describing the formation of the double-peaked profiles from discs [48]. However,

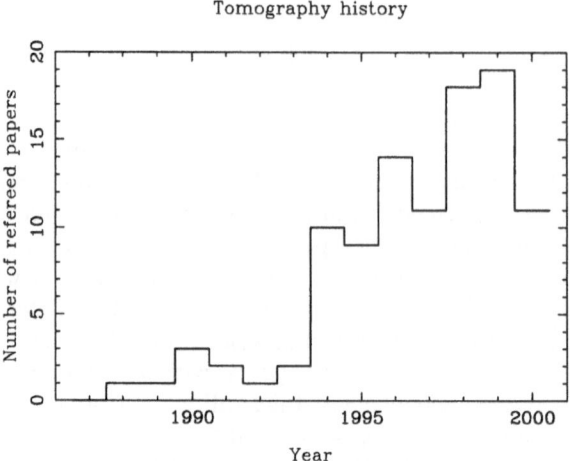

Fig. 2. Numbers of refereed publications using Doppler tomography versus year of publication.

the existence of a disc is not necessary for Doppler tomography. For instance, one of the most spectacular applications of tomography has been to the AM Her stars or polars [67], systems in which the white dwarf has such a strong magnetic field that there is no disc. Therefore, in this instance, I will attempt to describe tomography from a more general perspective.

The key to tomography is first to consider a point-like source of emission in a binary. Assuming this has a motion parallel to the orbital plane of the binary, line emission from such a source will trace out a sinusoid around the mean velocity of the system. Assuming one observed such a sinusoid, one could associate it with a particular velocity vector in the binary, depending upon its phase and amplitude. The trick of tomography is to cope with any number of such sinusoids, even when they are so overlapped and blended that one cannot distinguish one from another. To understand how this is possible, it is helpful to think of profile formation as a projection in the mathematical sense of integrating over one dimension of an N-dimensional space to produce an $N-1$ dimensional space.

3.1   Profile Formation by Projection

A given point in the binary can be defined by its spatial position, but, more usefully in this case, also by its velocity (V_x, V_y). One must be a little careful to define this velocity which is relative to the *inertial* rather than rotating frame. Inertial frame velocities are always changing as the binary rotates, and so in order to define unique values of V_x and V_y, they are measured at a particular orbital phase. Conventionally this is taken to be when the inertial frame lines up with the rotating frame. In the case of CVs it is usual to define the x-axis (in the rotating frame) to point from the white dwarf to the mass donor, and the y-axis to point in the direction of motion of the mass donor. With this convention, and defining orbital phase zero to be the moment when the donor star is closest to us, the radial velocity of the point in question at orbital phase ϕ is

$$V_R = \gamma - V_x \cos 2\pi\phi + V_y \sin 2\pi\phi, \tag{3}$$

where γ is the mean or systemic velocity of the star. The use of a single value of γ is equivalent to assuming that all motion is parallel to the orbital plane as mentioned above.

With these definitions, an "image" of the system can be defined as the strength of emission as a function of velocity, $I(V_x, V_y)$. That is, the flux observed from the system that comes from the velocity element bounded by V_x to $V_x + dV_x$, V_y to $V_y + dV_y$ is given by $I(V_x, V_y)\, dV_x\, dV_y$. I will refer to this as an "image in velocity space" or more simply a "velocity-space image". The relation of this image to the conventional image will be discussed below.

The line flux observed from the system between radial velocities V and $V + dV$ at orbital phase ϕ can now be obtained by integration over all regions of the image that have the correct radial velocity:

$$\int_{-\infty}^{\infty} \int_{-\infty}^{\infty} I(V_x, V_y)[g(V - V_R)\, dV]\, dV_x\, dV_y, \tag{4}$$

where g is a function (of velocity) representing the line profile from any point in the image, including instrumental blurring. I assume here that g is the same at every point, although it is possible to allow it to vary. The velocity width is divided out to obtain a flux density of course, and so the line profile can be expressed as

$$f(V, \phi) = \int_{-\infty}^{\infty} \int_{-\infty}^{\infty} I(V_x, V_y)g(V - V_R)\, dV_x\, dV_y. \tag{5}$$

Ideally g is narrow, best of all a delta function, thus this equation picks out all regions of the image close to the line

$$V = V_R = \gamma - V_x \cos 2\pi\phi + V_y \sin 2\pi\phi.$$

This is a straight line in V_x, V_y coordinates. Different values of V define a whole family of parallel straight lines across the image, with a direction dependent upon the orbital phase. With the definition of velocity and orbital phase described above, orbital phase 0 corresponds to a collapse in the positive V_y direction, phase 0.25 corresponds to the positive V_x direction, phase 0.5 to the negative V_y direction etc, with the angle rotating clockwise. Thus the formation of the line profile at a particular phase can be thought of as a projection (or collapse) of the image along a direction defined by the orbital phase. Note that if this model is correct, two line profiles taken half-an-orbit apart should be mirror images of one another. The extent to which this is not the case is one measure of violations of the basic assumptions made.

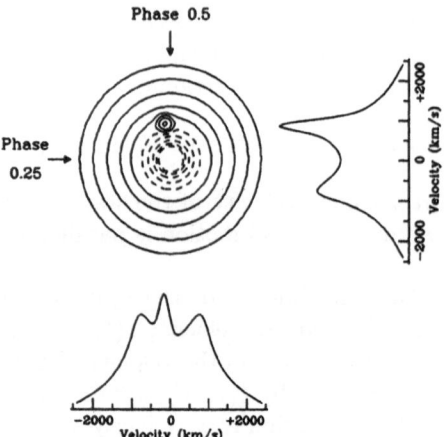

Fig. 3. A model image and the equivalent profiles formed by projection at angle appropriate to orbital phases 0.25 (right-most profile) and 0.5 (lower profile).

Fig. 3 shows a pictorial representation of this process for two projection angles. The artificial image has been created with a spot which can be seen to project into different parts of the profile at different phases. Tracing back from the peaks along the projection directions leads to the position of the original spot. This is in essence how line profile information can be used to reconstruct an image of the system.

A series of line profiles at different orbital phases is therefore nothing more than a set of projections of the image at different angles. The inversion of projections to reconstruct the image is known as "tomography", the case of medical X-ray imaging being perhaps the most famous, although it occurs in many other fields too. I now look at the two methods that have been applied in the case of Doppler tomography.

3.2   Inversion Methods

The mathematics of the inversion of projections dates back to the work of Radon in 1917 [60]. If one knows the function (in my notation) $f(V, \phi)$ for all V and ϕ, a linear transformation – the Radon transform – can produce the desired end product, $I(V_x, V_y)$. In reality, things are not so easy, and we never have the luxury of knowing the line profiles at all orbital phases, although one can get close in some cases. With the advent of fast computers and the development of medical imaging, interest in the implementation of Radon's transform increased greatly in the 1970s, and one particular method, that of "filtered back-projection" found favour [63]. The original paper on Doppler tomography used an alternative method inherited from eclipse mapping, that of maximum entropy regularisation. In this section I describe these two methods, which are both in use today. Each has its pros and cons, which I discuss at the end of the section.

Filtered Back-Projection. The mathematical inversion of Eq. 5 is detailed in the Appendix A. The process can be summarized in the following two steps. First the line profiles are filtered in velocity to derive modified profiles, $\tilde{f}(V,\phi)$. The filter is applied through a Fourier transform, multiplication by $|s|/G(s)$, where $G(s)$ is the Fourier transform over V of $g(V)$ and s is the frequency in inverse velocity units, and finally an inverse Fourier transform. It can be applied to one spectrum at a time and is a fairly fast process.

The second step is that of *back-projection*:

$$I(V_x, V_y) = \int_0^{0.5} \tilde{f}(\gamma - V_x \cos 2\pi\phi + V_y \sin 2\pi\phi, \phi)\, d\phi. \qquad (6)$$

An intuitive understanding of back-projection is extremely useful when trying to make sense of Doppler maps. There are two ways of imagining the process (see Fig. 4). The first one is perhaps the most obvious from Eq. 6 which implies that each point in the image can be built by integration along a sinusoidal path through "trailed" spectra (spectra viewed in 2D form with axes of phase versus velocity). The particular sinusoid is exactly that which a spot at the particular place in the image would produce in the trailed spectrum. This view is illustrated on the left of Fig. 4.

However, back-projection is named for another, perhaps more useful, way of regarding this operation shown on the right of Fig. 4. In effect, Eq. 6 means that the image is built up by smearing each filtered profile along the same direction as the original projection which formed it. This way of looking at back-projection shows very clearly why small numbers of spectra cause linear artifacts in Doppler

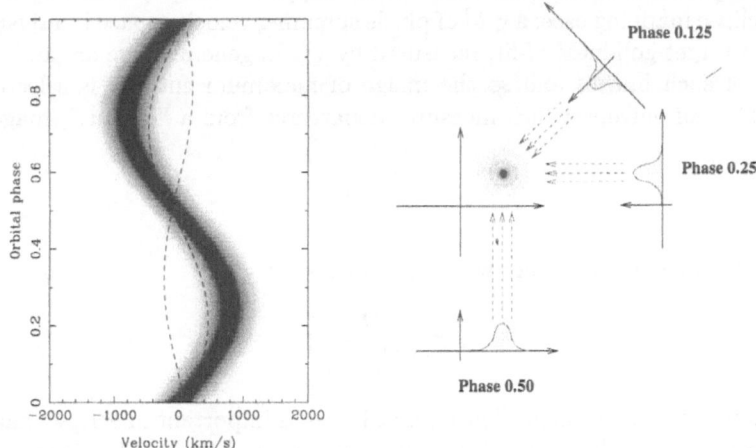

Fig. 4. Two view of back-projection: on the left are paths of integration through trailed spectra (spectra plotted as greyscale images with time running upwards and wavelength from left to right). A track close to a sinusoidal component gives a spot in the final image. On the right three profiles are smeared back along their original projection directions to give a spot.

maps, and one should always be wary of such features. Similarly, any anomalies, such as unmasked cosmic rays, dead pixels, flares or unmasked eclipses are liable to cause streaks across Doppler maps.

If the local line profile $g(V)$ is gaussian then so too is $G(s)$, dropping to zero at large s. Thus the filter $|s|/G(s)$ will strongly amplify high frequencies, and the image will be corrupted by noise. This may be familiar when it is realised that division by $G(s)$ is just the standard (and noise-sensitive) Fourier deconvolution; the presence of $|s|$ in this case only exacerbates the problem. One can just remove the $G(s)$ term and thus make no attempt to de-convolve the image. Typically one goes further still and the filter applied is $|s|W(s)$, where $W(s)$ is a (typically gaussian) "window" function to cut off high frequencies and therefore limit the propagation of noise into the final image. The penalty for this is that the final image is a blurred version of the true image. A similar trade-off will become apparent in the maximum entropy inversion which I turn to now.

Maximum Entropy Inversion. In the original paper presenting Doppler tomography, the focus was upon an alternative to filtered back-projection using maximum entropy regularisation [48]. This stemmed in part from the earlier development of eclipse mapping but also because I originally developed Doppler tomography in spatial coordinates in which the projection becomes one over a set of curves rather than straight lines [47]. However, velocity space is nearly universally used now, and the linear inversion has become more commonly used on the whole. Is another inversion method useful? I think the answer is yes, and will discuss why in detail below. First of all, I describe the maximum entropy method in some more detail.

The application of maximum entropy to Doppler tomography is very similar to the eclipse mapping case: a grid of pixels spanning velocity space is adjusted to achieve a target goodness-of-fit, measured by χ^2. In general there are an infinite number of such images and so the image of maximum entropy is selected. A refined form of entropy which measures departures from a "default" image [39] is used:

$$S = -\sum_{i=1}^{M} p_i \ln \frac{p_i}{q_i}. \tag{7}$$

Here all symbols are as before with the addition of

$$q_i = \frac{D_i}{\sum_{j=1}^{M} D_j}, \tag{8}$$

where $\mathbf{D}$ is the default image. The default image is important in eclipse mapping (see Baptista, this volume) but less so in Doppler tomography. Usually a moving default is used, computed as a blurred version of the image. This constrains the map to be smooth on scales shorter than the blurring, but fixed by the data alone on larger scales. For reasonable data, the choice of default appears to have little effect. Indeed, it would be my guess that neither does the form of S, and that its most important role in this case is to allow a unique solution to be found.

Relative Merits of the Two Inversions. Table 1 compares the Maximum Entropy Method (MEM) and Filtered Back-Projection (FBP) inversion methods; plus and minus signs indicate pros and cons respectively. First of all, χ^2 in

Table 1. Comparison of maximum entropy and filtered back-projection for Doppler tomography.

Characteristic	MEM	FBP
Controlling parameter	χ^2	FWHM of noise filter
Processor time	$-$ (22 sec)[†]	$+$ (0.7 sec)
Comparison with data	$++$	$--$
Consistency of noise level	$-$[‡]	$+$
Flexibility	$++$	$--$

[†]100 by 100 image, 100 by 50 data, 300 MHz Pentium II.

the maximum entropy method (MEM) is identified as the equivalent of the window filter in the filtered back-projection (FBP). For instance a low χ^2 forces the image to become highly structured in order to fit the data better. Conversely, a high χ^2 allows the image to become smooth and blurred. Next I compare processing time, which is perhaps the major disadvantage of MEM. In the example given, MEM took 30 times longer than FBP. Even so, for single images the absolute amount of time taken is not large, especially when compared with the steps taken to get the data in the first place, although it can become more significant when trying to estimate noise, as I will discuss later. Nevertheless, I still regard it as a relatively minor disadvantage, and award a single minus. The next entry "Comparison with data" refers to the central role that χ^2 plays in the MEM reconstructions which allows one to compare the predicted data directly to observations. Filtered back-projection does not try to achieve a good fit to the data, leaving one uncertain as to how much better the fit could have been; I regard this as a significant disadvantage of the method.

Propagation of noise into MEM images can be problematical when one is comparing, say multiple datasets. The reason is that if one sets the same χ^2 level for each inversion, in one case this might be easy to reach and a rather smooth looking map is the result, while in the next may have a hard time reaching the desired level at all, resulting in a noisy map. FBP on the other hand seems to give an images of similar appearance for the same window filter. It was for this reason that Marsh & Duck [52] used FBP in their tomography of the DQ Her star, FO Aqr. It is possible that this problem could be fixed by iterating towards a fixed entropy and minimising χ^2. While this may seem a bit *ad hoc*, it may provide a more accurate representation of how one actually sets χ^2 in practice: it is rather rare to achieve statistically acceptable values of χ^2 for data of reasonably good signal-to-noise ratio, and often χ^2 is set so that the image is neither too smooth nor too corrupted by noise.

Flexibility, the final entry in Table 1, is another major plus point of MEM. For instance, it is very easy to adapt it to the common case of blended lines

[48,53]. Steeghs (this volume) presents a nice extension of MEM to allow for variation of flux with orbital phase. Another more minor example of this is that it is easy in MEM to mask out bad data without the need to interpolate.

For straightforward cases, I think that there is relatively little to choose between the two methods, although the speed of the filtered back-projection may be advantageous when many maps are being computed. MEM has the edge in difficult cases where modifications of the standard model are needed.

3.3   Noise in Doppler Images

In the penultimate part of this section, I look at the propagation of noise into Doppler maps. To an extent Doppler images carry their own uncertainty estimates in the degree of fluctuation that one sees in the background, and perhaps this has motivated the lack of a more rigorous treatment to date. Moreover, it is often the case, as I remarked above, that Doppler tomography cannot achieve a good fit to the data, and one must assume that systematic errors are dominant. However, there is still a need to understand noise, with a common case being the question of the reality of a certain feature. Appealing to the level of background noise is not always good enough – for example, any map will have a highest point, but how is one to judge whether it is significant?

It has been said that noise cannot be propagated into Doppler maps [84]. It is in fact straightforward to do so. However, the important point to understand is that noise in Doppler images is *correlated*. In more detail it is correlated on short scales but less so on large scales. This correlation means for example that single pixel variances are more-or-less useless in defining the amount of noise on a Doppler image. The correlation is positive on short scales and so Doppler maps are effectively noisier than an uncorrelated image with the same variance per pixel; the difference is significant.

One of the best ways to appreciate the correlation is to view animations running simulated images in movie form. Since this is not possible in print, I illustrate the consequence by plotting the scatter in circular apertures in Fig. 5 In this figure, although the RMS scatter of the uncorrelated images was adjusted to match the MEM images pixel-by-pixel, the scatter at first grows more quickly with radius in the FBP and MEM images, leading to an RMS about 3 times larger at large radii. This illustrates the positive correlation at small separations. Ignoring this correlation would lead to a very significant underestimate of the true noise. In this case for radii larger than ≈ 4 pixels, all three lines are roughly parallel, demonstrating the weak correlation on large scales. The final point is the very similar behaviour of the FBP and MEM plots: the two methods lead to a very similar propagation of noise into the image.

The exact pattern of noise is dependent upon the controlling parameter, χ^2 or the filter FWHM. While driving down χ^2 may force a better fit and higher resolution, a penalty is paid in increased noise. Fig. 6 shows this effect. In this case as χ^2 is lowered from 1.1 to 0.9, the noise at the smallest scales increases by a factor of two; large scales are almost unaffected. Very similar behaviour is seen for different filter widths for filtered back-projection.

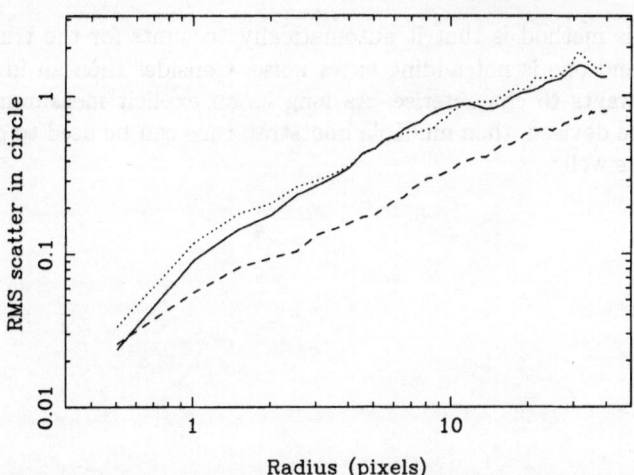

Fig. 5. The figure shows the RMS scatter in simulated Doppler images measured in circular apertures, as a function of the aperture radius for three sets of simulations. The solid line shows results of MEM inversions with $\chi^2 = 1.0$; the dotted line shows the FBP results with a filter of FWHM = 1.0 in terms of the Nyquist frequency; the dashed line shows what the result would have been if the noise were uncorrelated with the same variance per pixel as the MEM simulations.

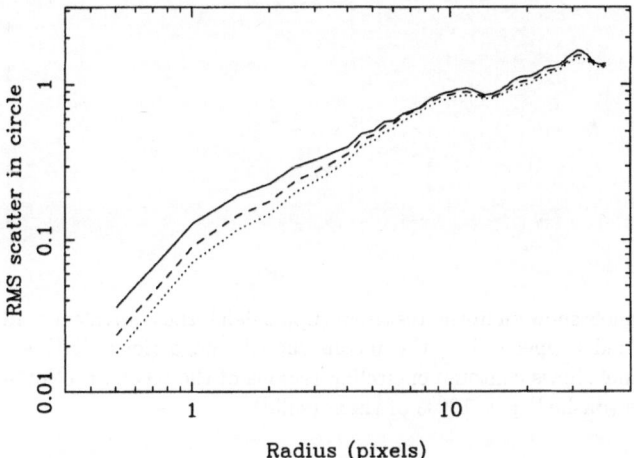

Fig. 6. The figure shows the RMS scatter in MEM simulations measured in circular apertures for $\chi^2 = 0.9$ (solid line), 1.0 (dashed), and 1.1 (dotted).

Figures 5 and 6 were made by adding noise to a simulated dataset and then reconstructing images. A similar method was employed by Hessman & Hopp in their analysis of GD 552 [35]. In practice, the "bootstrap" method [15] is preferable. In this technique, real data are used to generate artificial data by randomly selecting, *with replacement*, N new points from N old points. The beauty of this method is that it automatically accounts for the true statistics of the data and one is not adding extra noise. Consider then an image with a feature one wants to characterise. As long as an explicit measurement of the feature can be devised, then multiple bootstrap runs can be used to generate an uncertainty as well.

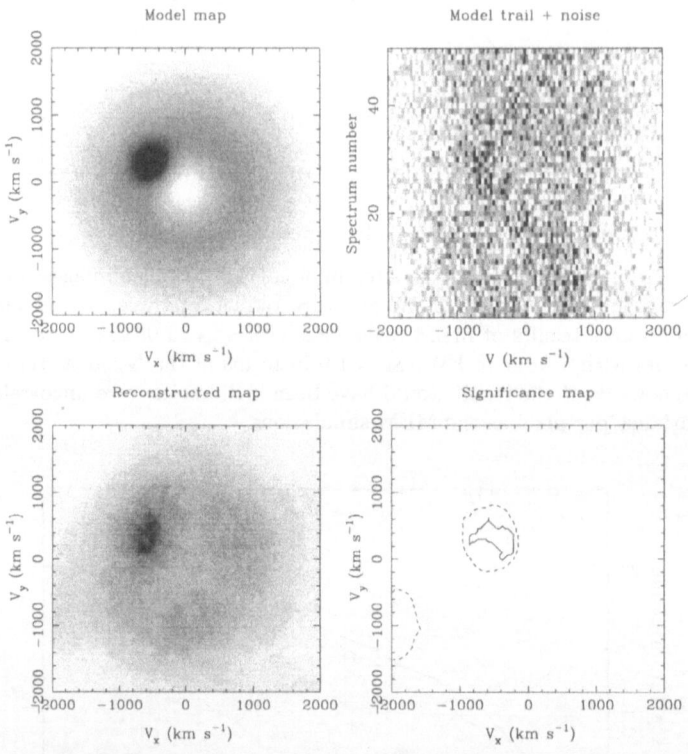

Fig. 7. The panels show an initial test map (upper-left), the equivalent trailed spectrum with noise added (upper-right), the reconstructed image (lower-left) and finally the lower-right panel shows contours encircling regions of the reconstruction above 99% of the 1000 trials (dashed) and 100% of them (solid).

As an example of how one might use the bootstrap method, consider an image showing evidence for a bright-spot, with some question as to the significance of the feature (Fig. 7). First one needs a method of measuring the strength of such a feature. The method I use here is first to subtract off the symmetric part of the image and then to measure the flux in a circular aperture centred on the spot. I then use the following procedure to determine the significance of the feature:

1. Generate a large number (~ 1000) of new datasets from the original data by bootstrap resampling (a fast procedure).

2. Compute maps for each of these in the same manner as for the true map (here the CPU penalty of MEM is paid in full).
3. Subtract the true map from each boot-strapped map to obtain difference maps.
4. Measure the flux of each difference map in the same way as before i.e. by subtracting the symmetric part of each image and computing the flux in a circular aperture.
5. Finally, rank the observed flux relative to the fluxes measured from the difference maps.

This procedure generates a set of values showing the stochastic noise level in the circular aperture. If carried out for apertures centred on every pixel in the image, one can generate a "significance map" which shows the fraction of fluctuation values that the observed value exceeds. The result of such a scheme is shown in Fig. 7 where an artificial image shown in the top-left led, after the addition of gaussian noise, to the trailed spectrum shown in the top-right. The "true map" referred to above is displayed in the lower-left and shows evidence for a spot in the upper-left quadrant. Carrying out the bootstrap computations with 1000 trials leads to the significance map of the lower-right. This shows that the region of the spot is higher than all 1000 of the simulated datasets, whereas no other part of the image is. In this case the circular aperture used has a radius of 200 km s^{-1} , and so can be fitted $\sim$ 100 times over into the image. Taking these to be independent, then there is $\sim$ 10% chance that one region will exceed all 1000 trials. More trials would be needed to establish the reality of the feature more firmly, but the principle is clear.

Similar measurements are easily imagined, for instance, one could perhaps fit the position of the spot [35], and subsequently obtain uncertainty estimates from bootstrapping. All that is necessary is that the measurement is precisely defined and applied in the same way to the true and bootstrap maps.

3.4 Axioms of Doppler Tomography

Doppler tomography rests on certain approximations to reality that are, at best, only partially fulfilled. Violation of these approximations does not mean that the resulting maps are useless, but everyone who carries out Doppler tomography should be aware of its limitations. Thus in this final part of this section, I list the "axioms" that underly the method:

1. All points are equally visible at all times.
2. The flux from any point fixed in the rotating frame is constant.
3. All motion is parallel to the orbital plane.
4. All velocity vectors rotate with the binary star.
5. The intrinsic width of the profile from any point is negligible.

Exceptions exist to each of these. For instance, emission on the mass donor violates the first axiom, while outbursts clearly contravene (2). Doppler tomography is an interpretation of the data within a specific model of binary systems and only applies inasmuch as the model itself does.

4   The Application of Doppler Tomography to CVs

I now move on to discuss Doppler tomography in the context of cataclysmic variable stars. I will mainly focus upon results, but before doing so I consider the interpretation of Doppler maps in the case of CVs.

4.1   Understanding Doppler Maps

On the basis of the standard model presented in section 2 one can easily predict the locations of the various components in velocity-space; Fig. 8 shows some of the key components of a CV represented in velocity space. The donor star

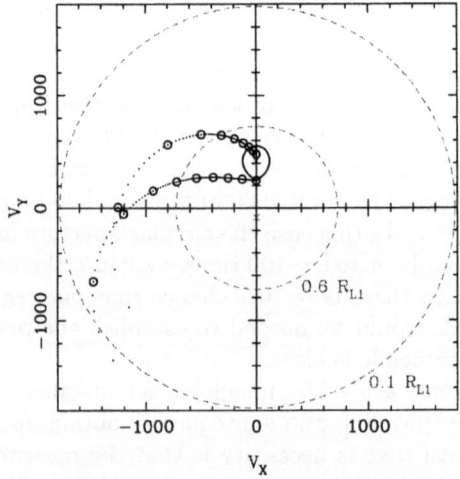

Fig. 8. A schematic of some key components in velocity coordinates.

is assumed to co-rotate with the binary, which means that it appears with the same shape in velocity as it does in position coordinates, although rotated by 90° owing to the relation $v = \Omega \wedge r$ between velocity and position for "solid-body" rotation. This reassuring property is somewhat misleading, since the disc, which is very definitely not co-rotating with the binary, ends up being turned inside out so that the inner disc is at large velocities while the outer disc appears as a ring at low velocity. The gas stream is plotted twice: once with its true velocity and once with the velocity of the disc along its path; one can also imagine intermediate cases. The positions of all these components is fully specified if the projected orbital velocities of the two stars, K_1 and K_2, and the orbital phase are known. The overall scale is set by $K_1 + K_2$; their ratio, which is the mass ratio $q = K_1/K_2 = M_2/M_1$, defines the detailed shape of the stream and Roche lobe. The orbital phase sets the orientation of the image, and if it is not known the image will be rotated by an unknown amount relative to the "standard" orientation shown in Fig. 8. This is not uncommon: for instance if the orbital phase is based upon emission line measurements, it is typically delayed by 0.05 – 0.1 cycles with respect to the true ephemeris. This causes an anti-clockwise rotation of the image by an equivalent number of turns. The published map of LY Hya [78] is a nice example of this phenomenon.

Although velocity coordinates simplify the picture of line profile formation, it is simple enough to invert into position coordinates – indeed this is how I originally computed Doppler images [47]. All that is required is a specification of the velocity at every point in the system. However, I abandoned this approach for two reasons. First, the translation between velocity and position is often not known. In fact, perhaps it is never known, given that it is likely that deviations from keplerian flow occur. This means that position maps would require recomputation each time system parameters were updated. Second, the same place in the system can produce emission at more than one velocity. This is not an abstract possibility, but happens in almost every system that has been imaged. There are many examples of bright-spot emission from the gas stream while the disc at the same location produces emission at a completely different velocity. If such data is imaged into position coordinates on the basis of keplerian rotation, a spot of emission would be produced at a spurious location in the disc. Sticking to velocity coordinates is a reminder of these potential difficulties of interpretation. Only in eclipsing systems is there potential for disentangling such effects.

While we cannot translate the data into position space, there is no difficulty in translating any theoretical model into velocity coordinates. Indeed, ideally, the theory–data comparison should be made by predicting trailed spectra, doing away with the need for Doppler maps altogether. However, Doppler maps still have a rôle in that theoretical models are not good enough to predict all the peculiarities of real systems, and comparison is easier in the half-way house of velocity space.

The idea of translating to velocity space also applies to how one should think about Doppler maps. Rather than trying to translate features of maps mentally from velocity to position coordinates, one should try to think of various components and imagine where they would appear in velocity space. The difference may seem slight, but it is a significant one. With that said, I now turn to look at some results.

4.2 Doppler Imaging Results

There are now a large number of examples of Doppler tomography, covering CVs along with other types of binary as well, such as Algols and X-ray transients, the latter being very similar to CVs in many ways [51,8]. Rather than spend space covering these in detail when the original references do so already, as do other contributions in this volume, I have decided to devote this section mainly to highlights based upon a literature survey of as many published Doppler images as I could find. The results of this survey are tabulated in Tables 2, 3 and 4 contained in Appendix B where I list systems with published maps, which lines were mapped, the spectral resolution, the outburst state and some indication of the appearance of the maps.

Spiral Shocks. The discovery by Steeghs et al. [77] of spiral shocks in the dwarf nova IP Peg is perhaps the most significant result from Doppler tomography applied to CVs. These shocks appear to be present in all outbursts of IP Peg,

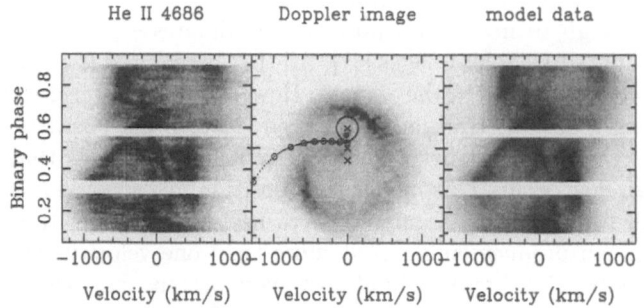

Fig. 9. Spiral shocks in IP Peg [27].

including (with hindsight!) pre-discovery outburst data [50]; Fig. 3 shows one example. There is corroborating evidence from other systems such as SS Cyg [76], V347 Pup [82] and EX Dra [42], although none of these are as convincing as IP Peg (Steeghs, this volume).

Stream Emission in Polars. Beautiful work by Schwope (this volume) and others [66,67,30,68,71] has revealed the gas stream in the polar class of cataclysmic variables in which the white dwarf is magnetically locked to the mass donor star. Although initially ballistic, there is evidence for the influence of the field upon the stream and some emission from the gas as it hurtles down towards the white dwarf (see Fig. 10). Such work has tremendous scope for teaching us about the stream/magnetosphere interaction. Changes in the appearance of maps between high and low states have been seen in HU Aqr [67] and further observation of differing states should tell us how the plasma/field interaction varies with accretion rate.

Bright-Spots. Many systems show bright-spots in Doppler maps, classic cases being WZ Sge [75] and GP Com [53]. The locations of the spots are interesting. In some cases they line up with the stream's velocity [50], while others are closer to the disc's velocity [93]. Still others adopt a position half-way between the disc and stream velocities [49,75] (see also Fig 11). There is some evidence for extended cooling following the spot [75] and there is potential for examining this with lines of differing excitation.

Emission from the Secondary Star. Many Doppler maps show emission from the secondary star [50,49,23,57,66]. Some of this may be related to magnetic activity of these rapidly rotating stars, but there is little doubt that much of it

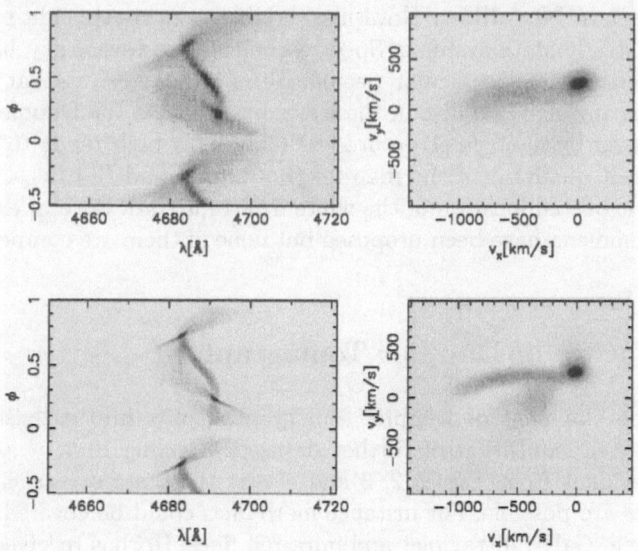

Fig. 10. Stream-stripping in HU Aqr [30]. The upper panels show the HeII 4686 line and Doppler map; the lower panels show simulated data based upon a simple model in which gas is pulled of the stream and threaded onto the magnetic field.

Hα   HeI 6678

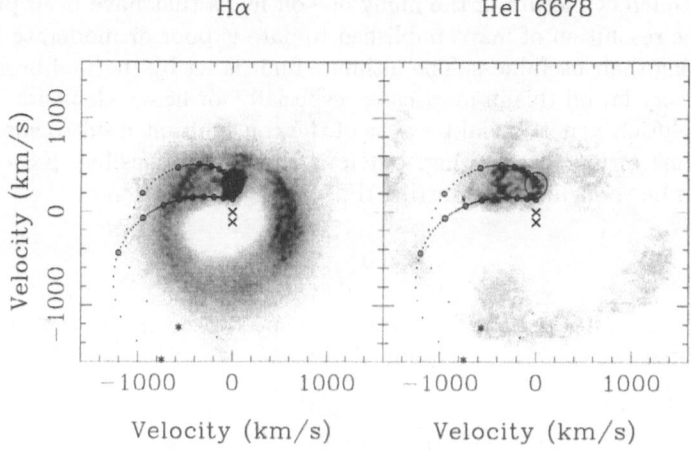

Fig. 11. The peculiar bright-spot of U Gem observed in January 2000 (Unda et al. in prep).

is caused by irradiation by the inner disc. The disc systems are interesting in this respect since one can expect the disc to cast a shadow over the equator of the mass donor. This seems indeed to be the case [28,57] and perhaps has the potential to tell us about the vertical structure of the disc.

Missing Discs in Novalikes. Nova-like variable stars are thought to be classic examples of steady-state systems. Spectroscopically however they have proved hard to understand. A particular peculiarity of these stars is that it is often difficult to see any sign of a disc in these systems. Instead the Doppler maps are often dominated by a single structure-less blob. This blob tends to be located in the lower-left quadrant of the map for the Balmer and HeI lines, but rather closer to the expected location of the white dwarf for HeII. Several explanations for these phenomena have been proposed but none of them are compelling in my view.

5   The Future of Doppler Tomography

The future development of Doppler tomography splits into extensions of the method and the acquisition of further datasets. Dealing first of all with the latter, it is evident from Tables 2, 3 and 4 that there are several areas where improvements are possible. For instance more lines could be covered, especially metal lines (e.g. CaII), ultraviolet and infra-red lines. $H\alpha$ has received relatively little attention, but when it has been looked at, often appears peculiar [81,76]. In the case of dwarf novae, the discovery of spiral shocks makes extended coverage of outbursts of considerable interest.

Multi-epoch tomography is probably the most gaping hole because it is hard to make a coherent picture of the many one-off maps that have been published to date. The resolution of maps published to date is poor or moderate in many cases, limiting their usefulness. The ultimate limit is set by thermal broadening, but we are very far off this in most cases, especially for heavy elements. Pushing to high resolution is not trivial because of the concomitant need to shorten the exposure time owing to smearing, but it is thoroughly feasible. Exposures of length t can be thought of as blurring the image by a rotation of

$$360° \frac{t}{P}$$

where P is the orbital period. For a feature a speed K from the centre of mass, this will match the spectral resolution ΔV when

$$\frac{t}{P} \sim \frac{1}{2\pi} \frac{\Delta V}{K}.$$

Thus if ΔV is lowered by a factor of two, the exposure time must also be reduced by a factor two. As a concrete example, consider trying to image the mass donor in a system where $K_2 = 400\,\mathrm{km\ s^{-1}}$. Typical equatorial velocities in CVs are $\approx 100\,\mathrm{km\ s^{-1}}$, so we may attempt to obtain data with $\Delta V = 10\,\mathrm{km\ s^{-1}}$. We would then need $\Delta\phi = t/P < 0.004$, equivalent to $t < 30\,\mathrm{s}$ for a system of $P = 2$ hours. For many CVs, this will require 8m-class telescopes.

Various extensions are possible, and have already been developed in some cases. Standard Doppler tomography does not treat the geometry of the donor star correctly. It is straightforward to fix this [64]. Bobinger et al. [3] describe a method for simultaneous Doppler and eclipse mapping of emission lines in which a single image is computed to fit both spectra and light curves of the lines, with a keplerian velocity field used to translate between position and velocity space. It is difficult to evaluate whether the spectra or fluxes dominate the final maps, but it is clear that spectral information does alleviate the degenerate nature of eclipse mapping. Of course, the need to assume a particular velocity field is a disadvantage. An attempt has been made to avoid this by simultaneously adjusting a spatial image and a position–velocity map to fit spectra of eclipsing systems [1]. In this method, spectra out of eclipse serve to fix the velocity space image as usual, which is then translated to position space through the eclipse information. The technique was able to recover a $V \propto R^{-1/2}$ relation from spectra of V2051 Oph, but as developed it could not handle the difficult case of the same place producing emission at more than one velocity. Finally, Steeghs (this volume) describes a new modification which allows orbital variability to be included in Doppler images. As this is so common, it has considerable potential.

6   Conclusions

I have reviewed the principles and practice of the analysis method of Doppler tomography which helps the interpretation of the complex line profile variations from close binary stars. The key discoveries from the application of this technique are the spiral shocks in outbursting dwarf novae and the stream/magnetic field interaction in the polar class of cataclysmic variable stars, but Doppler tomography has taught us much about the stream/disk interaction and irradiation of the donor star too.

For the future, efforts need to be made to acquire multi-epoch datasets for tomography as these are wholly lacking at present. Following that higher spectral and temporal resolution data are needed to exploit tomography to its limit.

References

1. Billington, I., 1995, PhD Thesis, Oxford University.
2. Billington, I., Marsh, T.R., Dhillon, V.S., 1996, MNRAS, , 278, 673–682.
3. Bobinger, A. et al., 1999, A&A, , 348, 145–153.
4. Burwitz, V. et al., 1998, A&A, , 331, 262–270.
5. Casares, J. et al., 1995, MNRAS, , 274, 565–571.
6. Casares, J., Charles, P.A., Marsh, T.R., 1995, MNRAS, , 277, L45–L50.
7. Casares, J. et al., 1996, MNRAS, , 278, 219–235.
8. Casares, J. et al., 1997, New Astronomy, 1, 299–310.
9. Casares, J. et al., 1997, New Astronomy, 1, 299–310.
10. Catalán, M. S., Schwope, A. D., Smith, R. C., 1999, MNRAS, , 310, 123–145.
11. Dhillon, V.S., Marsh, T.R., Jones, D.H.P., 1991, MNRAS, , 252, 342–356.
12. Dhillon, V.S., Jones, D.H.P., Marsh, T.R., Smith, R.C., 1992, MNRAS, , 258, 225–240.
13. Dhillon, V.S., Jones, D.H.P., Marsh, T.R., 1994, MNRAS, , 266, 859.

14. Dhillon, V.S., Marsh, T.R., Jones, D.H.P., 1997, MNRAS, , 291, 694.
15. Diaconis, P., Efron, B., 1983, Sci. Am. 248, 96–.
16. Diaz, M. P., Steiner, J. E., 1994, ApJ, , 425, 252–263.
17. Diaz, M. P., Steiner, J. E., 1994, A&A, , 283, 508–514.
18. Diaz, M. P., Steiner, J. E., 1995, AJ, , 110, 1816.
19. Diaz, M. P., Hubeny, I., 1999, ApJ, , 523, 786–796.
20. Dickinson, R. J. et al., 1997, MNRAS, , 286, 447–462.
21. Gaensicke, B. T. et al., 1998, A&A, , 338, 933–946.
22. Harlaftis, E.T., Marsh, T.R., Dhillon, V.S., Charles, P.A., 1994, MNRAS, , 267, 473.
23. Harlaftis, E.T., Marsh, T.R., 1996, A&A, , 308, 97–106.
24. Harlaftis, E. T., Horne, K., Filippenko, A. V., 1996, PASP, , 108, 762.
25. Harlaftis, E. T., Steeghs, D., Horne, K., Filippenko, A. V., 1997, AJ, , 114, 1170–1175.
26. Harlaftis, E., Collier, S., Horne, K., Filippenko, A. V., 1999, A&A, , 341, 491–498.
27. Harlaftis, E. T. et al., 1999, MNRAS, , 306, 348–352.
28. Harlaftis, E., 1999, A&A, , 346, L73–L75.
29. Hastings, N. C. et al., 1999, PASP, , 111, 177–183.
30. Heerlein, C., Horne, K., Schwope, A.D., 1999, MNRAS, , 304, 145–154.
31. Hellier, C., Robinson, E.L., 1994, ApJL, , 431, L107–L110.
32. Hellier, C., 1996, ApJ, , 471, 949.
33. Hellier, C., 1997, MNRAS, , 288, 817–832.
34. Hellier, C., 1999, ApJ, , 519, 324–331.
35. Hessman, F.V., Hopp, U., 1990, A&A, , 228, 387–398.
36. Hoard, D. W., Szkody, P., 1996, ApJ, , 470, 1052.
37. Hoard, D. W., Szkody, P., 1997, ApJ, , 481, 433.
38. Hoard, D. W. et al., 1998, MNRAS, , 294, 689.
39. Horne, K., 1985, MNRAS, , 213, 129–141.
40. Horne, K., 1991, Fundamental Properties of Cataclysmic Variable Stars: Proc. 12th North American Workshop on CVs and Low Mass X-Ray Binaries, ed. A. W. Shafter (San Diego: San Diego State Univ.), , 23–.
41. Howell, S. B., Ciardi, D. R., Dhillon, V. S., Skidmore, W., 2000, ApJ, , 530, 904–915.
42. Joergens, V., Spruit, H. C., Rutten, R. G. M., 2000, A&A, , 356, L33–L36.
43. Joergens, V. et al., 2000, A&A, , 354, 579–588.
44. Kaitchuck, R. H. et al., 1994, ApJS, , 93, 519–530.
45. Kaitchuck, R. H., Schlegel, E. M., White, J. C., Mansperger, C. S., 1998, ApJ, , 499, 444.
46. Littlefair, S. P., Dhillon, V. S., Howell, S. B., Ciardi, D. R., 2000, MNRAS, , 313, 117–128.
47. Marsh, T.R., 1985, PhD Thesis, Cambridge University.
48. Marsh, T.R., Horne, K., 1988, MNRAS, , 235, 269–286.
49. Marsh, T. R. et al., 1990, ApJ, , 364, 637–646.
50. Marsh, T.R., Horne, K., 1990, ApJ, , 349, 593–607.
51. Marsh, T.R., Robinson, E.L., Wood, J.H., 1994, MNRAS, , 266, 137.

52. Marsh, T.R., Duck, S.R., 1996, New Astronomy, 1, 97–119.
53. Marsh, T.R., 1999, MNRAS, , 304, 443–450.
54. Martell, P. J., Horne, K., Price, C. M., Gomer, R. H., 1995, ApJ, , 448, 380.
55. Mennickent, R. E., Diaz, M., 1996, A&A, , 309, 147–154.
56. Mennickent, R. E., Diaz, M. P., Arenas, J., 1999, A&A, , 352, 167–176.
57. Morales-Rueda, L., Marsh, T.R., Billington, I., 2000, MNRAS, , 313, 454–460.
58. Nogami, D., Masuda, S., Kato, T., Hirata, R., 1999, PASJ, 51, 115–125.
59. North, R. C. et al., 2000, MNRAS, , 313, 383–391.
60. Radon, J., 1917, Ber. Verh. Sächs. Akad. Wiss. Leipzig Math. Phys. K1, 69, 262–277.
61. Ratering, C., Bruch, A., Diaz, M., 1993, A&A, , 268, 694–704.
62. Robinson, P.F.L., Marsh, T.R., Smak, J., 1999, Accretion Disks in Compact Stellar Systems, Edited by J. Craig Wheeler. World Scientific, ISBN 981-02-1273-9 (1993)., , 75–116.
63. Rowland, S.W., 1979, Image Reconstruction from Projections, , 8–79.
64. Rutten, R.G.M., Dhillon, V.S., 1994, A&A, , 288, 773–781.
65. Schmidt, G. D. et al., 1999, ApJ, , 525, 407–419.
66. Schwope, A.D., Mantel, K.-H., Horne, K., 1997, A&A, , 319, 894–908.
67. Schwope, A. D. et al., 1998, Wild stars in the old west: Proceedings of the 13th North American Workshop on Cataclysmic Variables and Related Objects, eds. Howell, S. Kuulkers, E., Woodward, C., astro-ph/9708228, , 44–59.
68. Schwope, A. D. et al., 2000, MNRAS, , 313, 533–546.
69. Shafter, A. W., Veal, J. M., Robinson, E. L., 1995, ApJ, , 440, 853.
70. Shahbaz, T., Wood, J. H., 1996, MNRAS, , 282, 362–372.
71. Simic, D. et al., 1998, A&A, , 329, 115–130.
72. Smak, J., 1971, Acta Astronomica, 21, 15.
73. Smak, J., 1979, Acta Astronomica, 29, 309.
74. Smith, D.A., Dhillon, V.S., Marsh, T.R., 1998, MNRAS, , 296, 465–482.
75. Spruit, H.C., Rutten, R.G.M., 1998, MNRAS, , 299, 768–776.
76. Steeghs, D., Horne, K., Marsh, T.R., Donati, J.F., 1996, MNRAS, , 281, 626–636.
77. Steeghs, D., Harlaftis, E.T., Horne, K., 1997, MNRAS, , 290, L28–L32.
78. Still, M.D., Marsh, T.R., Dhillon, V.S., Horne, K., 1994, MNRAS, , 267, 957.
79. Still, M. D., Dhillon, V. S., Jones, D. H. P., 1995, MNRAS, , 273, 863–876.
80. Still, M. D., Dhillon, V. S., Jones, D. H. P., 1995, MNRAS, , 273, 849–862.
81. Still, M. D., 1996, MNRAS, , 282, 943–952.
82. Still, M. D., Buckley, D. A. H., Garlick, M. A., 1998, MNRAS, , 299, 545–553.
83. Still, M.D., Duck, S.R., Marsh, T.R., 1998, MNRAS, , 299, 759–767.
84. Still, M. D., Steeghs, D., Dhillon, V. S., Buckley, D. A. H., 1999, MNRAS, , 310, 39–42.
85. Szkody, P., Armstrong, J., Fried, R., 2000, PASP, , 112, 228–236.
86. Tovmassian, G. H. et al., 1997, A&A, , 328, 571–578.
87. Tovmassian, G. H. et al., 1998, A&A, , 335, 227–233.
88. Tovmassian, G. H. et al., 2000, ApJ, , 537, 927–935.
89. Warner, B., Nather, R. E., 1971, MNRAS, , 152, 219.

22   T.R. Marsh

90. Welsh, W. F., Horne, K., Gomer, R., 1998, MNRAS, , 298, 285–302.
91. White, J. C., Honeycutt, R. K., Horne, K., 1993, ApJ, , 412, 278–287.
92. White, J. C., Schlegel, E. M., Honeycutt, R. K., 1996, ApJ, , 456, 777.
93. Wolf, S. et al., 1998, A&A, , 332, 984–998.
94. Wood, J. et al., 1986, MNRAS, , 219, 629–655.

Appendix A

In this appendix I show that, as stated in section 3.2, Eq. 5 can be inverted by application of the filter $|s|/G(s)$ followed by the back-projection of Eq. 6. I define the Fourier transform $F(s)$ of a function $f(x)$, and its inverse by

$$F(s) = \int_{-\infty}^{\infty} f(x)e^{-i2\pi sx} \, dx.$$

$$f(x) = \int_{-\infty}^{\infty} F(s)e^{i2\pi sx} \, ds.$$

The frequency s here is measured in cycles per unit x. Now take the Fourier transform over V of the line profile equation, 5:

$$F(s, \phi) = \int_{-\infty}^{\infty} f(V, \phi)e^{-i2\pi sV} \, dV \tag{9}$$

$$= \int_{-\infty}^{\infty} \int_{-\infty}^{\infty} I(V_x, V_y) \int_{-\infty}^{\infty} g(V - V_R)e^{-i2\pi sV} \, dV \, dV_x \, dV_y \tag{10}$$

$$= G(s) \int_{-\infty}^{\infty} \int_{-\infty}^{\infty} I(V_x, V_y)e^{-i2\pi sV_R} \, dV_x \, dV_y. \tag{11}$$

Dividing through by $G(s)$, multiplying by $|s|$ and taking the inverse Fourier transform gives the filtered line profiles

$$\tilde{f}(V, \phi) = \int_{-\infty}^{\infty} \frac{|s|F(s, \phi)}{G(s)} e^{i2\pi sV} \, ds$$

$$= \int_{-\infty}^{\infty} \int_{-\infty}^{\infty} I(V_x, V_y) \int_{-\infty}^{\infty} |s|e^{-i2\pi s(V - V_R)} \, ds \, dV_x \, dV_y. \tag{12}$$

Finally, back-project these filtered profiles according to Eq. 6, that is compute the integral

$$\int_0^{0.5} \tilde{f}(V_R, \phi) \, d\phi,$$

where

$$V_R = \gamma - V_x \cos 2\pi\phi + V_y \sin 2\pi\phi. \tag{13}$$

Putting dashes on various symbols to avoid confusion later, then the back-projection integral becomes

$$\int_0^{0.5} \tilde{f}(V_R, \phi) \, d\phi = \int_{-\infty}^{\infty} \int_{-\infty}^{\infty} I(V_x', V_y') \int_0^{0.5} \int_{-\infty}^{\infty} |s|e^{-i2\pi s(V_R - V_R')} \, ds \, d\phi \, dV_x' \, dV_y'$$

$$= \int_{-\infty}^{\infty} \int_{-\infty}^{\infty} I(V_x', V_y') \int_0^1 \int_0^{\infty} s e^{-i2\pi s(V_R - V_R')} \, ds \, d\phi \, dV_x' \, dV_y'$$

$$= \int_{-\infty}^{\infty} \int_{-\infty}^{\infty} I(V_x', V_y') \delta(V_x' - V_x) \delta(V_y' - V_y) \, dV_x' \, dV_y'$$

$$= I(V_x, V_y). \tag{14}$$

The third line above follows from the second after transforming from polar coordinates s and ϕ to cartesian $s_x = s \cos 2\pi\phi$ and $s_y = s \sin 2\pi\phi$, and using Eqs. 13 so that

$$\int_0^1 \int_0^{\infty} s e^{-i2\pi s(V_R - V_R')} \, ds \, d\phi = \int_{-\infty}^{\infty} \int_{-\infty}^{\infty} e^{-i2\pi[-(V_x - V_x')s_x + (V_y - V_y')s_y]} \, ds_x \, ds_y$$

and then the integrals over s_x and s_y separate to give the two Dirac δ-functions of the penultimate line of Eq. 14 since

$$\delta(x) = \int_{-\infty}^{\infty} e^{\pm i2\pi s x} \, ds.$$

This justifies the assertions of section 3.2.

Appendix B

Table 2. Doppler maps of CVs and X-ray novae published in refereed journals as of September 2000.

Object	Type	State	Res. km s^{-1}	Line(s)	Features	Ref.
V616 Mon	BH	Q	80	Hα, β	1; 2a; 3	51
GU Mus	BH	Q		Hα	3	9
V2107 Oph	BH	Q	120	Hα	1; 3	25
V518 Per	BH	O	100 (Hβ); 35 (Hα)	Hα, β, HeII	1; 2?; 3	5
"	"	Q	120	Hα	1	26
QZ Vul	BH	Q	200	Hα	1; 2a	6
"	"	Q	120	Hα	1; 2a	24
AR And	DN	Q, O	180 (Q); 130 (O)	Hα (Q); Hβ,γ (O)	1; 7	69
AE Aqr	DN	F	50	Hα	2b; 7	90
VY Aqr	DN	Q	300	Pβ	1?; 2b	46
OY Car	DN	O	80	Hβ	1; 2 or 4; 3	23
AT Cnc	DN	SS	140	Hα	2; 7	58
EM Cyg	DN	SS	35	Hα	1; 3	59
SS Cyg	DN	O	35	Hα, β, γ, HeI, HeII	1; 3; 4?; 6	76
EX Dra	DN	?	80	Hα, HeI, HeII	2ab; 3	2
"	DN	Q, O	100 – 250	Hα – δ, HeI, CII	1; 2b	43
"	DN	O	35	Hα,β, HeI, HeII	1; 3; 4	42
U Gem	DN	Q	170	Hβ, HeI, HeII	1; 2a; 3	49, 44
V2051 Oph	DN	Q	170	Hβ, HeI	1; 7	44
IP Peg	DN	Q, O	150	Hβ, γ, HeII	1; 2ab; 3; 4	50
"	"	Q	170	Hβ	1; 7	44
"	"	Q	140	Hα	1; 3	22
"	"	O	35	Hα, HeI	1; 3; 4; 6	76
"	"	O	35	Hα, HeI	1; 3; 4	77
"	"	Q	70	Hα, β, γ	1; 2a, c; 3	93, 3
"	"	O	54	HeII, HeI, MgII	1; 3; 4; 6	27, 28
"	"	O	100	HeII, HeI, MgII, CII	1; 3; 4	57
KT Per	DN		200	Hβ	1	61
WZ Sge	DN	Q	170	Hβ, HeI	1; 2a	44
"	DN	Q	90	Hα	1; 2a	75
"	DN	Q	300	Pβ	1; 2b	46
CU Vel	DN	Q	320	Hα	1	55

(See end of Tab. 4 for notes)

Table 3. Doppler maps of CVs and X-ray novae published in refereed journals as of September 2000.

Object	Type	State	Res. km s^{-1}	Line(s)	Features	Ref.
GD 552	DN?	Q	70	Hβ	1; 2	35
LY Hya	DN	Q	120	Hβ, γ, δ	1; 2	78
GP Com	IBWD	–	70	HeI, HeII	1; 2; 6	53
FO Aqr	IP	–	80	HeII	2a; 3?	52
"	IP	–	80	HeII		34[†]
BG CMi	IP	–	115	Hβ, HeII	2	33, 34[†]
PQ Gem	IP	–	115	Hβ, HeII	2	33, 34[†]
EX Hya	IP	–	170	Hβ	1; 2a	44
"	IP	–	65	Hβ		34[†]
AO Psc	IP	–	80	HeII		34[†]
V405 Aur	IP	–	50	Hα,β, HeI, HeII	1; 2b, d	83
"	"	–	120	Hα,γ, HeII	2b, d	85
RX0757+63	IP	–	280	Hβ	1; 2a	87
RX1238-38	IP	–	?	Hβ		34[†]
RX1712-24	IP	–	?	HeII		34[†]
DQ Her	N	–	170	Hβ, HeI, HeII	1; 6	44
"	"	–	120	Hγ, HeI, HeII, CaII	1; 3	54
BT Mon	N	–	170	Hβ,γ, HeII	2b; 6; 7	44, 92
"	"	–	100, blue; 36, red	Hα,β, HeI, HeII	2b; 6	74
GQ Mus	N	–	220	HeII	1; 3	16
CP Pup	N	–	170	Hβ, HeII	1	44, 91
PX And	NL	–	90	Hα	1; 2b	31
"	"	–	45	Balmer, HeII	2b, c; 7	79
V1315 Aql	NL	–	65	Hβ, HeII	2b; 6; 7	11
"	"	–	170	Hβ, HeI, HeII	2b; 6; 7	44
"	"	–	115	Hβ, HeII	2b; 6; 7	32
UU Aqr	NL	–	170	Hβ	1; 2ab	44, 45
"	"	–	115	Hα,β,γ, HeI, HeII	1; 2ab	38
V363 Aur	NL	–	170	Hβ, HeII	2bc; 7	44
WX Cen	NL	–	130	Hβ, HeII	1; 3; 7	18
AC Cnc	NL	–	170	Hβ, HeII	2bd; 6; 7	44
V795 Her	NL	–	60	Hβ	1	20
"	"	–	70	Hα,β,γ, HeI, HeII	1	7
BH Lyn	NL	L	75	Hβ,γ,δ, HeI	2b,c; 7	12
"	"	–	120	Hβ	1; 2c	37

For codes see Table 4.

[†]These maps were computed on the spin phase of the white dwarf rather than the standard orbital phase and thus I have not attempted to describe their features.

Table 4. Doppler maps of CVs and X-ray novae published in refereed journals as of September 2000.

Object	Type	State	Res. km s^{-1}	Line(s)	Features	Ref.
BP Lyn	NL		120	Hα,β,γ	1; 2ad	36
"	"		80	Hα, HeI	1; 2bc; 3	81
V347 Pup	NL	–	120	Balmer	1; 3; 4?	82
"	"	–	120	Hβ	2ab; 6	19
LX Ser	NL	–	170	Hβ, HeI, HeII	2a, b; 6	44
SW Sex	NL	–	170	Hβ, HeI	1; 2a; 6	44
"	"	–	75	Hβ,γ,δ, HeI, HeII	1; 2a; 6	14
VZ Scl	NL	–	170	Hβ	2ab; 7	44
RW Tri	NL	–	170	Hβ, HeI, HeII	1, 3	44
"	"	H	50	Hβ,γ	3	80
DW UMa	NL	–	170	Hβ, HeI, HeII	2b; 6	44
"	"	L	75	Balmer	3	13
UX UMa	NL	–	170	Hβ, HeII	1; 6	44
HU Aqr	P	H	110	Hγ, HeII	3; 5	66, 30
AM Her	P	H	130	NV, SiIV, CIII (UV)	2b; 3; 7	21
V884 Her	P	H	200	Hα,β, HeI, HeII	2ab	29
BL Hyi	P	H	160	Hα, HeI	2a	56
ST LMi	P	–	70	NaI, CaII	3	70
"	"	L	350	NaI	3	41
V2301 Oph	P	H	80	Hα,β,γ, HeI, HeII	3; 5?	71
VV Pup	P	H	90	Hα	3; 5	17
MR Ser	P	–	70	NaI, CaII	3	70
QQ Vul	P	H	70	NaI, MgII, HeII, CI	3	10
"	"	H	100	HeII	3, 5	68
AR UMa	P	M	100	Hα,β, HeI, HeII, MgI, NaI	3, 5	65
RX0719+65	P	H	300	Hβ, HeII	2ab; 7	86
RX1015+09	P	H	?	HeII	3; 5	4
RX2157+08	P	H	180	Hβ, HeII	3	88

Notes:
References are amalgamated when they refer to the same data.
Type codes: BH = black-hole system; DN = dwarf nova; N = old nova; NL = nova-like variable; IBWD = interacting binary white dwarf; IP = intermediate polar; P = polar.
State codes (where relevant): Q = quiescence; O = outburst; SS = stand-still; F = flaring; H = high; L = low; M = middle.
Feature codes: (1) Ring (which may be from a disc), (2) Spot, (3) Secondary star, (4) Spiral shocks, (5) Gas stream, (6) Low velocity emission, (7) Little structure or low signal-to-noise.
Type (2) = "spot" does not necessarily imply stream/disc impact,but just refers to the appearance of the image. Entries such as 2a refer to the quadrant the spot is located in (if the orbital phase is known). The quadrants start at the upper-left with "a", and then go anti-clockwise from there. A combination such as 2ab implies a spot located on the boundary of the upper-left and lower-left quadrants. Type (6) = "low-velocity emission" refers to such features as the slingshot prominences seen in IP Peg [76] and the emission at low velocities commonly seen in nova-like variables.

Mapping the Peculiar Binary GP Com

L. Morales-Rueda, T.R. Marsh, and R.C. North

University of Southampton, Southampton SO17 1BJ, UK

Abstract. We present high resolution spectra of the AM CVn helium binary GP Com at two different wavelength ranges. The spectra show the same flaring behaviour observed in previous UV and optical data. We find that the central spike contributes to the flare spectra indicating that its origin is probably the compact object. We also detect that the central spike moves with orbital phase following an S-wave pattern. The radial velocity semiamplitude of the S-wave is $\sim 10\,\mathrm{km\,s^{-1}}$ which lies very close to the white dwarf. The Stark effect seems to significantly affect the central spike of some of the lines, suggesting that it forms in a high electron density region. This again favours the idea that the central spike originates in the white dwarf. We present Doppler maps obtained for the emission lines which show three clear emission regions.

1 Helium Rich Binaries

GP Com belongs to a group of binaries called AM CVn systems. In these binaries, a white dwarf accretes material from the stripped-down core of a giant star. Only 6 known systems belong to this group although they are predicted to have a space density about a factor of 2 higher than that of cataclysmic variables (CVs) of which some 700 are known [14]. They show properties similar to those of CVs but with some peculiarities, for instance ultra short orbital periods – between 15 and 46 min – which indicate that the two stars are very close together, and a complete lack of hydrogen in their spectra. Some of the systems show strong flickering indicating the presence of mass transfer. AM CVn and EC 15330 show only absorption lines. These lines have large widths indicating either rapid rotation, or pressure broadening by compact stars. CR Boo, CP Eri and V803 Cen show absorption lines when they are bright and emission lines when faint. This absorption/emission line behaviour has been compared to that of CVs, as most of them show emission lines when in quiescence and absorption lines when in outburst. GP Com only shows emission lines and therefore could be considered as always being in quiescence. For a summary of the properties of AM CVn systems see [15].

2 GP Com's Spectra

We took spectra of GP Com in two wavelength ranges, $\lambda\lambda 6600 - 7408\text{Å}$ and $\lambda\lambda 4253 - 5058\text{Å}$ with the 4.2 m William Herschel telescope during two consecutive nights. For details on the observations and their reduction see [8].

The average spectrum of GP Com consists of strong emission lines, mainly He I, on a weak continuum, see the top panels of Figs. 1 and 2. The spectra look very similar to previously published spectra [9,4], but the intensity of the central spike is stronger indicating that GP Com shows long term variability. The emission lines consist of a double-peaked profile superposed with a narrow line component that moves between the red and blue peaks with orbital phase, and a central narrow spike near the rest wavelength. The central spike seems to be independent of the double peaks, which are associated with an accretion disc around the compact object [11]. The narrow line component probably has its origin in the region of impact between the accretion stream and the accretion disc, i.e. the bright spot [9]. The origin of the central spike has been a puzzle for a long time. It was suggested that it came from a surrounding nebula [9] but subsequent searches were unsuccessful [12]. The central spike seems to participate in the flaring activity shown by GP Com which would suggest that it is associated to the compact object [4]. We present results that encourage us to suggest that its origin is the compact object in the binary and not a surrounding nebula.

2.1 Flares

GP Com shows strong flaring activity at UV [7] and optical [4] wavelengths probably driven by X-ray variability. By using the method described in [4] we obtain the characteristic flare spectra of GP Com for both wavelength ranges,

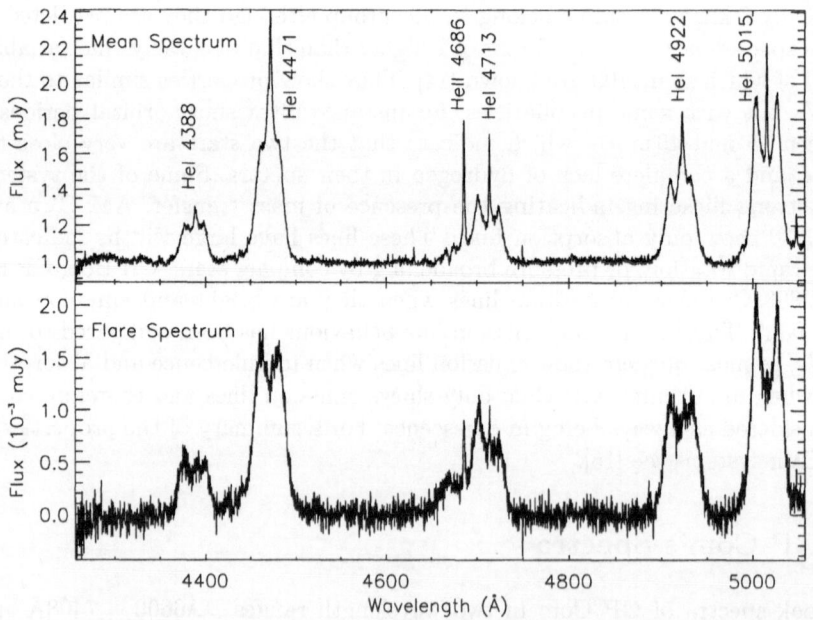

Fig. 1. The panels show the mean GP Com normalised spectrum (**top**) and the flare spectrum (**bottom**)

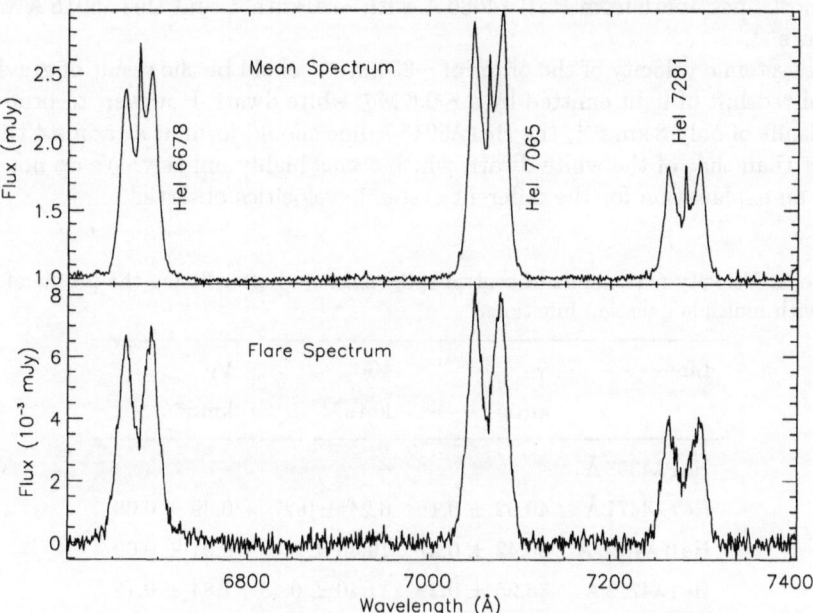

Fig. 2. Same as for Fig. 1 but for a different wavelength range

see bottom panels of Figs. 1 and 2. The first thing we notice is that the central spike is not present in the flare spectra for most emission lines but it is strong for He II $\lambda4686$ Å and present, although weak, for He I $\lambda4713$ Å. We are certain that the central spike contributes to the flare spectrum of GP Com and therefore suggest that it must have its origin somewhere in the binary and not in a nebula around it. Another important feature of the flare spectra is that lines are slightly broader than in the mean spectra, indicating that they are formed mainly in the regions of the disc that rotate faster, i.e the inner regions, whereas the lines that contribute to the mean spectrum are formed in lower velocity regions.

2.2   Central Spike

When we plot all the spectra together versus orbital phase we notice that the central spike behaves like an S-wave with a radial velocity semiamplitude of about $10\,\mathrm{km\,s^{-1}}$. This S-wave completes a whole cycle in an orbit indicating that its origin is somewhere in the binary and not in a nebula around it (as that would produce a stationary peak). We carried out multi-gaussian fittings to the profiles of the lines and fitted the velocities associated with the gaussian corresponding to the central spike by the $\gamma - V_X \cos 2\pi\phi + V_Y \sin 2\pi\phi$ function. The parameters of the fit are shown in table 1. We could not fit He I $\lambda4388$ Å and He I $\lambda4922$ Å accurately as the central spike is double peaked. The radial velocity semiamplitudes measured range between $\sim$6–12 $\mathrm{km\,s^{-1}}$ similar to the $\sim$10 $\mathrm{km\,s^{-1}}$ semiamplitude measured by [9] and [4]. The systemic velocity γ is $\sim$40 $\mathrm{km\,s^{-1}}$

for most lines apart from He II λ4686 Å with $\sim$20 km s^{-1} and He I λ5015 Å with $\sim$8 km s^{-1}.

A systemic velocity of the order of $\sim$35 km s^{-1} could be the result of gravitational redshift of light emitted by a $\sim$0.6 M$\odot$ white dwarf. However, to produce a redshift of only 8 km s^{-1}, the He I λ5015 Å line should form at a radius 4 times larger than that of the white dwarf, which seems highly unlikely. We do not yet have an explanation for the different systemic velocities observed.

Table 1. Velocity parameters of central spike measured after fitting the profile of the line with multiple gaussian functions

Line	γ km s^{-1}	V_X km s^{-1}	V_Y km s^{-1}
He I λ4388 Å	–	–	–
He I λ4471 Å	49.02 ± 0.18	6.24 ± 0.27	0.39 ± 0.06
He II λ4686 Å	20.42 ± 0.27	10.46 ± 0.40	0.91 ± 0.09
He I λ4713 Å	38.25 ± 0.29	11.46 ± 0.43	1.84 ± 0.13
He I λ4922 Å	–	–	–
He I λ5015 Å	7.64 ± 0.33	8.55 ± 0.47	0.89 ± 0.12
He I λ6678 Å	42.98 ± 0.23	10.33 ± 0.35	-0.67 ± 0.07
He I λ7065 Å	46.20 ± 0.24	6.56 ± 0.35	-1.10 ± 0.11
He I λ7281 Å	42.59 ± 0.43	10.70 ± 0.63	-0.35 ± 0.11

3   Doppler Maps

We modified the orbital ephemeris given by [4] so the modulation of the central spike corresponded to the motion of the white dwarf in the binary. Using the maximum entropy technique we obtained Doppler maps for the emission lines and show them, for 4 of the lines, in Fig. 3. For maps of all the lines see [8]. The top panels present the spectra binned in orbital phase and plotted twice. The central spike is the strongest feature in almost all the lines. In some cases it appears to be double-peaked (He I λ4388 Å and He I λ4922 Å) which we suggest is the result of Stark broadening. The Stark effect does not affect helium lines in the same way as hydrogen lines. In hydrogen the effect is symmetrical whereas in helium it results in some forbidden transitions being allowed. What we believe we are seeing is two of those forbidden transitions which happen to lie very close to He I λ4388 Å and λ4922 Å, see [1–3] for details. This indicates that the central spike must form in a high electron density region, which favours the white dwarf as its origin.

Hel 4388   Hel 4922   Hel 5016   Hel 7065

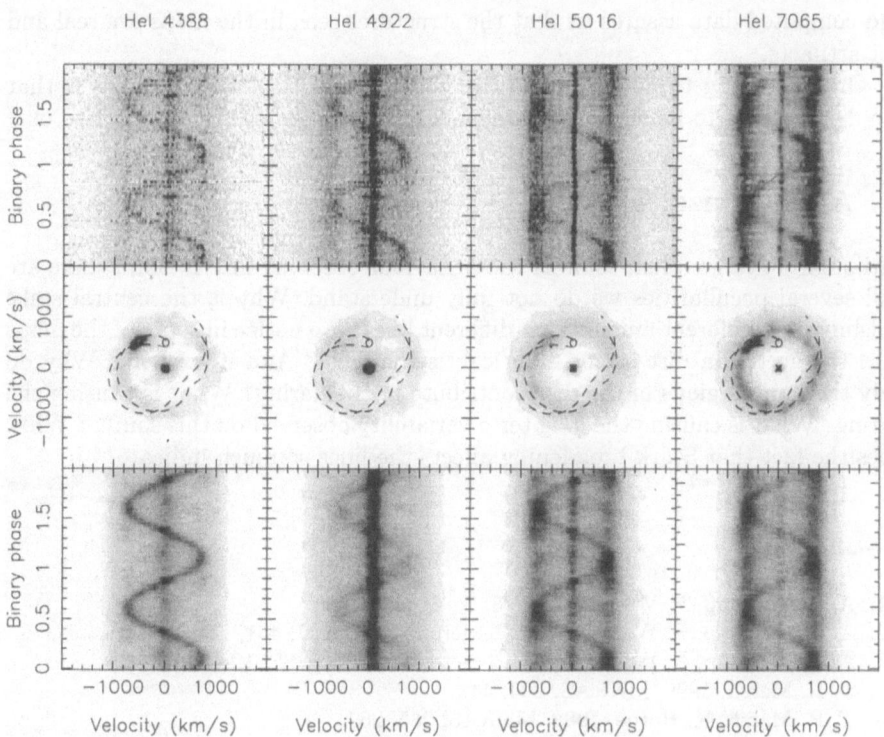

Fig. 3. Trail spectra of 4 helium emission lines (**top**), Doppler maps calculated from the spectra (**middle**) and spectra computed back from the maps (**bottom**). The donor star, the gas stream velocity and the Keplerian velocity along the stream are also plotted on the Doppler maps. The position of the white dwarf is marked by a cross

Also clear in the trails are the two peaks equidistant from the rest wavelength that correspond to the accretion disc, and a sinusoidal component that moves between the two peaks. The middle panels are the Doppler tomograms obtained using MEMSYS techniques [28] (Marsh, this volume). The emission in the centre of the map, therefore at low velocities, corresponds to the central spike. The red and blue peaks of the lines map into a ring around the centre of mass that corresponds to the accretion disc. The sinusoidal component maps into an emission region on the top left quadrant of the map. This is the position in the map where we expect to find any emission coming from the bright spot. We observe that the bright spot shows a complex structure, stretched along the accretion disc, for some lines. This behaviour had been observed previously [4]. When Marsh [4] measured the radial velocity of the spot at different orbital phases, he realised that it moved in a semi-sinusoidal fashion, the values of the velocity always being between the radial velocities of the stream and the disc at the bright spot position. The bottom panels of Fig. 3 present the trails of the spectra computed back from the Doppler maps. The agreement between the real and computed data assures us that the structures seen in the maps are real and not artifacts.

One last thing to notice in the He I λ5015 Å and He I λ7065 Å maps is that the disc appears to be slightly elliptical. The reason for this is not clear to us.

4   A Few Puzzles Still

Although we can explain some of the behaviour observed in GP Com, there are still several peculiarities we do not fully understand. Why is the central spike redshifted by different amounts for different lines? We notice in some of the maps that the accretion disc seems elliptical: is that real? And if so, why? Why do only the inner regions of the disc contribute to the flaring? What is causing the flaring? What is causing the long term variability observed on this source? What does the fact that Stark broadening affects the lines so much indicate?

References

1. A. Beauchamp, F. Wesemael, 1998, ApJ, 496, 395
2. A. Beauchamp, F. Wesemael, P. Bergeron, 1995, ApJ, 441, L85
3. A. Beauchamp, F. Wesemael, P. Bergeron, 1997, ApJSS, 108, 559
4. T. R. Marsh, 1999, MNRAS, 304, 443
5. T. R. Marsh, K. Horne, 1988, MNRAS, 235, 269
6. T. R. Marsh, K. Horne, S. Rosen, 1991, MNRAS, 366, 535
7. T. R. Marsh, J. H. Wood, K. Horne, D. Lambert, 1995, MNRAS, 274, 452
8. L. Morales-Rueda, T. R. Marsh, R. C. North, J. H. Wood, MNRAS, *in preparation*
9. R. Nather, E. Robinson, R. Stover, 1981, ApJ, 244, 269
10. K. Osawa, 1959, ApJ, 130, 159
11. J. Smak, 1975, Acta Astr, 25, 227
12. R. Stover, 1983, PASP, 95, 18
13. M. Tsugawa, Y. Osaki, 1997, PASJ, 49, 75
14. A. Tutukov, L. Yungelson, 1996, MNRAS, 280, 1035
15. B. Warner, 1995, *Cataclysmic Variable Stars* (Cambridge University Press, Cambridge)

Hα-Emission Doppler Tomography
of Long-Period Cataclysmic Variable Stars

R.C. North[1], T.R. Marsh[1], C.K.J. Moran[1], U. Kolb[2],
R.C. Smith[3], and R. Stehle[4]

[1] University of Southampton, Department of Physics & Astronomy,
    Highfield, Southampton, SO17 1BJ, UK
[2] Open University, Department of Physics, Milton Keynes, MK7 6AA, UK.
[3] University of Sussex, Astronomy Centre, Brighton, BN1 9QJ, UK
[4] University of Leicester, Astronomy Group,
    University Road, Leicester, LE1 7RH, UK

Abstract. We present Doppler maps of Hα (6562.76Å) emission of 4 well-known dwarf
novae, SS Cyg, AH Her, EM Cyg and V426 Oph. All 4 systems were in quiescence during
our observations. All of them have visible mass donor stars allowing us to establish
precise orbital phases. None of them show what is often thought of as the classic
pattern of symmetric disc plus bright-spot at the gas stream/disc impact. Instead they
have regions of emission at low velocities in the area below the point representing
the velocity of the white dwarf. In addition, emission with a velocity consistent with
an origin on the heated face of the mass donor can be seen. We consider possible
explanations for these peculiar images.

1 Long-Period Dwarf Novae

Using Doppler tomography to obtain maps of the emission from the accretion discs in cataclysmic variables (hereafter CVs) has now been widely accepted
as a technique to analyse these systems [4], Marsh, this volume. CVs are binary
stars in which one of the stellar components, a main-sequence star like our Sun, is
overfilling its Roche lobe and losing material onto its companion, a white dwarf,
through the inner Lagrangian point. Dwarf novae are a sub-group of CVs. They
spend time in two distinct states; one when the disc is bright (outburst), and
the other when the disc is dim (quiescence). The change in brightness of the
system from one state to the other is of the order 2–5 magnitudes. These objects
have periods ranging from less than an hour to a few days, however most of
them are gathered in two distinct groups, one with $\sim 1 \le P_{\rm orb} < 2.2\,{\rm hr}$ and the
other with $P_{\rm orb} > \sim 3\,{\rm hr}$. Long-period dwarf novae belong to the second group.
They generally have a K or M spectral-type mass donor. The larger mass donors
manifest themselves in the spectra through a set of absorption lines, which are
broadened compared to those in field stars of similar spectral type due to the
high rotational velocities ($\sim 100\,{\rm km\ s^{-1}}$). These lines can be used to derive
accurate orbital parameters for the mass donor, and therefore, the binary.

Four long-period dwarf novae ($P_{\rm orb} > 6\,{\rm hr}$) were observed over a full orbital
period whilst in their quiescent state with the 2.5 m Isaac Newton Telescope on

the island of La Palma. All four objects are double-lined spectroscopic binaries. Spectra with a width of 400 Å centred on ≈ 6450 Å were obtained. Each exposure was a few hundred seconds in duration so that orbital smearing was minimised. Covering this wavelength range ensured that the Hα emission line and absorption lines from the mass donor were observed. The He I line at 6678Å was unfortunately just out of range. Figure 1 shows the average spectrum of SS Cyg as an example.

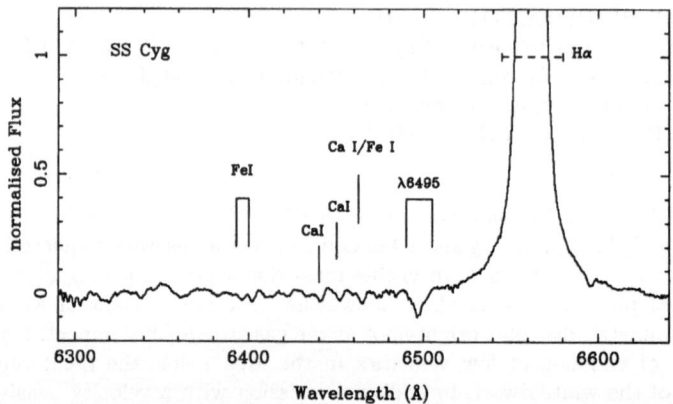

Fig. 1. Average spectrum of SS Cygni

Like all dwarf novae in quiescence, Fig. 1 has prominent Hα emission (mainly from the accretion disc), with a width of thousands of km s^{-1} . Between wavelengths 6400 & 6540 Å the absorption lines from the mass donor can be seen. These have widths of ~ 100 km s^{-1} and can be used to determine the radial velocities for the mass donor. Producing radial velocities in this way allows accurate orbital phases to be derived from the amplitude of the motion, and measurement of the projected rotational broadening of the lines allows a value of the mass ratio, q, to be calculated without having to depend on any measurements made using emission lines – inherently error-prone. The radial velocities calculated for each spectrum (using a cross-correlation method involving radial-velocity standard stars [10]) were fit with a sinusoid of the form

$$V = \gamma + K \sin \frac{2\pi(t - t_0)}{P_{\mathrm{orb}}} \qquad (1)$$

where γ is the centre-of-mass velocity of the binary, K is the semiamplitude of the sine curve, and t_0 is the HJD at the zero phase point, and then the best fit radial velocity curve was deduced. All the orbital periods were fixed at previously determined values.

An example of the results of this process, using the absorption lines in order to obtain an accurate radial velocity curve for each object, is shown in Fig. 2. Following this procedure enabled accurate orbital phases to be determined for the construction of the Doppler maps.

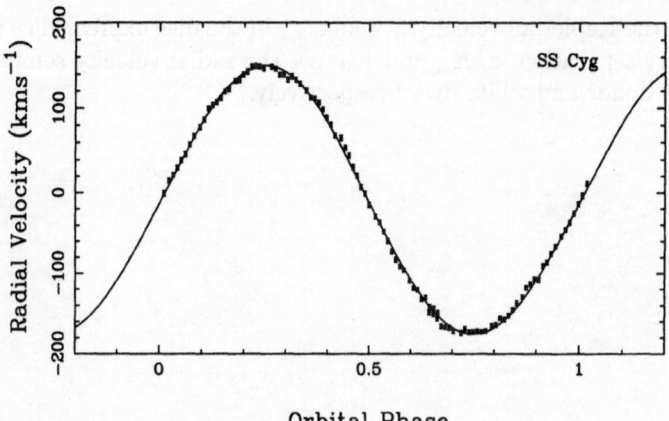

Fig. 2. The radial velocity curve obtained for SS Cygni, showing the data points and the fit (full line)

2 The Hα Doppler Maps

Doppler maps are basically images in velocity space. Reconstructing the image in velocity space means that no assumptions have to be made about the nature of the velocity field in the binary, and so emission from the gas stream and red star (neither of which are expected to obey a Keplerian relation) can be identified. The frame of reference used for the basis of the coordinate system is one which co-rotates with the binary. Since the mass donor is synchronously rotating with the binary, the shape of its Roche lobe is conserved during the transformation to velocity space, and so any emission sites due to the mass donor become obvious, appearing as spots at velocities equivalent to its semiamplitude on the positive y-velocity axis, in the top-centre of the image.

We would expect a 'classic' Doppler map at Hα to have an annulus of emission centred on the velocity of the white dwarf, due to the double-peaked line emission from the rotating accretion disc [3]. Then, if emission from the mass donor is significant, we see a spot at the velocity of the mass donor, as explained previously. A bright-spot, forming where the gas stream hits the outer edge of the disc, will be seen in the top-left quadrant of the map somewhere along the velocity path calculated for the stream. In the Doppler maps plotted here, the lower path from the mass donor indicates the direct velocity along the gas stream, and the upper path represents Keplerian velocities. The points marked onto those paths are at $0.1\,\mathrm{R}_{L_1}$ intervals. In addition, on the Doppler maps we have plotted a circle at the Keplerian velocity of the outer disc edge, if the disc had a radius of $0.8\mathrm{R}_{L_1}$. This velocity was calculated using the equation

$$v_d = \sqrt{K_R(K_R + K_W)}\left(\frac{a}{r_d}\right)^{\frac{1}{2}} \tag{2}$$

where v_d is the Keplerian velocity at radius r_d in the disc expressed as a function of the binary separation, a. K_R and K_W are the radial velocity semiamplitudes of the mass donor and white dwarf respectively.

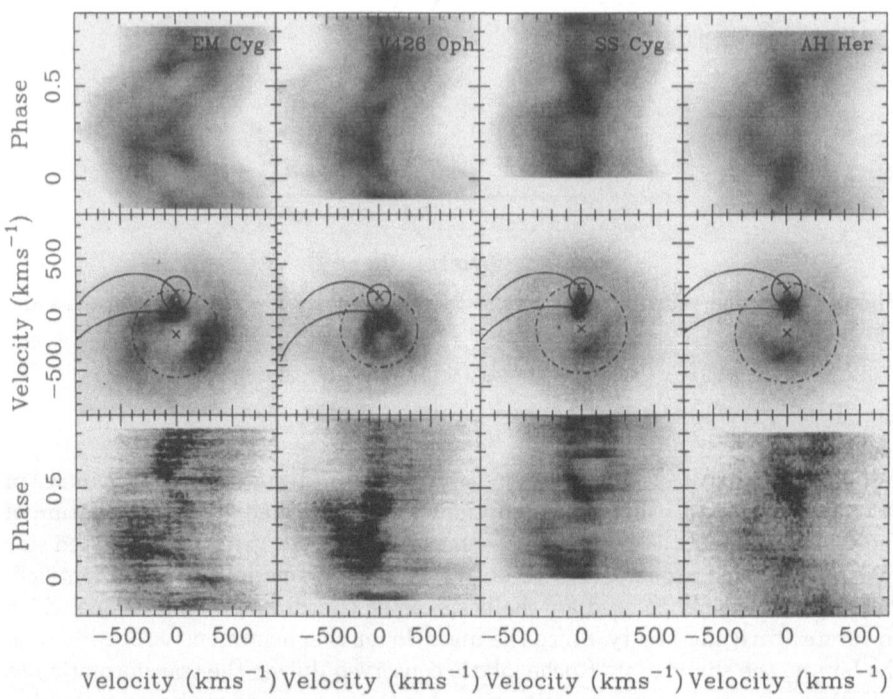

Fig. 3. Hα Doppler maps of the four dwarf novae. From top to bottom: Reconstructed data, Doppler map and trailed spectrum.

Figure 3 is a plot of all four of the Hα Doppler maps (centre panels), bracketed by the observed trail (bottom panel) and the trail reconstructed from the Doppler map (top panel) using a MEMSYS implementation (Marsh, this volume). The appearance of the reconstructed trail is a good indicator of how well the maximum-entropy method fitted the data. The Doppler maps will be discussed in turn, from left-to-right in the figure.

2.1   EM Cygni

EM Cyg shows the most 'normal' Doppler map at the Hα wavelength. The Doppler map of EM Cyg is the closest to what would be expected from a dwarf nova in quiescence. There appears to be evidence of a ring-like formation, however most of this emission is at sub-Keplerian velocities (as indicated by its position inside the circle which marks the calculated outer-disc edge). Errors on the system parameters do not resolve this sub-Keplerian behaviour. From [5] the system appears to be at, or coming down from a standstill to, a quiescent level at the time of the observations. There appears to be some emission situated on the hemisphere of the mass donor facing the white dwarf. There is also absorption at the inner Lagrangian point. This is a systematic effect due to the presence of an extra source of absorption in Hα, most likely an additional late-type star [6]. The trailed spectrum in the bottom panel of Fig. 3 show clearly the 'Z-wave' effect caused by the disc eclipse around phase zero. The outer disc appears at around $300\,\mathrm{km\,s^{-1}}$, which indicates a radius larger than the Roche lobe of the white dwarf, which is impossible. Of course, if the Hα emission is not confined to the plane of the binary then this may well explain why we are observing emission at apparently sub-Keplerian velocities in the Doppler map.

2.2   V426 Ophiuchi

Column 2 shows the results for V426 Oph. This system has been classified as a dwarf nova of the Z Cam sub-type [7], however questions have been raised as to whether it belongs to the magnetic group of CVs instead [9] [1]. There is no indication in the literature that any eclipses occur. However, in the trail of the actual data, around phase 0 there appears to be an eclipse. The Doppler map (as can be seen in the middle panel) has no real evidence for the annular signature of an accretion disc. There is a general emission background apparent out to $\sim 700\,\mathrm{km\,s^{-1}}$. Superimposed on top of it is a sausage-shaped region of enhanced emission apparently situated between the inner Lagrangian point and the centre of the white dwarf. Slingshot prominences [8] have been suggested as an explanation for emission in this location, however the idea was used to explain features in outburst-state Doppler maps. At high negative velocities on the Doppler map (lower-left quadrant), emission is present at velocities which could indicate stream overflow and appear to agree with the theory.

2.3   SS Cygni

SS Cyg is one of the well-studied dwarf novae, due to its brightness. Column 3 in Fig. 3 shows the data, map and reconstructed data, in that order from bottom to top. The first thing to notice is that there doesn't appear to be the expected emission distribution from an accretion disc. There is instead two spots of emission, one located at the mass donor, and the other at apparently sub-Keplerian velocities moving in phase with the white dwarf. The two spots can be seen as sinusoidal components in the trailed data. One sine wave starts around zero velocity and moves to positive velocities, with a semiamplitude of around $150\,\mathrm{km\,s^{-1}}$ – this behaviour pins it's origins to the mass donor. The other sine wave starts out at negative velocities, reaches a maximum and moves back towards zero velocity which it appears to reach around phase 0.5, with a semiamplitude of $\sim 350 - 400\,\mathrm{km\,s^{-1}}$. Horne [2] proposed that this type of Doppler map would appear due to the presence of a *magnetic propeller* – allowing blobs of gas to be ejected from the system due to the interaction between the angular momentum of the gas and a magnetic field on the inner accretion disc. Could it be that the low velocity material in the Doppler map is an indication of this effect?

2.4   AH Herculis

Finally, column 4 shows the results from AH Her. The Doppler map is similar to that for SS Cyg. There are two S-waves visible in the trailed spectra, in anti-phase with each other. The reconstructed data shows enhanced emission at phases 0 & 0.5; the observations only show it at phase 0.5, when we see the side of the mass donor facing the white dwarf, and it does appear on the sine wave attributable to it, indicating an origin there. This sine wave does map onto the Doppler image as an emission region at velocities equivalent to the hemisphere of the mass donor facing the white dwarf. As in SS Cyg, there is an emission region in the bottom-centre of the Doppler map, but in AH Her it spills over into the bottom-left quadrant of the map. The semiamplitude of its motion is again $\sim 350\,\mathrm{km\,s^{-1}}$ and it is phased with the accretion disc/white dwarf.

3   Conclusions

Four Hα Doppler maps of dwarf novae with orbital periods greater than six hours have been presented. None of them show the classic disc-emission/bright-spot pattern we would expect from these objects. Various reasons have been given for the appearance of these Hα maps. There is a possibility that emission out of the binary orbital plane exists which would affect the Doppler images. Why are these Doppler maps so different to what is expected, and observed in other spectral lines, for objects like these in quiescence?

References

1. C. Hellier, D. O'Donoghue, D. Buckley, A. Norton, 1990, MNRAS, 242, P32
2. K. Horne, 1999, 'Disk-Anchored Magnetic Propellers – A Cure for the SW Sex Syndrome'. In: *Annapolis Workshop on Magnetic Cataclysmic Variables, 13–17 July 1998*, ed. by C. Hellier, K. Mukai, (A.S.P. conference series Vol. 157), pp. 349–356
3. K. Horne, T.R. Marsh, 1986, MNRAS, 218, 761
4. T.R. Marsh, K. Horne, 1988, MNRAS, 235, 269
5. J.A. Mattei, 1997, *Observations from the AAVSO international database*, priv. comm.
6. R.C. North, T.R. Marsh, C.K.J. Moran, U. Kolb, R.C. Smith, R. Stehle, 2000, MNRAS, 313, 383
7. H. Ritter, U. Kolb, 1998, A&AS, 129, 83
8. D. Steeghs, K. Horne, T.R. Marsh, J.F. Donati, 1996, MNRAS, 281, 626
9. P. Szkody, 1986, ApJ, 301, L29
10. J. Tonry, M. Davis, 1979, AJ, 84, 1511 (1979)

Doppler Tomography of the Dwarf Nova IY UMa During Quiescence

D.J. Rolfe[1], T.M.C. Abbott[2], and C.A. Haswell[1]

[1] Department of Physics and Astronomy, The Open University,
   Walton Hall,
   Milton Keynes, MK7 6AA, UK
[2] Nordic Optical Telescope, Roque de Los Muchachos & Santa Cruz de La Palma,
   Canary Islands, Spain

Abstract. Quiescent Doppler tomography of the newly discovered deeply-eclipsing SU UMa system IY UMa reveals properties of the region where the accretion stream from the donor impacts the edge of the disc. A very strong bright spot is produced and the Keplerian disc emission in the impact region is disrupted or obscured. The differing properties of Hα, Hβ and He I emission will allow physical parameters of the converging flow region to be studied.

1   Introduction

IY UMa was observed for the first time with a superoutburst in January 2000 identifying it as a SU UMa type dwarf nova cataclysmic variable [11] with or-

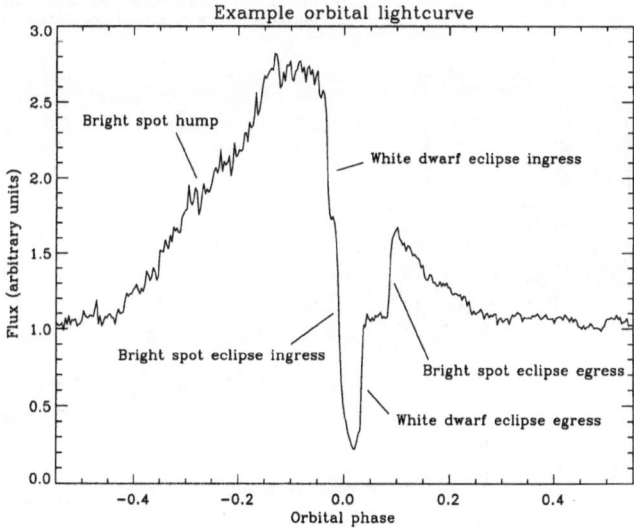

Fig. 1. Example orbital lightcurve of IY UMa close to quiescence. Note the strong orbital hump as the stream-disc impact region comes into view, the eclipse of this region and the eclipse of the white dwarf. Taken from [7]

bital period 1.77 hours. It exhibits deep eclipses, with the eclipse of the white dwarf and the stream-disc impact region being clearly identifiable in quiescence (Fig. 1); this behaviour is similar to that of the other eclipsing SU UMa systems OY Car [8] and Z Cha [12]. The orbital parameters have been estimated [6] as $M_{wd} = 0.93 \pm 0.14 M_\odot$, $M_{donor} = 0.12 \pm 0.03 M_\odot$ and inclination $i = 87° \pm 3°$.

2   Spectroscopy

2.1   Observations

We obtained 3 orbits of coverage on 19th March 2000 using ALFOSC on the Nordic Optical Telescope in La Palma. There are 80 spectra with resolution ~4Å covering wavelength range 3900Å to 6850Å.

2.2   Average Spectra

Figure 2 shows average spectra covering different sections of the orbit. The top spectrum (a) is out of eclipse data; the middle spectrum includes phases where the disc is eclipsed but the white dwarf is **not**; (c) is the average during white dwarf eclipse. The spectra show double-peaked Balmer, He I and Fe II emission, the double-peaked structure signalling that this emission comes from the accretion disc. There are also strong broad Balmer absorption wings in Hβ, Hγ and

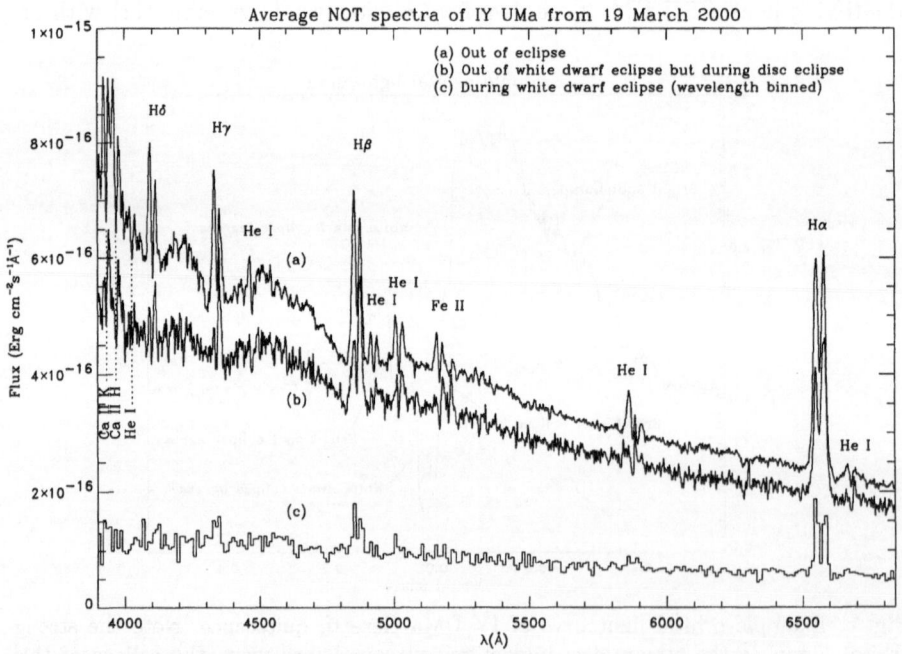

Fig. 2. Average spectra of IY UMa on 19th March 2000

Hδ in spectra (a) and (b) which only disappear during white dwarf eclipse (c), telling us that these features come from close to the white dwarf, if not from the white dwarf itself. There is also deep core absorption clearly seen in Hδ, He I and Fe II. These features are almost identical to those seen in OY Car [2]. Z Cha also shows very similar features in quiescence [4].

2.3 Systemic Velocity

The radial velocity of the Hα line was measured for each spectrum by fitting a double gaussian profile and using the velocity of the midpoint. A sinusoid of the form $V = V_0 - V_1 sin2\pi(\phi - \phi_0)$ was fitted to those velocities outside eclipse. We obtain systemic velocity $V_0 = 15.8 \pm 1.3$ km s^{-1}, velocity amplitude $V_1 = 100.4 \pm 1.4$ km s^{-1} and $\phi_0 = 0.119 \pm 0.003$. The large phase shift, ϕ_0 relative to white dwarf mid-eclipse, tells us that the emission does not follow the motion of the white dwarf and so V_1 cannot be treated as the white dwarf velocity. This phase shift is the same as that seen in the unusual dwarf nova WZ Sge [5] and also similar to those in SU UMa systems OY Car (measurements summarized in [2]) and Z Cha [4]. Similar phase shifts are seen in quiescent low mass X-ray transients e.g. V616 Mon [1].

2.4 Trailed Spectra

In Fig. 3 we show trailed spectra of Hα, Hβ and He I 5876Å. In all three lines we see the eclipse beginning in the blue and ending in the red as first the side of

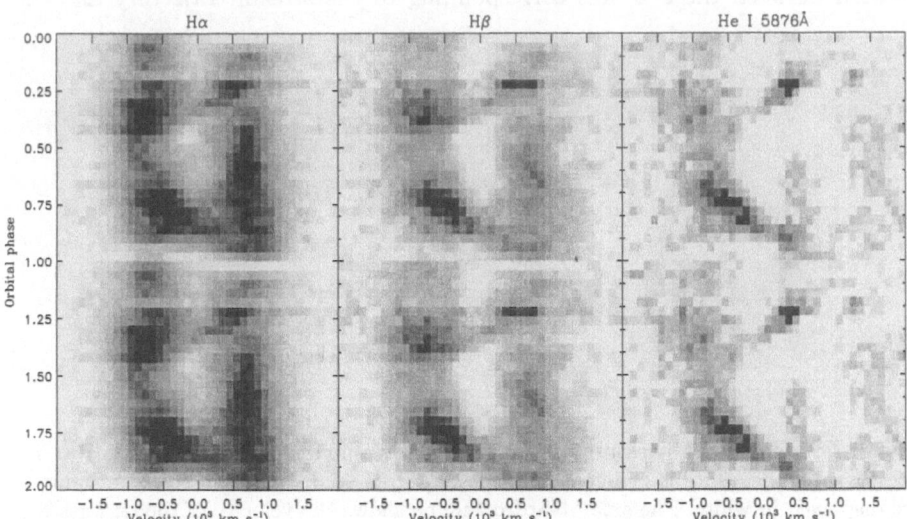

Fig. 3. Phase-folded, velocity-binned and continuum-subtracted trailed spectra. Black corresponds to the maximum flux, while white is the continuum level. Absorption (values below continuum) are displayed in white

the disc coming towards us and then the side of the disc moving away is occulted by the donor star. In Hα we clearly see the double-peaked emission component from the disc, with average peak-to-peak separation in the phase range 0.4–0.6 (where the bright spot is on the far side of the disc and so will have minimal effect on the measurement) of 1440 km s^{-1} corresponding to a radius of 0.45a assuming a Keplerian velocity field (a is the orbital separation). There is also a strong S-wave component corresponding to a localized region of emission. The S-wave is weakest around orbital phase 0.5. The structure in Hβ is very similar, except that the disc emission is fainter compared to the S-wave. He I shows very little evidence of disc emission, but again exhibits a strong S-wave which is weakest around phase 0.5. Without the complication of the disc component, we see that the brightness of the S-wave closely follows the orbital hump in the continuum lightcurve. The He I emission reveals that the eclipse of the S-wave is late, placing it in the correct region of the disc to be the stream-disc impact region. The strong low velocity absorption is present.

2.5 Doppler Maps

We used the Fourier-filtered back-projection method [3] to obtain maps of the velocity-space distribution of emission in each line, shown in Fig. 4. Only out-of-eclipse spectra were used to generate the Doppler maps, since occultation by the donor star violates a core assumption of the Doppler tomography method, which assumes that all regions have the same visibility at all orbital phases.

The Hα map (Fig. 4a) shows a ring of disc emission with Keplerian velocities corresponding to locations within the tidal radius. The stream-disc impact should be seen between the two arcs corresponding to the stream trajectory and its

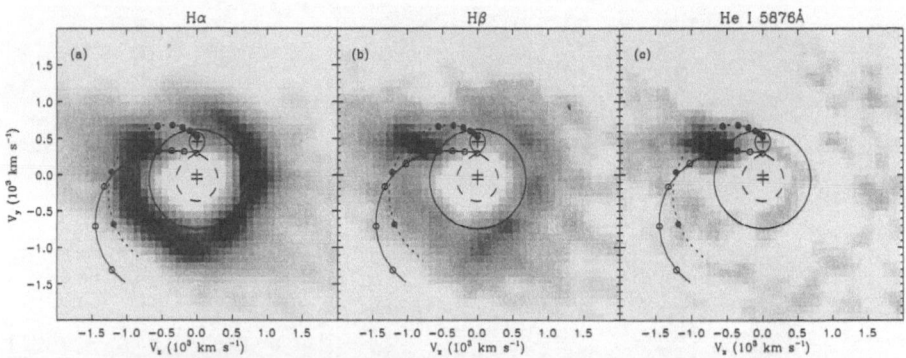

Fig. 4. Fourier-filtered back-projections. Black corresponds to the maximum flux in each map, while white is the continuum level. Absorption (values below continuum) are displayed in white. The small teardrop shape shows the velocity of the donor star, the arc with unfilled circles shows the velocity along the stream trajectory with circles denoting steps of 0.1a along the stream. The arc with solid circles is the Keplerian velocity along the stream trajectory. The circle is the Keplerian velocity at the tidal truncation radius

Keplerian velocity in the top left of the map, but there is no significantly brighter region here in the $H\alpha$ map. The disc emission just to the left of the donor star is very much weaker than elsewhere. Assuming a Keplerian velocity field, this corresponds to weaker emission from the region of the disc marked with a thick black outline in Fig. 1, and is coincident with the stream-disc impact. We do not expect a Keplerian velocity field where the stream and disc merge, and any underlying Keplerian emission could be obscured by an optically thick region around the impact. It is therefore no surprise that we see weaker disc emission in this velocity region. At low velocities (less than $\sim$500 km s^{-1}) we see strong absorption. This absorption is of similar strength to the disc emission.

The $H\beta$ map (Fig. 4b) again shows the Keplerian emission from the disc and the strong low-velocity absorption. Most notable, however, is the bright spot located exactly where we expect to see the stream-disc impact. Combined with spatial information provided by the late eclipse of the S-wave which corresponds to this hot spot, and the fact that the variation in brightness of the S-wave also closely follows the orbital hump, we conclude that this emission is coming from stream-disc impact. The position along the stream trajectory at which the $H\beta$ bright spot is brightest corresponds to the grey region in Fig. 1, just within the disc radius deduced from the peak-peak separation of the $H\alpha$ disc emission.

The He I 5876Å map (Fig. 4c) shows no disc emission at all. It has the strong low-velocity absorption and also the bright spot due to the stream-disc impact.

The behaviour of the $H\alpha$ and $H\beta$ maps is the same as that seen in WZ Sge [9] and also very similar to that in HE 1047 [10]. The strong disc component in both lines, and the increase in the relative strength of the stream-disc impact to the disc in $H\beta$ is seen in both of the systems, while the weaker region of disc emission in $H\alpha$ between the hot spot and donor is seen in WZ Sge but not in HE 1047.

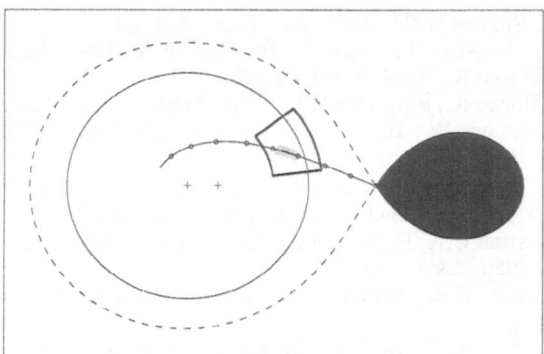

Fig. 5. Spatial geometry interpreted from Doppler map. Stream from donor star (with circles at separation $0.1a$ marked along it) first impacts disc at hot spot (grey region). Circle marks radius at which Keplerian velocity is equal to half the $H\alpha$ peak separation

2.6 Conclusions

The deeply-eclipsing dwarf nova IY UMa has average optical spectra during quiescence similar to those of the two similarly high inclination dwarf novae OY Car and Z Cha. Time-resolved spectroscopy and the method of Doppler tomography reveals 'classic' accretion flow behaviour in this system. The stream from the donor star impacts the edge of the Keplerian accretion disc, dissipating its energy in a fairly concentrated region (the bright spot seen in Figs. 4b and 4c). This stream-disc impact disrupts and/or obscures the Keplerian emission leading to the weak disc emission in the top left of Fig. 4a.

Acknowledgements

The data presented here have been taken using ALFOSC, which is owned by the Instituto de Astrofisica de Andalucia (IAA) and operated at the Nordic Optical Telescope under agreement between IAA and the NBIfA of the Astronomical Observatory of Copenhagen. The Nordic Optical Telescope is operated on the island of La Palma jointly by Denmark, Finland, Iceland, Norway, and Sweden, in the Spanish Observatorio del Roque de los Muchachos of the Instituto de Astrofisica de Canarias. The authors thank Joe Patterson and Jonathan Kemp for the data used to produce Fig. 1. The Doppler maps were produced using the software package MOLLY by Tom Marsh. We acknowledge the data analysis facilities at the Open University provided by the OU research committee and the OU computer support provided by Chris Wigglesworth. DJR is supported by a PPARC studentship. CAH acknowledges support from the Leverhulme Trust F/00-180/A.

References

1. Haswell C.A., Shafter A.W., 1990, ApJ Lett., 359, L47
2. Hessman F.V., Koester D., Schoembs R., Barwig H., 1989, A&A, 213, 167
3. Marsh T.R., Horne K., 1988, MNRAS, 235, 269
4. Marsh T.R., Horne K., Shipman H.L., 1987, MNRAS, 225, 551
5. Mason E., Skidmore W., Howell S., Ciardi D., Littlefair S., Dhillon V.S., 2000, MNRAS, 318, 440
6. Patterson J., Thorstensen J., Kemp J., Jensen L., Vanmunster T., Abbott T., Skillman D., Martin B., Fried R., 2000, PASP, submitted
7. Rolfe D.J., Haswell C.A, Patterson J., 2000, astro-ph/0009420
8. Schoembs R., 1986, A&A, 158, 233
9. Skidmore W., Mason E., Howell S.B., Ciardi D., Littlefair S., Dhillon V.S., 2000, MNRAS, 318, 429
10. Skidmore W., Mason E., Still M., Horne K., 2000, poster presented at 'Astro-Tomography', Brussels
11. Uemura M., Kato T., Matsumoto K., Takamizawa K., Schmeer P., Jensen L.T., Vanmunster T., Novák R., Martin B., Pietz J., Buczynski D., Kinnunen T., Moila-nen M., Oksanen A., Cook L.M., Watanabe T., Maehara H., Itoh H., 2000, PASJ, in press
12. Wood J.H., Horne K., Berriman G., Wade R., O'Donoghue D., Warner B., 1986, MNRAS, 219, 629

Spiral Waves in Accretion Discs – Observations

D. Steeghs

Physics & Astronomy, Southampton University, Southampton SO17 1BJ, UK

Abstract. I review the observational evidence for spiral structure in the accretion discs of cataclysmic variables (CVs). Doppler tomography is ideally suited to resolve and map such co-rotating patterns and allows a straightforward comparison with theory. The dwarf nova IP Pegasi presents the best studied case, carrying two spiral arms in a wide range of emission lines throughout its outbursts. Both arms appear at the locations where tidally driven spiral waves are expected, with the arm closest to the gas stream weaker in the lines compared to the arm closest to the companion. Eclipse data indicates sub-Keplerian velocities in the outer disc. The dramatic disc structure changes in dwarf novae on timescales of days to weeks, provide unique opportunities for our understanding of angular momentum transport and the role of density waves on the structure of accretion discs. I present an extension to the Doppler tomography technique that relaxes one of the basic assumptions of tomography, and is able to map modulated emission sources. This extension allows us to fit anisotropic emission from, for example, spirals shocks, the irradiated companion star and disc-stream interaction sites.

1 Accretion Discs and Angular Momentum

The energetic phenomena associated with a wide range of accreting systems rely on the efficient conversion of potential energy into radiation and heat. In close binaries, the deep potential well of the compact object leads to mass transfer and accretion once the companion star evolves and Roche lobe overflow commences. The efficiency of accretion is proportional to the compactness, M/R, of the accreting compact star with mass M and radius R. As matter spills over near the first Lagrangian point, it sets off on a ballistic trajectory towards the accretor. Its potential energy is converted into kinetic energy, but its net angular momentum, due to orbital motion of the mass donor, prevents a straightforward path to the accretor. The natural orbit for such matter is a circular Keplerian orbit corresponding to its specific angular momentum. Instead of dumping material directly onto the compact star, the primary Roche lobe is slowly filled with a near Keplerian disc. Angular momentum needs to be dispersed within this accretion disc in order to allow gas to spiral inwards towards the compact star [7].

It is the detailed process of angular momentum transport that determines the structure of this accretion disc and therefore the rate at which gas, supplied from the mass donor, is actually accreted by the compact object. Although so fundamental to the process of accretion through discs, our understanding of angular momentum dispersal is very limited. We can roughly divide the possible physical mechanisms that may provide the required angular momentum transport into two classes. Those that work on a local scale and exchange angular momentum among neighbouring parcels in the disc, and those that rely on global, large scale structures in the disc. The local processes are commonly referred to as 'viscous processes' even though it was clear that the molecular viscosity of the accretion disc material itself was many orders of magnitudes too small [14]. Viscous interaction in the sheared Keplerian disc allows some material to spiral inwards, losing angular momentum, while excess momentum is carried outwards by other parts of the flow. In the famous α-parameterisation of Shakura & Sunyaev [22], this viscosity was replaced by a single dimensionless constant, which allows one to solve the structure equations for thin, viscously heated accretion discs [19].

A very different way of transporting the angular momentum is via density waves in the disc. In self-gravitating discs, the ability of density waves to transport angular momentum is a direct result from purely gravitational interaction between the wave and the surrounding disc material [31,4]. Waves of this type can still transport angular momentum in the absence of self gravity, provided some mechanism exists that exchanges momentum between the wave and the fluid. Sawada et al. [21] conducted numerical simulations of mass transfer via Roche lobe overflow of inviscid, non self gravitating discs, and witnessed the development of strong spiral shocks in the disc which were responsible for the bulk of the angular momentum transport throughout the flow. Such trailing spiral patterns are the natural result of a tidal deformation of the disc that is sheared into a spiral pattern by the (near) Keplerian rotation profile of the disc material [20].

In this review I aim to give an overview of the observational efforts to study such spiral structures in the discs of cataclysmic variables (CVs). Tomography is ideally suited for the study of disc structure, and the detection of a spiral structure in the disc of IP Pegasi [27], 11 years after the work by Sawada et al., triggered a renewed interest into such models. H. Boffin, in this volume, will focus on the theoretical side of the issue and the comparison between the observations and numerical simulations of spiral waves in discs.

2   Prospect for Detecting Spiral Waves in Discs

Tidally generated spiral waves are the result of tidal torques of the companion star on the orbiting disc material. Initially triggered at the outer edge of the disc, where the tidal interaction between the disc and the companion star is strongest, they take the form of trailing spirals because the azimuthal velocity of the disc material is supersonic and increases monotonically with decreasing distance from the compact object ($v_\phi \propto r^{-1/2}$). Both the density as well as the disc temperature are much higher at the location of the spirals, and continuum and line emission from such spirals can thus be expected to be in clear contrast with the surrounding areas of the disc. This contrast depends on the density contrast in the wave, i.e. the strength of the shock, as well as local radiative transfer conditions.

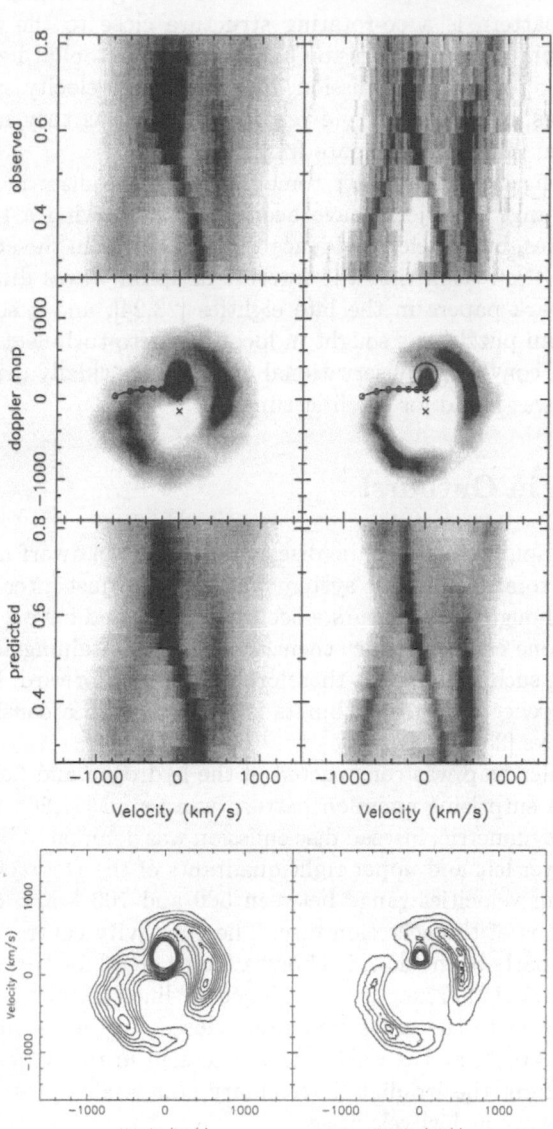

Fig. 1. Top; the observed Hα and HeI emission line profiles during the early stages of an IP Pegasi outburst. The maximum entropy Doppler tomograms (second row) reveal a prominent spiral pattern in the disc as well as secondary star emission. Third row of panels are the predicted line profiles given the calculated maps. Bottom row is a contour plot of the Doppler tomograms. From [27].

Although predicted in the 80s, observational evidence for spirals relies on the ability to spatially resolve the accretion discs in interacting binaries. Indirect imaging methods are thus required to search for such global disc asymmetries. Since the wave pattern is a co-rotating structure close to the orbital plane, Doppler tomography of strong emission lines is the ideal tool at hand. Although providing an image of the line emission distribution in velocity space, and not spatial coordinates, spirals should be readily identified as they maintain their spiral shape in the velocity coordinate frame.

Since the application of Doppler tomography to the discs in CVs (Marsh, this volume), a range of objects have been imaged, showing a rich variety of accretion structures, but no clear evidence for spiral arms in the accretion discs. This was part of the reason that the interest in spiral waves diminished after a series of landmark papers in the late eighties [23,24], and a solution to the angular momentum puzzle was sought in local magneto-turbulent processes [1]. In 1997, however, convincing observational evidence for tidally driven spirals in the disc of a CV was found for the first time.

3   IP Pegasi in Outburst

As part of a programme to study the disc evolution of the dwarf nova IP Pegasi through Doppler tomography, the system was observed just after rise to one of its outbursts. Although such outbursts occur regularly and have a characteristic recurrence time, one cannot predict them accurately. Obtaining scheduled telescope time during such outbursts is therefore not straightforward. For IP Pegasi, the average recurrence time for outbursts is 88 days, with a considerable RMS variation of 18 days [25].

When a Doppler map was constructed of the hydrogen and helium emission from IP Pegasi, a surprising emission pattern was found [27,28]. The accretion disc was far from symmetric, instead disc emission was dominated by a two armed pattern in the lower left and upper right quadrants of the Doppler map (Figure 1). The spiral arm velocities range between 500 and 700 km/s, corresponding to the outer regions of the accretion disc. The emissivity contrast between the spirals and other parts of the disc, is about a factor of ∼3 for Hα, and ∼5 in the case of HeI6678 emission. There is no evidence for line emission from the bright spot. The spiral arm in the lower quadrant, closest to the secondary, extends over an angle of ∼100°, and is weaker than the arm in the opposite quadrant. Strong emission from the irradiated secondary star is also present, producing the prominent S-wave at low velocities.

The location of this pattern corresponds closely to the radii and azimuths where tidally driven spiral waves are expected. Although the Doppler map is in velocity coordinates, and not Cartesian position coordinates, a spiral pattern in position space corresponds to a spiral pattern in velocity space. In Figure 2, a simple disc model is shown containing an accretion disc around the white dwarf with a purely geometrical spiral pattern and a Keplerian velocity field. The two armed trailing spiral in position coordinates, maps into a similar two armed pattern in velocity space. Although this model is a purely geometric pattern, the location of the spiral roughly corresponds to the location in the disc were

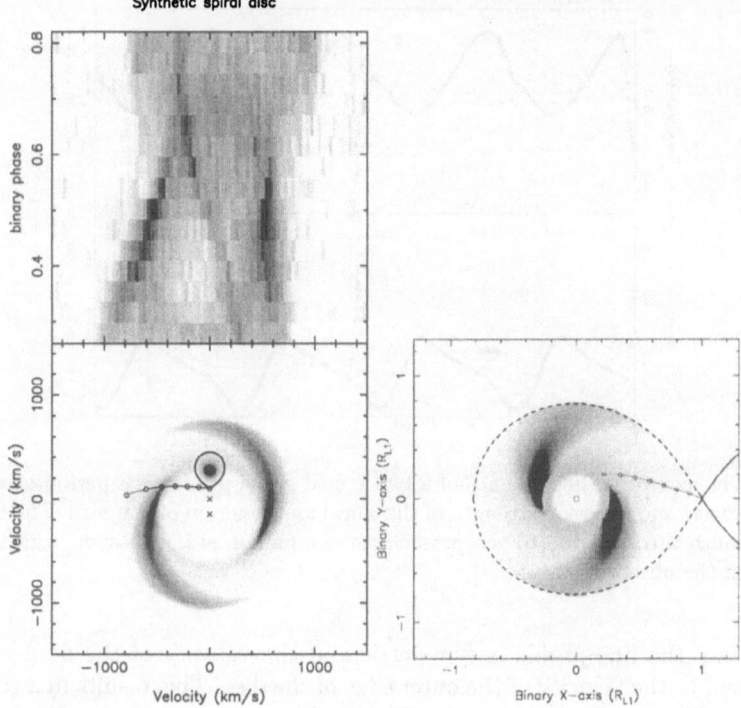

Fig. 2. A simple model of a disc carrying a two armed spiral pattern. Bottom right the distribution of emission in XY coordinates, bottom left in Doppler coordinates and top is the predicted line profiles from such a disc. From [27]

tidally driven waves are expected from model simulations. Even such a simple model already captures most of the structure we observe in IP Pegasi, indicating that most of the disc emission in IP Peg during outburst is indeed localised in a two armed pattern.

Let us look at the main signatures of spiral arms in the observable line profiles based on our simple model. In Figure 3, the velocity and emissivity of the model disc is plotted as a function of azimuth in the Doppler tomogram. Azimuth 0 corresponding to the positive V_y axis, increasing clock wise. The defining feature of a trailing spiral is the change in velocity as a function of azimuth in the tomogram. This corresponds to a change in radius R as a function of azimuth θ in position coordinates that is determined by the opening angle of the spiral, $dR/d\theta$. The velocity field of the disc relates this geometrical opening angle to a particular velocity-azimuth relation in the Doppler map. In the spiral arm dominated line profiles, this produces a regular motion of the double peaks, since azimuth in position coordinates translates to orbital phase in the observed spectrogram. Near orbital phases 0.25 and 0.75, the two arms cross over, producing a sudden jump in the position of the double peaks. For a symmetric Keplerian disc on the

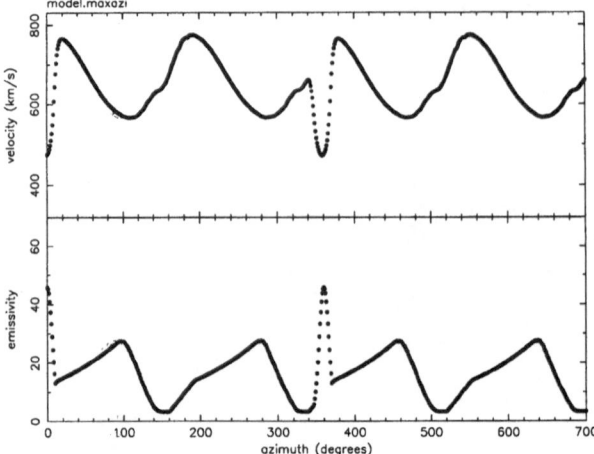

Fig. 3. The azimuthal dependence of a two armed spiral produces a periodic variation in the velocity and emission strength of the spiral as a function of azimuth. The features near azimuth zero are due to the presence of companion star emission, which is also present in the observed data.

other hand, the line profile is symmetric and the velocity of the double peaks correspond to the velocity of the outer edge of the disc. This results in a constant double peak separation as a function of orbital phase and a circular disc image in the Doppler map. With a two armed spiral, the double peaks move, varying their separation considerably and sharply across the binary orbit in a particular way. Although an identification in Doppler maps is perhaps more straightforward, spirals may thus also be identified in the observed line profiles directly.

A week later, during the same outburst, more spectroscopy was obtained before the decline of the outburst had started. The Doppler image (Figure 4) shows that the spiral pattern persists throughout the outburst, and the secondary star now makes a considerably larger contribution (from 6% to 10%) to the line flux. The arm in the upper right quadrant is still stronger, and the location of the arms have not changed, although the upper right arm appears shorter. To highlight the spiral arms in the line profiles, the S-wave of the secondary star was subtracted from the data in the following manner. A Roche lobe shaped mask was used to select all the emission from the secondary, which was then projected in order to produce the trailed spectrogram of the red star emission. This was subtracted from the observed data, and a Doppler image of the disc emission only was constructed (Figure 4, right). The dramatic phase dependence of the double peaks is now clearly visible, and is directly related to the two spiral arms in the disc. This also confirms that the presence of a strong S-wave due to the mass donor does not distort the reconstructed disc structure, since they occupy different locations in the velocity plane.

original data   disc emission only

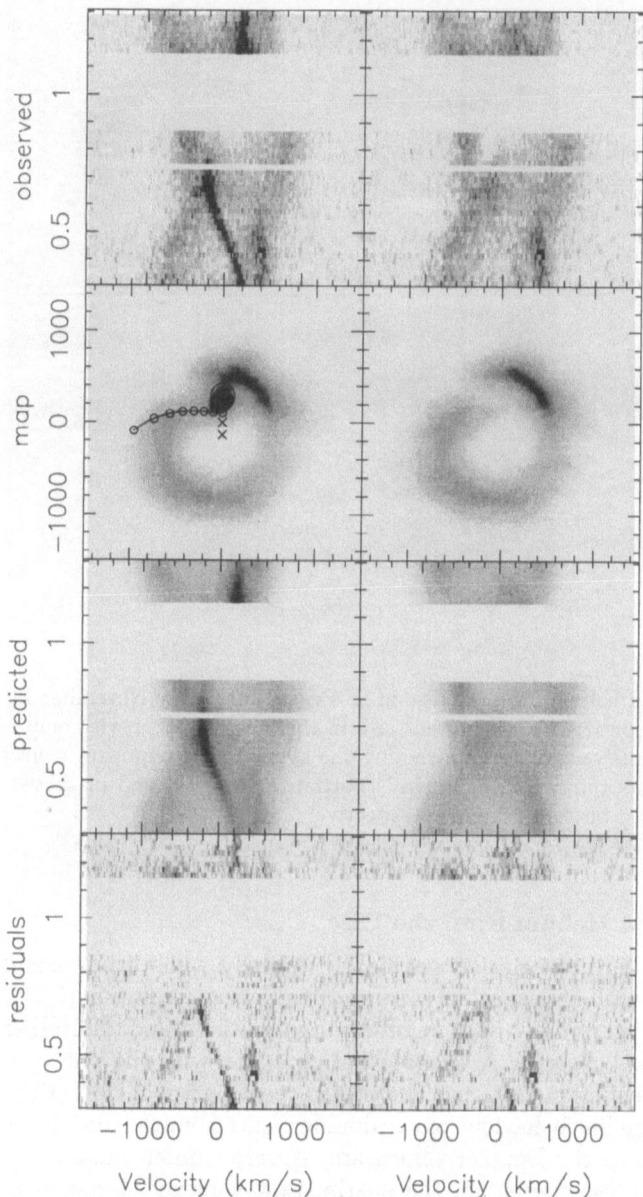

Fig. 4. The spiral pattern persists throughout the outburst, above are HeI6678 data towards the end phases of outburst maximum. From top to bottom, observed data, Doppler map, predicted data and residuals. For the right hand panels, the contribution from the secondary star has been subtracted. From [25]

He II 4686   Doppler image

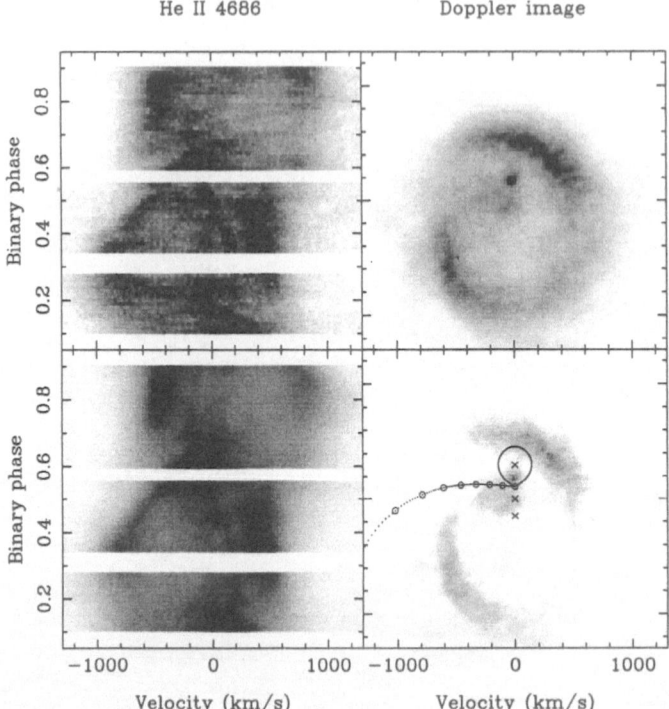

Fig. 5. Ionised helium from the disc in IP Pegasi during the November 1996 outburst. The familiar spiral arms are present also in this line, together with emission from the front side of the secondary. Subtracting the symmetric component from the Doppler map, highlights the two armed spiral (bottom). Observed and predicted data are in the top left and bottom left panel respectively.

3.1   Ionised Helium from the Disc

In a different outburst Harlaftis et al. [11] secured a whole orbit, including eclipse, of IP Pegasi with high spectral (40 km/s) and phase resolution (0.01). This time, the outburst started two days before the observations, and the HeII emission line at 4686Å was observed to provide a comparison with the previously observed Hα and HeI emission patterns. The trailed spectrogram of HeII (Figure 5), again shows the familiar behaviour of the double peaks from the disc, leading to a two armed spiral in the Doppler tomogram. A very similar emission pattern from the disc was also present in the nearby Bowen blend consisting of CIII/NIII emission and the Helium I line at 4471Å. The symmetric component of the map was calculated, using the velocity of the white dwarf as the centre of symmetry, and subtracted in order to highlight the location of the two arms (Figure 5, bottom right).

In order to characterise the properties of the spirals quantitatively, we determined the position of the two spiral arms as a function of velocity in the re-

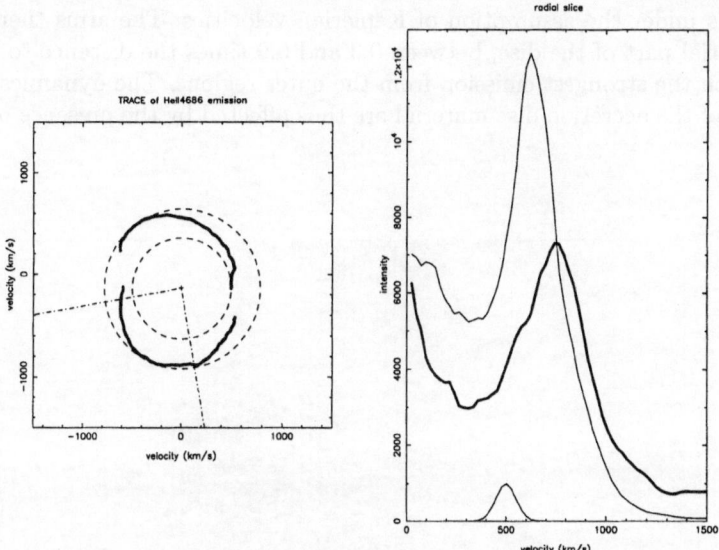

Fig. 6. Tracing the spiral arms in the Doppler map through cross correlation with a Gaussian. Two radial slices through the Doppler map are plotted in the right panel, revealing the radial extent and position of the spiral arms. The instrumental profile (lowest curve) is plotted for comparison. The fitted positions of the spiral arms in the velocity plane are shown in the left panel. Circles deno te the maximum/minimum velocity of the spirals and the two lines indicate the azimuth of the two slices plotted on the right.

constructed tomogram. A slice through the map, starting from the white dwarf at $V_y = -137$ km/s, was made for each azimuth in order to extract the radial profile. Each radial profile was then cross correlated with a Gaussian in order to determine the velocity of the spiral at that particular azimuth. Figure 6 plots the fitted positions of the spirals in the velocity plane together with two radial slices across the map. The two spirals can be traced for almost 180 degrees, and the velocity of the arms varies from 495 km/s to 780 km/s, indicated by the two dashed circles. Near the azimuths where the spiral cross over, the fitting assumptions break down as there is a sudden transition between the velocity of one arm and the next. Unfortunately, the contrast of the arms is too low to follow this switching with our cross correlation method, or a double Gaussian. The widths of the arms are significant, and even change as a function of azimuth. We are thus resolving an intrinsically broad feature. Such an analysis will clearly profit from even higher resolution and signal to noise data since at peak intensity the width of the arm is comparable to our resolution element, and the maximum entropy constraint will tend to broaden features as much as is allowed by the signal to noise of the data.

A tidally distorted disc will not have a pure Keplerian velocity field, and we will see later that the observations indeed indicate this is not the case in IP Pegasi. However, we can estimate the radii corresponding to the velocity of the spirals under the assumption of Keplerian velocities. The arms then cover a substantial part of the disc, between 0.3 and 0.9 times the distance to the L_1 point, with the strongest emission from the outer regions. The dynamics of the majority of the accretion disc material are thus affected by the presence of these spirals.

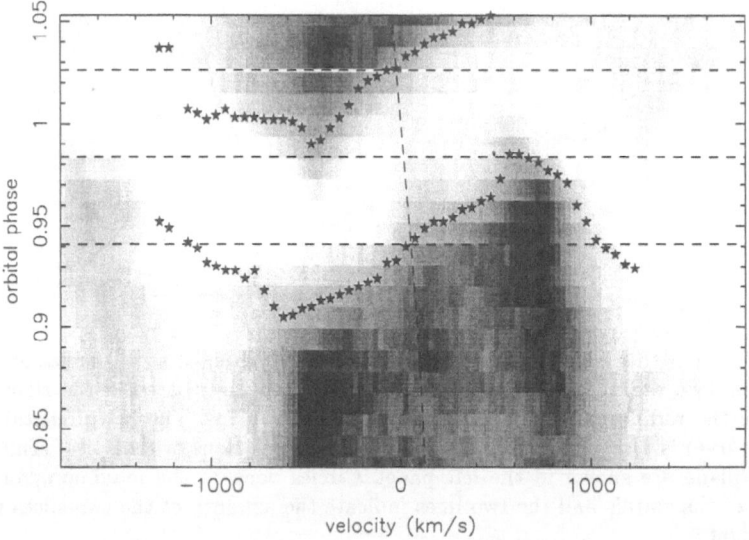

Fig. 7. Measurements of the phases of half depth across the HeII eclipse are denoted by asterisks. Horizontal, dashed lines are the average post-eclipse half depth (top), pre-eclipse half-depth (bottom) and mid-eclipse. The orbital phases are based on the conjunction of the white dwarf.

The eclipses of the lines are also affected by the large accretion disc asymmetry. In Figure 7, the eclipse of the Helium II emission line is plotted. For each velocity bin, the orbital phases where half the light is eclipsed are marked by asterisks. The outer disc on the blue-shifted peak is eclipsed first, followed by emission at higher, blue-shifted velocities. After the blue side of the disc is covered, the red-shifted is emission is progressively eclipsed, and during egress, the situation reverses. This is the classical pattern of the eclipse of a pro-gradely rotating disc, where the disc velocity increases with decreasing radius. A strictly Keplerian accretion disc would result in a smooth eclipse, symmetric around orbital phase 0. Mid-eclipse for the disc emission from IP Peg occurs considerably earlier (orbital phase 0.987) compared to the continuum, and most of the light is eclipsed well before white dwarf ingress. A large asymmetry in the outer disc is thus corroborated by the eclipse. The distorted disc eclipse is a combination of the spiral asymmetry as well as deviations from Keplerian velocities. Although these eclipse phases are not fitted by Doppler tomography codes, high quality eclipse data has the potential to reveal departures from Keplerian velocities and disentangle these two effects.

Eclipse mapping of this data in both the lines and continuum, reveals the presence of a two armed asymmetry in the outer regions of the disc, at the azimuths corresponding to the spiral arms in the Doppler maps [2] (see also Baptista, this volume). The velocities of the spirals as derived from the emission lines, can be compared with the position of the spirals from the eclipse map. This indicates that velocities near the spirals deviate considerably from the local Keplerian value. The two imaging methods thus nicely complement each other and both support a tidal interpretation of the spirals.

Apart from the data sets discussed here, Morales-Rueda et al. [18], also recovered prominent spiral waves in the disc of IP Pegasi during outburst in a range of emission lines. The disc structure during their observations, about 5 days after the start of the outburst, is very similar to the Doppler maps presented here, with a slightly stronger arm in the upper right quadrant. They also note the shifted emission line eclipses and shielding of the irradiated companion star by the geometrically thick disc.

Doppler tomography of IP Pegasi during outburst thus invariably shows the presence of spiral shaped disc asymmetries. The two armed spiral dominates the disc emission from the start of the outburst maximum and persists for at least 8 days, corresponding to about 50 binary orbits. The spirals are present in a range of emission lines from neutral hydrogen to ionised helium, the latter indicating that the gas concerned has to be hot, although it is not clear if we are looking at direct emission from the shock, or recombination emission from the spiral arms. The asymmetry between the two arms that was observed in the discovery data, is also present at other epochs and in other lines. The disc structure is co-rotating with the binary, and corresponds to the velocities were tidally driven spiral arms are expected. A more detailed comparison between theory and observations is discussed by Boffin, this volume. We shall see that the observed properties of the spirals fit recent hydro-dynamical simulations of such discs in detail [26].

Although the disc emission of the different data sets is in general terms the same, differences between the various data sets should tell us something about the evolution of the spiral across the outburst (and therefore their origin). In addition, comparing lines with different excitation potential allows us to investigate the physical conditions of the emitting gas directly. Unfortunately, the various data sets have been obtained with different telescopes and instruments, different resolutions, and different orbital phase coverage. Disentangling those systematic effects from true variations of the spiral in various lines is therefore not straightforward. Ideally one would observe a range of emission lines simultaneously with the same instrumental setup, while covering a substantial part of the disc outburst to study its evolution. As a first step, we have obtained echelle spectroscopy of IP Pegasi during a recent outburst with the NTT (Steeghs & Boffin, in preparation). Covering a large part of the optical spectrum at high resolution, ensures that we will be able to compare the properties of the spiral among a large set of emission lines, obtained under identical circumstances. Fig-

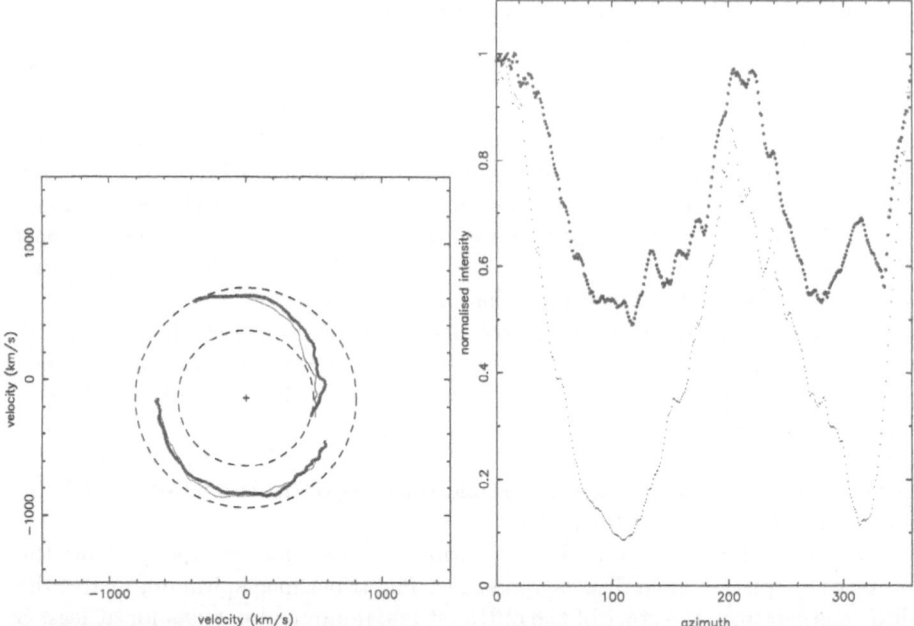

Fig. 8. Comparing the properties of the spirals in two emission lines from the same data set. The position and intensity of the spirals were traced as a function of azimuth in the Hα (thick line) and HeI6678Å tomograms constructed from NTT echelle spectroscopy during the August 1999 outburst of IP Peg. The intensities are normalised with respect to the maximum emissivity values of each line.

ure 8 compares the properties of the spirals between just two lines covered, and indicates a very similar position for the spirals in terms of velocity as a function of azimuth for the Hα and HeI6678Å lines. The intensity modulation, on the other hand is significantly different.

4 The Quiescent Disc

In stark contrast with the prominent spiral waves observed during outburst, Doppler mapping of IP Pegasi during quiescent epochs does not reveal such features [15,13,10,33]. Although significant disc asymmetries are observed even during quiescence, it is clear that the open spiral pattern the disc carries in outburst is not present. The disc radii of dwarf novae are expected to vary significantly across the outburst cycle. In the disc instability picture, a dwarf nova disc is in a cool state during quiescent phases, when most mass transferred from the mass donor is not accreted by the white dwarf. The disc density builds up until a radical opacity change due to the ionisation of hydrogen flips the disc from a neutral, cool state to a hot, ionized state. Angular momentum transport in this hot state is increased by a factor of 10 or so, which results in a rapid expansion of the disc at the onset of the outburst. During outburst, the disc is depleted as more mass is accreted than is transferred from the mass donor, until the density in the disc drops below the critical value again and the system returns to quiescence.

Disc radius variations are indeed observed in dwarf novae, and the disc in IP Pegasi fills between half of the primary Roche lobe during quiescence up to most of the Roche lobe during outburst [56,32,25]. Tidally driven waves are thus much more likely to have an effect on the disc in outburst, since the tidal torques are a steep function of the distance to the secondary star. The fact that spirals are not prominent in quiescence, is therefore what one expects if they are due to the tidal torques of the companion star [26]. The detailed properties of the spirals depend upon disc temperature, radius and the mass ratio of the binary. The most likely discs carrying prominent open-armed spirals are thus dwarf novae in outburst and the nova-like variables, where the disc is always in the hot, ionised state. Dwarf novae are likely to display highly variable spiral arm structures, depending on the state of the system, whereas nova-like variables would provide a more persistent tidal pattern since large disc radius variations are relatively rare.

Marsh, this volume, has compiled an up to date overview of all the systems for which Doppler maps have been published. One would perhaps expect that such spirals should have been found before if they are a common phenomenon. The number of Doppler maps of dwarf nova in outburst are very rare. And even in quiescence, some well known systems have not yet been mapped using good quality data. Unfortunately, in the case of nova-likes were one would have the potential of observing a persistent spiral pattern, Doppler mapping has shown complicated emission geometries. In those systems, line emission is not dominated by a disc component, instead prominent spots dominate the tomograms whose interpretation is unclear. However, IP Pegasi is no longer the only system with evidence for distortions of a spiral nature. The next section will discuss a small number of other systems which display very similar disc behaviour.

5 Disc Asymmetries in Other Systems

5.1 EX Dra

The dwarf nova EX Dra was the second object to show similar disc behaviour. This is another eclipsing system above the period gap and is in many ways very similar to IP Pegasi [3,6]. Joergens et al. [12] present Doppler images of EX Dra during outburst, and the similarity of the Doppler maps with those of IP Pegasi is obvious (Figure 9). Although perhaps not as convincing as in IP Pegasi in all the mapped lines, the disc carries a two armed asymmetry during outburst in the same quadrants of the Doppler map. The appealing characteristic of this object is its very short outburst recurrence time. Although its orbital period and mass ratio is very similar to IP Peg, EX Dra's outbursts recur every 23 days, and catching such a system in outburst is thus more likely. It would be an ideal target for a longer campaign, with the prospect of securing the disc evolution

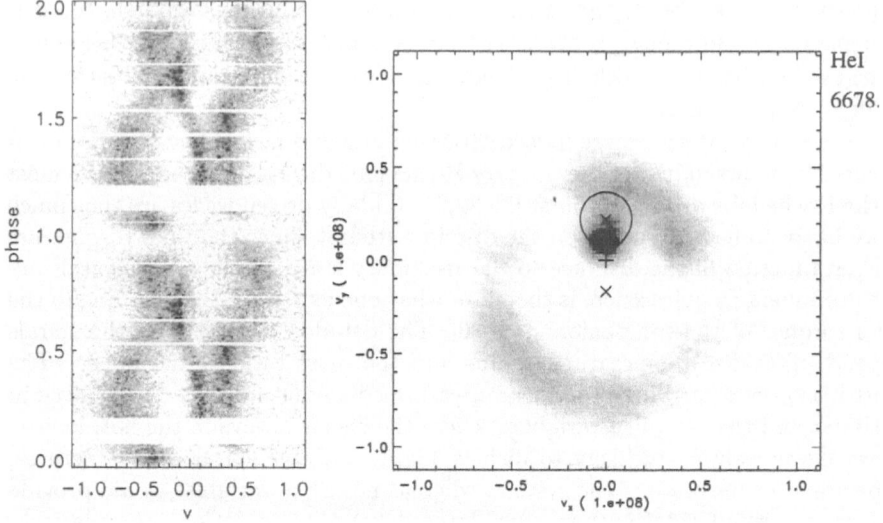

Fig. 9. Doppler tomography of EX dra in outburst [12]. Left the observed HeI 6678Å emission and its tomogram on the right. Apart from the irradiated secondary, the disc contains a two armed asymmetry with a shorter arm in the top right and a more extended arm in the lower left quadrant.

across the whole outburst cycle, and performing time lapsed tomography in order to construct a movie of the accretion disc along its outburst cycle. A photometric campaign lasting over 6 weeks was conducted during the summer of 2000, just after this workshop. Although no Doppler tomography will be possible, two outburst cycles were covered, and eclipse mapping methods will reveal the development of disc asymmetries and radial temperature variations.

5.2   U Gem

U Gem is one of the brightest and best studied CVs, reaching a magnitude of V~8.5 during outburst. Surprisingly, no outburst Doppler tomography of this system was available until very recently, Groot observed the system during the March 2000 outburst [17]. Figure 10 shows the HeII tomogram of U Gem in outburst, revealing a prominent two armed spiral pattern. The spirals are strong even towards the late phases of the outburst maximum, and weaken once decline sets in. This supports a tidal interpretation for these structures since the spirals are there when the disc is hot and large, but weaken once the disc shrinks back towards quiescence like in IP Pegasi. The rise and decline phases of the outbursts are particularly useful for testing the tidal interpretation of spiral waves. During those phases the spiral arm geometry will change dramatically, and the exact nature of the evolution will tell us if indeed the spirals behave as tidally driven spirals would do, and how they affect the angular momentum exchange in the accretion disc.

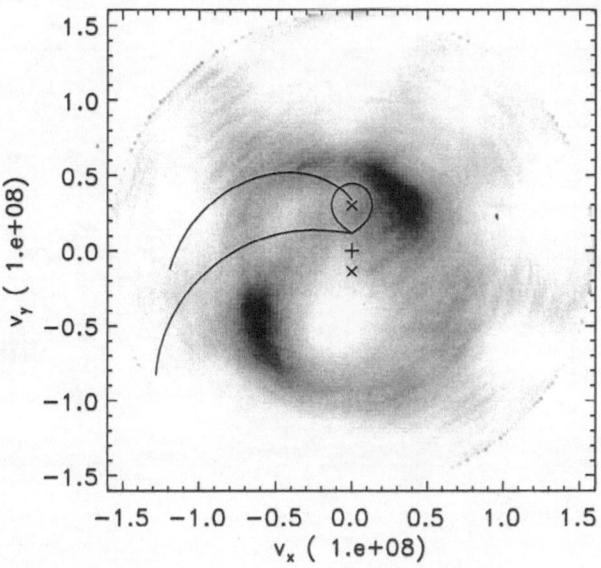

Fig. 10. Doppler tomogram of the HeII 4686Å emission from U Gem in outburst [17]. Two striking spiral arms are revealed, matching the properties of a tidal spiral pattern.

5.3 SS Cyg

During outburst, SS Cygni is the brightest CV in the sky, with an almost continuous light curve available for the last 100 years thanks to amateur astronomers. Its outburst history has been well studied and compared with disc instability predictions [5]. Phase resolved spectroscopy reveals emission from the secondary star as well as a distorted accretion disc, both in quiescence and outburst [16,17]. Doppler tomography during outburst revealed enhanced emission from the disc in the top right and lower left quadrant of the tomograms in the Balmer, HeI and HeII lines [29]. MEM maps of two emission lines are presented in Figure 11. Again, the disc has a two armed asymmetry, but on the other hand differs significantly from a simple two armed spiral pattern. SS Cyg is an intermediate inclination system, and our line of sight thus penetrates deeper into the disc atmosphere, producing broad absorption wings from the optically thick disc during outburst. This may complicate the comparison between models of spiral arms and the line emission from them because of more intricate, and poorly understood, radiative transfer conditions in the vertically stratified disc. However, these differences need an explanation and given the brightness of SS Cyg, justifies high resolution phase resolved spectroscopy of this system in quiescence and outburst.

SS CYG HeII 4686   SS CYG HeI 6678   V347 PUP HeI 4921   V347 PUP HeI 5015

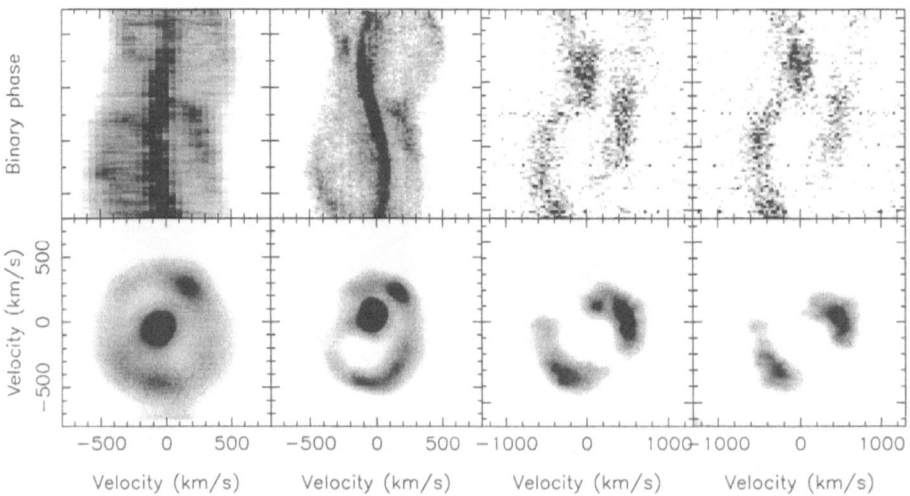

Fig. 11. Two other cases of distorted accretion discs with a two armed asymmetry. Left the HeII and HeI emission of SS Cygni during outburst, right two panels the eclipsing nova-like V347 Pup.

5.4   V347 Pup

The accretion discs of nova-likes are in many ways regarded as dwarf nova in permanent outburst, since the high mass transfer rate in the disc keeps the disc gas ionized and stable against thermal disc instabilities. Unfortunately the properties of the emission lines of these systems remains a puzzle. The atlas of Doppler maps from [13] contains many nova-like systems, for which evidence for an accretion disc is absent in most cases. One of the few nova-likes that appears to reveal a clear signature of an accretion disc is the eclipsing nova-like V347 Pup [30]. Doppler mapping revealed the presence of an accretion disc, containing a two armed asymmetry. However, the properties of some of the emission lines at a different epoch several years later were quite different (Steeghs et al., in preparation). The Balmer lines were dominated by a spot of emission in the lower left side of the tomogram rather than the disc, a typical feature of nova-like variables. However, the HeI emission (Figure 11) appears to be originating from the disc only, and displays a two armed asymmetry like that was observed in the Balmer lines by Still et al. [30] The opportunities to investigate the tidal structures of a high mass transfer rate accretion disc are ideal in this eclipsing object, and follow up NTT data is under investigation.

5.5   A Systematic Picture?

Phase resolved spectroscopy of CVs in outburst, suitable for Doppler mapping of the accretion disc, is still limited to a small number of objects. It is very encouraging that on the occasions that such data are available and can be compared, very similar accretion disc distortions are seen. This suggests a common physical cause for these large scale asymmetries and tidally driven waves appear to offer the most likely explanation. It is interesting to note that all the dwarf novae that have indicated spiral-type distortions are long period systems above the period gap. One does expect the structure of the spiral arms to depend on the mass ratio of the binary, with higher mass ratio binaries corresponding to a heavier secondary star that induces stronger tidal torques. On the other hand selection effects work against us since the outburst of the short period systems are usually short, and suitable outburst spectroscopy of short period dwarf novae during outburst maximum is extremely rare. Tomography of OY Car in outburst [9], reveals an extended arm on the side of the gas stream impact, but not a two armed spiral pattern. Limited resolution and signal to noise in this case prevents a strong case for or against spirals in SU UMa systems.

The important question is whether spiral waves affect the structure of accretion discs under a wide range of conditions, or are merely a dynamical side effect of the expanding disc during dwarf novae outbursts. Clear progress requires a more balanced observational picture covering the observational parameter space in terms of orbital periods, mass ratios, outburst behaviour, etc. This relies on our ability to obtain a considerable number of data sets of outbursting dwarf novae, in particular during the rise and decline phases. Flexible, or even robotic, scheduling of telescope time would be highly beneficial to such projects.

Doppler tomography has proven to be an invaluable tool for the discovery and study of spiral waves in the discs of CVs. With better resolution and signal to noise data becoming available with large aperture telescopes, however, there is also room for improvements to the technique itself. In the next section I describe an extension to the Doppler tomography technique that aims to improve our ability to fit to data sets containing anisotropic emission sources, a rather common situation.

6   Modulation Mapping

Doppler tomography provides a velocity resolved image of the accretion flow in the corotating frame of the binary. Since it relies on only a few basic assumptions, such images can be recovered in a model independent way. They then provide a perfect frame in which to compare data with models. One of those assumptions is that the flux from any point fixed in the rotating frame is constant (Marsh, this volume). However, observations show the presence of anisotropic emission sources, that modulate their emission as a function of the orbital phase. Some general examples are the irradiated front of the mass donor, the hot spot, and the anisotropic emission from spiral shocks. Doppler tomography can still be used in that case, since the Doppler map serves to present a time averaged image of the distribution of line emission. However, one will not be able to fit the data very well, and the phase dependent information contained in the observed line profiles is lost. The remainder of this review will present a straightforward extension to the Doppler tomography method that tries to remove the above mentioned assumption from tomography.

6.1   Extending Doppler Tomography

Rather than assuming that the flux from each point in the binary frame is constant, I include modulations of the flux on the orbital period. The line flux from each location (in terms of its position in the $V_x V_y$ plane of the tomogram) is not just characterised by the average line flux, but also the amplitude and phase of any modulation on the orbital period (Figure 12). We thus have two additional parameters to describe the line flux form a specific velocity vector. The flux f from a modulated S-wave as a function of the orbital phase ϕ can be written as;

$$f(\phi) = I_{avg} + I_{cos} \cos \phi + I_{sin} \sin \phi$$

where I_{avg} is the average line flux for the s-wave, and I_{cos} and I_{sin} the cosine and sine amplitudes (Figure 12). In other words, the amplitude of the modulation is $\sqrt{I_{cos}^2 + I_{sin}^2}$, and its phase is $\tan^{-1}(I_{cos}/I_{sin})$. The velocity of the S-wave is defined in terms of the vector $V = (V_x, V_y)$ in the usual manner (Marsh, this volume);

$$v(\phi) = \gamma - V_x \cos 2\pi\phi + V_y \sin 2\pi\phi$$

With γ the systemic velocity of the binary. In this prescription, we need three images describing the values of I_{avg}, I_{cos} and I_{sin} for each velocity (V_x, V_y) instead of the conventional one describing I_{avg}. In other words the projection from

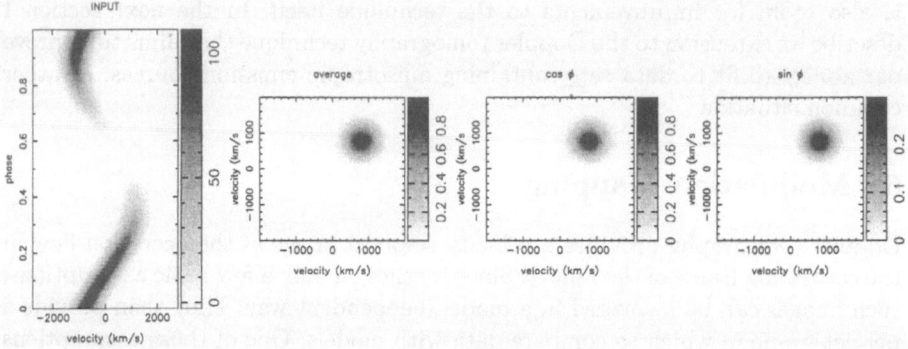

Fig. 12. A spot that modulates its emission across the orbital phase is not only characterised by its position in the $V_x V_y$ plane but also the amplitude and phase of its modulation. Two additional velocity images are used to store the cosine and sine amplitudes of such S-waves. All modulations on the orbital periods can then be described as the sum of the image values in the three velocity images.

Doppler map to trailed spectrogram is then;

$$F(v,\phi) = \int \left(I_{avg}(V_x, V_y) + I_{cos}(V_x, V_y) \cos\phi + I_{sin}(V_x, V_y) \sin\phi \right) g(V-v) dV_x dV_y$$

with $g(V-v)$ describing the local line profile at a Doppler shift of $V-v$, which is assumed to be a Gaussian convolved with the instrumental resolution.

Keith Horne implemented a similar extension to the filtered back-projection code. However, cross talk among the three terms results in significant artifacts in the back-projected maps. I choose to implement the extension using a maximum entropy optimiser in order to reconstruct artefact free images. Since the entropy is only defined for positive images, and the cosine and sine amplitudes can be either positive or negative, the problem was implemented using 5 images to characterise the data. One (positive) average image, and two (positive) images for both the cosine and sine amplitudes. For those two images one image reflects positive amplitudes, the other negative amplitudes. The image entropy for each image is defined in the usual manner relative to a running default image, and the data is fitted to a requested χ^2 while maximising the entropy of the 5 images.

6.2   A Test Reconstruction

In order to test the method, and check for possible artefacts, fake data was generated and noise was added to that data to see how well the input images can be reconstructed by the code. In Figure 13, I show an example of an input data set consisting of a constant emission from a disc as well as various anisotropic contributions from 3 spots. The large modulation of the S-waves corresponding to the spots results in a trailed spectrogram that cannot be fit using the conventional Doppler tomography technique. However, using the above described extension, the modulation mapping code is able to fit the input data to a χ^2 of 1, without introducing any spurious features in any of the images, or leaving systematic residuals to the fit. As expected, the higher the signal to noise of the input data, the more reliable and accurate the reconstruction is. Most importantly though, there is no cross-talk between the various images. A series of tests were performed to confirm that the problem is well constrained and that maximum entropy ensures that no image structure is added unless the data dictates it. Two different types of default were used. For the average image, the default is set to a Gaussian blurred version of the average image after each iteration, while for the modulation images we have tried both Gaussian blurring as well as steering the images to zero. Both converge easily to the maximum entropy solution. Of course this method suffers from the same limitations as Doppler tomography with regards to errors in the assumed systemic velocity of the binary, limited image structure due to poor signal to noise, and artefacts due to limited phase sampling.

6.3   Application to Real Data

As an example of applying the method to real data, we use the previously discussed data of SS Cygni during outburst (Figure 11). Figure 14 looks at the

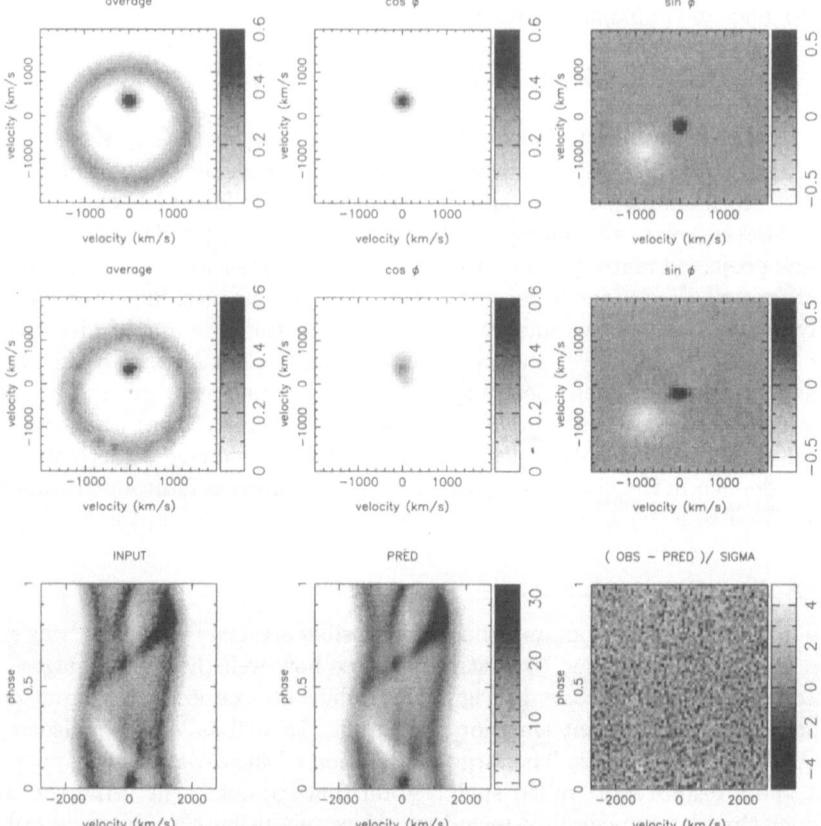

Fig. 13. A fake data set in order to test the reliability of modulation mapping. Top row plots the input images consisting of emission from a ring and various discrete modulated spots. The data corresponding to those images is plotted in the lower left panel (with random noise added). Middle row are the reconstructed images, bottom rows compares observed and predicted data, together with a plot of the normalised residuals.

residuals between predicted and observed data when conventional MEM Doppler tomography was applied to the HeI emission line data. Although recovering the general features in the data, such as the S-wave from the irradiated secondary and the two armed disc asymmetry, residuals are significant and the best χ^2 that can be achieved is 3.2. In particular the anisotropic emission from the secondary star and parts of the disc leave large residuals due to their phase dependence. Doppler tomography tries to reproduce this phase dependence as best as it can, but is fundamentally unable to describe such emission sources adequately.

If the same data is passed to the modulation mapping code, a much better fit to the data can be obtained (Figure 15). The χ^2 value of the fit is 1, as good as one may expect, and leaves much reduced systematic residuals between data and

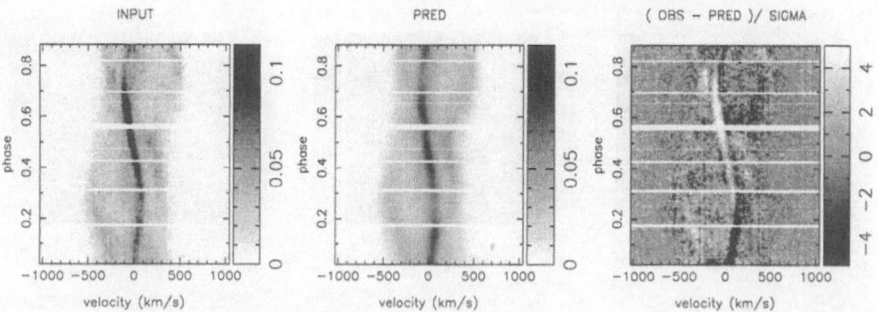

Fig. 14. The result of Doppler tomography of SS Cyg in outburst. The observed trailed spectrogram of HeI6678 is on the left, the predicted data of the reconstructed tomogram in the middle panel. The tomogram itself was shown in Figure 11. Significant residuals (right panel) exist due to the anisotropic emission from the secondary st ar and parts of the disc.

fit. The structure of the average image is very similar to that obtained with the conventional method, indicating that the majority of the emission is relatively unmodulated. In the two modulated images, one can identify in particular the emission from the secondary and some areas of the disc. However, the two images tell us not only where the modulated emission is coming from, but also the phase of the modulation. The secondary star emission shows a left-right asymmetry in the cosine image, and a front-back asymmetry in the sine image, indicating that the emission is indeed beamed away from the Roche lobe surface. Although one ideally applies Roche tomography in order to model the contribution of the secondary, modulation mapping ensures that strong anisotropic S-waves originating from the mass donor are fitted well, and that they do not leave artefacts or limit the goodness of fit that can be achieved.

Obvious other emission sources where the phasing and beaming of the emission can be quantified with this method are the emission from the bright spot, and emission from anisotropic spiral shocks. In SS Cygni, the emission from the disc seems to be only weakly modulated (Figure 16), which may not be too surprising because of its low orbital inclination i. Shear broadening, for example, leads to anisotropic emission and is and proportional to $\cos^{-1} i$. Its effect is thus considerably reduced for inclinations below 60° [26]. Work is in progress to apply this method to the emission of the spirals in IP Peg and other objects in order to quantify the anisotropy of the emission from the spiral shocks.

Modulation mapping provides a straightforward extension to Doppler tomography, with a wide range of applications. The code as such does not need to make any assumptions about the nature of the modulations, except that the period of modulation is the orbital period. It thus relaxes one of the fundamental assumption of Doppler tomography, in order to make the technique more versatile. Many data sets already exist that could benefit from this extension, but it will be of particular use for the high resolution data sets that will be obtained with

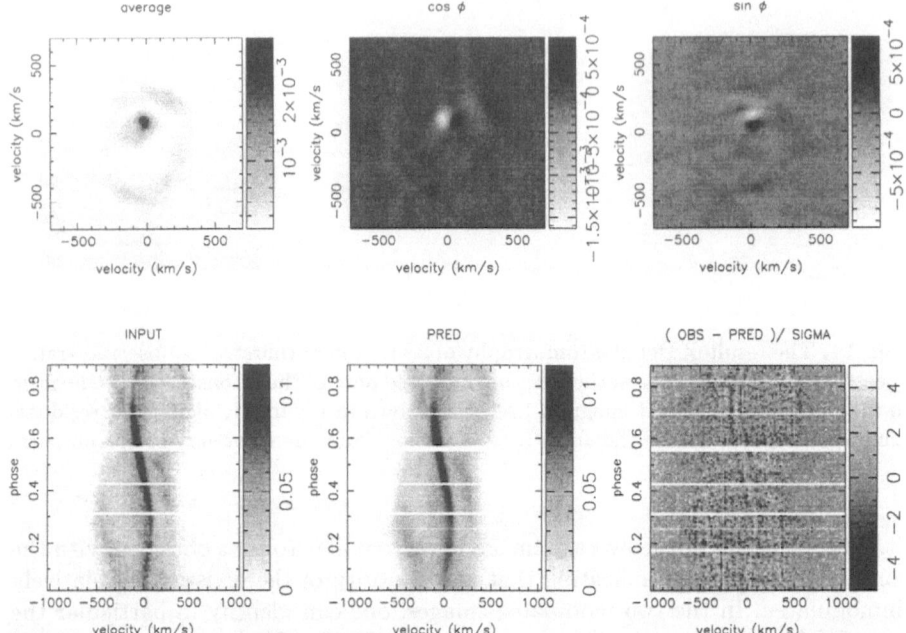

Fig. 15. The same SS Cygni data now processed with the modulation mapping code. Top row are the reconstructed images, with the observed and predicted spectrogram in the bottom row.

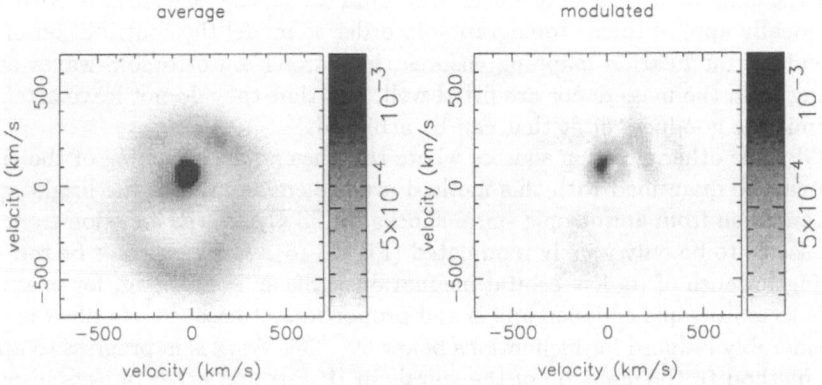

Fig. 16. A closer look at the derived tomograms with modulation mapping. Left the average image, right the total amplitude in the modulated images. Most prominent is the modulated emission from the secondary star, as well as a small contribution from the disc.

the new large aperture telescopes. An even more general prescription can easily be envisaged that attempts to model modulations on other periods as well.

7   Conclusions

Doppler tomography of the dwarf nova IP Peg has revealed a remarkable two armed spiral structure in its accretion disc that dominates the emission from a range of emission lines over a large range of radii. The pattern is observed from the start of the outburst up to the later stages of the outburst maximum, and is fixed in the binary frame. Its location in the tomograms fits remarkably well with the two armed spiral shocks that are expected to be generated by the tidal torques of the companion star. A tidal origin is also supported by the fact that the structure is corotating with the binary for at least 50 orbital periods, and the fact that the asymmetry is not visible during quiescence, when tidal torques on the much smaller accretion disc are significantly reduced.

In order to appreciate the relevance of tidally driven spiral waves for the structure of accretion discs in general, a more varied observational picture is required, spanning a range of objects and source states. Since the detection of spirals in IP Pegasi, a handful of other systems have displayed very similar disc structures, but a systematic picture is still difficult to extract. A second important restriction of the current data sets, is the limited coverage we have of each outburst. In the CV sub-class of dwarf novae we have a truly unique opportunity to track the evolution of the disc in real-time through time-lapsed tomography. What is needed is a focused campaign that aims to obtain spectroscopy across a significant fraction of the outburst cycle. Such a data set would be an extremely valuable test bed for both disc instability models as well as the question of angular momentum transport associated with density waves. Although the observed spirals appear to fit to detailed simulation in surprising detail, it is still not clear what the impact of such prominent spiral arms is on the angular momentum budget of the disc and how it relates to the local shear viscosity. This requires improved observations and realistic simulations in order to test quantitatively the impact of such waves (Boffin, this volume).

I also discussed an extension to Doppler tomography, with the aim of mapping modulated emission sources in emission line data. This is a rather common situation and will not only benefit the observational study of spiral waves, but a range of other emission sources commonly observed. I demonstrated that artefact free reconstructions can be calculated from phase resolved spectroscopy using maximum entropy regularisation.

References

1. Balbus S. A., Hawley J. F., 1991, ApJ, 376, 214
2. Baptista R., Harlaftis E. T., Steeghs D., 2000, MNRAS, 314, 727
3. Billington I., Marsh T.R., Dhillon V.S., 1996, MNRAS, 278, 673
4. Binney J., Tremaine S., 1987, Galactic Dynamics. Princeton Series in Astrophysics, Princeton University Press
5. Cannizzo, J. K., Mattei, J. A., 1998, ApJ, 505, 344
6. Fiedler H., Barwig H., Mantel K. H., 1997, A&A, 327, 173

7. Frank J., King A., Raine D., 1985, Accretion Power in Astrophysics, Cambridge Astrophysics Series, Cambridge University Press
8. Groot P., 2000, ApJL, submitted
9. Harlaftis, E.T, Marsh, T., 1996, A&A, 308, 97
10. Harlaftis E. T., Marsh T. R., Dhillon V. S., Charles P. A., 1994, MNRAS, 267, 473
11. Harlaftis E. T., Steeghs D., Horne K., Martin E., T. M., 1999, MNRAS
12. Joergens V., Spruit H. C., Rutten R. G. M., 2000, A&A, 356, 33
13. Kaitchuck R. H., Schlegel E. M., Honeycutt R. K., Horne K., Marsh T. R., White J. C. I., Mansperger C. S., 1994, ApJS, 93, 519
14. Livio M., 1994, in "Interacting Binaries", Saas-Fee Advanced Course 22. Lecture Notes 1992. Swiss Society for Astrophysics and Astronomy, XVI, Springer-Verlag Berlin Heidelberg New York
15. Marsh T. R., Horne K., 1990, ApJ, 349, 593
16. Martinez-Pais I. G., Giovannelli F., Rossi C., Gaudenzi S., 1994, A&A, 291, 455
17. Martinez-Pais I. G., Giovannelli F., Rossi C., Gaudenzi S., 1996, A&A, 308, 833
18. Morales-Rueda L., Marsh T., Billington I., 1999, MNRAS,
19. Pringle J. E., 1981, ARA&A, 19, 137
20. Savonije G.J., Papaloizou J.C.B., Lin D.N.C., 1994, MNRAS, 268, 13
21. Sawada K., Matsuda T., Hachisu I., 1986, MNRAS, 219, 75
22. Shakura N. I., Sunyaev R. A., 1973, A&A, 24, 337
23. Spruit H. C., Matsuda T., Inoue M., Sawada K., 1987, MNRAS, 229, 517
24. Spruit H. C., 1987, A&A, 184, 173
25. Steeghs D., 1999, PhD thesis, University of St.Andrews (available at http://www.astro.soton.ac.uk/~ds/thesis.html).
26. Steeghs D., Stehle R., 1999, MNRAS, 307, 99
27. Steeghs D., Harlaftis E. T., Horne K., 1997, MNRAS, 290, L28
28. Steeghs D., Harlaftis E. T., Horne K., 1998, MNRAS, 296, 463
29. Steeghs D., Horne K., Marsh T. R., Donati J. F., 1996, MNRAS, 281, 626
30. Still M. D., Buckley D. A. H., Garlick M. A., 1998, MNRAS, 299, 545
31. Toomre A., 1969, ApJ, 158, 899
32. Wolf S., Mantel K. H., Horne K., Barwig H., Schoembs R., Baernbantner O., 1993, A&A, 273, 160
33. Wolf S., Barwig H., Bobinger A., Mantel K.-H., Simic D., 1998, A&A, 332, 984
34. Wood J., Horne K., Berriman G., Wade R., O'Donoghue D., Warner B., 1986, MNRAS, 219, 629

Spiral Waves in Accretion Discs – Theory

H.M.J. Boffin

Royal Observatory of Belgium, Av. circulaire 3, B-1180 Brussels
Henri.Boffin@oma.be

Abstract. Spirals shocks have been widely studied in the context of galactic dynamics and protostellar discs. They may however also play an important role in some classes of close binary stars, and more particularly in cataclysmic variables. In this paper, we review the physics of spirals waves in accretion discs, present the results of numerical simulations and consider whether theory can be reconcilied with observations.

If, in the course of the evolution of a binary system, the separation between the stars decreases, there comes a point where the gravitational pull of one of the stars removes matter from its companion. There is mass transfer, through the so-called Roche lobe overflow. This is what is believed to happen in cataclysmic variable stars (CVs; see Warner [40] for a review). In these, a white dwarf primary removes mass from its low-mass late-type companion which fills its Roche lobe. In this particular case, the decrease in the separation is due to a loss of angular momentum by magnetic braking or by gravitational radiation.

1 Roche Lobe

Consider a binary system with a primary white dwarf of mass M_1, and a companion of mass M_2 with a mean separation a. We can define the mass ratio, $q = M_1/M_2$, which, for CVs, is generally smaller than 1. From Kepler's law, the orbital period, P_{orb}, is then:

$$P_{\mathrm{orb}}^2 = \frac{4\pi^2 a^3}{G(M_1 + M_2)}, \qquad (1)$$

G being the gravitational constant, and the masses being expressed in unit of the solar mass.

In the reference frame rotating with the binary and with the center of mass at the origin, the gas flow is governed by Euler's equation:

$$\frac{\partial v}{\partial t} + (v \cdot \nabla)v = -\nabla \Phi_r - 2\omega \times v - \frac{1}{\rho}\nabla P, \qquad (2)$$

where ω is the angular velocity of the binary system relative to an inertial frame, and is normal to the orbital plane with a module $\omega = 2\pi/P$, ρ is the density and P is the pressure. The last term in the right hand side of this equation is thus the gas pressure gradient, while the second is the Coriolis force.

Here, Φ_r is the Roche potential, and includes the effect of both gravitational and centrifugal forces :

$$\Phi_r = -\frac{GM_1}{|r - r_1|} - \frac{GM_2}{|r - r_2|} - \frac{1}{2}(\omega \times r)^2, \tag{3}$$

where r_1, r_2 are the position vectors of the centres of the two stars. The equipotential surfaces of Φ_r are shown in Fig. 1. It can be seen that there is a particular equipotential which delimits the two Roche lobes of the stars. The saddle point where the two lobes join is called the inner Lagrange point, L_1. These Roche lobes are the key to understanding mass transfer in close binary systems. Indeed, if one of the two stars fills its Roche lobe, then matter can move into the Roche lobe of its companion and be gravitationnally captured by it. Mass transfer occurs via the so-called Roche lobe overflow mechanism. The size of the Roche lobes, R_{L1} and R_{L2}, in unit of the separation is only a function of the mass ratio, $R_{L2} = a\ f(q)$ and $R_{L1} = a\ f(q^{-1})$, where an approximate value for $f(q)$ is given by

$$f(q) = \begin{cases} 0.38 + 0.20 \log q & (0.3 \leq q < 20), \\ 0.462 \left(\frac{q}{1+q}\right)^{1/3} & (0 < q < 0.3). \end{cases} \tag{4}$$

From Eq.(1) and (4), one can see that the mean density of a star which fills its Roche lobe is a function of the orbital period only [40]:

$$\bar{\rho} \simeq \frac{107}{P_{\rm orb}^2}\ \text{gcm}^{-3}, \tag{5}$$

if $P_{\rm orb}$ is expressed in hours. Thus, for the typical orbital periods of CVs, from 1 to 10 hours, the mean density obtained is typical of lower main-sequence stars.

2   Disc Formation

Matter which is transferred through the L_1 point to the companion has a rather high specific angular momentum with respect to the later, $b_1^2 \omega$. Here, b_1 is the distance of the inner Lagrangian point to the centre of the primary and can be obtained with the following fitted formula :

$$b_1 = (0.500 - 0.227 \log q)\ a. \tag{6}$$

Once the gas comes close to the primary, it will mainly feel its gravitational force and follow a Keplerian orbit, for which the circular velociy is given by

$$v_\phi(r) = \sqrt{\frac{GM_1}{r}}, \tag{7}$$

and the associated angular momentum, $r\ v_\phi(r)$. If we equal this to the angular momentum of the gas transfered through the Roche lobe, we can define the

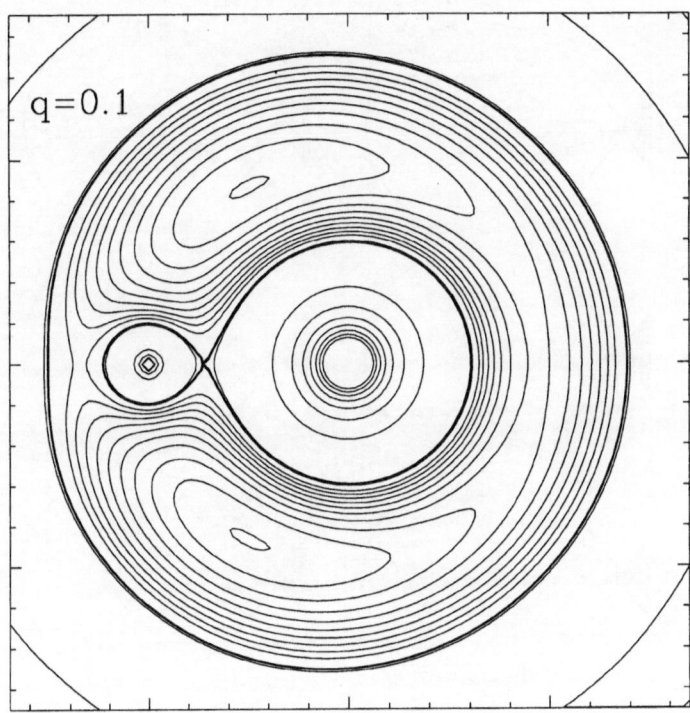

Fig. 1. Roche equipotentials in a binary system with a mass ratio, $q = 0.1$. The Roche lobes are shown with the heavy lines. The primary is in the middle

circularization radius $R_{\text{circ}} = (1 + q)(b_1/a)^4 a$. Gas coming from the companion will thus form a ring of radius approximately R_{circ}, provided this is larger than the radius of the accreting object. For cataclysmic variable stars, we typically have $P_{\text{orb}} = 1 - 10$ hours, $M_1 \approx M_\odot$ and $q < 1$. Therefore, $a \lesssim 3R_\odot$ and $R_{\text{circ}} \approx 0.1 - 0.3\ a$. By comparison, the primary white dwarf has a radius about $0.01\ R_\odot$. Thus, in non-magnetic CVs, the mass transfer will always lead to the formation of such a ring. This ring will then spread out by viscous processes: because energy is lost, the gas will move deeper into the gravitational well until it reaches the primary and accretion will occur (see [6]). To conserve angular momentum, the outer part will have to move further away. An accretion disc forms. The formation of an accretion disc is pictured in Fig. 2 as obtained from numerical simulations.

In principle, the disc could expand forever. In a binary system, this is not possible however because of the torque exerted by the companion. The radius of the disc is therefore limited by the tidal radius, where the tides induced by

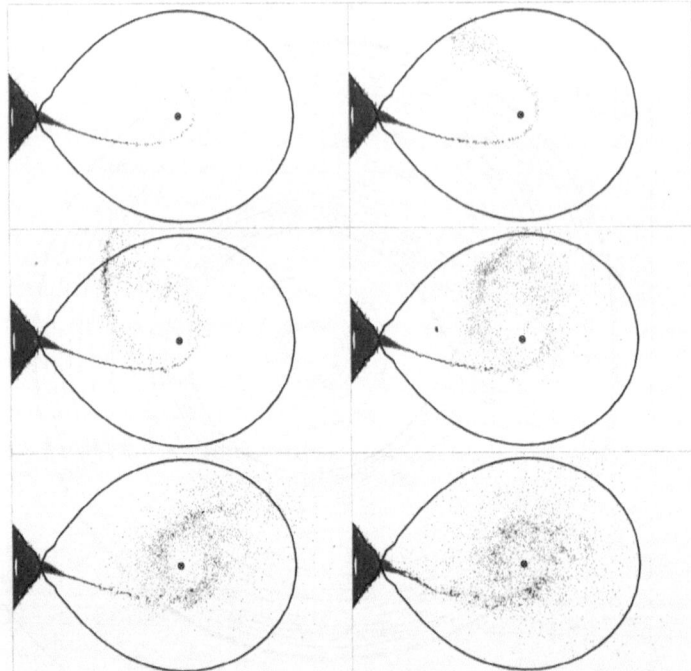

Fig. 2. The formation of an accretion disc by Roche lobe overflow. The low mass transferring secondary is at the left. Mass flows through the inner Lagrange L_1 point towards the white dwarf primary

the secondary star truncate the disc. The angular momentum is then transferred back to the orbital motion. Paczynski [24] has computed the maximum size of a disc in a binary system by following the orbits of a set of particles and deriving the largest non-intersecting orbit. This orbit would represent the maximum size of an accretion disc, when neglecting the effect of viscosity and pressure. This is called the disc's tidal truncation radius. Typically, the disc radius is limited to 0.7-0.9 the Roche lobe radius.

3 Viscosity

As we have seen, viscous processes are at play to explain to formation of accretion discs. It is this viscosity which will also ensure that matter transferred by the companion can be accreted onto the primary. In the thin disc approximation, i.e. when the disc half thickness H is much smaller than the radius r at each radius, $H/r \ll 1$, and in a steady state,

$$\nu\Sigma = \frac{\dot{M}}{3\pi}\left[1 - \left(\frac{R_{\text{wd}}}{r}\right)^{1/2}\right], \tag{8}$$

where ν is the effective kinematic viscosity, Σ the surface density and $\dot{M}$ the mass accretion rate (e.g. [6]). The origin of this viscosity is as yet unknown. Shakura & Sunyaev [32] used a parametric formulation to hide in a parameter, α, our lack of knowledge:

$$\nu = \alpha c_{\text{s}} H, \tag{9}$$

with c_{s} being the sound speed. Note that in the thin disc approximation,

$$\frac{H}{r} \sim \frac{c_{\text{s}}}{v_\phi} \equiv \frac{1}{\mathcal{M}}, \tag{10}$$

with $\mathcal{M}$ the Mach number in the disc. In the framework of the turbulent viscosity mechanism, the Shakura & Sunyaev prescription can be understood in writing the kinematic viscosity as the product of the turbulent velocity, v_{t} and the typical eddy size, l_{ed}: $\nu \sim v_{\text{t}} l_{\text{ed}}$. The eddy size cannot be larger than the disc scale height, thus $l_{\text{ed}} \lesssim H$. Moreover, in order to avoid shocks, the turbulent velocity must be subsonic, $v_{\text{t}} \lesssim c_{\text{s}}$. Thus, α must be smaller or equal to 1. There is no reason however for α to be constant throughout the disc.

By using Eq. (8) and the direct relation between the mass accretion rate and the luminosity of the disc, it is possible to have an estimate of the amount of viscosity present in the accretion disc of cataclysmic variables. In a typical dwarf nova, the disc is observed to brighten by about five magnitudes for a period of days every few months or so. According to the thermal instability model, which is the most widely accepted model to explain these outbursts, the disc flips from a low accretion, cool state in quiescence to a high accretion hot state at outburst (e.g. [39]). It is therefore generally believed that $\alpha \simeq 0.01$ in quiescent dwarf nova, while α is typically 0.1-0.3 in dwarf novae in outburst or in nova-like stars. This is clearly too large by several order of magnitudes for standard molecular viscosity. More serious candidates are therefore turbulent viscosity, magnetic stresses and the Balbus-Hawley and Parker magnetic instabilities.

Although the common mechanims invoked for this anomalous viscosity are thought to be local, hence the Shakura & Sunyaev prescription, it may be possible that some global mechanism is acting as a sink for angular momentum. In this case, one could still use the above prescription if we now use an *effective* α parameter. This will be the case for spiral shocks which we will discuss in the rest of this review.

4   Spiral Shocks

Although it was already known that accretion discs in close binary systems can lose angular momentum to an orbiting exterior companion through tidal interaction [12,25], it is Sawada, Matsuda & Hachisu [10,11] who showed, in their 2D inviscid numerical simulations of accretion discs in a binary of unit mass ratio, that spiral shocks could form which propagate to very small radii. Spruit [13,35] and later Larson [11] made semi-analytical calculations which were followed by numerous - mostly 2D - numerical simulations [3,7,9,14,15,18–21,26,27,31,37,45]. An historic overview can be found in Matsuda et al. [22].

Savonije, Papaloizou & Lin [28] presented both linear and non-linear calculations of the tidal interaction of an accretion disc in close binary systems. The linear theory normally predicts that spiral waves are generated at Lindblad resonances in the disc. A $m : n$ Lindblad resonance corresponds to the case where n times the disc angular speed is commensurate to m orbital angular speed: $n\Omega = m\omega$, i.e. $r = \left(\frac{1}{1+q}(\frac{n}{m})^2\right)^{1/3} a$. In fact, in their study of tidal torques on accretion discs in binary systems with extreme mass ratios, Lin & Papaloizou [13] already observed the spiral pattern. In their case, this pattern was indeed due to the 2:1 Lindblad resonance which can fall inside the Roche lobe and inside the disc if the mass ratio is small enough. However, the typical mass ratios and disc radius of cataclysmic variable stars does not allow the centre of such resonances to be in the disc. For the 2:1 resonance to lay inside the disc requires $q < 0.025$, a value too small for most cataclysmic variables altough possible for some low-mass X-ray binaries. The 3:1 resonance can be located inside the disc for $q < 0.33$, hence for most of SU UMa stars (see e.g. [43]).

In their study, however, Lin & Papaloizou [13] showed that the resonant effect is significant over a region

$$\Delta x \sim \left(\frac{\nu\, r}{v_\phi\, r_s^2}\right)^{1/3} r, \qquad (11)$$

where r_s is the position of the resonance. Using $\nu = \alpha c_s H$, this leads to $\Delta x \propto c_s^{2/3}$. Thus, as also found by Savonije et al. [28], even in CVs with larger mass ratios, the centre of the 2:1 resonance can still be thought of as lying in the vicinity of the boundaries of the disc and because the resonance has a finite width that increases with the magnitude of the sound speed, it can still generate a substantial wave-like spiral response in the disc, but only if the disc is large, inviscid and the Mach number is smaller than about 10. For larger Mach number, more typical of cataclysmic variables, however, Savonije et al. consider that wave excitation and propagation becomes ineffective and unable to reach small radii at significant amplitude.

We will look more closely to this later on. As for now, we will follow Spruit [13] to show why spiral waves can be thought of as an effective viscosity. Spiral waves in discs have been sudied extensively in the context of galactic dynamics and of protostellar discs. Such waves carry a negative agular momentum. Their dissipation leads to accretion of the fluid supporting the waves onto the central object.

Consider some disturbance at the outer disc edge. As the generated wave propagates inward, it is being wound up by the differential Keplerian rotation

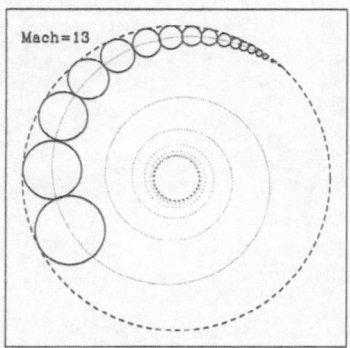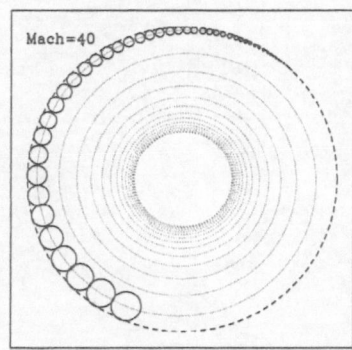

Fig. 3. Illustration of the fact that the angle of the spiral pattern depends on the Mach number. In the hot disc (left), the perturbance propagates faster and the spiral is more open than in the cold disc (right)

in the disc into a trailing spiral pattern (Fig. 3). The wave frequency σ for an azimuthal wavenumber m in the comoving frame is $\sigma = \omega - m\Omega \simeq -m\Omega$, because we can neglect the much lower orbital frequency, ω. The conserved wave action is given by

$$S_{\mathrm{w}} = \frac{1}{2} \int \frac{\rho v_{\mathrm{w}}^2}{\sigma} dV \tag{12}$$

where v_{w} is the amplitude of the wave and the integration is carried out over the volume of the wave packet. The angular momentum of the wave, given by $j = mS_{\mathrm{w}}$, is conserved, giving:

$$j \simeq -\frac{1}{2} \int \frac{\rho v_{\mathrm{w}}^2}{\Omega} dV. \tag{13}$$

Hence, it is negative. This can be understood because the tidally excited spiral pattern rotates with the binary angular speed which is smaller than the angular speed of the gas in the disc.

Because of the differential rotation, the amplitude of the wave increases as it propagates inwards. Indeed, the radial extent of the wave packet, ΔR, is proportional to the sound speed, $\Delta R \propto c_{\mathrm{s}}$, while the volume of the wave packet is $V = 2H \times 2\pi R\Delta R$. Now, because j is conserved, Spruit obtains

$$v_{\mathrm{w}} \propto \left(\frac{\Omega}{\rho r H c_{\mathrm{s}}} \right)^{1/2} \propto r^{-11/16}, \tag{14}$$

where in the last approximation, we made use of the relations valid for a thin disc. The wave therefore steepens into a shock. Dissipation in the shock lead to a loss of angular momentum in the disc, hence to accretion.

The opening angle of the spirals is related directly to the temperature of the disc as, when the shock is only of moderate strength, it roughly propagates at

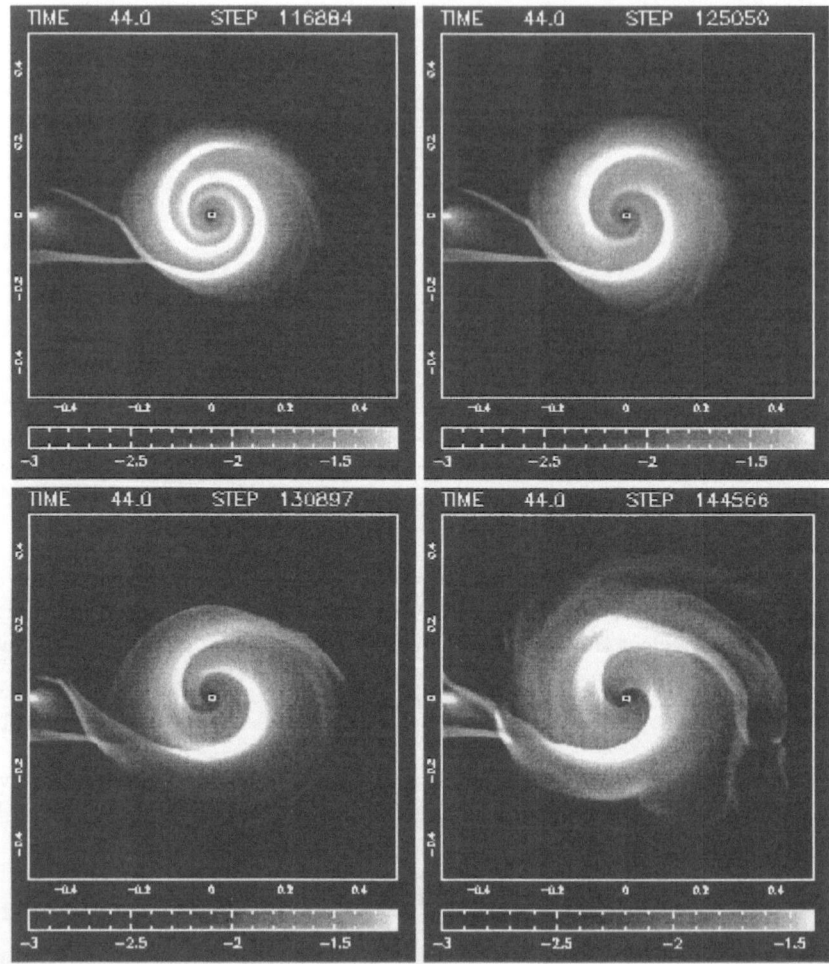

Fig. 4. Density plots of 2D finite-difference simulations (in the case of a mass ratio of 1) for different value of $\gamma = 1.01$ (upper left), 1.05, 1.1, 1.2 (lower right). The scale is logarithmic and the results are shown at about 7 orbital periods [16]

sound speed. This is shown schematically in Fig. 3 for two values of the Mach number, where we have assumed that the perturbation propagates radially at a speed of 1.3 times the Mach number [2,10,13]. Thus the angle between the shock surface and the direction of the orbital motion is of the order $\tan \theta = c_s/v_\phi = 1/\mathcal{M}$. In the approximation of an adiabatic equation of state, often used in numerical simulations, where the temperature soon reaches a value given by $T = 0.5(\gamma - 1)T_{\mathrm{vir}}$, with T_{vir} being the virial temperature, the angle becomes $\sqrt{0.5\gamma(\gamma - 1.)}$. Thus, low γ discs will have more tightly wound spirals than large

γ ones. This is indeed was is shown by two-dimensional simulations ([16,22]; see also Fig. 4).

5   Observational Facts

The most prevalent and successful model to explain dwarf nova outbursts is the disc instability model based on a viscosity switch related to the ionisation of hydrogen in the disc. It is an hysteresis cycle, in which the disc switches back and forth between a hot optically thick high viscosity state - the outburst - and a cool optically thin low viscosity state - quiescence. The disc radius increases at outburst and then after maximum, decreases exponentially. In U Gem for example, a clear increase in the radius is seen at outburst : the radius of the disc is of the order of $0.4a$ at maximum and then it decreases on a timescale of tens of days to $0.28a$ [40]. While for EX Dra, Baptista & Catalan [1] found the disc radius to be $0.30a$ in quiescence and $0.49a$ in outburst.

Global disc evolution models reproduce the main properties of observed outbursts, but only if the efficiency of transport and dissipation is less in quiescence than during outburst [5]. This seems in agreement with spiral shocks. Indeed, the tidal force is a very steep function of r/a, hence the spirals rapidly become weak at smaller disc sizes. Spiral shocks are produced in the outer regions of the disc by the tides raised by the secondary star. During the outburst, the disc expands, and its outer parts feel the gravitational attraction of the secondary star more efficiently, leading to the formation of spiral arms.

Having these general ideas in mind, we can now summarise the results from observations (see the review by Danny Steeghs in this volume):

- The spiral pattern has been established in several cataclysmic variables for a wide range of emisson lines
- The spiral arms appear right at the start of the outburst and persist during outburst maximum for at least 8 days, i.e. several tens of orbital periods
- The structure is fixed in the corotating frame of the binary and corresponds to the location of tidally driven spiral waves
- During quiescence, the spiral pattern is no longer there but the disc remains asymmetric

These results can be exactly interpreted in the spiral shock theory. They do not prove necessarily, however, that spiral shocks are the main viscosity mechanism in the disc. There is indeed a problem similar to the egg and the hen: who begon? In order to have well developed spirals, one needs a large disc, which is due to an increase in viscosity !

6   Numerical Simulations

As can already be seen from Fig. 3 and 4, in order to explain the widely open spirals seen in the Doppler maps of IP Peg [38], one require a rather hot disc.

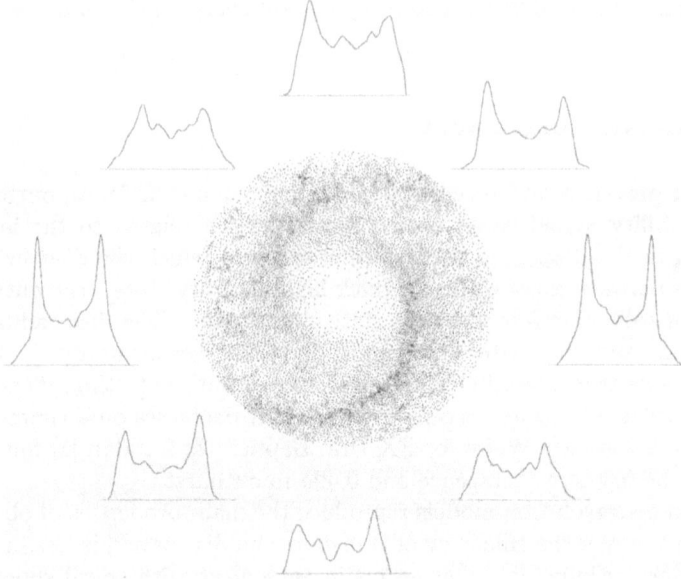

Fig. 5. An accretion disc showing spiral arms as obtained by SPH simulations. When viewed from different angles (*i.e.* at different orbital phases), the emission-line profile is different. When these lines are recorded at several orbital phases, one can then construct a spectrogram which can be inverted, using a maximum entropy method, to give a doppler tomogram

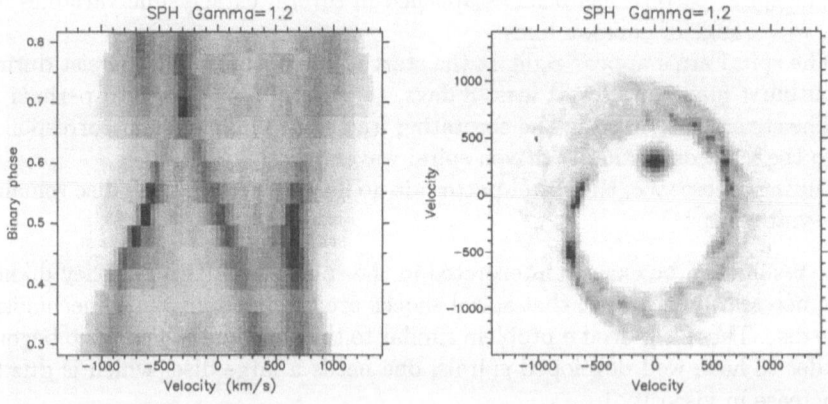

Fig. 6. Binary phase-velocity map (spectrogram) and Doppler map corresponding to the simulation of Fig. 5

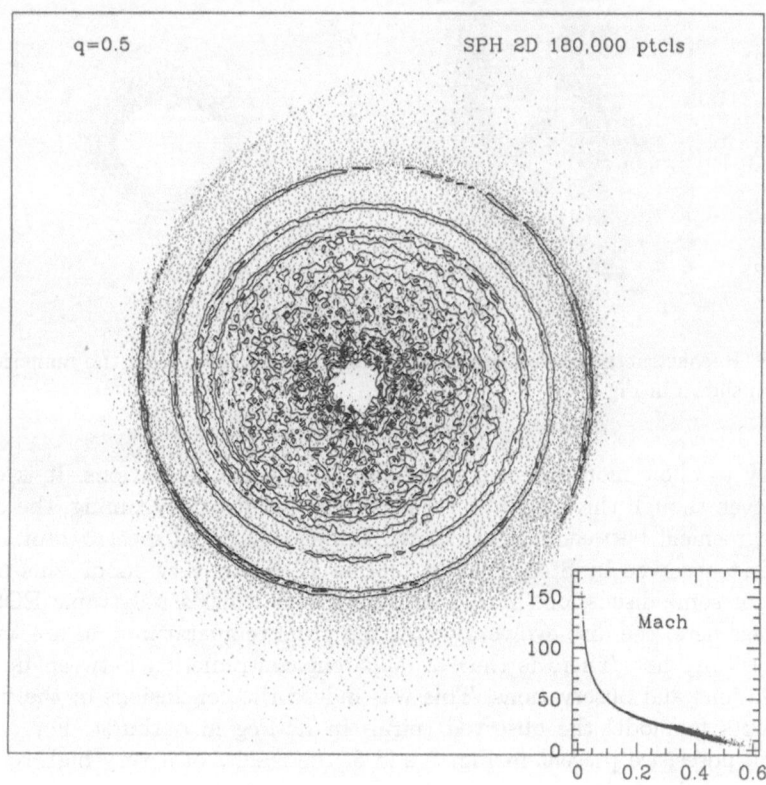

Fig. 7. Result of a high-resolution 2D SPH simulation of an isothermal ($c_s = 0.05$) accretion disc in a binary system with a mass ratio of 0.5. The density contours are plotted over the particles positions. The inset shows the variation of the Mach number with the distance from the white dwarf.

Figure 2 shows for example the results of a 2D numerical simulation, using the Smoothed Particle Hydrodynamics (SPH) method, of an accretion disc in a binary system with a mass ratio of 0.5, as observed for IP Peg. Here, a polytropic equation of state (EOS) has been used with a polytropic index, $\gamma = 1.2$. In Fig. 2, we show the different emission lines profiles corresponding to the various orbital phases for such a disc. Those profiles can then be presented in a trailed spectrogram as seen in Fig. 6. With numerical simulations, we have the advantage of having all the dynamical information of the flow and we can therefore easily construct a Doppler map which can then be compared to those observed. The Doppler map corresponding to this $\gamma = 1.2$, 2D simulation is also shown in Fig. 6. Note that we have artificially added a spot at the location of the sec-

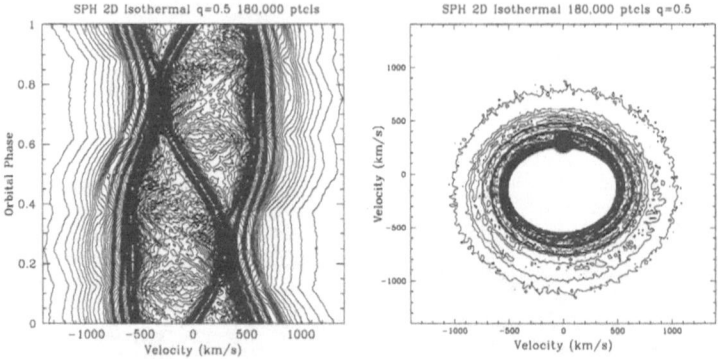

Fig. 8. Reconstructed trailed spectrogram and Doppler map from the numerical simulation shown in Fig. 7

ondary to allow more direct comparisons with real observations. It is obvious that even though this simulation was not the result of any tuning, the qualitative agreement between this calculated Doppler map and spectrogram, and the one first observed by Steeghs et al. [38] for IP Peg is very good. This however calls for some discussion. Indeed, in simulations using a polytropic EOS, as is the case here, the disc evolve towards the virial temperature and are therefore unphysically hot. There is thus an apparent contradiction between numerical simulations and observations. This was indeed the conclusions of the first few attempts to model the observed spirals in IP Peg in outburst. For comparison purposes, we present in Fig. 7 and 8, the results of a very high resolution isothermal simulation.

Godon, Livio & Lubow [5] presented two-dimensional disc simulations using Fourier-Chebyshev spectral methods and found that the spiral pattern resembles the observations only for very high temperatures. This was already the conclusion of Savonije et al. [28] who claimed that spirals could not appear in the colder disc of cataclysmic variables.

We have therefore run another set of simulations, with our SPH code, where we use an "isentropic" equation of state, instead of a polytropic one. By "isentropic", we mean a barotropic EOS, $P = K\rho^\gamma$, and keeping K constant. The heating due to the viscous processes is supposed to be instantaneously radiated away. In this case, the temperature remains always very close to the initial value. The results of isentropic runs are compared with those of polytropic runs using $\gamma=1.2$ and two values for the initial sound speed: 0.1 and 0.04 times the orbital velocity (Fig. 9). While, in the two polytropic runs, spiral structures are clearly present with a very similar pitch angle, there is a clear distinction between the two isentropic runs: the 0.1 case produces well defined spiral arms which are more tightly wound than in the equivalent polytropic case, while in the 0.04 case, spiral structure can hardly be seen at all. The difference clearly lies in the Mach number of the flow: in the polytropic cases, the Mach number is well below 10 for both a sound speed of 0.04 and 0.1, while in the isentropic cases, the Mach

c=0.04 polytropic   c=0.1 polytropic

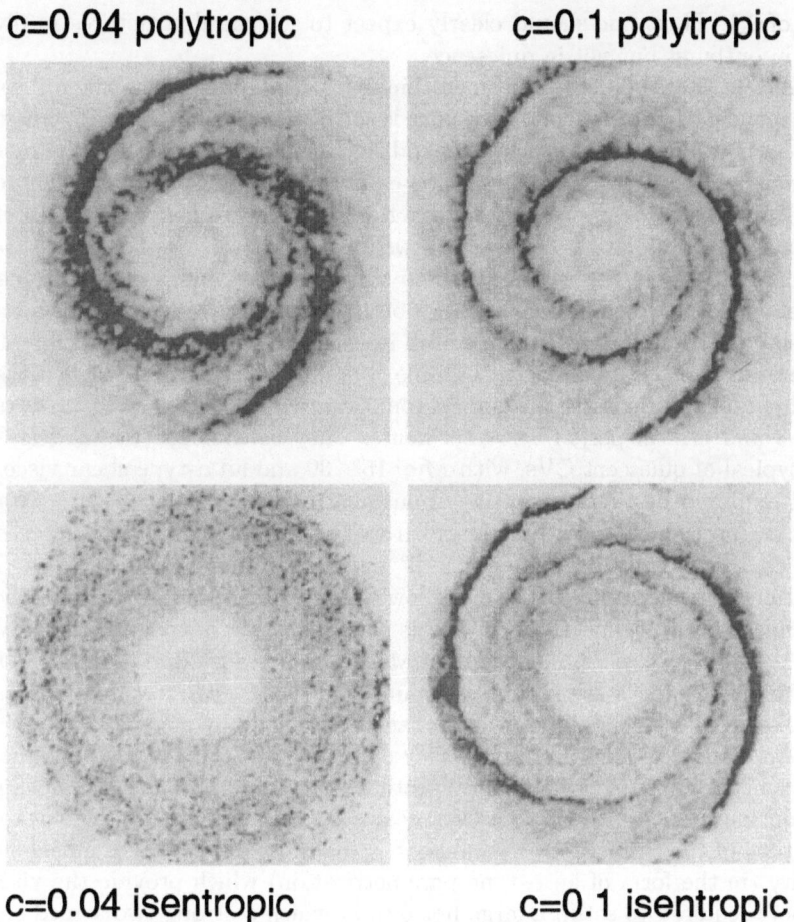

c=0.04 isentropic   c=0.1 isentropic

Fig. 9. Comparison between our SPH results for the isentropic and polytropic equation of state for two values of $c_o=0.1$ and $c_o=0.04$

number goes from 20 at the outer edge of the disc to more than 100 inside when the sound speed is 0.04, but goes from 10 to only 30 when the sound speed is 0.1. Therefore, the observations seem to require a hot disc with a wide spiral. But, are dwarf novae discs hot or cold ? During quiescence, the disc will have a temperature of the order of 10,000 K. This corresponds to a sound speed of the order of 15 km/s, or in our units 0.03. Thus, one should compare observations with our isentropic case with sound speed 0.04, and we therefore predict that no spiral should be seen. In outburst however, a dwarf nova will have a temperature profile corresponding to that of the steady-state solution for a viscous disc (e.g. [40]), and the temperature will be closer to 10^5 K, *i.e.* a sound speed of roughly 45 km/s (in our units, 0.1). In this case, the Mach number will be around 20 and one is allowed to compare observations with the isentropic case with a sound speed of 0.1. Thus, one should clearly expect to see spiral structures in dwarf novae in outburst but not in quiesence.

Similarly, Stehle (1998) performed thin disc calculations where the full set of time dependent hydrodynamic equations is solved on a cylindrical grid. The disc thickness is explicitly followed by two additional equations in a one-zone model, allowing the disc to be vertically in non-equilibrium. The spatial and temporal evolution of the disc temperature follows tidal and viscous heating, the latter in the α-ansatz of Shakura & Sunayev, as well as radiation from the disc surfaces. The surface temperature is connected to the disc mid-plane temperature using Kramers opacities for the vertical radiation transport. In this sense, it is a much more elaborated model than the classical isothermal or adiabatic approximation for the equation of state. Steeghs & Stehle (1999) use the grid of disc calculations of Stehle (1998) to construct Doppler tomograms for a binary with mass ratio $q = 0.3$ and orbital period $P = 2.3$ hours. They considered two models, one more typical of quiescent CVs, with $M \simeq 15 - 30$ and an α-type shear viscosity of 0.01, and another, representative of outbursting discs, with $M \simeq 5 - 20$ and $\alpha = 0.3$. Effective temperature ranged from less then 10^4 K in the outer part of the cold disc, to values between $2\,10^4$ and $5\,10^4$ K for the hotter disc. The cold disc was rather small (due to the low viscosity), varying in radius between 0.55 and 0.65 r_{L1}, while the hot disc is pushed by the increased viscosity to larger radii of 0.6-0.8 r_{L1}. For the high Mach number accretion discs, they find that the spiral shocks are so tightly wound that they leave few fingerprints in the emission lines, the double peaks separation varying by at most 8%. For the accretion disc in outburst, however, they conclude that the lines are dominated by the emission from an $m = 2$ spiral pattern in the disc, resulting in converging emission line peaks with a cross over near phases 0.25 and 0.75. It has to be noted that in the simulations of Stehle, it is the presence of a large initial shear viscosity (in the form of an α-type parametrization) which provide the viscous transport required to setup a large hot disc in which a strong two armed spiral patter forms. This remark will take all its relevance in view of the discussion in Sect. 8.

For example, Godon et al. [5] used a value of $\alpha = 0.1$, so that viscous spreading was less efficient compared to the simulations of Stehle. As a consequence, their discs were smaller by up to 50 % compared to the hot disc model of Steeghs & Stehle, hence the tidal effect was less severe.

7   Angular Momentum Transport

As noted above, the parameter α introduced by Shakura and Sunayev [32] refers to some local unknown viscosity. In the case of spiral shocks, which is a global phenomenon, we can still refer to an *effective* α that would give the same mass accretion rate in an α-disk model that one finds in numerical simulations.

The standard α-disk theory gives the mass accretion rate in terms of α as $\dot{M} = 3\pi\alpha c_s H \Sigma$ [see Eq. (8) and (9)]. Using Eq.(10), Blondin [2] finds an equation for α:

$$\alpha_{\text{eff}} = \frac{2}{3} \frac{v_\phi \langle v_r \rangle}{c_s^2}, \tag{15}$$

where $\langle v_r \rangle$ is a density-weighted average of the radial velocity:

$$\langle v_r \rangle = \frac{\int_0^{2\pi} \Sigma v_r d\phi}{\int_0^{2\pi} \Sigma d\phi}.$$

It has to be noted that the accretion time scale can be estimated (e.g. [36]) :

$$\tau_\alpha \sim \frac{1}{\alpha_{\text{eff}} \Omega} \left(\frac{v_\phi}{c_s}\right)^2.$$

Thus this time scale can only be followed for rather hot discs with high value of α_{eff} and therefore, for cold discs, the evolution can usually be followed until the wave pattern is staionary but not long enough for the accretion process to reach a steady state. The wave pattern on the other hand is stable in only a few sound crossing time.

From his study of self-similar models of very cool discs, Spruit [13] obtains $\alpha_{\text{eff}} = 0.013 \left(\frac{H}{r}\right)^{3/2}$. As in cataclysmic variables, $\frac{H}{r} \leq 0.1$, this leads to very low values $(10^{-4} - 10^{-3})$, which was the reason why several people dismissed spiral shocks as a viable efficient accretion mechanism. Spruit himself, however, in his paper, insists that this must not be the final word: "For most common mass ratios $0.1 < q < 1$, the forcing by the companion is so strong however that the resulting spiral shock therefore have a strength much above the self similar value over a large part of the disc". In their inviscid - but adiabatic - simulations, Matsuda et al. (1987) obtained values of the effective α up to 0.1, hence large enough to explain mass accretion observed in outbursting cataclysmic variables. In fact, Spruit [13] and Larson [10] obtained a relation between the effective α and the radial Mach number M_r at disc mid-plane. Larson [10], for example, obtains $\alpha_{\text{eff}} \simeq 0.07(M_r^2 - 1)^3$ for isothermal discs. Note that for an isothermal disc, M_r^2 is equal to the compression ratio. With M_r being of the order of 1.3 to 1.5, this typically implies values of $\alpha_{\text{eff}} \simeq 0.02 - 0.14$. Numerical simulations by Blondin [2] seem to confirm this, as in his isothermal simulations, values as high as 0.1 are obtained near the outer edge of the disc. For very cold discs, he even found values close to 1. The accretion efficiency however decreases very sharply and reaches value below 10^{-3} in the part of the disc closer than $0.1a$. For a hotter disc, if the value of α_{eff} is about 0.1 in the outer part of the disc, it stays above 0.01 well inside the disc. Larson [10] goes on to predict that the strength of tightly wound spiral shocks in a cold, thin disk should be proportional to $(\cos\theta)^{3/2}$, where θ is the trailing angle of the spiral wave. Estimating $\cos\theta \approx 1.5c_s/v_\phi$, Larson [10] finds an effective α of

$$\alpha \approx 0.026 \left(\frac{c_s}{v_\phi}\right)^{3/2}. \tag{16}$$

The prediction for an isothermal disk with $c_s = 0.25$ is $\alpha \approx 3.2 \times 10^{-3} r^{3/4}$, orders of magnitude below that found in the numerical simulations (e.g. [2,18]).

But in these simulations, the spirals are not tightly wound and the formula of Larson may not be applicable.

Blondin [2], for example, finds that the maximum value of α, found near the outer edge of the disk, remains roughly independent of sound speed, while the radial dependence of $\alpha(r)$ steepens with decreasing sound speed. In fact, because the radial decay of α is so steep, he found steady mass accretion only for radii above $r \approx 0.1a$ in the coldest disk with an outer Mach number of ~ 32. A further result of his simulations is the relative independence of spiral waves on the mass ratio in the binary system. The strength of the two-armed spiral shocks at their origin near the outer edge of the disk was fairly constant in all of his models. Blondin believes this has the consequence that, despite all else, one can be confident that the effective α in the outer regions of an accretion disk in a binary system is (at least) ~ 0.1.

8   Spirals in Quiescence?

By using results from shearing-box simulations as an input for a global numerical model designed to study disc instabilities, Menou [23] shows that as the disc goes into quiescence, it suffers a runaway cooling. Hence, in quiescence, the disappearance of self-sustained MHD turbulence is guaranteed. As accretion is known to occur during quiescence in cataclysmic variables, another transport mechanism must operate in the discs. The rapid disc expansion observed during the outbursts of several dwarf novae is consistent with MHD-driven accretion because it shows that disc internal stresses dominate transport during this phase [23]. On the other hand, the same discs are observed to shrink between consecutive outbursts, which is a signature that transport is dominated by the tidal torque due to the companion star, at least in the outer regions of the disc during quiescence. When looking at a sample of 6 well studied SU UMa stars, Menou [23] finds an anti-correlation of the recurrence times with mass ratios. This, he interprets, is evidence that tidal torques dominate the transport in the quiescent discs. Indeed, the recurrence times of dwarf novae represent the time-scales for mass and angular momentum redistribution in the quiescent discs. For his small sample, the correlation is significant for normal and super-outbursts. No correlation is however found for U Gem type dwarf novae but this could be because the correlation is masked by other effects. Menou [23] proposes thus the concept of accretion driven by MHD turbulence during outburst and by tidal perturbations during quiescence.

This is, at first sight, a rather astonishing result. Indeed, we have seen that the spirals will be more developed and more effective when the disc is hotter and larger, i.e. in outburst. If the main source of viscosity during quiescence are the spiral shocks, then one might expect that their contribution during outburst cannot be negligible. But as stated before, one needs first to make the disc large for the tidal effect to become more important. The answer may lie - like always - in a combination of processes. Spiral arms are indeed very good at transporting angular momentum in the outer part of the disc. But they do not succeed generally to penetrate deep into the disc. This is were MHD turbulence may provide the main source of viscosity. This scenario clearly needs to be further developed and tested.

9   Spiral Shocks in Other Stellar Objects

Even if spiral shocks are not the main driving viscosity mechanism in accretion discs, Murray et al. (1998) argue that they could have another observational consequence in intermediate polars. Indeed, the presence of spiral waves break the axisymmetry of the inner disc and tells the accreting star the orbital phase of its companion. This could put an additional variation in the accretion rate onto the white dwarf, a variation dependent on the orbital period. This could explain the observed periodic emission in intermediate polars.

Savonije, Papaloizou & Lin [28] note that although tidally induced density waves may not be the dominant carrier of mass and angular momentum throughout discs in CVs, they are more likely to play an important role in protostellar discs around T Tauri binary stars, where the Mach number is relatively small. In GW Ori for example, both circumstellar and circumbinary discs have been inferred from infrared excesses [17] and the disc response to the tidal disturbance might be significant. In the same line of ideas, Boffin et al. [4,41,42,44] found large tidal waves in large protostellar discs being induced by the tidal interaction of a passing star. In this case, the perturbation might be large enough in these self-gravitating discs to lead to the collapse of some of the gas, thereby forming new stars as well as brown dwarfs and jovian planets.

Another class of objects where spiral arms may play a role is X-ray binaries (see the review by E. Harlaftis in this volume). There also, because the discs are rather hot, the Mach number is much smaller than in CVs, hence their effect might be more pronounced. An observational confirmation of this would be most welcome. The work of Soria et al. [33] is maybe such a first step. These authors studied optical spectra of the soft X-ray transient GRO J1655-40. They claim that, during the high state, the Balmer emission appears to come only from a double-armed region on the disc, possibly the locations of tidal density waves or spirals shocks.

References

1. Baptista R., Catalan M.S., 2000, ApJ, 539, L55
2. Blondin, 1999, New Astronomy, 5, 53
3. Boffin H.M.J, Haraguchi K., Matsuda T., 1999, In: *Disk Instabilities in Close Binary Systems — 25 Years of the Disk-Instability Model*, Mineshige S., Wheeler J.C., eds., Universal Academy Press, p. 137
4. Boffin H.M.J., Watkins S.J., Bhattal A.S., Francis N., Whitworth A.P., MNRAS, 300, 1189
5. Cannizzo J.K., 1993, In: *Accretion disks in Compact Stellar Systems*, Wheeler J.C., ed., p. 6
6. Frank J., King A.R., Raine D.J., 1992, *Accretion Power in Astrophysics*, 2nd ed., Cambridge University Press

7. Godon P., 1997, ApJ, 480, 329
8. Godon P., Livio M., Lubow S., 1998, MNRAS, 295, L11
9. Haraguchi K., Boffin H. M. J., Matsuda T., 1999, In: *Star Formation*, p. 241
10. Larson R. B. 1989, In: *The Formation and Evolution of Planetary Systems*, H. A. Weaver & L. Danly, eds., Cambridge Univ. Press, p. 31
11. Larson R.B., 1990, MNRAS, 243, 358
12. Lin D.N.C., Pringle J. E., 1976, In: *Structure and Evolution of Close Binary Systems*, Eggleton P., Mitton S., Whelan J., eds., Proc. IAU Symp. 73, Reidel, p. 237
13. Lin D.N.C., Papaloizou J., 1979, MNRAS, 186, 799
14. Makita M., Matsuda T., 1999, In: *Numerical Astrophysics*, Miyama S.M., Tomisaka K., Hanawa T., eds., Kluwer, p. 227
15. Makita M., Yukawa H., Matsuda T., Boffin H.M.J., 1999, In: *Disk Instabilities in Close Binary Systems — 25 Years of the Disk-Instability Model*, Mineshige S., Wheeler J.C., eds., Universal Academy Press, p. 147
16. Makita M., Miyawaki K., Matsuda T., 2000, MNRAS, 316, 906
17. Mathieu R.D., Adams F.C., Latham D.W., 1991, AJ, 101, 2184
18. Matsuda T., Inoue M., Sawada K., Shima E., Wakamatsu K., 1987, MNRAS, 229, 295
19. Matsuda T., Sekino N., Shima E., Sawada K., Spruit H., 1990, A & A, 235, 211
20. Matsuda T., Makita M., Boffin H.M.J., 1999, In: *Disk Instabilities in Close Binary Systems — 25 Years of the Disk-Instability Model*, Mineshige S., Wheeler J.C., eds., Universal Academy Press, p. 129
21. Matsuda T., Makita M., Yukawa H., Boffin H.M.J., 1999, In: *Numerical Astrophysics*, Miyama S.M., Tomisaka K., Hanawa T., eds., Kluwer, p. 207
22. Matsuda T., Makita M., Fujiwara H., Nagae T., Haraguchi K., Hayashi E., Boffin H.M.J., 2000, Ap&SS, 274, 259
23. Menou K., 2000, Science, 288, 2022
24. Paczyński B., 1977, ApJ, 216, 822
25. Papaloizou J.C.B., Pringle J.E., 1977, MNRAS, 181, 441
26. Różyczka M., Spruit H., 1989, In: *Theory of Accretion Disks*, Meyer F., Duschl W.J., Frank J., Meyer-Hefmeister E., eds, Kluwer, Dordrecht, p.341
27. Różyczka M., Spruit H. C., 1993, ApJ, 417, 677
28. Savonije G.J., Papaloizou J.C.B., Lin D.N.C., 1994, MNRAS, 268, 13
29. Sawada K., Matsuda T., Hachisu I., 1986, MNRAS, 219, 75
30. Sawada K., Matsuda T., Hachisu I., 1986, MNRAS, 221, 679
31. Sawada K., Matsuda T., Inoue M., Hachisu I., 1987, MNRAS, 224, 307
32. Shakura N.I., Sunyaev R.A., 1973, A&A, 24, 337
33. Soria R., Wu K., Hunstead R.W., 2000, ApJ, 539, 445
34. Spruit H., 1987, A&A, 184, 173
35. Spruit H.C., 1989, In: *Theory of Accretion discs*, Meyer F., Duschl W.J., Frank J., Meyer-Hofmeister E. (eds.), Kluwer, Dordrecht, p. 325
36. Spruit H.C., 2000, In: *The neutron star black hole connection*, C. Kouveliotou et al., eds., NATO ASI series
37. Spruit H., Matsuda T., Inoue M., Sawada K., 1987, MNRAS, 229, 517
38. Steeghs D., Harlaftis E.T., Horne K., 1997, MNRAS, 290, L28
39. Tout C.A., 2000, New Astronomy Reviews, 44, 37
40. Warner B., 1995, *Cataclysmic variable stars*, Cambridge University Press
41. Watkins S.J., Bhattal A.S., Boffin H.M.J., Francis N., Whitworth A.P., MNRAS, 300, 1205

42. Watkins S.J., Bhattal A.S., Boffin H.M.J., Francis N., Whitworth A.P., MNRAS, 300, 1214
43. Whitehurst R., King A., 1991, MNRAS, 249, 25
44. Whitworth A.P., Boffin H.M.J., Francis N., 1998, A&G, 39, 6.10
45. Yukawa H., Boffin H.M.J., Matsuda T., 1997, MNRAS, 292, 321

Spiral Shocks in an Inviscid Simulation of Accretion Flow in a Close Binary System

M. Makita[1], H. Fujiwara[2], T. Matsuda[2], and H.M.J. Boffin[3]

[1] Department of Astronomy, Kyoto University, Sakyo-ku, Kyoto 606-8502, Japan
[2] Department of Earth and Planetary Sciences, Kobe University,
   Nada-ku, Kobe 657-8501, Japan
[3] Royal Observatory of Belgium, 3 av. Circulaire, B-1180 Brussels, Belgium

Abstract. We perform 2D and 3D numerical simulations of an accretion disc in a close binary with a mass ratio of 0.5, corresponding to the dwarf nova IP Pegasi. We do not include artificial nor α-viscosity and supply gas from the inner L1 point. We construct Doppler maps and artificial trailed spectrograms. In the 2D calculation, the results agree very well with those observed in IP Peg. In the 3D calculations, the stream from the L1 point penetrates into the accretion disc. The resulting flow pattern, therefore, is different from the 2D results.

1 Introduction

Spiral shock waves in an accretion disc were found in two-dimensional numerical simulation by Sawada, Matsuda and Hachisu [10,11]. Spruit [13] obtained self-similar solutions of spiral shocks in a semi-analytic manner. Since then, many researchers have performed numerical simulations and confirmed the presence of spiral shocks in the accretion disc both in 2D and 3D simulations.

On the observational side, eleven years after the numerical discovery of spiral shocks, Steeghs, Harlaftis & Horne [14], using the Doppler tomography technique, found the first convincing evidence for spiral structures in the accretion disc of the dwarf nova IP Peg during outburst. Since then, spiral structures have been found successively in other accretion discs: SS Cyg, V347 Pup, EX Dra, U Gem. (Detailed discussions about the spiral shock model and its history are written by H.M.J. Boffin and D. Steeghs in this volume.)

The observational rediscovery of spiral structures in accretion discs led to renewed interest in the topic and several group performed numerical studies to investigate the observed spiral waves. Godon, Livio and Lubow [5] performed 2D numerical calculations of the tidal interaction between the companion star and the disc around the primary, in a cataclysmic variable system with parameters appropriate for IP Peg, using time-dependent hybrid Fourier-Chebyshev spectral method. They attempted to reproduce the observations of IP Peg. They found, however, that the spiral pattern resembles the observations only for very high temperature. Armitage and Murray [1] calculated the evolution of the disc outburst using Smoothed Particle Hydrodynamics (SPH) scheme to construct Doppler tomograms. They claimed that the obtained Doppler tomograms are in close agreement with the observations if the spiral pattern arises as a transient feature when the disc expands viscously at the start of the outburst. Steeghs and Stehle [15] used the grid of hydrodynamics accretion disc calculations of Stehle [16] to construct orbital phase-dependent emission-line profiles of thin discs carrying spiral density waves. They confirmed that the observed spiral pattern can be reproduced by tidal density waves in the accretion disc and demanded the presence of a large, hot disc, at least in the early outburst stages.

2 Model and Calculations

We calculate 2D as well as 3D flows in a compact binary system with a mass ratio of 0.5, using the Simplified Flux vector Splitting (SFS) scheme [7]. The numerical accuracy of the calculations is second-order both in space and in time. We use almost the same assumptions and conditions as in [7]. Only the region surrounding the white dwarf is considered. The origin of the coordinates is at the center of the white dwarf, and the computational region is $-0.57 \leq x, y \leq 0.57$ in 2D and $-0.57 \leq x, y \leq 0.57$ and $0 \leq z \leq 0.25$ in 3D. The region is divided into 228×228 grid points in 2D and $228 \times 228 \times 50$ in 3D. We assume an ideal gas, which is characterized by the specific heat ratio γ, assumed to be 1.2 for 2D calculations and, 1.2 and 1.01 for 3D calculations. The gas is injected from a small hole at the L1 point, which is at $x = -0.57, y = 0$, and $z = 0$. In the 3D calculations, the sound speed of the gas to be injected into the computational region is $0.02\Omega a$, where Ω is the orbital angular frequency and a is the binary separation. This value of the sound speed is smaller than in [7], in order to improve the boundary condition at the L1 point.

The differences between our calculations and previous similar calculations are the following:

- We do not include any artificial viscosity nor α-viscosity.
- Gas is supplied from the inner L1 point.

3 Results of 2D Calculations

Here, we show the results from our 2D hydrodynamic calculations. Figure 1 shows our calculated Doppler map (a), artificial trailed spectrograms (b), and calculated density distribution (c). Clear spiral shocks can be seen in Fig. 1(c). Based on the density distributions in the x-y space, we construct a Doppler map, showing the density distributions in the velocity space, (v_x, v_y). In constructing the Doppler map, a contribution due to the companion star is added at its position. The contribution from the outer region of the accretion disc, $r > 0.22$, is omitted in order to avoid the effect of the stream from the L1 point. This contribution is apparently not seen in the observation, conducted during outburst. The calculated Doppler map and line flux shown in Figs. 1(a,b) agree very well with observations found in [14].

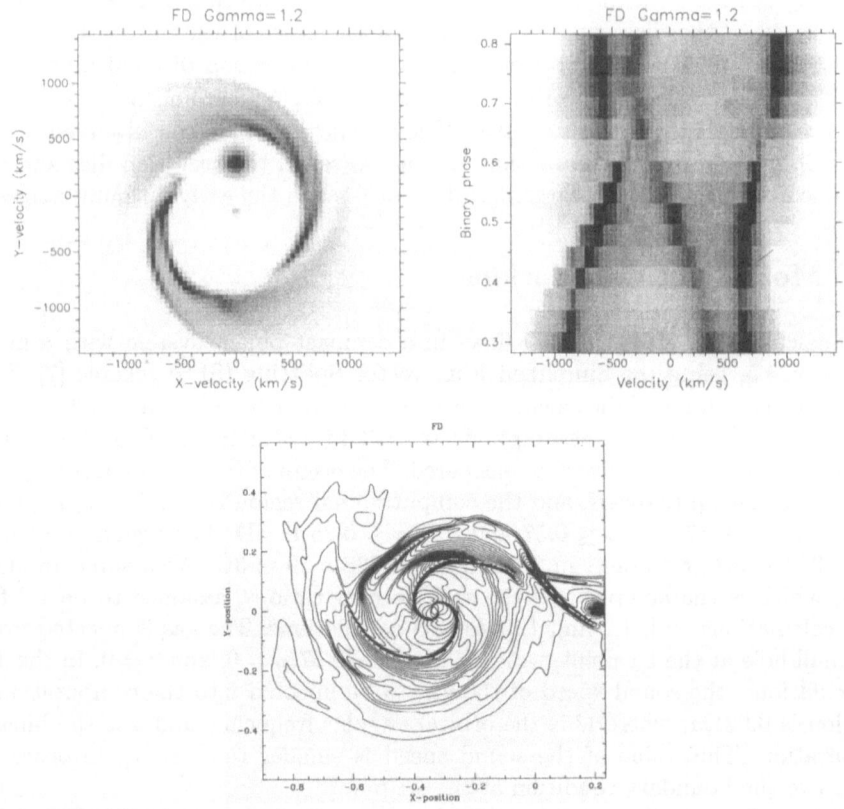

Fig. 1. Result of 2D calculations. (a)Top left: calculated Doppler map. The abscissa and the ordinate depict v_x and v_y, respectively. (b)Top right: calculated line flux as a function of the binary phase (ordinate), i.e. artificial trailed spectrograms. The abscissa represents the radial velocity. (c)Bottom: the density distribution (After [6,8])

Here, some remark is necessary. In this calculation, the obtained Mach number of the flow is always below 10, which means that the disc has a considerably high temperature. It is well known, however, both from theoretical studies as from observations, that the accretion discs of cataclysmic variables have much lower temperatures with Mach numbers of 20-30. For such high Mach numbers, spiral shocks would tightly wind, and would therefore not agree with observations. This problem has been already pointed out [1,5].

We also construct an eclipse light curve. We use an inclination of 80° also corresponding to the case of IP Peg. There is a little bit of asymmetry in Figure 2, but the effect of the presence of spiral shocks can not be seen. This effect is not as clear as seen in [2].

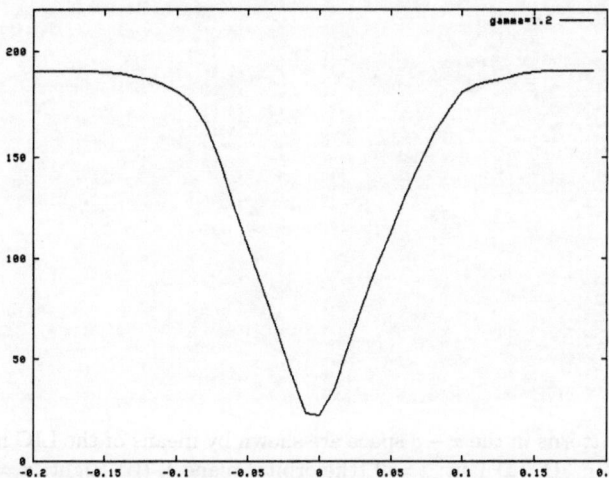

Fig. 2. Eclipse light curve. The abscissa depicts orbital phase

4   Results of 3D Simulations

Figure 3 presents 3D iso-density surfaces. Spiral structures are clearly seen. Comparing these pictures with Makita et al. [7], spiral waves wind more tightly in the case of $\gamma = 1.01$ in this simulation because of the lower temperature at the L1 point. The flow from L1 point does not expand, so that it does not form an under-expanded jet which was observed in previous results [7].

Figure 4 shows flow patterns at $z = 0$ (a), that is the orbital plane, and $z = 0.02$ (b), by means of the Line Integral Convolution (LIC) method, in the case of $\gamma = 1.2$. The stream from the L1 point directly penetrates into the accretion disc in Fig. 4(a). On the other hand, at $z = 0.02$, gas from the disc

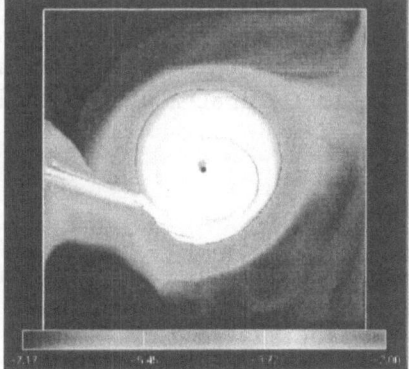

Fig. 3. 3D view of iso-density surfaces at $\log \rho = -4.0$. Gray scale of the density in the orbital plane is also shown. (a)Left: $\gamma = 1.01$. (b) Right: $\gamma = 1.2$

Fig. 4. Flow patterns in the $x-y$ space are shown by means of the LIC method in the case of $\gamma = 1.2$ in 3D. (**a**) Left: $z = 0$ (the orbital plane). (**b**) Right: $z = 0.02$

rotates around the primary star in Fig. 4(**b**). These pictures show that the stream from the L1 point penetrates into the accretion disc rather than being blocked by it and that the rotating gas in the disc collides with this penetrating flow to form a bow shock [3]. This shock can be called a hot line rather than a hot spot. Rotating gas overflows the hot line, so that the edge of the hot line may be seen like a hot spot. This penetration occurs because the L1 stream has a much larger density than the accretion disc. The density of the accretion disc is still increasing at the end of these calculations. The penetration, therefore, may be a transient feature before the disc gas is piled up. (See [4,9] for the detailled discussion of the effect of penetration.) The penetration is, for example, observed in the accretion disc of WZ Sge [12].

It is hard to prepare a correct, meaningful Doppler map based on the results of three-dimensional calculations, since the distribution observed is that on the photosphere of the accretion disc. The preparation thus requires complex calculations of radiative transfer, which is beyond the scope of the present paper. Doppler maps based on the density at some z value are a little bit different from the ones in 2D because of the presence of the penetrating flow [6]. We also comment that the Mach number in the accretion disc in 3D is again as small as less than 10.

5   Summary

We performed 2D and 3D numerical simulations with a mass ratio of 0.5, which corresponds to the dwarf nova IP Pegasi. We obtained the following results:

1. A computer synthesized Doppler map and trailed spectrogram based on 2D result are compared with observation and a good agreement is obtained.
2. The constructed eclipse light curve shows a small asymmetry. The effect of the presence of the spiral structure, however, is not seen.

3. In 3D, the stream from the L1 point penetrates into the accretion disc rather than being blocked by it. The gas rotating in the disc collides with this penetrating stream and forms a bow shock. This shock can be called a hot line rather than a hot spot. The rotating gas overflows the hot line, so that the edge of the hot line may be seen like a hot spot.

4. In 3D, the presence of the penetrating flow makes the structure of the accretion disc complicated.

5. The Mach numbers in our accretion discs are less than 10 both in 2D and in 3D. The discs in our numerical simulation are therefore rather hot compared with what is ordinarily assumed for dwarf novae. This is an important problem to be solved.

References

1. Armitage P.J., Murray J.R., 1998, MNRAS, 297, L81
2. Baptista R., Harlaftis E.T., D. Steeghs D., 2000, MNRAS, 314, 727
3. Bisikalo D.V., Boyarchuk A.A., Chechetkin V.M., Kuznetsov O.A., Molteni D., 1998, MNRAS, 300, 39
4. Fujiwara H., Makita M., Nagae T., Matsuda T., 2000, PASJ, submitted
5. Godon P., Livio M., Lubow S., 1998, MNRAS, 295, L11
6. Makita M., Yukawa H., Matsuda T., H. M. J. Boffin H.M.J., 1999, In: *Disk Instabilities in Close Binary Systems, the Disk-Instability Workshop on "Disk Instabilities in Close Binary Systems" at Kyoto, Japan, October 27-30, 1998,* ed S. Mineshige, J. C. Wheeler (Universal Academy Press, Tokyo) pp. 147-150
7. Makita M., Miyawaki K., Matsuda T., 2000, MNRAS, 316, 906
8. Matsuda T., Makita M., Yukawa H., Boffin H.M.J., 1999, In: *Numerical Astrophysics, International Conference on Numerical Astrophysics 1998 at Tokyo, Japan, March 10-13, 1998,* ed. by S. M. Miyama, K. Tomisaka, T. Hanawa (Kluwer, Dordrecht) pp. 207-210
9. Matsuda T., Makita M., Fujiwara H., Nagae T., Haraguchi K., Hayashi E., Boffin H.M.J., 2000, Ap & SS, in press
10. Sawada K., Matsuda T., Hachisu I., 1986, MNRAS, 219, 75
11. Sawada K., Matsuda T., Hachisu I., 1986, MNRAS, 221, 679
12. Skidmore W., Mason E., Howell S.B., Ciardi D.R., Littlefair S., Dhillon V.S., 2000, MNRAS, in press
13. Spruit H., 1987, A& A, 184, 173
14. Steeghs D., Harlaftis E.T., Horne K., 1997, MNRAS, 290, L28
15. Steeghs D., Stehle R., 1999, MNRAS, 307, 99
16. Stehle R., 1999, MNRAS, 304, 687

Imaging the Secondary Stars in Cataclysmic Variables

V.S. Dhillon and C.A. Watson

Department of Physics and Astronomy, University of Sheffield, Sheffield S3 7RH, UK

Abstract. The secondary, Roche-lobe filling stars in cataclysmic variables (CVs) are key to our understanding of the origin, evolution and behaviour of this class of interacting binary. We review the basic properties of the secondary stars in CVs and the observational and analysis methods required to detect them. We then describe the various astro-tomographic techniques which can be used to map the surface intensity distribution of the secondary star, culminating in a detailed explanation of Roche tomography. We conclude with a summary of the most important results obtained to date and future prospects.

1 Introduction

CVs are semi-detached binary stars consisting of a Roche-lobe filling secondary star transferring mass to a white dwarf primary star via an accretion disc or magnetically-channelled accretion flow. The CVs are classified into a number of different sub-types, including the *novae, recurrent novae, dwarf novae,* and *novalikes,* according to the nature of their cataclysmic (i.e. violent but non-destructive) outbursts. A further sub-division into *polars* and *intermediate polars* is also made if a CV accretes via magnetic field lines. Figure 1 depicts the five principal components of a typical non-magnetic CV: the primary star, the secondary star, the gas stream (formed by the transfer of material from the secondary to the primary), the accretion disc and the bright spot (formed by the collision between the gas stream and the edge of the accretion disc). The distance between the stellar components is approximately a solar radius and the orbital period is typically a few hours. For a detailed review of these objects, see [44].

1.1 The Nature of the Secondary Stars in CVs

The spectral type and luminosity class of the secondary star in CVs can be estimated from basic theory, as follows. The mean density, $\bar{\rho}_2$, of the secondary star is given by

$$\bar{\rho}_2 = \frac{M_2}{\frac{4}{3}\pi R_2^3},\tag{1}$$

where M_2 and R_2 are the mass and volume radius (i.e. the radius of a sphere with the same volume as the Roche lobe) of the secondary star, respectively. The volume radius of the Roche lobe can be approximated by

$$\frac{R_2}{a} = 0.47\left(\frac{q}{1+q}\right)^{1/3}\tag{2}$$

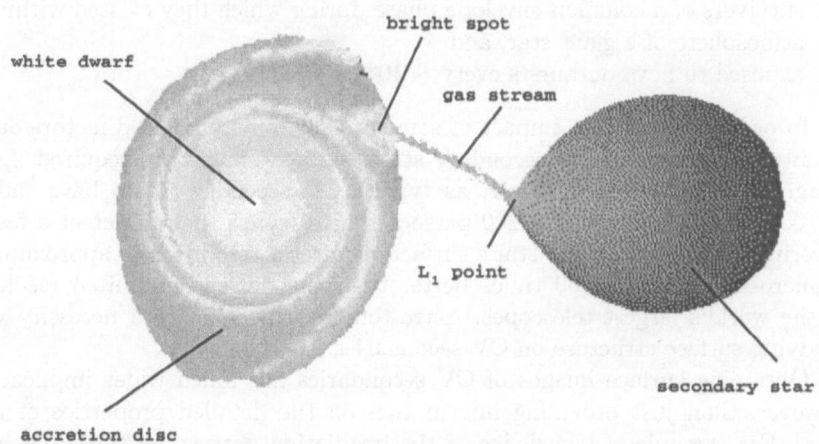

white dwarf

bright spot

gas stream

L_1 point

accretion disc

secondary star

Fig. 1. A non-magnetic CV.

[37], where $q = M_2/M_1$ and a is the distance between the centres of mass of the binary components. Newton's generalisation of Kepler's third law can be written as

$$\frac{4\pi^2 a^3}{GP^2} = M_1 + M_2 = M_2\left(\frac{1+q}{q}\right), \tag{3}$$

where P is the orbital period of the binary. Combining equations 1, 2 and 3 gives the mean density-orbital period relation,

$$\bar{\rho}_2 = 105\,P^{-2}\;(\text{h})\quad\text{g cm}^{-3}, \tag{4}$$

which is accurate to ~6 per cent [15] over the range of mass ratios relevant to most CVs ($0.01 < q < 1$). Most CVs have orbital periods of $1.25\,\text{h} < P < 9\,\text{h}$, resulting in mean densities of $67\,\text{g cm}^{-3} < \bar{\rho}_2 < 1.3\,\text{g cm}^{-3}$. Such mean densities are typically found in M8V–G0V stars [1] and hence the secondary stars in CVs should be similar to M, K or G main-sequence dwarfs. With a few caveats, this prediction is largely confirmed by observation [37],[3].

1.2   Why Image the Secondary Stars in CVs?

Even though we have just shown that most CV secondary stars are similar to lower main-sequence stars in their gross properties, it is not clear that they should share the same detailed properties. This is because CV secondaries are subject to a number of extreme environmental factors to which isolated stars are not. Specifically, CV secondaries are:

1. situated $\sim 1R_\odot$ from a hot, irradiating source (see [40]);
2. rapidly rotating ($v_{rot} \sim 100\,\text{km s}^{-1}$);
3. Roche-lobe shaped;
4. losing mass at a rate of $\sim 10^{-8} - 10^{-11}M_\odot\,\text{yr}^{-1}$;

5. survivors of a common-envelope phase during which they existed within the atmosphere of a giant star, and
6. exposed to nova outbursts every $\sim 10^4$ yr.

In order to study the impact of some of these environmental factors on the detailed properties of CV secondary stars, surface images are required. Direct imaging is impossible, however, as typical CV secondary stars have radii of 400 000 km and distances of 200 parsecs, which means that to detect a feature covering 20 per cent of the star's surface requires a resolution of approximately 1 micro-arcsecond, 10 000 times better than the diffraction-limited resolution of the world's largest telescopes. Astro-tomography is hence a necessity when studying surface structure on CV secondaries.

Obtaining surface images of CV secondaries has much wider implications, however, than just providing information on the detailed properties of these stars. For example, a knowledge of the irradiation pattern on the inner hemisphere of the secondary star in CVs is essential if one is to calculate stellar masses accurate enough to test binary star evolution models (see section 3.3.4). Furthermore, the irradiation pattern provides information on the geometry of the accreting structures around the white dwarf (see section 3.3.4). Perhaps even more importantly, surface images of CV secondaries can be used to study the solar-stellar connection. It is well known that magnetic activity in isolated lower-main sequence stars increases with decreasing rotation period (e.g. [28]). The most rapidly rotating isolated stars of this type have rotation periods of ~ 8 hours, much longer than the synchronously rotating secondary stars found in most CVs. One would therefore expect CVs to show even higher levels of magnetic activity. There is a great deal of indirect evidence for magnetic activity in CVs – magnetic activity cycles have been invoked to explain variations in the orbital periods, mean brightnesses and mean intervals between outbursts in CVs (see [44]). The magnetic field of the secondary star is also believed to play a crucial role in angular momentum loss via magnetic braking in longer-period CVs, enabling CVs to transfer mass and evolve to shorter periods. One of the observable consequences of magnetic activity are star-spots, and their number, size, distribution and variability, as deduced from astro-tomography of CV secondaries, would provide critical tests of stellar dynamo models in a hitherto untested period regime.

2 Detecting the Secondary Stars in CVs

Detecting spectral features from the secondary stars in CVs is not easy. There are two main reasons for this. First, CVs are typically hundreds of parsecs distant, rendering the lower main-sequence secondary very faint. Second, the spectra of CVs are usually dominated by the accretion disc and the resulting shot-noise overwhelms the weak signal from the secondary star. The result is that, in the 1998 study of Smith and Dhillon [37], only 55 of the 318 CVs with measured orbital periods had spectroscopically identified secondary stars.

The best secondary star detection strategy depends very much on the orbital period of the CV, as this approximately determines the spectral type of the secondary via the empirical relation

$$Sp(2) = 26.5 - 0.7 \ P(\text{h}), \ P < 4\,\text{h}$$
$$\pm \ 0.7 \pm 0.2$$

$$= 33.2 - 2.5 \ P(\text{h}), \ P \geq 4\,\text{h}$$
$$\pm \ 3.1 \pm 0.5$$

(5)

[37], where $Sp(2)$ is the spectral type of the secondary; $Sp(2) = 0$ represents a spectral type of G0, $Sp(2) = 10$ is K0 and $Sp(2) = 20$ is M0. Longer period CVs therefore have earlier spectral-type secondaries, which generally contribute a greater fraction (typically >75 per cent) of the total optical/infrared light than the secondary stars found in shorter period CVs, which usually contribute only ~10–30 per cent [10]. Equation 5 can then be used in conjunction with a black-body approximation to the wavelength, $\lambda_{\max}$, of the peak flux, f_ν, for a star of effective temperature T_{eff},

$$\lambda_{\max} = 5100/T_{\text{eff}} \quad \mu\text{m},$$

(6)

to obtain a crude idea of the optimum observation wavelength required to detect the secondary star in a CV of known orbital period. With T_{eff} ranging from ~6000–2000 K for the G–M dwarf secondary stars found in most CVs, $\lambda_{\max}$ ranges from ~0.8–2.5 μm, i.e. the optimum wavelength always lies in the optical and near-infrared. In practice, when observing the secondary stars in longer-period CVs it is generally best to observe the numerous neutral metal absorption lines in the R-band around Hα (e.g. [13]; bottom-right, figure 2), whereas short and intermediate-period secondaries are best observed via the TiO molecular bands and Na I absorption doublet in the I-band (e.g. [43]; top-right, figure 2). If optical spectroscopy fails to detect the secondary star in a CV, as is often the case in the shortest-period dwarf novae and the nova-likes (which have very bright discs), it is possible to use near-infrared spectroscopy in the J-band [19] and K-band (e.g. [10]); the brighter background in the near-infrared is offset to some extent by the greater line flux from the secondary star at these wavelengths (e.g. [20]; top-left, figure 2). In addition to absorption-line features, emission-line features from the secondary star are sometimes visible in CV spectra, especially in dwarf novae during outburst and novalikes during low-states (e.g. [9]; bottom-left, figure 2). These narrow, chromospheric emission lines originate on the inner hemisphere of the secondary star and are most probably due to irradiation from the primary and its associated accretion regions. The emission lines are usually most prominent in the Balmer lines, but have also been observed in He I, He II and Mg II (e.g. [17]).

2.1 Skew Mapping

If the secondary star is not visible in a single spectrum of a CV, one might naturally think that co-adding additional spectra will increase the signal-to-

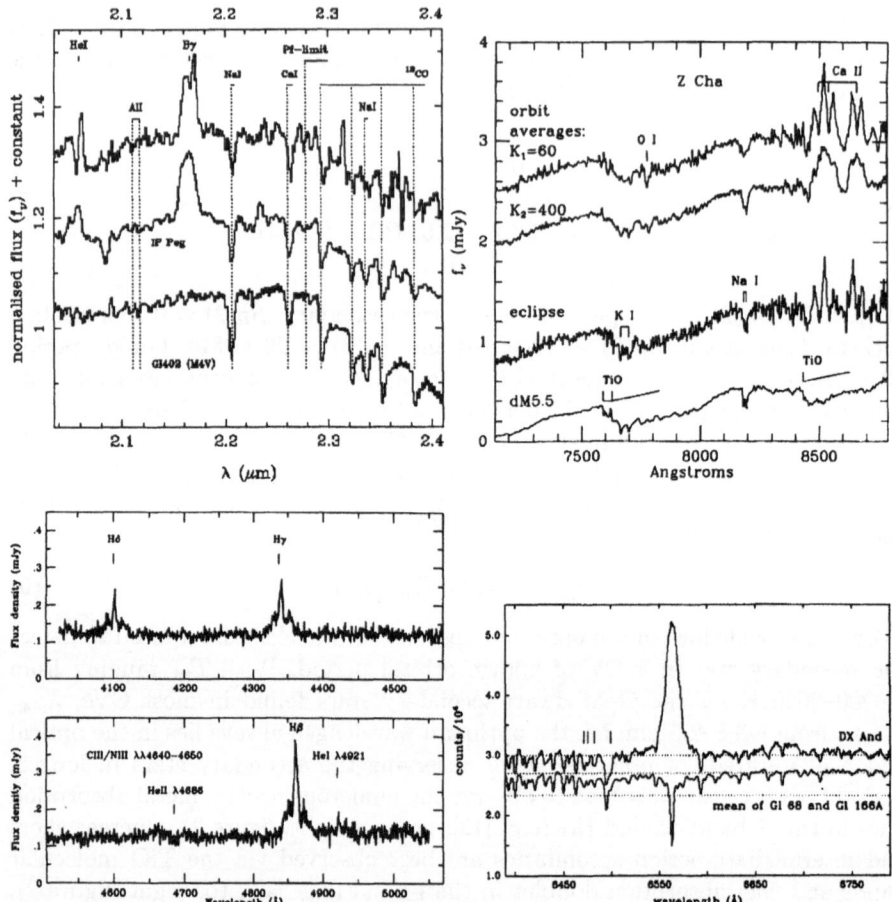

Fig. 2. Secondary star features in CV spectra. Clockwise from top-left: IP Peg [20], Z Cha [43], DX And [13] and DW UMa [9].

noise and hence increase the chances of a detection. This is true, but only if the additional spectra are first shifted to correct for the orbital motion of the secondary star, as otherwise the weak features will be smeared out. The problem is that the orbital motion is not known in advance; the solution is to use a technique known as *skew mapping* [39].

The first step is to cross-correlate each spectrum to be co-added with a template, usually the spectrum of a field dwarf of matched spectral type, yielding a time-series of cross-correlation functions (CCFs). If there is a strong correlation, the locus of the CCF peaks will trace out a sinusoidal path in a 'trailed spectrum' of CCFs, in which case plotting the velocity of each of the CCF peaks versus orbital phase allows one to define the secondary star orbit. More often, however, the CCFs are too noisy to enable well-defined peaks to be measured. Instead, the trailed spectrum of CCFs is back-projected in an identical manner

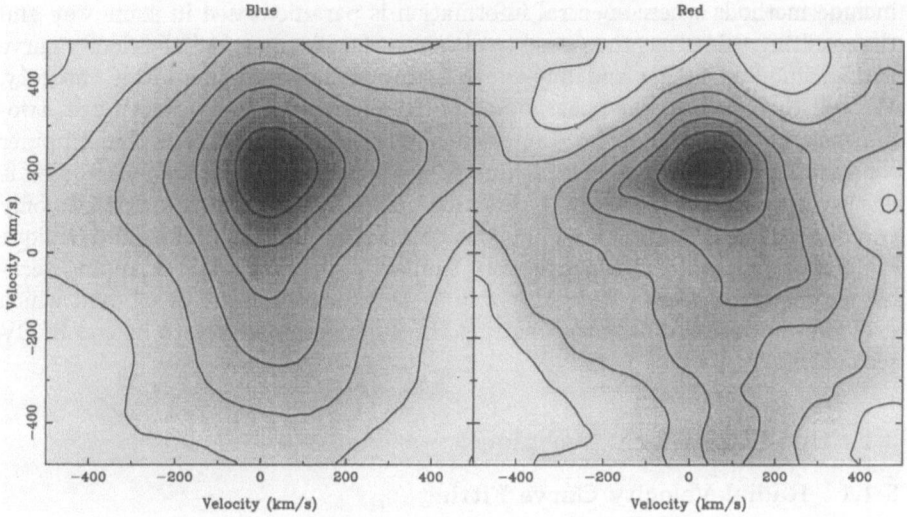

Fig. 3. Skew maps of the old nova BT Mon [38].

to that employed in standard Doppler tomography ([22]; Marsh, this volume) to produce what is known as a *skew map*. Any noisy peaks in the trailed spectrum of CCFs which lie along the true sinusoidal path of the secondary star will reinforce during the back projection process, resulting in a spot on the skew map at $(0, K_2)$, where K_2 is the radial-velocity semi-amplitude of the secondary star.

Figure 3 shows an example of the successful use of the skew mapping technique applied to the old nova BT Mon [38]. The value of K_2 determined from the skew map (205 $\mathrm{km\,s^{-1}}$) was used to produce a co-added spectrum which enabled the rotational broadening and spectral type of the secondary star to be accurately determined. These system parameters were then used to calculate the distance and the component masses of BT Mon, which provide fundamental input to thermonuclear runaway models of nova outbursts. Although a powerful technique which is invaluable in the detection of faint secondary stars in CVs, skew mapping does not, however, provide surface images. To do this, other techniques are required, which we shall turn to now.

3 Mapping the Secondary Stars in CVs

The secondary stars in CVs are three-dimensional objects. Time-series of astronomical data, however, are effectively one-dimensional (in the case of light curves) or two-dimensional (in the case of spectra). The problem of obtaining surface images of CV secondaries is hence poorly constrained (see [34]), especially if one is restricted to the so-called one-dimensional case. We will begin the description of CV secondary mapping techniques by looking at the one-dimensional techniques. These include photometric light curve fitting, but also include methods where spectral information is parameterised in some way and the resulting values are then used to obtain surface images; radial-velocity curve fitting, line-flux fitting and line-width fitting all fall into this latter category. We will then look at the special case of Doppler tomography, which uses two-dimensional data to map two-dimensional structures. This means that Doppler tomography is fully constrained, but it also means that the secondary star is effectively compressed along the direction defined by the rotation axis into only two dimensions. Finally, we will look at potentially the most powerful technique – Roche tomography – which is very similar to the single-star mapping techniques described elsewhere in this volume by Cameron and Donati, and which uses two-dimensional data to construct three-dimensional surface images of CV secondaries.

3.1   One-Dimensional Techniques

3.1.1   Radial-Velocity Curve Fitting

If the centre-of-light and centre-of-mass of the secondary are not coincident, the star's radial-velocity curve will be distorted in some way from the pure sine wave which represents the motion of the centre-of-mass. The observed radial-velocity curves of CV secondaries suffer from this distortion, due to both geometrical effects caused by the Roche-lobe shape and non-uniformities in the surface distribution of the line strength due to, for example, irradiation. Davey and Smith [7],[8] introduced an inversion technique which exploits this effect and uses the asymmetries present in observed radial-velocity curves to produce surface images of CV secondaries. Their 'one-spot' model assumed a single region of heating on the inner hemisphere of the secondary star which was allowed to vary in strength, size and position, i.e. there were 3 free parameters. The position of the spot, however, was restricted to vary in only the longitudinal direction, and hence the resulting maps assume symmetry about the orbital plane and are only two-dimensional.

Davey and Smith have successfully applied their technique to four dwarf novae and one polar [7],[40],[8]. The surface image they derived of the polar AM Her is shown in the right-hand panel of figure 4. It can be seen that there is a general reduction in the line strength on the inner hemisphere of the secondary, a result consistent with the effects of heating caused by irradiation [4]. It can also be seen that the line absorption is stronger on the leading hemisphere (the lower edge of the map in figure 4) than it is on the trailing hemisphere (the upper edge of the map in figure 4). This asymmetry has been interpreted as due to the gas stream blocking the radiation produced in the magnetic accretion column close to the white dwarf, thereby shielding an area on the leading hemisphere of the secondary star from irradiation. A similar asymmetry, but in the opposite sense (i.e. the trailing hemisphere has stronger line absorption than the leading hemisphere), has also been seen in the dwarf novae mapped by Davey and Smith [40]. In these objects, however, there is an accretion disc instead of an accretion column. Irradiation by the bright-spot is apparently insufficient to account for the

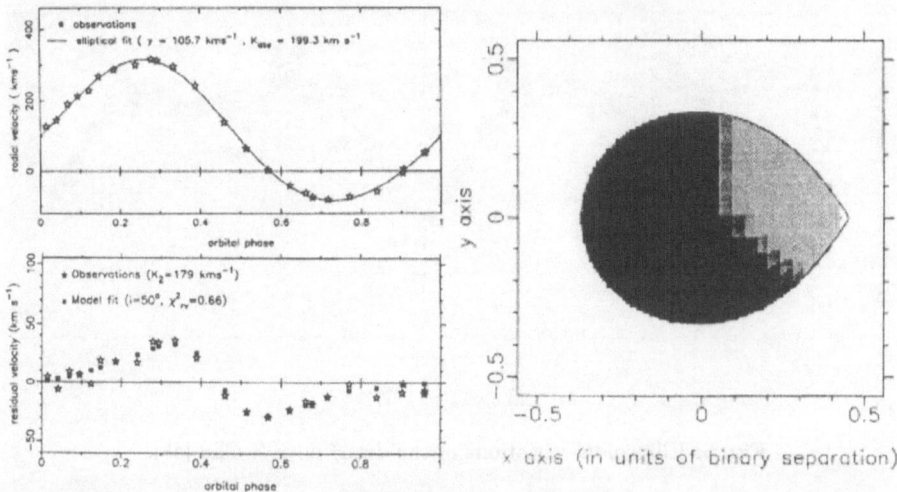

Fig. 4. Radial-velocity curve fitting of the polar AM Her. Upper-left: Radial-velocity curve of the Na I absorption doublet around 8190Å [7]. Lower-left: Residual-velocity curve, i.e. the motion of the centre-of-mass of the secondary has been removed [8]. Right: Surface map of AM Her (top view) – the light region is where the Na I doublet is weakest [8].

observed asymmetry [7], and so Davey and Smith speculated that circulation currents induced by the irradiation spreads the heating over a larger area, but preferentially towards the leading hemisphere due to Coriolis effects. This idea was later confirmed by SPH modelling [25].

3.1.2   Light-Curve Fitting

Light-curve fitting is now a well-established tool in CV research, having found particular success in the study of accretion discs via the eclipse mapping method ([18]; Baptista, this volume). The brightness of the secondary star in eclipse mapping is, however, either completely ignored or included in the fit as a single nuisance parameter [29]. In order to obtain surface images of CV secondaries via light-curve fitting, it is necessary to extend the eclipse mapping method by constructing a secondary star grid of tiles in addition to, or instead of, an accretion disc grid and fit the whole light curve rather than just the eclipse portion. Each element in the grid is assigned an intensity, which is weighted according to its projected area and limb darkening at each orbital phase. A model light curve is then derived by summing the weighted intensities as a function of orbital phase. By comparing the model light curve for some grid brightness distribution with the observed light curve, the element intensities can be iteratively adjusted until the observed light curve is optimally fitted, usually using the maximum-entropy approach described in section 3.3.1.

Such an approach has been adopted by Wade and Horne [43], who used a secondary star grid, but no accretion disc grid, to fit the TiO band-flux light

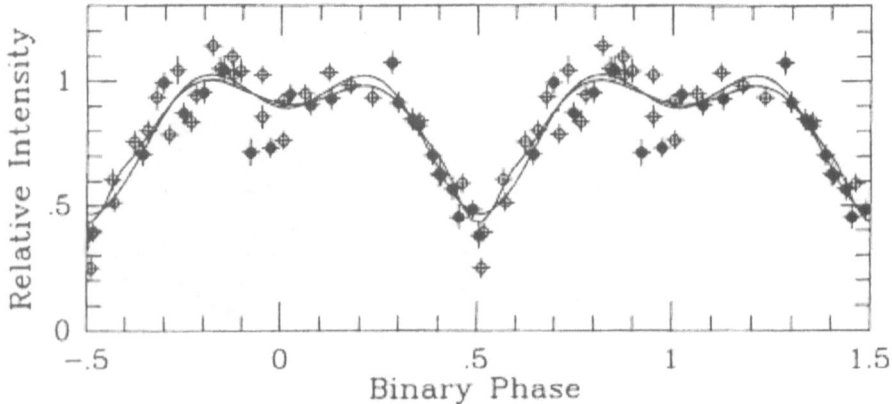

Fig. 5. Ellipsoidal variations in the dwarf nova Z Cha [43].

curve of the dwarf nova Z Cha. The observed data is represented by the points in figure 5 and their fits to the data are represented by the solid lines. The observed data show a double-humped variation, known as an *ellipsoidal modulation*. This is due to the changing aspect of the secondary's Roche-lobe which presents the largest projected area, and hence highest flux, at quadrature (phases 0.25 and 0.75) and the smallest project area, and hence lowest flux, during conjunction (phases 0 and 0.5). The best fit to the data was obtained with a TiO distribution which has a minimum around the L_1 point and rises smoothly to a level 3 times higher on the hemisphere facing away from the white dwarf. This surface variation is consistent with the effects of irradiation [4] and was used to correct the radial-velocity curves for the effects of a mis-match between the secondary's light centre and its centre-of-mass in order to derive accurate stellar masses.

Wade and Horne [43] did not include an accretion disc grid in their light-curve fits because the disc does not contribute to the TiO band-flux. Broad-band light curves, however, especially those obtained in the infrared, have approximately equal contributions from the disc and secondary, which means that any light-curve fitting must include both a secondary star grid and an accretion disc grid. Rutten [52] has developed such a technique, known as *3D eclipse mapping*. The upper-left panel of figure 6 shows a typical accretion disc and secondary star grid used in 3D eclipse mapping. A secondary star with a luminous inner hemisphere and dark outer hemisphere, combined with a standard $T_{\text{eff}}^4 \propto R_{\text{disc}}^{-3}$ disc [44], produces the light curve shown in the right-hand panel of figure 6. Fitting this light curve produces the map of the system shown in the lower-left panel of figure 6, which is a good representation of the original intensities assigned to the grid elements. 3D eclipse mapping has only recently been applied to real data [16], however, and awaits full exploitation in secondary star studies.

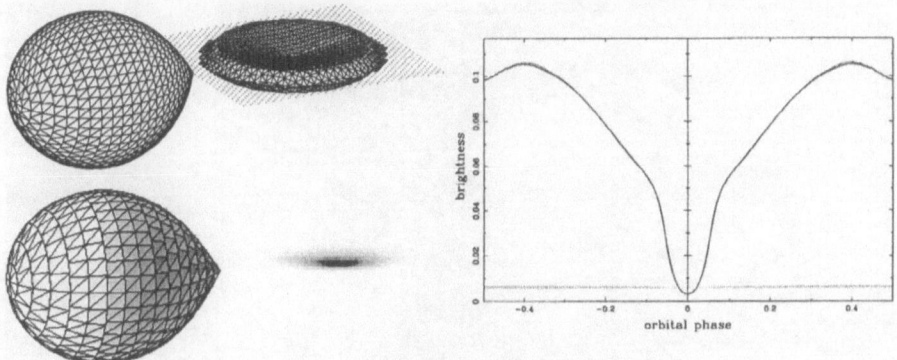

Fig. 6. 3-D eclipse mapping [52].

3.1.3   Line-Width Fitting

We have just seen how variability in two of the 'integral' properties of a line profile, the line strength and the line position, can be used to map the secondary stars in CVs. There is, however, a third integral property of line profiles which can be used to deduce surface images – the variability in line width, or more generally, in line shape. The reason line width varies with orbital phase can be understood by recalling that the secondary stars in CVs are synchronously rotating and geometrically distorted, which results in a variation in the projected radius of the secondary as a function of orbital phase. For a given orbital period, the larger the projected radius of the secondary the broader the line profile will be, which means that the rotational broadening varies with orbital phase. At the conjunction phases, the projected radius and rotational broadening are at a minimum, while at quadrature they are at a maximum.

The observed variations in rotational broadening also depend on the inclination of the binary plane and the surface distribution of line strength. Casares et al. [5] exploited this effect in order to determine the inclination of the magnetic novalike AE Aqr. By constructing a secondary star with essentially no surface structure, other than that due to an assumed gravity darkening law, they calculated a grid of model rotational-broadening curves which were then matched to the observed curve using a χ^2-test. Shahbaz [34], also motivated by a desire to determine inclination angles, added an extra dimension by fitting the line shape rather than just the line width. Both of these techniques are examples of *model fitting*, in which any surface structure must be added to the model in an ad-hoc manner prior to fitting. A better approach to mapping surface structure is *image reconstruction*, in which the intensity of each image pixel is a free parameter; Roche tomography is an example of such a technique (see section 3.3).

3.2   Doppler Tomography

Doppler tomography uses a time-series of emission line profiles spanning the binary orbit to construct a two-dimensional map of the system in velocity space.

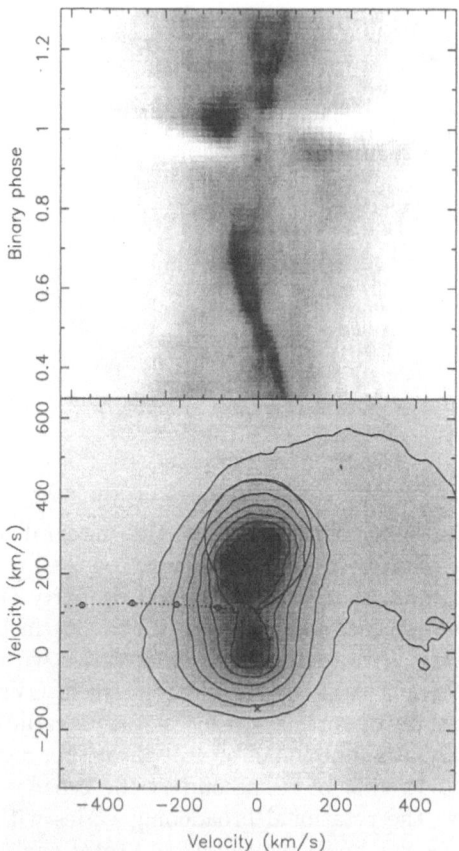

Fig. 7. A slingshot prominence and accretion disc shadow in the dwarf nova IP Peg [41].

Although primarily used as a tool to study the accretion regions in CVs, Doppler tomography has also proved to be remarkably successful in secondary star studies. This is because Doppler tomography provides a way of cleanly separating the emission component due to the secondary star from the broader and often stronger emission originating in the accretion regions. The technique and some of its most important achievements to date are reviewed elsewhere in this volume by Marsh and Schwope. In this section, we highlight one particularly important example of the application of Doppler tomography to the study of CV secondaries – the work of Steeghs et al. [41] on the dwarf nova IP Peg during outburst (figure 7).

The upper panel of figure 7 shows the observed trailed spectrum of the Hα emission line in IP Peg. Note the stationary component at 0 $\mathrm{km\,s^{-1}}$ running the entire length of the time-series and the narrow sinusoidal component which crosses from red-shifted to blue-shifted around phase 0.5. The sinusoidal emission component is mapped onto the inner hemisphere of the secondary star's Roche lobe (lower panel, figure 7). This emission is almost certainly being powered by irradiation from the white dwarf and the hot, inner regions of the outbursting accretion disc, a conclusion supported by the fact that the emission appears to be stronger around the poles than it is around the L_1 point, implying that obscuration of the inner disc by the flared, outer disc might be occurring. The stationary component is mapped onto the centre-of-mass of the binary, indicated by the cross at zero velocity in the lower panel of figure 7. This component is much more difficult to explain as there is no obvious part of the binary system which is at rest. One possible interpretation is that the emission is due to material trapped in a 'slingshot prominence', a magnetic loop originating on and co-rotating with the secondary. With this model, Steeghs et al. [41] used the Doppler map in figure 7 to estimate the magnetic field strength of the secondary star ($\sim$ kG) and the temperature (2×10^4 K) and total mass ($\sim 10^{18}$ g) of the material in the prominence, and found good agreement with the parameters deduced from similar prominences observed in isolated dwarf stars (e.g. [5]).

3.3 Roche Tomography

Roche tomography [30],[45] takes as its input a trailed spectrum, i.e. a time-series of spectral-line profiles, and provides on output a map of the secondary star in three-dimensions. The main advantage of Roche tomography over the one-dimensional techniques described in section 3.1 is that the one-dimensional techniques each extract and then use just one piece of information about the line profile (e.g. the variation in its flux, radial velocity or width) to map the secondary star, whereas Roche tomography uses all of the information in the line profile, as described in section 3.3.1.

Roche tomography is very similar to the Doppler imaging technique used to map single stars (e.g. Cameron, Donati, this volume), contact binaries [21] and detached secondaries in pre-CVs [26]. In fact, Roche tomography and Doppler imaging of single stars differ in only two fundamental ways. First, the secondary stars in CVs are tidally-distorted into a Roche-lobe shape and are in synchronous rotation about the binary centre-of-mass; isolated stars rotate only about their own centre-of-mass and are symmetric about their rotation axis. Second, the continuum is ignored in Roche tomography, whereas it is included in Doppler imaging. This is because of the variable and unknown contribution to the spectrum of the accretion regions in CVs, forcing Roche tomography to map absolute line fluxes. The data are therefore slit-loss corrected and then continuum-subtracted prior to mapping with Roche tomography, whereas when Doppler imaging the spectra need not be slit-loss corrected and the continuum is divided into the data.

Although the system parameters are generally better constrained in CVs than they are in single stars, Roche tomography is a much harder task than Doppler imaging. This is because CV secondaries are usually both fainter and more rapidly rotating than isolated stars, resulting in surface images of a much lower quality than are routinely obtained via Doppler imaging (see section 3.3.4).

3.3.1   Principles and Practice

In Roche tomography, the secondary star is modelled as a grid of quadrilateral surface elements of approximately equal area lying on the critical potential surface which defines the Roche lobe. Each surface element, or tile, is then assigned a copy of the local specific intensity profile convolved with the instrumental resolution. These profiles are then scaled to take into account phase dependent effects, such as variations in the projected area, limb darkening and obscuration[1] and Doppler-shifted according to the radial velocity of the surface element at a particular phase. Simply summing up the contributions from each element gives the rotationally broadened profile at any particular phase. An example of this 'forward' process is shown in figure 8, which shows the trailed spectrum resulting from a secondary star with a uniformly radiating inner hemisphere.

By iteratively varying the strengths of the profile contributed by each tile, it is possible to perform the 'inverse' process and obtain the map which fits the observed data. How well the map fits the observed data is defined by a *consistency statistic*, given by the *reduced chi-squared*:

$$\chi^2 = \frac{1}{n} \sum_{i=1}^{n} \left(\frac{p_i - o_i}{\sigma_i} \right)^2, \qquad (7)$$

where n is the number of data points, p_i and o_i are the predicted and observed data, and σ_i is the error on o_i. Fitting the data as closely as possible (i.e. minimising χ^2) is not a good approach, however, as noise will dominate the resulting map. A better approach is to reduce χ^2 until the observed and predicted data are consistent, i.e. $\chi^2 \sim 1$. Such a condition is satisfied by many maps, however, and so a *regularisation statistic* is employed to select just one of them. Following Horne [18], we select the 'simplest' map, which is given by the map of *maximum entropy*. The definition of image entropy, S, that we use is

$$S = \sum_{j=1}^{k} m_j - d_j - m_j \ln\left(m_j/d_j\right), \qquad (8)$$

where k is the number of tiles in the map, and m_j and d_j are the map and the so-called 'default map', respectively. Equation 8 shows that the entropy is a measure of the similarity of the map to the default map and hence it is the default map which defines what we mean by the 'simplest' image. The choice of default map is therefore of great importance. We have experimented with a number of different prescriptions for the default map [45]. The two most successful have been a uniform default, where every tile in the default is set to the average value in the map, and a smoothed version of the map, achieved using a Gaussian blurring function. In the former case, we are selecting the *most uniform map consistent with the data*, which constrains large-scale surface structure, and in

---

[1] Note that this list does not include gravity darkening, which is not phase-dependent and hence need not be included in the algorithm; if any gravity darkening is present in the data, it will be reconstructed in the maps.

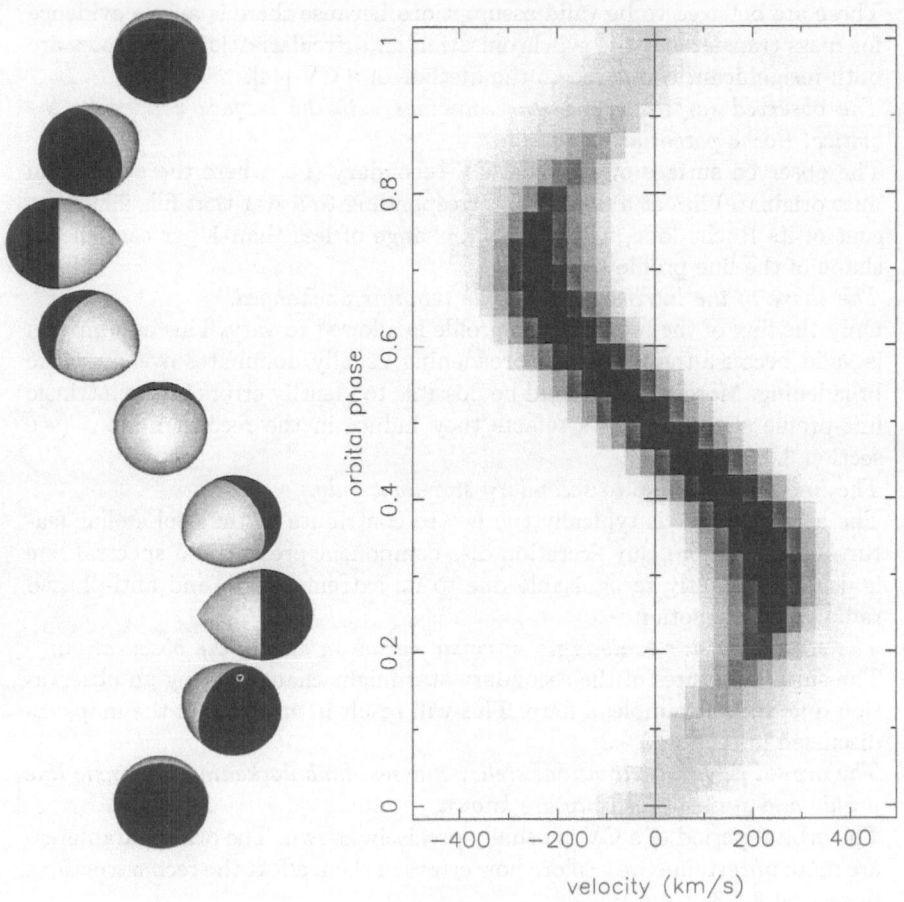

Fig. 8. The principles of Roche tomography.

the latter case we are selecting the *smoothest map consistent with the data*, which constrains short-scale structure. An efficient algorithm for the task of maximising entropy subject to the constraint imposed by χ^2 is given by Skilling and Bryan [23] and has been implemented by them in the FORTRAN package MEMSYS.

3.3.2 Assumptions

The basic assumptions which underlie Roche tomography are:

1. *The secondary star is Roche-lobe filling, synchronously rotating and in a circular orbit.*

These are believed to be valid assumptions because there is ample evidence for mass transfer and the synchronisation and circularisation timescales are both insignificant compared to the lifetime of a CV [44].

2. *The observed surface of the star coincides with the surface defined by the critical Roche potential.*

   The observed surface of a typical CV secondary (i.e. where the absorption lines originate) lies at a potential corresponding to a star that fills 99.97 per cent of its Roche lobe, which gives a change of less than 1 per cent in the shape of the line profile [34].

3. *The shape of the intrinsic line profile remains unchanged.*

   Only the flux of the intrinsic line profile is allowed to vary. This assumption is valid because the rotational broadening usually dominates over intrinsic broadening. Moreover, it should be possible to identify errors in the intrinsic line-profile shape by the artefacts they induce in the reconstructions (see section 3.3.3).

4. *The line profile is due to secondary star light only.*

   The accretion disc is typically too hot to contribute to the cool stellar features. Furthermore, any accretion disc component present in a spectral line is usually instantly recognisable due to its extreme width and anti-phased radial-velocity motion.

5. *The secondary star exhibits no intrinsic variation during the observation.*

   The surface features of the secondary star might change during an observation due to, for example, a flare. This will result in artefacts in the maps, as discussed in section 3.3.3.

6. *The orbital period, inclination, stellar masses, limb darkening, intrinsic line profile and systemic velocity are known.*

   The orbital period of a CV is usually precisely known. The other parameters are more uncertain – we explore how errors in them affect the reconstructions in section 3.3.3.

7. *The final map is the one of maximum entropy (relative to an assumed default image) which is consistent with the data.*

   The data constrains the final map through the consistency statistic, χ^2. The default map constrains the final map through the regularisation statistic, S. If the data are noisy, the data constraints will be weak and the map will be strongly influenced by the default map. If the data are good, however, the image will not be greatly influenced by the default and the choice of default makes little difference to the final map. The default may thus be regarded as containing prior information about the map, and the map will be modified only if the observations require it. The importance of this assumption therefore depends on the quality of the data.

3.3.3   Errors

The maps resulting from any form of astro-tomography are prone to both systematic errors, due to errors in the assumptions underlying the technique, and statistical errors, due to measurement errors on the observed data points. It is essential that the effects of these errors on the reconstructions are quantified in order to properly assess the reality of any surface structure present. This is especially true of Roche tomography, for which the input data is generally noisier due to the faintness of CV secondaries, a problem further exacerbated by the fact that noise in Roche tomograms can mimic the appearance of star-spots [45].

We begin by discussing our approach to statistical error determination [45]. The maximum-entropy technique is non-linear, in the sense that each image value of the final map is not a linear function of the data values. This makes it very difficult to propagate the statistical error on a data point through the maximum-entropy process in order to calculate the statistical error on each tile of the map. Furthermore, the statistical errors on each tile will not be independent due to, for example, the projection of bumps in the line profiles across arcs of constant radial velocity on the secondary star (see figure 9). The simplest approach to error estimation, in this case, is to use a Monte Carlo-based simulation.

Monte-Carlo techniques rely on the construction of a large (typically hundreds, in the case of Roche tomography) sample of synthesized datasets which have been effectively drawn from the same parent population as the original dataset, i.e. as if the observations have been repeated many hundreds of times. This large sample of synthesized datasets is then used to create a large sample of Roche tomograms, resulting in a probability distribution for each tile in the map. The main difficulty with this technique, aside from the demands on computer time, lies in the construction of the sample of synthesized datasets. One approach (e.g. [29]) is to 'jiggle' each data point about its observed value, by an amount given by its error bar multiplied by a number output by a Gaussian random-number generator with zero mean and unit variance. This process adds noise to the data, however, which means that the synthesized datasets are not being drawn from the same parent population as the observed dataset – noise is being added to a dataset which has already had noise added to it during the measurement process. In practice, this means that fitting the sample datasets to the same level of χ^2 as the original dataset is either impossible (i.e. the iteration does not converge) or results in maps dominated by noise, which overestimates the true error on each tile.

A much better approach to creating synthesized datasets is the *bootstrap method* [14], which we have implemented as follows [45]. From our observed trailed spectrum containing n data points, we select, *at random and with replacement*, n data values and place them in their original positions in the new, synthesized trailed spectrum. Some points in the synthesized trailed spectrum will be empty, in which case they will be omitted from the fit; in practice, this is achieved by setting the error bars on these points to infinity. Other points in the synthesized trailed spectrum will have been selected once or more, in which case their error bars are divided by the square root of the number of times they were picked. The advantage of bootstrap resampling over jiggling is that the data is not made noisier by the process as only the errors bars on the data points are manipulated. It is therefore possible to fit the synthesized trailed spectra to the same level of χ^2 as the observed data, giving a much more reliable estimate of

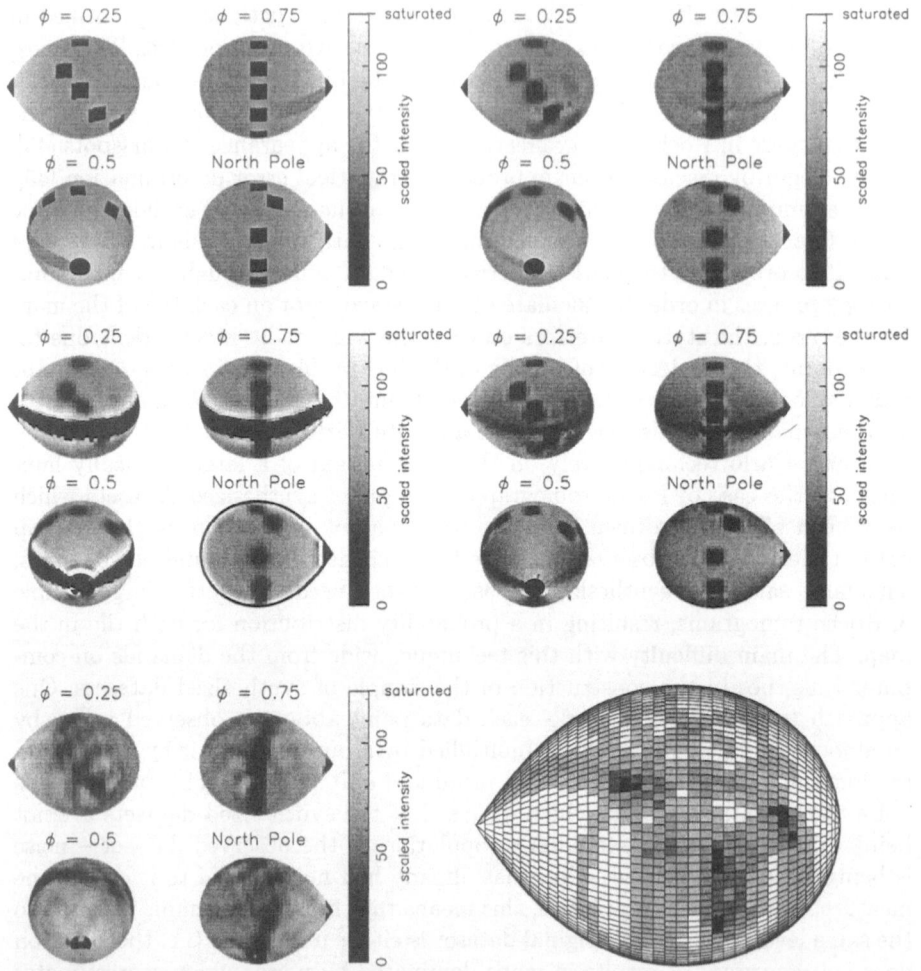

Fig. 9. Systematic errors in Roche tomography [45]. Top row: The test image (left) and the best fit (right). Middle row: Equatorial banding due to the effect of an incorrect systemic velocity (left) and incorrect limb darkening (right). Bottom row: Ring-like streaks due to the effects of phase undersampling (left), which correspond to lines on the secondary star with the same radial velocities as the spots (right).

the statistical errors in the maps. The bootstrap method has been shown to work very effectively with Roche tomography [45], and we present an example of its application in figures 11 and 13.

The determination of systematic errors requires a completely different approach to that employed for the determination of statistical errors. We have approached systematic errors in Roche tomography [45] in much the same way that Marsh and Horne [22] explored them in Doppler tomography. We first constructed a test image (top-row, left, figure 9) of the secondary star containing all of the surface features we might expect to find when mapping real data. These include irradiation, shadowing by the gas stream and star spots covering a range of latitudes and longitudes. We used the test image to create a test trailed spectrum which we then fit using Roche tomography to reconstruct the surface image of the star. As can be seen from figure 9 (top-row, right), the best fit reproduces all of the features in the test image. We then explored the effects of systematic errors on the reconstruction by varying the input parameters to Roche tomography and re-fitting the test trailed spectrum. In this way we explored how variations in the systemic velocity, limb darkening, inclination, velocity smearing, phase undersampling, noise, intrinsic line profile, resolution, default map, stellar masses and stellar flares affect the reconstructions [45].

We find that systematic errors in Roche tomography generally result in only two major artefacts in the Roche tomograms: ring-like streaks and equatorial banding [45]. Equatorial banding results whenever there is a mis-match between the maximum velocities present in the line profile and the maximum velocities available on the secondary star grid, which occurs whenever the wrong systemic velocity, limb darkening, inclination, velocity smearing, intrinsic line profile or stellar masses are used in the reconstruction. The banding occurs around the equator of the star because this is where the highest radial velocities are found. Some examples of equatorial banding patterns in Roche tomograms are shown in figure 9, where the maps have been reconstructed using incorrect values for the systemic velocity (middle row, left) and limb darkening (middle row, right).

Ring-like streaks appear in Roche tomograms when one or a few of the spectra in the input data dominate. This occurs when there is phase undersampling, for example, or when there is a flare in the data at a particular phase. The effect is analogous to the streaks observed in Doppler tomography [22] and can be understood by considering that lines of constant radial velocity on the secondary star at a particular phase can be integrated along to construct a line profile. These lines of constant radial velocity are ring-like in shape and if there are only a few phases, or if the profile is particularly bright at a certain phase, the streaks will not destructively interfere, leaving ring-like artefacts on the Roche tomogram. An example of a Roche tomogram reconstructed using only 5 orbital phases is shown in figure 9 (bottom row, left). The streaks present in this image correspond to lines on the secondary star with the same radial velocity as the spots, as shown at phase 0.25 in figure 9 (bottom row, right).

The error experiments presented here show that any feature on a Roche tomogram must be subjected to two tests before its reality can be confirmed. The first test is to determine whether the feature is statistically significant and is performed via a Monte-Carlo technique with bootstrap-resampling. The second test is to compare the feature with the appearance of known artefacts of the technique due to errors in the underlying assumptions, such as those presented in figure 9. If a surface feature survives both of these tests unscathed, it can be assumed to be real.

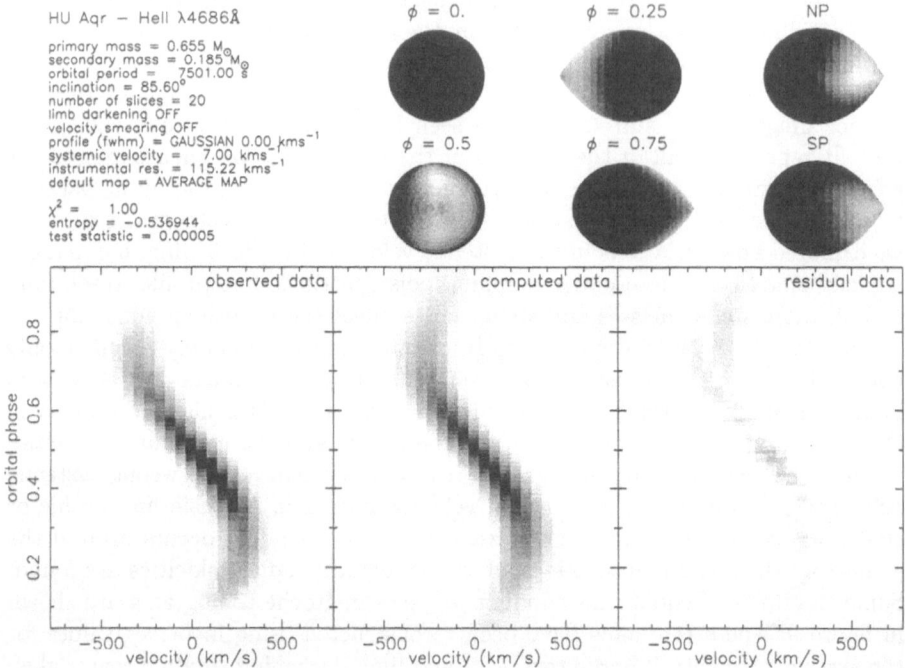

Fig. 10. Roche tomogram of the polar HU Aqr.

3.3.4 Results

Roche tomograms currently exist for only 3 CVs: the novalike DW UMa [30], the dwarf nova IP Peg [31] and the polar AM Her [8]. In this section we will present two new Roche tomograms – the polar HU Aqr and an improved study of IP Peg.

The Roche tomogram of HU Aqr in the light of the He II 4686Å emission line observed by Schwope et al. [33] is shown in figure 10, where bright regions in the map represent areas where the emission line flux is at its strongest. There are two main features in the Roche tomogram – the strong asymmetry between the inner (phase 0.5) and outer (phase 0) hemispheres of the secondary and a weaker asymmetry between the leading (phase 0.75) and trailing (phase 0.25) hemispheres of the secondary. The reality of the latter asymmetry can be assessed from figure 11, which shows a slice passing through the leading hemisphere (LH), north pole (NP), trailing hemisphere (TH) and south pole (SP) of the secondary star. The triangular points represent the map values, and it can be seen that the leading hemisphere is a factor of two fainter than the trailing hemisphere. The significance of this difference can be assessed from the curves in figure 11, which show confidence limits on the map values derived from 200 bootstrap resampling experiments; 67 per cent of the map values (measured relative to the mode of the distribution) lie within the range bounded by the solid curves

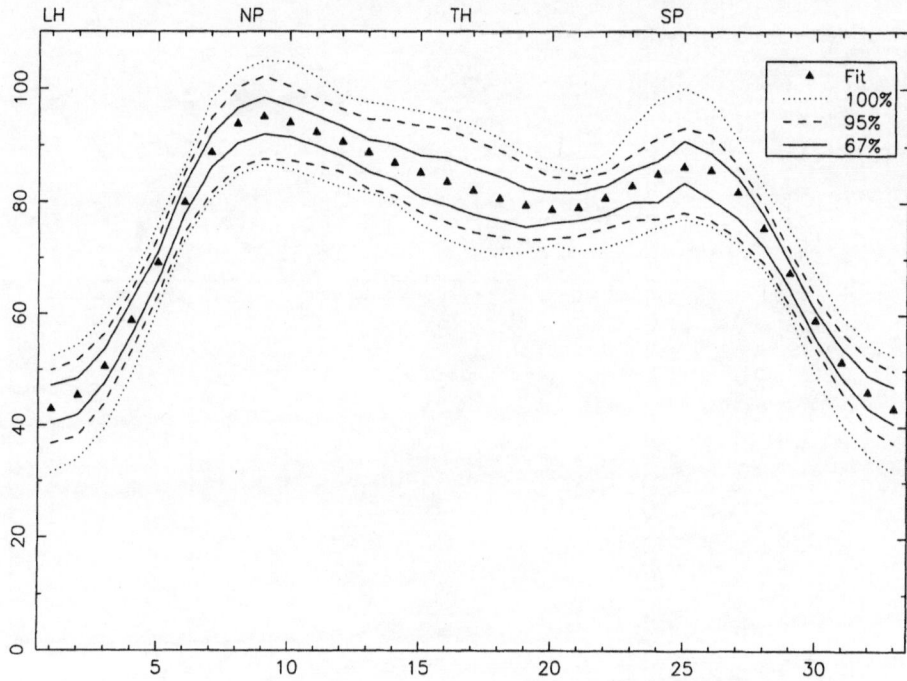

Fig. 11. A slice through the secondary star in HU Aqr.

and 100 per cent of the map values lie within the range bounded by the dotted curves. Figure 11 shows that we can be 100 per cent certain that the asymmetries between the trailing and leading hemispheres are not due to statistical errors. Furthermore, none of the systematic errors explored in section 3.3.3 result in asymmetries of the type observed in figure 10, implying that the asymmetries are real. Schwope et al. [33] reached a similar conclusion based on a model-fitting technique. They attributed the asymmetry between the inner and outer hemispheres to irradiation by the magnetic accretion column and the asymmetry between the leading and trailing hemispheres to shielding of this irradiation by the gas stream.

The Roche tomogram of IP Peg in the light of the Na I absorption doublet around 8190Å is presented in figure 12, where bright regions in the map represent areas where the flux deficit is at its largest, i.e. where the absorption line is at its strongest. The most noticeable feature in the map is the asymmetry between the inner (phase 0.5) and outer (phase 0) hemispheres, which is slightly skewed towards the leading edge (phase 0.75) of the secondary. As described in section 3.1.1, this can be attributed to the effect of irradiation by the accretion regions and the resulting circulation currents on the secondary. There is another feature of note in the tomogram of figure 12 – a spot on the leading edge of the secondary. The spot is bright, which means it is an area that exhibits stronger Na I absorption. This is not consistent with what one might expect from a star

114   V.S. Dhillon and C.A. Watson

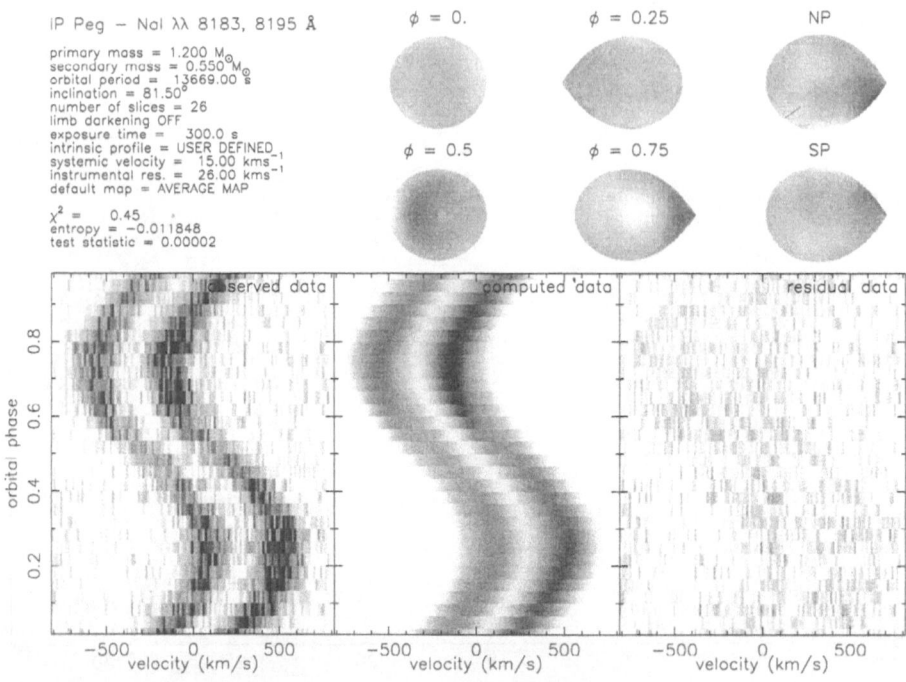

Fig. 12. Roche tomogram of the dwarf nova IP Peg.

spot, because it is known that the NaI flux deficit decreases with later spectral type [4] and star spots are generally believed to be of order 1000 K cooler than the surrounding photosphere [49] (although spots hotter than the photosphere have also been imaged [11]). An inspection of the confidence limits in a slice through the spot (figure 13), however, shows that the feature is not statistically significant; the spot is the hump at 'LH' in figure 13 and it can be seen that the 67 per cent confidence limits widen significantly around it.

Although we have not, unfortunately, imaged a star spot in IP Peg, we can use the Roche tomogram to derive accurate values for the masses of the stellar components. As discussed in section 3.3.3, using incorrect values for stellar masses results in equatorial banding, which increases the entropy of the reconstructions when using a uniform default map. The masses (and inclinations consistent with the masses and the observed eclipse width) can therefore effectively be included in the fit by constructing an *entropy landscape* [30],[31], which plots image entropy as a function of the primary and secondary masses. The entropy landscape for IP Peg is shown in figure 14. The triangle denotes the point of highest entropy, corresponding to primary and secondary masses of $M_1 = 1.2$ M$_\odot$ and $M_2 = 0.55$ M$_\odot$. These values are in approximate agreement with the mass determinations of Beekman et al. [2], Martin et al. [24] and Marsh [23], marked A, B and C in figure 14. Because the masses derived from an entropy

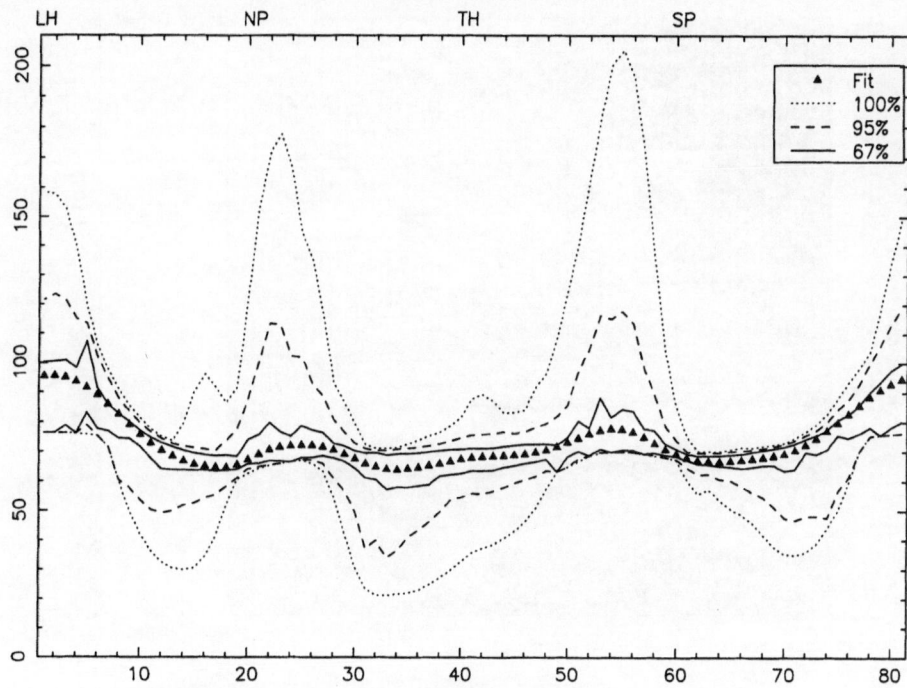

Fig. 13. A slice through the secondary star in IP Peg.

landscape automatically account for the geometrical distortion of the secondary and any non-uniformities in its surface structure, the technique provides very tightly constrained mass determinations, particularly in directions orthogonal to the diagonal feature in figure 14 (corresponding to mass ratios that allow the radial velocity of the secondary's centre-of-mass to remain constant).

4   Conclusions

The Roche tomogram of IP Peg presented in figure 12 represents the best that can be obtained with a 4-m class telescope when mapping a single spectral line. Yet even this does not appear to be good enough to image the star spots we might reasonably expect to be present on IP Peg's secondary. To provide routine imaging of star spots on CV secondaries it will therefore be necessary to move up to 8-m class telescopes and/or combine many spectral lines to increase signal-to-noise, using techniques such as *least-squares deconvolution* (LSD) [12]. An initial attempt at LSD using data on the dwarf nova SS Cyg is presented in figure 15 and appears to show the tell-tale signature of a star spot – the diagonal stripe moving from blue to red velocities between phases 0.5–0.8. These data require more careful reduction before they can be mapped, but it is clear that LSD is the way forward and we can expect the first star-spots on CV secondaries

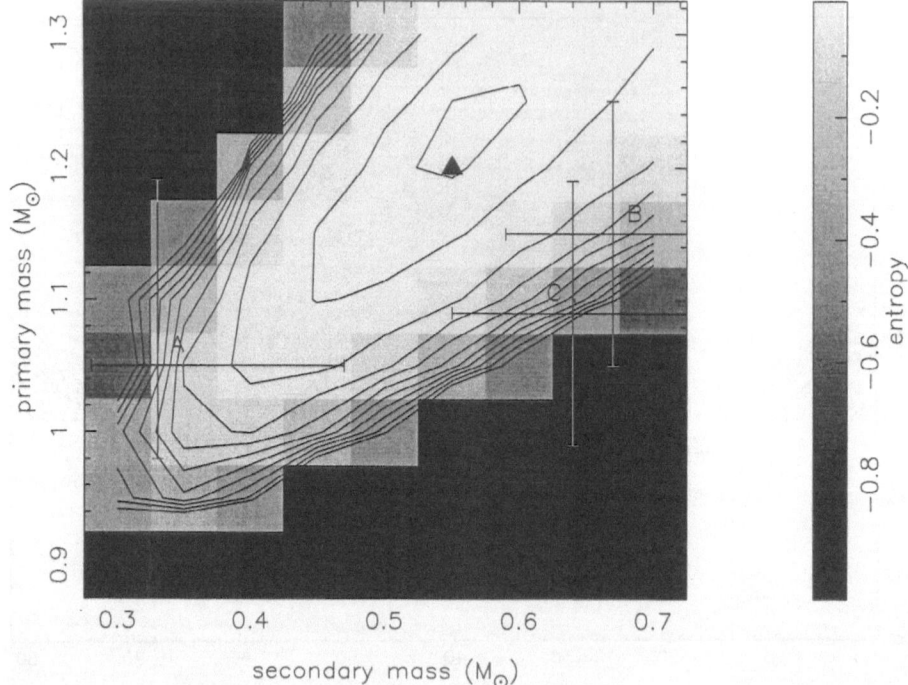

Fig. 14. Entropy landscape for IP Peg.

to have been unambiguously imaged by the time the next conference on astro-tomography convenes.

Acknowledgments

We thank Andrew Collier Cameron, Jean-Fançois Donati, Tom Marsh, René Rutten, Axel Schwope, Tariq Shahbaz and Danny Steeghs for allowing us to present their data in this review and for much useful advice on the art of astro-tomography.

References

1. C.W. Allen, 1976, *Astrophysical Quantities* (Athlone Press, London)
2. G. Beekman, M. Somers, T. Naylor, C. Hellier, 2000, MNRAS, 318, 9
3. K. Beuermann, I. Baraffe, U. Kolb, M. Weichhold, 1998, A&A, 339, 518
4. J.M. Brett, R.C. Smith, 1993, MNRAS, 264, 641
5. J. Casares, M. Mouchet, I.G. Martinez-Pais, E.T. Harlaftis, 1996, MNRAS, 282, 182
6. A.C. Cameron, R.D. Robinson, 1989, MNRAS, 238, 657
7. S.C. Davey, R.C. Smith, 1992, MNRAS, 257, 476

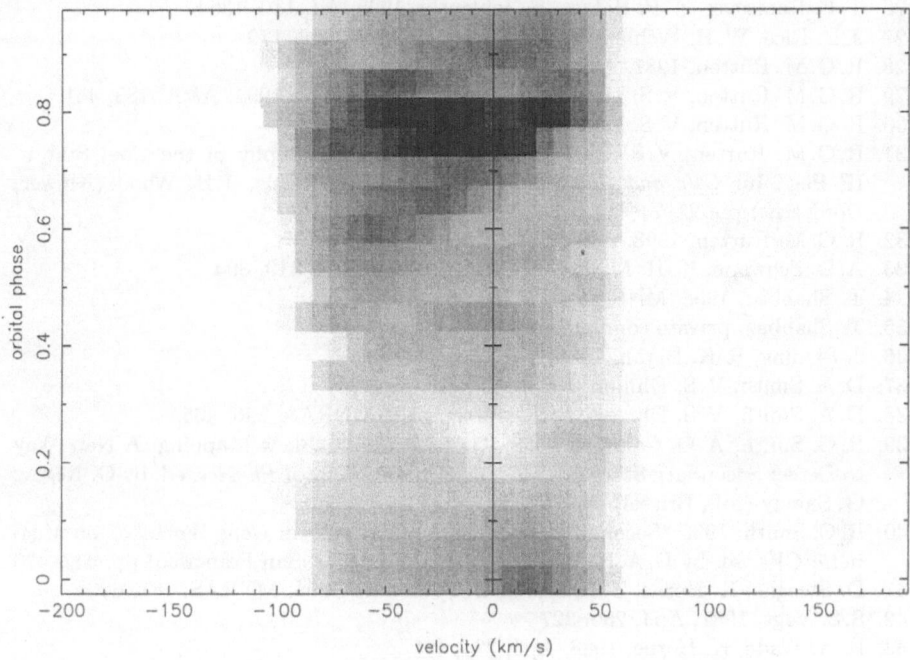

Fig. 15. Trailed spectrum of the combined line profile of the secondary in SS Cyg derived using LSD [53].

8. S. C. Davey, R. C. Smith, 1996, MNRAS, 280, 481
9. V. S. Dhillon, D. H. P. Jones, T. R. Marsh, 1994, MNRAS, 266, 859
10. V. S. Dhillon, S. P. Littlefair, S. B. Howell, D. R. Ciardi, M. K. Harrop-Allin, T. R. Marsh, 2000, MNRAS, 314, 826
11. J.-F. Donati, et al, 1992, A&A, 265, 682
12. J.-F. Donati, M. Semel, B. D. Carter, D. E. Rees, A. C. Cameron, 1997, MNRAS, 291, 658
13. J. E. Drew, D. H. P. Jones, J. A. Woods, 1993, MNRAS, 260, 803
14. B. Efron, 1982, *The Jackknife, the Bootstrap and Other Resampling Plans* (SIAM, Philadelphia 1982)
15. P. Eggleton, 1983, ApJ, 268, 368
16. P J. Groot, 1999, Optical Variability in Compact Sources. PhD Thesis, University of Amsterdam
17. E. T. Harlaftis, 1999, A&A, 346, L73
18. K. Horne, 1985, MNRAS, 213, 129
19. S. P. Littlefair, V. S. Dhillon, S. B. Howell, D. R. Ciardi, 2000, MNRAS, 313, 117
20. S. P. Littlefair, V. S. Dhillon, E. T. Harlaftis, T. R. Marsh, MNRAS, submitted
21. C. Maceroni, O. Vilhu, F. van 't Veer, W. Van Hamme, 1994, A&A, 288, 529
22. T. R. Marsh, K. Horne, 1988, MNRAS, 235, 269
23. T. R. Marsh, 1988, MNRAS, 231, 1117
24. J. S. Martin, D. H. P. Jones, M. T. Friend, R. C. Smith, 1989, MNRAS, 240, 519
25. T. J. Martin, S. C. Davey, 1995, MNRAS, 275, 31

26. T. F. Ramseyer, A. P. Hatzes, F. Jablonski, 1995, AJ, 110, 1364
27. J. B. Rice, W. H. Wehlau, V. L. Khokhlova, A&A, 208, 179
28. R. G. M. Rutten, 1987, A&A, 177, 131
29. R. G. M. Rutten, V. S. Dhillon, K. Horne, E. Kuulkers, 1994, A&A, 283, 441
30. R. G. M. Rutten, V. S. Dhillon, 1994, A&A, 288, 773
31. R. G. M. Rutten, V. S. Dhillon, 1996, 'Roche Tomography of the Cool Star in IP Peg'. In: *CVs and Related Objects*. ed. by A. Evans, J. H. Wood (Kluwer, Dordrecht) pp. 21–24
32. R. G. M. Rutten, 1998, A&AS, 127, 581
33. A. D. Schwope, K.-H. Mantel, K. Horne, 1997, A&A, 319, 894
34. T. Shahbaz, 1998, MNRAS, 298, 153
35. T. Shahbaz, private communication
36. J. Skilling, R. K. Bryan, 1984, MNRAS, 211, 111
37. D. A. Smith, V. S. Dhillon, 1998, MNRAS, 301, 767
38. D. A. Smith, V. S. Dhillon, T. R. Marsh, 1998, MNRAS, 296, 465
39. R. C. Smith, A. C. Cameron, D. S. Tucknott, 1993, 'Skew Mapping: A New Way to Detect Secondary Stars in CVs'. In: *CVs and Related Physics*. ed. by O. Regev, G. Shaviv (IoP, Bristol) pp. 70–72
40. R. C. Smith, 1995, 'Secondary Stars and Irradiation'. In: *Cape Workshop on Magnetic CVs*. ed. by D. A. H. Buckley, B. Warner (ASP, San Francisco) pp. 417–426
41. D. Steeghs, K. Horne, T. R. Marsh, J.-F. Donati, 1996, MNRAS, 281, 626
42. S. S. Vogt, 1981, ApJ, 250, 327
43. R. A. Wade, K. Horne, 1988, ApJ, 324, 411
44. B. Warner, 1995, *Cataclysmic Variable Stars* (CUP, Cambridge)
45. C. A. Watson, V. S. Dhillon, MNRAS, submitted

Statistics of Isolated Emission Sources in Cataclysmic Variables

C. Tappert[1,2] and R. Hanuschik[3]

[1] Dipartimento di Astronomia, Vicolo dell'Osservatorio 5, I-35122 Padova, Italy
[2] Astronomisches Institut, Ruhr-Universität, D-44780 Bochum, Germany
[3] ESO, Karl-Schwarzschild-Str. 2, D-85748 Garching, Germany

Abstract. The strong emission lines in cataclysmic variables are the primary source of information on physical parameters of and the accretion process in these systems. Many examples show that the lines are composed not only from (symmetrical) disc emission, but that additional contribution is provided by isolated, i.e. asymmetric emission sources, distorting the original line profile. In this paper, we study the distribution of these additional components with several system parameters.

1 Introduction: Why Do We (Have to) Study Line Profiles?

The research on cataclysmic variables (CVs) mainly focuses on two general topics: their long-term evolution and the accretion process. Both rely strongly on the investigation of the typical emission lines for those systems, especially of the Balmer and Helium series.

The importance of the examination of the line profile for the understanding of the accretion process has been emphasized in the recent years by the discovery of spiral shocks in CV disks, which are covered by Steeghs in this volume.

On the other hand, theoretical models on CV evolution can be tested only on a statistical basis, by the study of possible correlations between system parameters like masses, periods, etc. Especially for medium and low inclination systems, these values become only accessible by measuring the periodic radial velocity variations of spectral features. The parameters of the resulting sinusoidal curve can then be used to derive the system parameters (e.g. [13], p.99ff).

An important prerequisite for such an approach is the assumption that those features track the motion of the components. However, their direct signatures, i.e. the absorption lines of the secondary and the white dwarf, can only be measured in a very limited number of systems. One thus usually takes the strong and easily accessible emission lines as replacement. Those, however, only track the motion of the primary if the emission layer is distributed symmetrically around the white dwarf, rendering a symmetrical line profile [3], whose shape remains constant throughout the orbit. Still, numerous examples show the presence of additional emission components with different phase and amplitude (e.g. [10]). These components – often labelled 'S-wave' – stem from asymmetrically located emission distributions, which we therefore term 'isolated emission sources' (IES). Their presence distorts the line in such a way that radial velocity measurements which only treat the line as a whole and ignore the specific profile will yield incorrect phasing and amplitude, and consequently incorrect system parameters.

In this paper, which represents an extension and refinement of earlier work [11,12], we study the statistics of IES to examine their frequency and possible correlations with certain system parameters, which might provide clues to their physical origin.

2   Methods: How to Study Line Profiles?

There are a number of methods which are capable of indicating both the presence and a specific type of IES. Each has its strong and weak points, making in most cases the application of all of them necessary in order to extract the maximum information content from the line profile variations. These methods are mostly well-known and widely used, and we thus restrict ourselves to a short listing and refer the reader to the corresponding publications.

1. **Diagnostic Diagram:** Here the line profile is convolved with two identical Gauss functions. Varying the separation of the Gaussians measures the radial velocities of different parts of the line [7]. A plot of the parameters of the corresponding sinusoidal function vs. the separation yields the so-called diagnostic diagram [8]. The presence of IES is then indicated by the deviation from constant parameters. Under the assumption that the line wings are mostly undistorted, this method can furthermore be used (with caution) to obtain the parameters a pure disk profile would produce. It thus represents the only way – in the absence of further orbital phase information, e.g. through photometric eclipses – to determine the 'true' zero point of the radial velocity curve, and by comparison the phasing, and thus the type, of the IES. In this sense, the interpretation of all other methods relies on the results of the diagnostic diagram.
2. V/R **Plot:** The asymmetry of a line can be quantified by defining a blue and a red half of the profile and computing the ratio of the fluxes under both halfs as $V/R = \log(F_V/F_R)$ [11]. This is a more general approach than the often used ratio of the intensities of the blue and the red peak, as the latter is only applicable to double-peaked profiles. The variation of the asymmetry with the orbital phase then again indicates the location of the IES. The strength of this method lies in treating each spectrum individually. Its application – and to a certain extent its interpretation – is thus independent of the phase coverage or even the knowledge of the orbital period.
3. **Visual Inspection:** Either as a trailed and/or as a traced spectrum the visual approach gives qualitative information on multiple components, the extension of the wings and their possible distortion, and the variation of the line intensity, all of which influence the performance of the other methods.
4. **Doppler Tomography:** This method is described extensively in other publications of this volume (see the review by Marsh). It yields the best visual impression of the emission distribution and is able to indicate multiple and weak features.

3 Statistics: What Do We Find?

3.1 Classification

Several physical processes which produce IES are already well-known, such as emission from the hot spot region or from a possibly irradiated region from the secondary (e.g. [5,9]). In a Doppler map, they occupy coordinates $(-v_x, \pm v_y)$ and $(0, K_2)$, respectively, with K_2 being the apparent velocity of the secondary star. Although the different IES can also be distinguished with the other methods [11], Doppler maps yield the clearest overview and thus naturally suggest a geometrical approach for an IES classification, based on their location and appearance in the map. Avoiding an overclassification which would diminish the size of the sample and thus the significance of the statistics, we opted for five classes:

1. **Emission from the trailing side:** This concerns IES which appear on the left half of the Doppler map, i.e. at coordinates $(-v_x, \pm v_y)$. These components have later zero phases than the pure disk component $(0.5 < \varphi < 1.0)$, thus 'trailing' it in the orbital motion.
2. **Emission from the leading side:** Correspondingly, the coordinates covered here are $(+v_x, \pm v_y)$, the right half of the Doppler map.
3. **Emission from the secondary:** Note that this class contains all IES which have coordinates $(0, +v_y)$. Although we here term this 'secondary' emission, this does not a priori mean that the physical origin is really situated on the secondary star.
4. **Other locations:** This class includes different IES too low in number for a statistical analysis, like emission from the gas stream, from circumsystem material, or spiral shocks. It furthermore comprises systems with IES of uncertain type, i.e. when an additional component was clearly present, but its type could not be determined.
5. **None:** All systems where one or more of the above methods was applied to, and where no IES was found, although the quality of the data would have allowed its detection. This class does not include systems which only sometimes (i.e. at certain brightness states) lack the presence of IES.

3.2 The Sample

A prerequisite for the classification of an IES in a certain system is the phase information and thus the knowledge of the orbital period. We therefore used the catalogue of Ritter & Kolb [6] as starting point for our literature research. Subsequent studies were also included, e.g those presented during the Astro Tomography Workshop in July 2000 in Brussels (poster paper on "the new WZ Sge type dwarf nova candidate 1RXS J105010.3−140431" by Mennickent et al.) and those described by North & Marsh and Rolfe et al. in this volume. We restricted our research to systems which are supposed to inhabit accretion disks, i.e. to dwarf novae and nova-like variables.

Table 1. Distribution of the IES types with the system type. Abbreviations of the latter are explained at the bottom of the table. The total number of IES is larger than the number of systems (last row), as several CVs show more than one IES.

	total	Dwarf Novae							Nova-Likes				
		all	UG	ZC	SU	ER	WZ	DN	all	UX	VY	SW	NL
	(1)	(2)	(3)	(4)	(5)	(6)	(7)	(8)	(9)	(10)	(11)	(12)	(13)
trailing	37	22	6	1	10	0	3	2	15	6	1	7	1
leading	22	18	7	2	7	0	1	1	4	1	0	3	0
secondary	18	9	4	4	1	0	0	0	9	5	1	2	1
other	18	8	3	2	3	0	0	0	10	2	4	3	1
none	4	4	1	0	2	0	1	0	0	0	0	0	0
systems	68	43	11	7	18	0	5	2	25	10	5	9	1

UG = U Gem, ZC = Z Cam, SU = SU UMa, ER = ER UMa, WZ = WZ Sge, DN = dwarf nova of undefined subclass, UX = UX UMa, VY = VY Scl, SW = SW Sex, NL = nova-like of undefined subclass

In this manner we found 78 systems with published line profile studies. We subsequently excluded 10 systems with only ambiguous results, thus rendering a sample size of 68 systems. The last row of Table 1 shows that about 2/3 of them are dwarf novae. This is in mismatch to the sample of systems with known orbital period, where dwarf novae have a 3/4 majority [6]. A similar mismatch is observed within the dwarf novae alone, with the systems above the period gap (UG, ZC, and one of DN) making 44% of the sample, while they sum up to only 36% in the Ritter & Kolb catalogue. The reason for both mismatches is that several of the systems below the gap have been only observed photometrically, with the orbital period being accessible through the superhump phenomenon typical for these systems (e.g. [13], p.192ff). This is especially evident for the ER UMa subclass which has no published line profile study at all. However, as we investigate the IES statistics of dwarf novae and nova-likes separately, and examine possible correlations with the orbital period only within our sample, we consider our results to be not affected by these mismatches.

3.3 Correlations

Table 1 gives the distribution of the different IES types in general and with respect to the system type. The first data column shows that emission from the trailing side is dominant, but that also the corresponding one from the leading side is not an exceptional phenomenon, but is found in almost a third of the systems. Most interesting, however, is that only 4 systems were found to have no additional emission component in their spectra. We would furthermore like to point out that, apart from the orbital period, no system parameters are known of these objects (AR And, GW Lib, TY PsA, HE 2350–3908), which means that they are far from being well-studied.

A look at the distribution with the system type shows, by comparing columns 2 and 9 of Table 1, that the emission from the leading side appears to be mostly

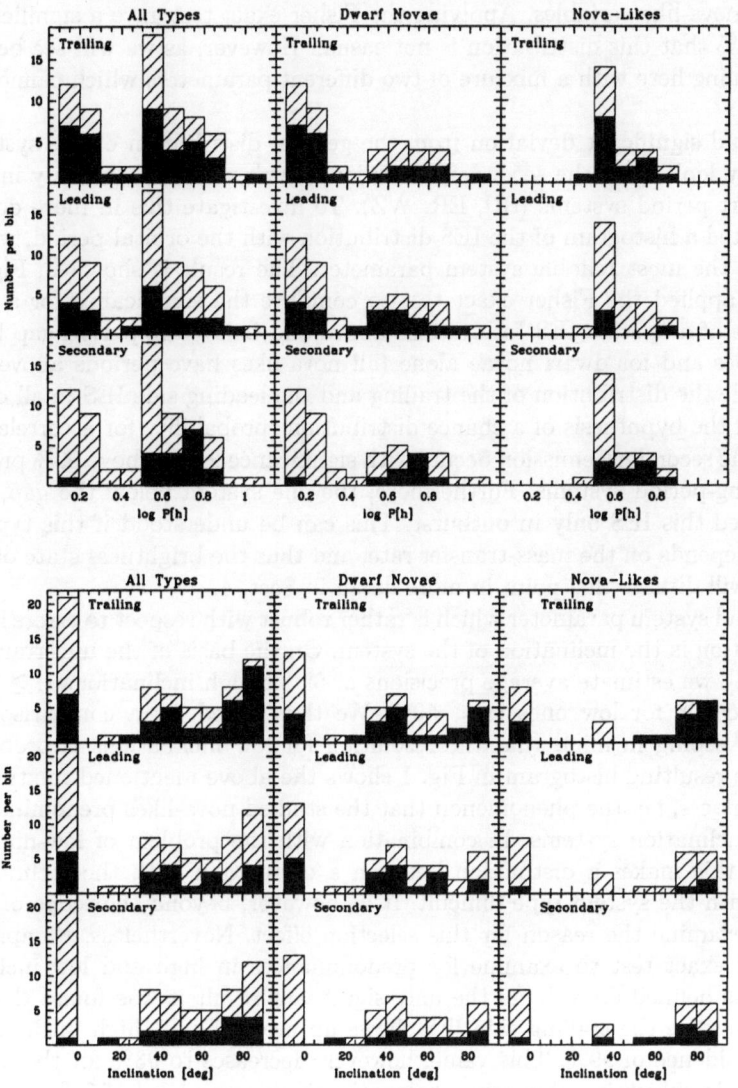

Fig. 1. Distribution of the emission types with system parameters. The hashed histogram represents the distribution of all investigated systems, the solid one refers to a specific emission type. The left plots show the distribution for all CV types, while the middle and the right ones regard specific types, as indicated. **Top:** Number per bin vs. the logarithm of the orbital period in hours. Each bin includes periods starting from a full tenth of $\log P$ up to the next one, e.g. the bin centred on $\log P = 0.55$ includes systems with $0.5 \leq \log P < 0.6$. **Bottom:** Number per bin vs. inclination. Each bin includes inclinations from a full ten up to the next one, e.g. the bin centred on $i = 55°$ includes systems with $50° \leq i < 60°$. Systems with unknown inclinations have been included at $i = -5°$, the highest bin also contains systems with $i = 90°$.

absent in nova-like variables. Applying the Fisher exact test gave a significance level of 96% that this distribution is not casual. However, as we will see below, we are dealing here with a mixture of two different parameters which cannot be separated.

A second significant deviation from the general distribution of the systems is found by looking at the secondary emission, which was detected only in one of the short period systems (SU, ER, WZ). To investigate this in more detail, we computed a histogram of the IES distribution with the orbital period, which represents the most reliable system parameter. The result is shown in Fig. 1. We again applied the Fisher exact test to compute the significance for a pre-dominance of a specific IES for systems above and below the period gap both for all types and for dwarf novae alone (all nova-likes have periods above the gap). While the distribution of the trailing and the leading side IES in all cases agree with the hypothesis of a chance distribution (probability for a correlation < 95%), the secondary emission occurs with significance levels above 99% prefer-ably in long-period systems. Furthermore, the one system below the gap, OY Car, showed this IES only in outburst. This can be understood if this type of emission depends on the mass-transfer rate, and thus the brightness state of the disk. We will discuss this point in more detail in Sect. 4.

A second system parameter which is rather robust with respect to errors in its determination is the inclination of the system. On the basis of the uncertainties given in [6], we estimate average precisions of 5° for high inclinations ($i \geq 70°$) and 15° to 20° for low ones ($i < 50°$). We thus opted for a comparison in 10° bins, keeping in mind that the resolution for low inclinations is probably lower. The resulting histogram in Fig. 1 shows the above mentioned mixture of two parameters, i.e. the phenomenon that the studied nova-likes predominantly are high inclination systems. In combination with the problem of low-number statistics, this makes a distinction between a correlation with the inclination and one with the system type difficult. It is, however, beyond the scope of this paper to examine the reason for this selection effect. Nevertheless, we applied the Fisher exact test to examine for predominances in high and low inclined systems (as defined above). As the only significant result it was found that in the total sample the trailing side IES shows up preferably at high inclinations with a confidence of 99%. This value, however, decreases to 93% for the dwarf novae sample, which is – if barely – below the significance level of 95%.

4   Conclusions: What Do We See?

When trying to draw conclusions from the results presented in Sect. 3, one has to be aware of several problems which affect the significance of our findings.

First, this study is still on the edge of small-number statistics. Already the addition of 7 new systems would represent an increase of 10%. This border is crossed in certain subsamples, e.g. the IES–inclination distribution for nova-likes only. A division in too many subsamples is thus not advisable. Consequently, the usual samples represent a mixture of lines and brightness states. However, the first is probably not very problematic, as most studies are restricted to Hα and Hβ lines, and although their appearences can be different in detail, our rather rough IES classification should secure in general same results regardless of the specific line. Lines with very different excitation levels, such as HeII, were not considered for our study. To take into account the second point, the brightness level, would have meant to decrease the sample size beyond significance, especially as respective information in the publications is often insufficient. However, it was checked by us for the secondary IES, where the period distribution showed clear evidence for a correlation with the mass-transfer rate.

This study also is built upon a mixture of methods, in the sense that not all data have been obtained by applying all methods. As pointed out in Sect. 2, they all possess a different quality with respect to the interpretation of the results. We would thus like to encourage our colleagues to also reanalyse their old data as new methods (such as Doppler mapping) come available.

Whereas the above mentioned points concern the methodical aspects of our study, there are also some problems with respect to the physical interpretation of the results. Specifically, it is not clear if a one-to-one relation between the physical process and the type of the IES exisits, or if different processes are able to produce at least similar phenomena. With respect to the rough classification we have chosen, this even appears quite probable. Last, not least, we are faced with the problem that we only observe 'snapshots' of the systems, in a specific (long-term) state. It is not known how persistent a specific IES remains at a certain brightness state. In the case of UX UMa and U Gem, Doppler maps taken in different years but at apparently similar brightness states show significant variations [4]. Unfortunately, only few studies enable such long-term comparisons, and a systematic investigation of related phenomena would certainly improve our understanding of the processes in these systems.

In spite of these problems our study produced several significant results:

1. The absence of IES is the very exception. This means that a careful study of the line profile is a necessity in order to derive meaningful system parameters.
2. The presence of IES from the leading side is *not* an exception. It is not the rule, either, but must nevertheless be regarded as a frequent phenomenon in CVs. The significance of this result lies in the fact that the 'classical' producers of isolated emission, like the hot spot or the secondary, are hardly able to occupy such velocity space. In our research, we have found no model to explain this IES convincingly. Unfortunately, the statistics show no significant correlation with any of the parameters, apart from the apparent lack of it in most nova-likes. The latter might mean that this process depends on the type of the accretion disk, but a more detailed study has to wait for an improved number of investigated systems.
3. The secondary IES is strongly correlated to the mass-transfer rate. This supports the general picture that this emission comes from an irradiated surface on the secondary star. Note, however, that the absence of this emission in systems with very low mass-transfer rates (i.e. below the gap) is not a priori obvious. The secondaries in those systems supposedly have spectral types

later than M4V [2]. Field stars of such type generally show chromospheric Balmer emission (e.g. [1]). It is not clear if this emission is simply not present in CVs or if it is too weak compared to the disk lines to be noticed. Note also that the simple equation 'high mass transfer = secondary IES' is not correct, as some nova-likes (e.g. DW UMa) exclusively show this emission in their low states.

4. Isolated emission from the trailing side appears to occur preferably in high inclined systems. Although the correlation for dwarf novae only is just below the significance level, we found no dependence on the system type for this emission, and have thus some confidence in the reality of this result. As the inclination is not a physical but a geometrical parameter, this means a) that this emission is probably mostly confined to the outer disk which dominates the information content in high inclination systems, and b) that it is present in basically all systems, although not always observable. While point a) makes the 'classical' hot spot region indeed the most likely location for this emission, point b) indicates that the properties of this shock front are similar for all CVs at least in the sense that the conditions for line emission are fulfilled.

Acknowledgements

This work made extensive use of the CDS-SIMBAD and NASA-ADS databases, whose presence and maintenance is gratefully acknowledged. We also would like to thank the Sternwarte Heidelberg, ESO-Garching, and ESO-Santiago, whose libraries enabled us to access some hard-to-find publications.

References

1. Bessell M.S., 1991, AJ ,101, 662
2. Beuermann K., Baraffe L., Kolb U., Weichhold M., 1998, A&A, 359, 518
3. Horne K., Marsh T.R., 1986, MNRAS, 218, 761
4. Kaitchuck R.H., Schlegel E.M., Honeycutt R.K., Horne K., Marsh T.R., White II J. C., Mansperger C.S., 1994, ApJS, 93, 519
5. Marsh T.R., Horne K., 1990, ApJ, 349, 593
6. Ritter H., Kolb U., 1998, A&AS, 129, 83
7. Schneider D.P., Young P., 1980, ApJ, 238, 946
8. Shafter A.W., 1983, ApJ, 267, 222
9. Spruit H.C., Rutten R.G.M., 1998, MNRAS, 299, 768
10. Stover R.J., 1981, ApJ , 248, 684
11. Tappert C., 1999, *Isolated Emisson Sources In Cataclysmic Variables*, PhD Thesis, Ruhr-Universität Bochum
12. Tappert C., Hanuschik R., Wargau W., 2000, in *Proceedings of PhD Conference on Variable Stars, Kécskemét, Hungary, August 1999* (in press)
13. Warner B., 1995, *Cataclysmic Variable Stars* (Cambridge University Press)

Tomography of Polars

A. Schwope

Astrophysikalisches Institut Potsdam,
An der Sternwarte 16, 14482 Potsdam, Germany

Abstract. Tomographic techniques of different flavour offer enormous diagnostic power for the analysis of magnetic cataclysmic binaries, particularly those of AM Herculis type, the so-called *polars*. The three main ingredients of such systems, the donor star, the accretor and the accretion stream between the two stars, are investigated by Doppler tomography, Roche tomography and eclipse mapping methods. Future applications currently in development will include Zeeman imaging, cyclotron imaging, as well as combined Eclipse Doppler Tomography. These techniques will reveal the magnetic field topology on the accreting white dwarf and the structure of the accretion curtain in a few carefully selected systems.

1 Introduction

The physics of magnetic cataclysmic binaries (MCVs) differs fundamentally from non-magnetic CVs due to the presence of a strong magnetic field. The field might prevent the formation of an accretion disc and dominate the internal dynamics of the binary. At sufficient low accretion rates and high magnetic field, the spin of the accreting white dwarf becomes synchronized with the binary's period. MCVs come in two flavours, the *polars* with almost strictly synchronized accretors (degree of asynchronism less than ∼1%), and the *intermediate polars (IPs)* with observed spin periods of 68% to 0.1% of the binary period.

While the polars do not show any sign of an accretion disc, the IPs might possess a truncated disc. In both cases, accretion onto the white dwarf occurs along magnetic field lines, either from the inner edge of the disrupted accretion disc, or directly from the ballistic accretion stream. Magnetic accretion gives rise to intense X-ray radiation in the soft and hard X-ray regime from the foot points of accreting field lines near the magnetic poles. X-rays are always pulsed with the spin frequency of the white dwarf due to self-occultation or fore-shortening of the X-ray emission region. In the polars, intense cyclotron radiation with sometimes resolved cyclotron harmonics dominates the optical spectra and yields additional information about the physical conditions in the accretion region. Most field strengths known today were derived by the identification of cyclotron harmonics.

Occasionally mass accretion ceases or is reduced substantially. During these episodes the optical spectra are dominated by the photospheres of the two stars. Particularly intriguing and challenging is the study of the photospheric Zeeman spectrum of the accretor. At the typical field strength encountered in polars,

$B = 10 - 100$ MG, Balmer Zeeman lines are split by several hundred Ångstroms, i.e. easily recognizable even in low-resolution spectra.

X-rays originating from the small accretion regions give rise to re-emission of high-excitation line radiation in the optical and the UV spectral range. The lines originate at all places in the binary systems exposed to the ionizing radiation, i.e. from the hemispheres of the donor star facing the white dwarf and the different parts of the accretion streams or accretion curtains. Hence, a detailed analysis of the emission lines can reveal the whole dynamics and kinematics of such a binary system.

The binary periods of MCVs are rather short, they range from 78 min to 480 min for polars and from 84 min to 2 days for IPs. Most (2/3) of the polars appear below the period gap between 2 and 3 hours, the periods of IPs cluster between 3 and 6 hours. Most polars are as faint as 17th or 18th magnitude, only 10 objects have magnitudes equal to or brighter than 15th mag. This explains why observations with sufficient high time and/or spectral resolution for the application of tomographic methods are published for only a handful of systems so far. Nevertheless, the insight gained so far from, e.g. the application of Doppler tomography, is breathtaking and the level of detailed structure already discovered and to be explored in the future utilizing 8m-class telescopes on a micro arcsec resolution spatial scale is highly promising.

In this review, I firstly describe the results of Doppler tomography using bright emission lines observed in polars, both synchronously and asynchronously rotating. Accretion Stream Mapping, ASM, an eclipse mapping technique applied to emission from the accretion stream, is described in Sect. 3. ASM complements Doppler tomography, the results so far are not completely agreeing, the prospects for application of the combined approach, Eclipse Doppler Tomography, are discussed in Sect. 4. In the following Section 5, I describe the success of Doppler tomography using absorption lines from the photosphere of the secondary star in order to resolve the stellar surface of the donor.

The remaining parts of the review are devoted to the accreting white dwarf. The efforts there are focussed on two different structures, the hot accretion spot (Sect. 6) and the structure of its magnetic field (Sect. 7). The hot spot is investigated with classical eclipse mapping methods using optical and/or X-ray light curves of eclipsing or self-eclipsing systems. Hot spot imaging using cyclotron radiation still needs to be developed, a first approach, the so-called Stokes imaging, is described somewhere else in this volume (contribution by S. Potter). The magnetic field of the white dwarf often displays drastic deviations from the simple configuration of a centered dipole. First attempts and suitable systems are presented, which will allow mapping of the field structure in an objective manner (Sect. 7).

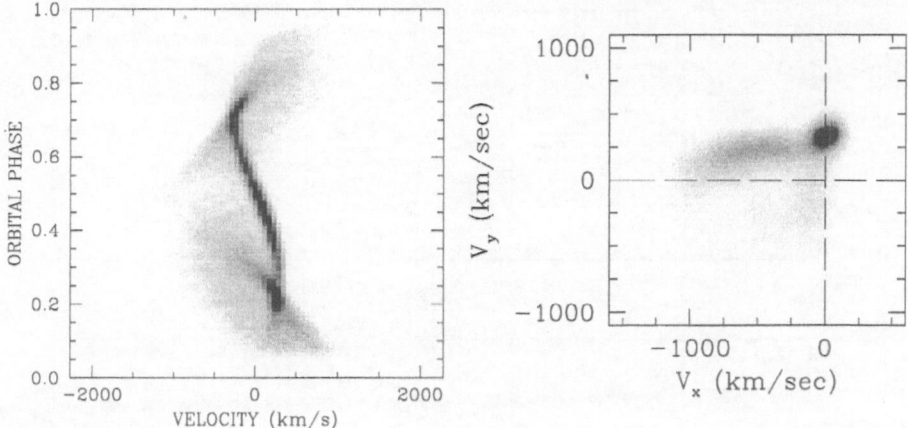

Fig. 1. HU Aqr: Trailed spectrogram of HeII λ4686 and MEM-based reconstruction of the Doppler image in the 1993 high accretion state.

2 Doppler Tomography of Polars

2.1 Technical Remarks

Doppler tomography implicitly assumes that all emission sites are visible at all orbital phases with velocities parallel to the orbital plane of the binary (see Marsh, this volume). The latter condition requires that systemic velocities have been removed from the data. Both prerequisites are violated in polars due to the presence of optically thick radiating or absorbing surfaces and the presence of out-of-plane velocities along magnetic field lines with $v_z \neq 0\,\mathrm{km\,s^{-1}}$. Both these effects and limitations were discussed in more detail in [42] and will not be reproduced here.

Published Doppler maps of polars are computed with either a Fourier-filtered back-projection algorithm (FFB) or a maximum-entropy minimization technique (MEM). In our Potsdam group we made experiments with the FFB-package kindly provided by K. Horne. It was originally developed for the analysis of disc systems, but was adapted to the analysis of non-disk systems and received an interface for MIDAS-input. A code for the construction of Doppler maps based on genetic optimization was developed in Potsdam, too, but converged very slowly compared to MEM-based algorithms. Also, Spruit's MEM-code [48], freely made available via his web-page, was adapted to our needs and extensively used by us. The IDL-based graphic was replaced by a `pgplot`-based graphic, his spectral normalization was removed and a *perl*-based interface now allows data in-/output using `fits`-files. Most of the Doppler maps shown in this review are calculated with the slightly changed code of Spruit.

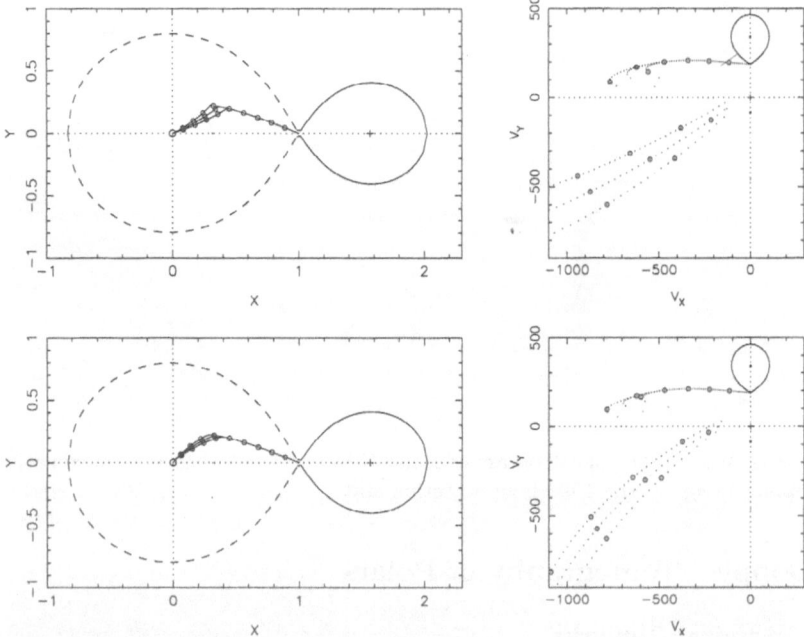

Fig. 2. Schematical representation of a simple accretion geometry in true spatial (left) and Doppler coordinates (right). The inclination of the dipole axis with respect to the rotation axis is 15°, in the upper row it is tilted towards the secondary star, in the lower row perpendicular to the line joining both stars.

2.2 Doppler Maps of Polars

The first Doppler maps of a polar, VV Pup, and a suspected polar, GQ Mus, were published in 1994 by Diaz & Steiner [11,12]. These maps were based on intermediate resolution spectroscopy of Hα and HeII λ4686 with a phase resolution slightly better than 0.1, i.e. with angular resolution of about 18°. Although not very high in resolution, they were showing the dramatic differences between Doppler maps of disc and discless accretors and the great potential for future discoveries. They also allowed a rough separation of emission line flux originating on the irradiated secondary star and from the accretion stream. The first well-resolved Doppler map of a polar, HU Aqr, a high-inclination eclipsing system, was presented by Schwope et al. [40]. The trailed spectrogram of HeII λ4686 used for the tomography experiment and the corresponding Doppler map are reproduced in Fig. 1.

The brightest emission lines of polars in their high accretion states are the Balmer lines and HeII λ4686, with the latter much less affected by optical depth effects than the Balmer lines. This makes the Balmer lines intrinsically broader and the corresponding Doppler maps show less structure. For this reason I am refering in this article mostly to HeII λ4686 maps.

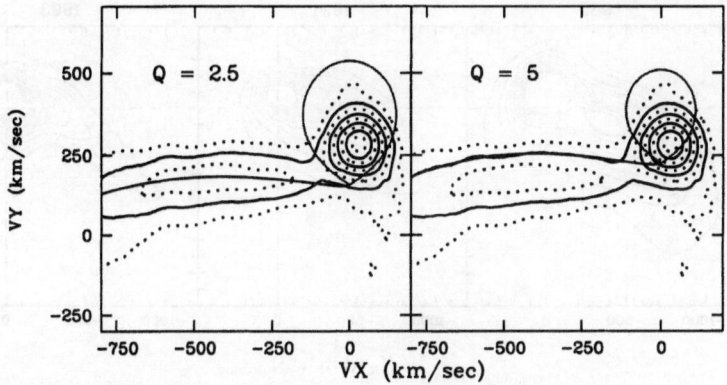

Fig. 3. Contour plot of the HeII λ4686 Doppler map of HU Aqr with Roche lobes and ballistic streams overlaid according to given mass ratio Q.

The Doppler map shown in Fig. 1 clearly shows three different structures which are not so easily recognizable in the trailed spectrogram. Easily interpretable is the emission spot at $(v_x, v_y) = (0, 300)\,\mathrm{km\,s^{-1}}$, which is the Doppler image of the pronounced s-wave emission of the dominant narrow emission line. A cometary tail linked to it stretching down to $v_x \simeq -1000\,\mathrm{km\,s^{-1}}$ and a diffuse patch of emission in the lower left quadrant are the two further structures mentioned. Parallel projections of elongated structures like the cometary tail yield very broad lines, FWHM up to $\sim 1000\,\mathrm{km\,s^{-1}}$, at certain phases (0.0 and 0.5 in the present case), whereas at phases 0.25 and 0.75 such projections result in rather narrow features of full width at half maximum (FHWM) of only about $200\,\mathrm{km\,s^{-1}}$. This means that one easily might loose track of certain features in trailed spectrograms and arrive at misleading conclusions if sinusoids are fitted to radial velocity curves of Gaussian-fitted emission lines.

A basic understanding of the Doppler maps of polars in general and of HU Aqr in particular can be reached by plotting the orthogonal components of the velocity vectors typically encountered in a polar system in a two-dimensional plane. One usually plots the projections of orbital and streaming velocities onto the orbital plane (Fig. 2). As usual, the x-axis of the coordinate system runs through the centres of both stars, the y-axis lies in the orbital plane and the z-axis parallel to the rotational axis.

Emission line radiation, particularly from species with a high ionization potential like HeII λ4686, is mainly of reprocessed photoionized origin. This kind of radiation is expected to be emitted from those places in the binary system, which are irradiated by X-rays from the accretion spot. Three distinctly different structures may be expected in a Doppler map of a polar, if one accepts the basic picture of the accretion geometry as shown in Fig. 2. Firstly, there is the heated front side of the secondary star, secondly the ballistic part of the accretion stream and finally the magnetically dominated part of the accretion stream. The comparison with Fig. 1 reveals that these structures indeed can be observed in nature. The computations which led to Fig. 2 assume that emission

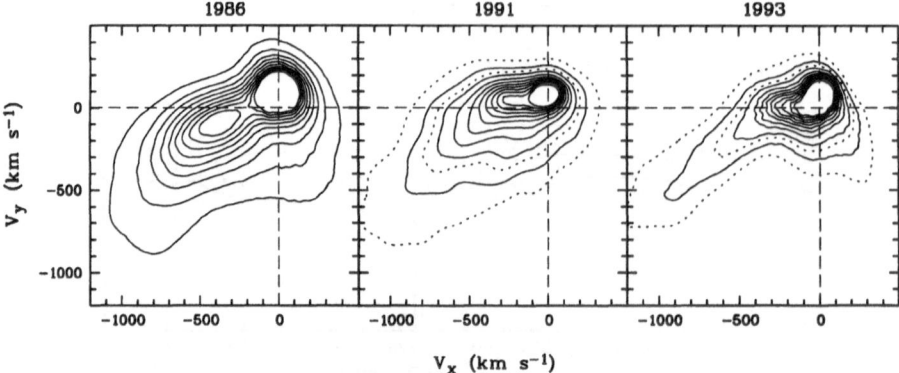

Fig. 4. QQ Vul: Comparison of Doppler maps in the line HeII $\lambda 4686$ obtained at epochs as indicated in the Figure.

from the secondary star originates from the Roche surface. The ballistic stream was defined by a single-particle trajectory. These assumptions can be tested by observations. The computations which led to the figure further assumed that the stream follows a particle trajectory until the magnetic pressure of the magnetic field overcomes the ram pressure in the stream. In that threading region the velocity component along the magnetic field is conserved, the other components are neglected. Depending on the assumed orientation of the dipole this then leads to significant jumps of the trajectory in Doppler space. The jumps become large, if the magnetic axis is roughly aligned with the rotation axis. The trajectory 'turns smoothly around the corner', if the dipole axis is highly inclined. In Fig. 2, I compare two situations with the same inclination of the dipole axis with respect to the rotation axis but with an azimuth differing by 90°. The velocity jump from the end of the ballistic stream at $(v_x, v_y) \simeq (-800, 100)\,\mathrm{km\,s^{-1}}$ to a velocity near the origin is almost the same in both cases, differences become obvious at higher velocities in the magnetic stream (lower left quadrant of the diagram). Under lucky circumstances the magnetic stream can be made visible in Doppler maps and the orientation of the field which guides matter down to the white dwarf can be determined. Further below I will discuss two such examples (UZ For, AR UMa).

The location of the secondary star and of the ballistic stream in the Doppler map are dependent on the mass ratio and the absolute masses of the stars in the binary. Doppler tomography thus may potentially be used to determine masses of the binary by locating the ballistic stream in a Doppler map. These expectations could not be fulfilled so far, the interpretation of the Doppler maps turned out out to be more complicated than suggested by the simple geometrical picture. This became evident already in HU Aqr, where the optimal mass ratio which reflects the location of the ballistic stream was only $Q = 2.5$ and the optimal mass ratio for the irradiated side of the mass-loosing star $Q = 5$ (Fig. 3).

Taking into account the detailed shape, i.e. length of the optical and X-ray eclipse and the radial velocity amplitude of the secondary star measured using the near infrared NaI $\lambda8183/8194$ lines (see Sect. 5), the mass ratio turned out to be $Q \simeq 4$. This means that in the case of HU Aqr neither the ballistic stream, which is, compared to other polars, an extraordinary distinct structure in the Doppler map, nor the narrow emission line from the secondary star fit straightforwardedly in the simple picture outlined in Fig. 2.

The long-period polar QQ Vul shows an even more drastic and variable disagreement between the simple picture and the observed location of the ballistic accretion stream. In Fig. 4, adapted from [43], the HeII $\lambda4686$ Doppler maps obtained at three different epochs are compared.

In 1986 no ballistic stream can be recognized, which in part can be explained by the fact that this particular data set has the lowest spectral and time resolution of the three sets shown. However, the maps of 1991 and 1993 clearly show structures which look like emission from the ballistic stream. But this kind of emission was observed at variable velocity v_y ($v_y = 45\,\mathrm{km\,s^{-1}}$ in 1991, $v_y = 0\,\mathrm{km\,s^{-1}}$ in 1993) which cannot happen, if it would originate from a single-particle trajectory starting at the inner Lagrangian point L_1.

There are two possibilities to cure the problem of dislocated 'ballistic' streams which are in all observed cases shifted towards lower than the nominal v_y-velocity for a given Q. Both possibilities involve magnetic fields as the likely cause, either that of the secondary or that of the primary star. The field of the secondary near L_1 is of the order of the field strength of the primary at that location. Hence it could have a large effect on the stream trajectory near the L_1 point. Matter near the L_1 point can either be deflected by the combined magnetic field of both stars or trapped in a slingshot prominence and released at a location and velocity very different than that of the L_1. Using HST-based Doppler tomography of bright UV lines in AM Herculis itself, Gänsicke et al. [15] showed the presence of highly ionized gas with low velocity dispersion corotating with the binary and located between L_1 and the centre of mass. This was explained as evidence for the occurrence of a magnetic slingshot prominence emanating from the secondary star, similarly to the dwarf novae IP Peg and SS Cyg [49].

A different approach was followed by [46], who modelled the accretion flow as an ensemble of diamagnetic blobs interacting independently with the magnetosphere of the primary. Their interaction term takes the form of a velocity-dependent surface drag force per unit mass, $f_{\mathrm{drag}} = -kv_r$, with v_r the relative velocity of the blob across the field line and $k \propto \mu^2(\rho_b l_b)^{-1}$ the drag coefficient (μ: magnetic moment of the white dwarf, ρ_b: blob density, l_b: blob length scale). This approach allows the magnetic stresses to vary continuously as each blob accelerates towards the secondary and thus raising an accretion curtain in 3D. This step of modelling needs to be elaborated, the first trailed spectra presented by Sohl & Wynn [46] were calculated for fixed k for all blobs and, when applied to HU Aqr, left systematic residuals in the trailed spectrograms. Only after improvement by allowing e.g. k to be variable, the question can be answered if this model yields clues to the problem of dislocated ballistic streams.

Another accretion curtain model was developed by Heerlein et al. [19] for HU Aqr who raised it by assuming a two-dimensional Gaussian density distri-

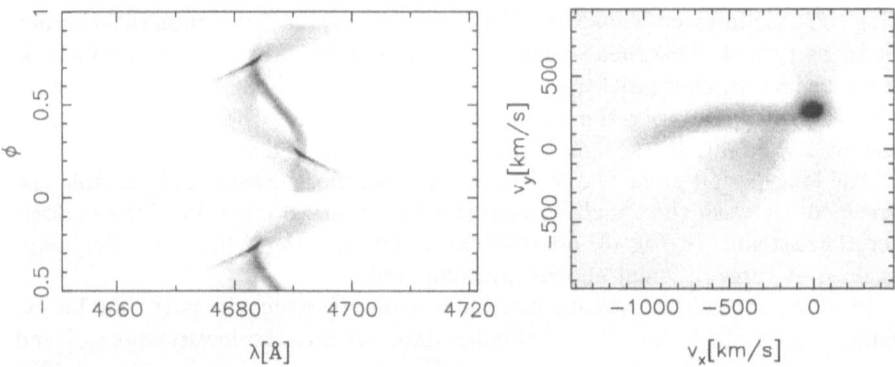

Fig. 5. Trailed spectrogram and Doppler map of HU Aqr according to Heerlein et al. [19]; to be compared with the observed trailed spectrum and Doppler map shown in Fig. 1.

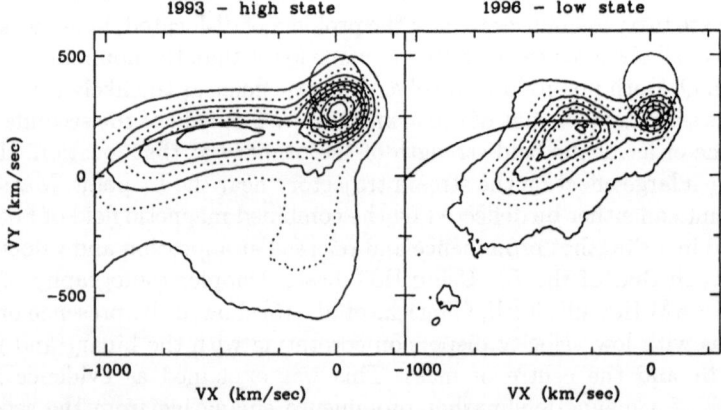

Fig. 6. Comparison of the 1993 high and the 1996 low state Doppler map of the eclipsing polar HU Aqr [41].

bution in the ballistic stream, which becomes stripped if the magnetic pressure at a given place exceeds the sum of ram and thermal pressures. Assuming a constant surface density in the curtain, the ballistic stream and on the secondaries surface, they optimized their solution with a χ^2-fit and solved mainly for geometrical parameters. Their modelled trailed spectrogram and Doppler map is shown in Fig. 5, when subtracted from observed data (Fig. 1) both, the trailed spectrogram and the Doppler map, show systematic residuals, which are particularly pronounced at the ballistic stream and the unshielded secondary star. This means that their dynamical model as well as the assumption of constant surface brightness along the ballistic and the magnetic stream are clearly much too simple.

The presence of an accretion curtain in HU Aqr, which makes proper modelling of the trailed spectrograms so difficult, is revealed indirectly also by the

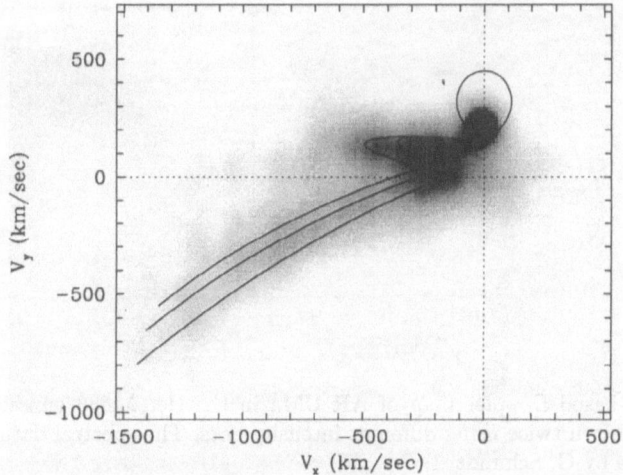

Fig. 7. Doppler map of UZ For in the HeII λ4686 emission line. Clearly visible are the irradiated front side of the secondary star, the ballistic stream and the magnetically coupled stream [41]

asymmetry of line radiation originating from the secondary star, both in high and low accretion states (see Fig. 6, adapted from [41]). In both Doppler maps, the centroid of light lies on the trailing (right) side of the Roche lobe, which means that the leading side is less affected by irradiation, i.e. shielded by an accretion curtain. This view is supported by soft X-ray observations of HU Aqr, which show attenuation of the X-ray flux prior to the eclipse, when the curtain blocks the view down to the hot accretion spot.

Exploring Doppler Maps of Polars: UZ For

UZ For is in many respects a twin system to HU Aqr, it has a period of 126.5 min, similar to the 125 min of HU Aqr, close to the lower edge of the period gap, it has a not too different magnetic field strength in the main accreting pole (53 MG compared to 36 MG) and, most important here, it is an eclipsing system with inclination of about 81° (compared to ∼85° for HU Aqr). The main difference is, that UZ For is a proven two-pole accretor, the second pole having a significantly higher field strength of ∼75 MG [35]. The existence of a second accreting pole is derived from the presence of cyclotron lines from this second region and from detailed eclipse photometry [1], which shows a two-step ingress and egress at optical wavelength attributed to two accretion regions. Interestingly, at X-ray wavelengths (ROSAT and EUVE observing windows) only one accretion region is evident, thus demonstrating the large temperature difference in the two regions.

The Doppler map of UZ For shown in Fig. 7 clearly shows three different emission structures, the irradiated front side of the secondary star, emission which is associated with the ballistic accretion stream and emission from the magnetically controlled part of the stream in the lower left quadrant. We modelled the

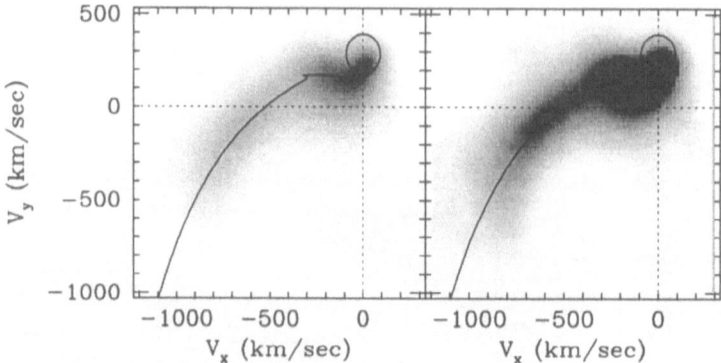

Fig. 8. MEM-based Doppler map of AR UMa in the HeII λ4686 emission line. The same map is shown twice using different intensity cuts. The spectral data were kindly made available by G. Schmidt.

stream emission with the same model which led to Fig. 2. With a co-latitude of the magnetic axis $\delta = 15° \equiv 165°$ and an azimuth $\varphi = 45°$, excellent agreement between the observed and modelled location of the stream can be reached. The three trajectories shown in Fig. 7 couple onto magnetic field lines $\varphi = 10° - 20°$ prior to eclipse centre. These parameters predict a location of the accretion spot at co-latitude $26°(\equiv 154°)$, azimuth 31°, and the occurrence of an X-ray absorption dip at phase 0.96. The spot co-latitude predicted is in good agreement, the azimuth and dip phase disagree with the detailed EUVE-observations by Warren et al. [53]. They observed the dip at phase 0.91 and the spot at an azimuth of 49°. This leaves us with three possibilities: (1) our modelling is insufficient, or (2) the stream that we are seeing in the Doppler maps feeds the secondary accretion spot which is not seen by EUVE and ROSAT, or (3) the disagreement is due to a pronounced re-arrangement of the accretion geometry. Without simultaneous observations in the X-ray and the optical wavelength range it will be difficult to discern between the different options. A further complication arises: For the computation of the model trajectories we assumed a mass ratio $Q = 3$, thus achieving a convincing fit. This mass ratio differs from the best-fit value, $Q = 5$, obtained from eclipse light curve modelling [2]. At higher mass ratio the secondary's Roche lobe is shifted upwards in Doppler space. Then the spot of emission originating from the secondary star is located near the L_1 and not on the irradiated surface, and the observed ballistic stream is dislocated with respect to the ballistic trajectory towards smaller v_y, similarly as in HU Aqr and in QQ Vul. Proper understanding and modelling of the Doppler map if UZ For requires an accurate determination of the mass ratio.

Exploring Doppler Maps of Polars: AR UMa

For more than a decade the measured field strength of polars were distributed between 10 MG and 70 MG and remained clearly below 100 MG. Then, in 1996,

Schmidt et al. detected in a low-resolution IUE-spectrum of the $P_{orb} = 115$ min *Einstein* source 1ES1113+432 (now called AR UMa) flux depressions near $L\alpha$ which were interpreted as Zeeman split $L\alpha$ absorption lines in a field of 230 MG, thus breaking the 100 MG barrier for the first time [33]. The interesting question arose, whether in such a high-field system accretion happens via the classical path with a ballistic and a magnetic accretion stream or whether matter leaving the secondary star would couple onto magnetic field lines directly at the L_1-point. This question was answered by Doppler tomography [34], showing the typical feature of a ballistic stream connected to the irradiated hemisphere of the secondary star.

A MEM-based Doppler map of the data presented by Schmidt et al. of the HeII $\lambda4686$ line is reproduced in Fig. 8. Besides the irradiated secondary which allowed proper phasing of the spectra the maps shows clearly some kind of ballistic stream and a magnetic stream.

Doppler tomography allows one to derive the far-field accretion geometry, i.e. far away from the white dwarf surface. The results achieved can be compared with results obtained from polarimetric observations which allow one to derive the near-field accretion geometry. Such a comparison is made in Fig. 8. The parameters used for computation of the single-particle shown in the figure are: orbital inclination $i = 50°$, co-latitude of the magnetic axis $\delta = 45°$, azimuth of the magnetic axis $\chi = 80°$, mass ratio $Q = M_1/M_2 = 3.8$ (fixed parameter). The mass of the white dwarf M_1 was fixed at 0.7 $M_\odot$.

Schmidt et al. [34] derived $i = 40° - 60°$, $\delta = 35° - 10°$, and $\chi = 90° \pm 7°$. For $i = 50°$, used for our example in Fig. 8, they predict $\delta \simeq 20°$ (depicted from their Fig. 5). Such a low value of δ is not compatible with our modelling of the Doppler map. For all possible values of i we had to assume higher values of δ (by 10°– 20°) than compatible with the results from polarimetry. Doppler tomography and polarimetry consistently confirm a very high azimuth of the magnetic axis.

A cautious remark is in place. In the high state of the system there are likely two streams present feeding the two poles, and this might confuse the tomogram analysis for the magnetically-controlled parts of the stream. Apart from these difficulties, Doppler tomography seems to help to further constrain the possible ranges of the mass ratio, the orbital inclination and the co-latitude of the magnetic axis.

Similarly to the systems discussed above, the structure which we refer to as 'ballistic stream' in the Doppler map is significantly broader (extended towards smaller v_y) than predicted by the single-particle model. Again, this seems to be indicative of an accretion curtain. Without further information other than provided by the Doppler tomogram (which is based primarily on gaseous optically thin radiation) we cannot judge whether the bulk of matter is transferred via the accretion curtain or via the magnetic stream, which is seen in the lower left quadrant of the map. But the clear existence of a structure like a magnetic stream supports the idea that even in a high-field system like AR UMa, the mass transfer between the two stars happens in a rather conventional way.

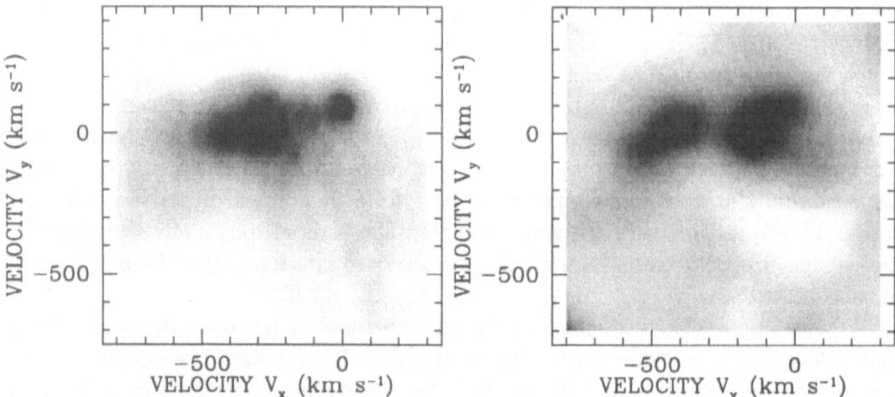

Fig. 9. Doppler maps of V1309 Ori obtained in November 1995 in the lines of HeII λ4686 (left) and HeIIλ8236 (right).

Exploring Doppler Maps of Polars: V1309 Ori

V1309 Ori (= 1RX J0515.41+0104.6) is the polar with by far the longest binary period, $P_{\text{orb}} = 7.98$ hours [52,44]. It contains a slightly evolved secondary of spectral type M0.5III. Fortunately, the system has a high binary inclination of about 78° (Staude et al., in preparation) and is eclipsing. The eclipses, which might last as long as 46 mins, display large variations of their length thus immediately demonstrating that the source of emission which becomes obscured is not a structure of fixed size, like e.g. the white dwarf, but a transient structure like an accretion stream or accretion curtain in the magnetosphere of the white dwarf primary.

Doppler maps of V1309 Ori using the two Helium lines (4686Å and 8236Å) are displayed in Fig. 9. The HeII λ4686 map shows the meanwhile well-known feature of an irradiated secondary and an elongated structure of emission from the accretion stream and/or curtain. Ballistic and magnetic stream do not emerge as separate structures. The ballistic stream is faint between $v_x = 0 \,\text{km s}^{-1}$ and $-200 \,\text{km s}^{-1}$. The broad, extended structure centred on $(-300, 0) \,\text{km s}^{-1}$ must be emission mainly from the magnetic dominated part of the accretion flow because of its low v_y velocity. The map of the near infrared line (HeII8236Å) is qualitatively different. Emission from the secondary is very weak and hardly separable from stream emission. The ballistic stream in HeII λ8236 is bright where the stream in HeII λ4686 is faint. A second isolated spot of emission appears at $(v_x, v_y) = (-420, 30) \,\text{km s}^{-1}$, where the ballistic stream probably reaches its final end and dissipative heating becomes important for excitation. No detailed modelling of the maps is available so far, but the maps constructed from different ionization stages and atomic species bear a great potential for modelling yet to be explored. However, such computations are rather difficult to be performed due to the high particle density in the stream, the unknown density structure

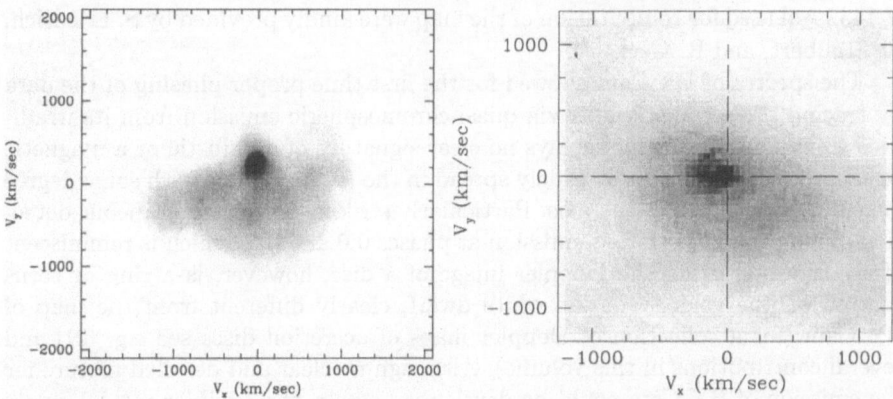

Fig. 10. Doppler maps of the asynchronous rotators BY Cam (left, Balmer Hβ) and V1432 Aql (1RX J1940-10, right, HeII λ4686) at one certain beat phase.

and the anisotropic absorption cross sections of matter in the stream due to the high flow velocities.

Exploring Doppler Maps of Polars:
The Asynchronous Rotators BY Cam and V1432 Aql

Most MCVs of AM Her subtype rotate, by definition, synchronously. There is, however, a small subgroup of presently only four confirmed systems (out of 67 polars known to me) which rotate slight asynchronously. Their spin and orbital periods typically differ by only about one per cent (see [6] for more details). There might be more such systems, the discovery of an asynchronism makes long photometric monitoring necessary and also the independent determination of the true orbital period, preferably by spectroscopy. Long-term photometry is lacking for most of the recently (ROSAT-)discovered and even for most systems known from the pre-ROSAT era.

It is not known how mass exchange works in asynchronously rotating polars, e.g. if always only one, or always both poles are accreting or if a pole-switching scenario applies. This question can be tackled by Doppler tomography and the results can be compared with spin-phase resolved tomography of the emission lines in IPs [20]. These show either one or two accretion curtains, some of them with opening angles of more than 100°.

Twice, in November 1998 and December 1999, we tried to follow the evolution of Doppler tomograms of BY Cam through a beat cycle between spin and orbital period, unfortunately at both occasions bad weather conditions allowed only a snapshot at one beat-phase to be taken (Schwarz et al. 2000, in preparation). The Hβ map of BY Cam obtained in November 1998 is shown in Fig. 10. The map is compared with a Doppler map of V1432 Aql, the only known magnetic CV which has a spin period longer than the orbital period. The spectral data of

V 1432 Aql used for computation of the map were kindly provided by S. Friedrich, R. Staubert, and R. Geckeler.

The spectra of BY Cam allowed for the first time proper phasing of the data by tracing the secondary star via quasi-chromospheric emission from its irradiated front side. The map displays no clear signature of a ballistic or a magnetic stream. Instead, emission is widely spread in the (v_x, v_y) plane with some degree of symmetry around the v_y-axis. Particularly striking is the simultaneous detection of red- and blue-shifted emission at phases 0.0 and 0.5, which is reminiscent of an accretion disc. The Doppler image of a disc, however, is a ring or torus centred on the velocity of the white dwarf, clearly different from the map of BY Cam (for a collection of Doppler maps of accretion discs see e.g. [21] and several contributions in this volume). Although no clear and detailed picture for the emission of BY Cam could be developped so far, the Doppler map suggests that some kind of extended accretion curtain is present with an opening angle between 180° and 360°.

The map of V1432 Aql, which has lower resolution and lower signal-to-noise, but shows a similar structure with red- and blue-shifted emission being simultaneously present at phases 0.0 and 0.5. Our understanding of these structures is still in its infancy. The similarity of the two maps suggest, that a common lesson can be learnt from a thorough study of asynchronously rotating polars.

3 Accretion Stream Mapping (ASM)

A second approach to model the brightness distribution along the accretion stream was developed recently by three independent groups. They make use of photometric data only (instead of spectrally resolved data). Classical eclipse mapping methods, which were originally applied to flat accretion discs, were modified by these groups and applied to a stream in three dimensional space. All methods invert a given light curve and map it onto an accretion stream with pre-defined geometry. This is in all recent cases a ballistic stream described by a single-particle trajectory and a dipolar field line connected to it. The methods differ mainly with respect to the kind of data which are used. Two systems, the twins HU Aqr and UZ For, were studied so far.

Hakala [16] and Harrop-Allin et al. [17,18] used broadband $UBVR$-photometry of HU Aqr in low and high accretion states (examples of such light curves are shown in Fig. 11). They made use of data around eclipse phase thus regarding the eclipse by the secondary star as the only element determining the visibility function.

Similarly, Kube et al. [23] used HST/FOS observations of UZ For around eclipse phase and mapped the light curve of the CIVλ1550 emission line. By using emission line data instead of broadband continuum data one avoids contaminations by the stellar photospheres, by cyclotron radiation from the hot accretion spot and by photospheric continuum radiation from the surroundings of the accretion spot. Cyclotron radiation and spot emission have angular-, and thus phase-dependent brightness, which leads to asymmetric eclipse light curves.

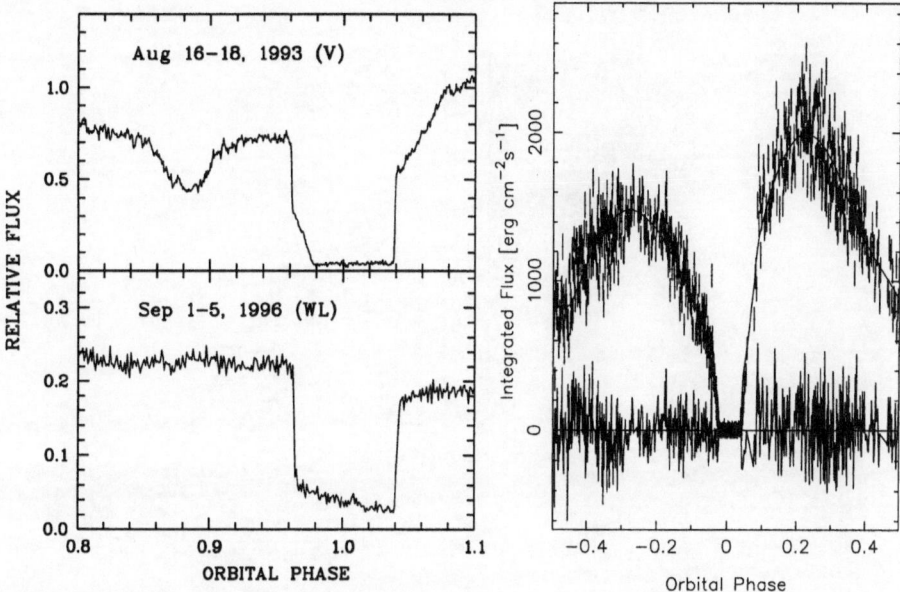

Fig. 11. *(left)* Optical light curves of HU Aqr in high (top) and low (bottom) accretion states. The high state light curve was obtained in the optical V-band with the high-speed multicolour photometer MCCP at the Calar Alto 2.2m telescope in August 1993 (time resolution 0.5 sec), the low state light curve taken in white, i.e. unfiltered light, was obtained with the 70cm telescope at the AIP in Potsdam-Babelsberg in September 1996 (time resolution 15 sec).
(right) Integrated light curve of HeII λ4686 of HU Aqr in its high accretion state. The spectral data were taken simultaneously with the high-speed photometry in the top left panel (adapted from [40]). The solid line through the data points is based on ASM by Vrielmann & Schwope [51].

This clearly complicates the mapping process and emission line data therefore should be used whenever available. On the other hand, broadband data are typically available with better signal-to-noise and higher time resolution than spectrally resolved data.

Finally, Vrielmann & Schwope [51] mapped Balmer (Hβ, Hγ) and HeII λ4686 emission line light curves of HU Aqr in its high accretion state and made use of the full orbital light curve. Contrary to the other teams, they treat the accretion stream as a true three-dimensional structure (a twelve-sided tube), thus allowing for different brightness on the irradiated and non-irradiated sides of the stream. This was found to be necessary because the observed emission line light curves displayed pronounced optical depth effects (see Fig. 11).

Results of the latter work are reproduced in Fig. 12. The brightness map shows a bright spot at the region, where the stream becomes redirected and couples onto magnetic field lines suggesting pronounced dissipative heating in the coupling region. A projection of the best-fit map with high phase resolution, however, shows that this feature seems to have exaggerated brightness in the

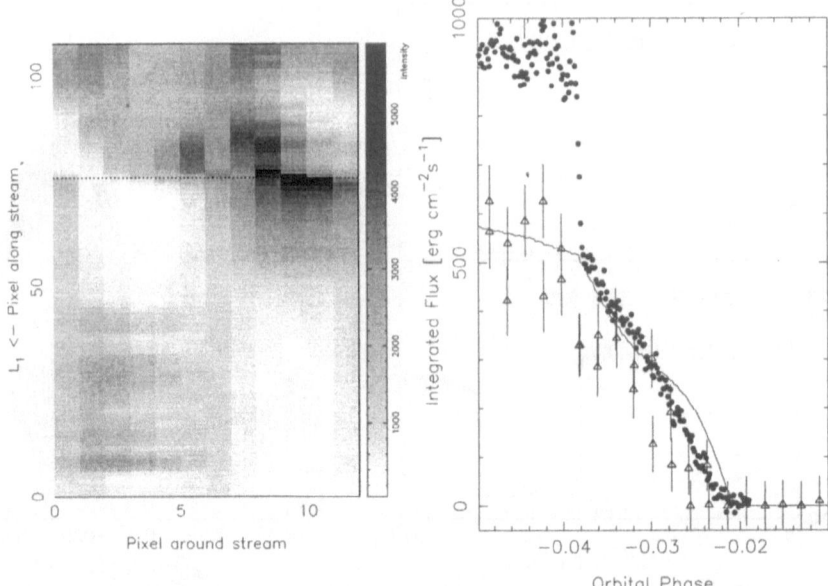

Fig. 12. *(left)* ASM-based brightness map along and around the accretion stream of HU Aqr in its high accretion state [51]. *(right)* Projected stream brightness map of HU Aqr in comparison with the input data (few data with large error bars) and with high time resolution broadband MCCP-data.

map. It is responsible for the hump in the predicted light curve just before the eclipse of the stream becomes complete (phase –0.027). Since the mapping algorithm stably found the same solution, irrespective whether noise was artificially added or only data from a restricted phase interval were used for the mapping experiment, this is suggestive of a more complex accretion geometry than used for the experiment. The existence of an accretion curtain in HU Aqr was mentioned above already. A numerical experiment performed by Vrielmann & Schwope in order to map the observed light curve onto an accretion curtain failed however, because the number of degrees of freedom in a curtain was found too large in order to constrain the fit. A complete picture of the stream thus makes higher time resolution emission line data with spectral resolution necessary.

Although investigating the same object, HU Aqr, at the same epoch, new moon in August 1993 (the observations were performed during the same nights from observatories in Spain and Texas), the maps by Vrielmann & Schwope differ clearly from those by Harrop-Allin et al., the latter showing a pronounced brightness increase towards the white dwarf. I ascribe this difference mainly to the use of the different kind of data: broadband photometry including contaminating radiation from the white dwarf by Harrop-Allin et al. vs. emission line flux from the stream only by Vrielmann & Schwope.

4   Eclipse Doppler Tomography

In the previous section the difficulties preventing us from successfully mapping an observed light curve on an accretion curtain were mentioned. The final reason for the failure is the lack of data with high time and spectral resolution all along the eclipse of the curtain and the stream. This situation will be changed in the era of 8-10m class telescopes. Instead of inverting the observed light curve of a spectral line in integral light, it will become possible to make use of the simultaneously recorded velocity information, i.e. to perform ASM in a velocity bin. A first attempt in this direction was undertaken recently by Bobinger et al. [3,4] who investigated the accretion disc structure in IP Peg. They performed a *double dataset eclipse mapping*, using trailed spectra outside eclipse and broadband photometry at eclipse phase.

The approach proposed here foresees usage of one dataset only, namely the phase-dependent flux $F_{v,\varphi}$ at velocity v in a selected spectral line which can be expressed as

$$F_{v,\varphi} = f(d) \sum_{j=1}^{M} I_j(v)\, \xi_{j,\varphi}\, a_{j,\varphi} \tag{1}$$

with $f(d)$ being a scaling factor (i.e. distance-dependent), $I_j(v)$ the intensity at the pixel j with velocity v in the curtain, M the number of the pixels, $\xi_{j,\varphi}$ the visibility function, $a_{j,\varphi}$ the projected area of the pixel and φ the orbital phase. The geometry of the curtain has to be defined, the inversion of a trailed spectrogram then will yield a map $I_j(v)$.

One difficulty one will encounter is the unknown velocity field in the curtain (contrary to disc systems, where a Keplerian velocity field can be assumed). One may think of different options which can be tested. The first conserves the velocity component along the local magnetic field in the coupling region for a given stream element (as was done above in Fig. 2), the second dissipates the kinetic energy of a stream element completely and uses as initial velocity along a given field line the mean kinetic energy in the coupling region. Despite these difficulties *Eclipse Doppler Tomography* is the logical next step in order to overcome the shortages of mapping experiments making use of light curves in integral light. One preferentially uses for such an experiment a long-period polar with reasonably long ingress and egress phases, like e.g. V1309 Ori or EUVE J1429-38.

5   Doppler Maps of Donor Stars in MCVs

Polars show in their spectra features of the donor star if the contribution of that star to the summed light makes a significant contribution, i.e. if it is not outshone by accretion-induced radiation. This happens in long-period systems with their big secondaries even in high states and it may happen in short-period systems if the accretion rate is reduced, i.e. if they have entered a low accretion state. If spectral signatures of the secondary can be traced through the orbital

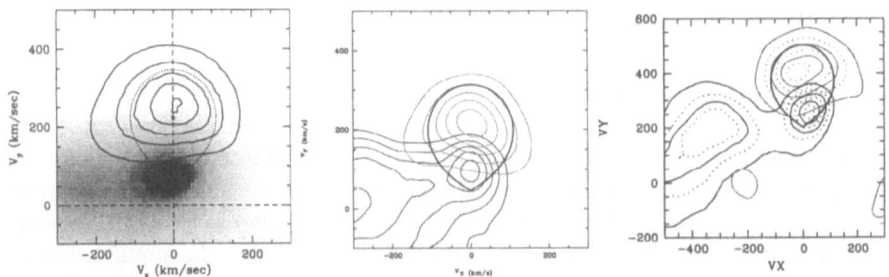

Fig. 13. Combined Doppler maps of NaI λ8183/8194 absorption lines and HeII λ4686 emission lines of the AM Herculis systems QQ Vul, V1309 Ori, and HU Aqr. The size of the Roche lobe of the donor star is drawn in each panel for the most likely mass ratio Q (left panel adapted from [43], middle panel from Staude et al. 2000 (submitted to A&A), right panel from Steeghs et al. 2000 (in preparation)).

cycle, the radial velocity information can be used for an accurate determination of the binary period and for the mass determination of the binary. In principle, both emission lines from the heated front-side and absorption lines from the photosphere, can be used for the mass determination, if the observed radial velocity amplitude is corrected from center-of-light to center-of-mass velocity. The crucial role of irradiation in this respect was firstly recognized in the prototypical system AM Her [9].

Fig. 13 shows the results of combined HeII λ4686 and NaI λ8183/8194 Doppler tomography of the long-period polars QQ Vul and V1309 Ori ($P_{\rm orb} = 3.71$ h and 7.98 h, respectively) observed in high accretion states and of the short-period polar HU Aqr ($P_{\rm orb} = 2.08$ h) which was observed in an intermediate to low accretion state [7,43]. In the high accretion state of HU Aqr the NaI λ8183/8194 absorption lines cannot be recognized due to their relative weakness. All four systems (the three shown in Fig. 13 and AM Her) show similar structures on the Roche surface of the secondary star with emission and absorption lines originating at mutually excluding sites.

The HeII λ4686 emission lines originate on the X-ray irradiated front-side of the donor star. A detailed comparison of the emission line radial velocities of these narrow emission lines in QQ Vul shows that the lines of different atomic species originate at different positions (i.e. radial velocity) in the atmosphere of the secondary. The higher ionization species like HeII λ4686 originate at the highest atmospheric levels, i.e. closest to the center of mass [7,43]. Whether this is a common feature of all polars in their high accretion states needs to be investigated in detail.

The absorption lines of NaI λ8183/8194 originate almost exclusively on the non-irradiated hemispheres of the donor stars away from the white dwarf. Although the luminosity of the irradiating source is higher than that of the exposed star, most of the X-ray photons are absorbed in upper atmospheric layers. This is due to the high column density above the photosphere which easily exceeds 10^{24} atoms cm^{-2} and the fact, that polars typically have soft X-ray spectra. As a consequence, an accretion-induced chromosphere is formed. Only the hard X-ray photons are able to penetrate down to sub-photospheric layers where their energy is finally deposited [22]. Hence, the depletion of absorption lines from the region around the L_1 point, the appearance of the donors as half stars in Doppler space, indirectly confirms the presence of a sufficiently powerful hard X-ray source at the time of the optical observation. This is true for all three systems presented here, although HU Aqr was encountered in a low accretion state.

Only at further reduced accretion rate (the lowest observed rates are of the order 10^{-13} $M_\odot$/yr the typical rate of a polar below the period gap is about 10^{-10} $M_\odot$/yr) the NaI $\lambda8183/8194$ radial velocity curve become apparently unaffected by irradiation [37,39].

The emission line Doppler maps of HU Aqr (Figs. 6 and 13) show the shielding effect of the accretion curtain on the trailing side of the secondary star. One could expect a corresponding mirrored asymmetry in the NaI $\lambda8183/8194$ map. This is not evident in the observed map (Fig. 13), probably due to the low signal-to-noise ratio of the input data. The NaI $\lambda8183/8194$ map of AM Her [9,10,47], clearly shows such an asymmetry, and it appears likely that this is also related to the shielding effect of an accretion curtain.

There are no model calculations available in the literature which may be applied directly to the X-ray irradiated hemisphere of a donor star in a polar. The physical scenario is similar to the HZ Her/Her X-1 system were several models with different level of sophistication were developed [26], but the type of the irradiated star and of the irradiating source are clearly different in the two types of stars. Brett & Smith [5] published a first model for irradiated donors in CVs, but neither X-ray irradiation, they used a 17000 K white dwarf atmosphere, nor non-LTE radiative transport were taken into account.

The formation and structure of an irradiation-induced quasi-chromosphere and the structure of a hard X-ray heated photosphere can be addressed only by proper numerical modeling of an X-ray irradiated atmosphere. Even then the problem will be solved only under strong simplification, e.g. neglection of meridional heating. In order to compare it with observations on the other hand, it needs geometric sophistication by e.g. proper treatment of shielding by an accretion curtain. A practical application of such computations would be a proper K-correction scheme for the correction of measured radial velocity amplitudes to give the true orbital velocity. The presently applied schemes [47,37] simply assume a strict black/white dichotomy between the irradiated and non-irradiated parts of the star.

Observationally, imaging of the donor stars is a challenging and rewarding task. The maps shown in Fig. 13 are based on data obtained at 4m-class telescopes under non-optimal observing conditions. With present-day instrumentation it will be possible, as model computations have shown, to determine the size of the Roche lobe of the donor star in a long-period polar in Doppler space with better than 10% accuracy by straight application of Doppler tomography. The next step then is to search for structures like star spots in these maps. Up to now star spot imaging was possible only for rapidly rotating single giant stars or those in RS CVn binaries (see e.g. [50] and references therein). The rotation velocities in CVs are much (factor 10) higher than in the giant stars investigated so far for star spot activity. This means that lower spectral resolution observations of CVs may yield similarly good resolution of the surface of the star. Even though they are much fainter than the giant stars mapped so far, the large light collecting area of the now operational 8m-telescopes will allow star spot imaging for donor stars in CVs (Dhillon, this volume). This way a new observational window will be opened in order to address the question of magnetic activity at the bottom of the main sequence.

6   The Hot Accretion Spot on the White Dwarf

The accretion spots on the white dwarfs near the polar regions are sources of emission from the infra-red to the X-ray spectral range. The hot accretion plasma, $kT \sim 10\,\mathrm{keV}$, cools via hard X-ray bremsstrahlung and cyclotron radiation. The latter is dominant in the infrared and at optical wavelengths. The X-ray spectrum is usually described with two components, the hard component of an optically thin cooling plasma and a soft component from the accretion-heated atmosphere with typical temperature $kT \sim 25\,\mathrm{eV}$. The soft component typically carries much more flux than the hard component, a phenomenon referred to as soft X-ray puzzle of AM Her stars and explained either in terms of particle or blob heating of the atmosphere instead of radiative heating by the cooling plasma [24].

Heating of the photosphere, either from above by irradiation or below by energy deposited below the photosphere, leads to an extended region around the accretion spot with enhanced temperature. These regions with temperatures somewhere between $10^4\,\mathrm{K}$ and some $10^5\,\mathrm{K}$, i.e. the temperatures of the photosphere and the X-ray emitting accretion spot, are dominating the ultraviolet spectral range.

The rich phenomenology offers, in principle, a wide field for the application of indirect imaging methods. The light curves in the different spectral ranges are markedly modulated by eclipses (by the companion star) and self-eclipses (by the white dwarf itself) and also by anisotropic emission. However, the following complications may arise if optical and X-ray light curves of radiation sources originating from the hot accretion spot are used for a mapping experiment:

1. The hot accretion region which emits X-ray and optical cyclotron radiation has vertical and lateral extent. Proper disentangling of the geometry is possible only if the latitude of the accretion spot can be reliably determined and if the lateral extent can be fixed. This means that only a high-inclination eclipsing system is promising for spot mapping.
2. The dominant soft X-ray component is subject to pronounced absorption in the binary system and in the very vicinity of the emission region. This is a serious constraint, it means that the visibility function is not only determined by the geometry of the emission region.

3. Cyclotron emission is angular- and frequency-dependent and changes from optically thick at long wavelengths to optically thin at short wavelengths with a transition region of badly determined plasma parameters. In addition, the observed cyclotron spectrum is always a superposition of high- and low-temperature regions with different spectral characteristics.

4. X-ray emission is highly variable on short (seconds to minutes) time-scales. A variable source of emission will inevitably produce spurious features in a brightness map, because a sudden brightness increase will be interpreted by the mapping algorithm as due to a spot of emission which just rotates into view. The only way to reduce the effect of variability is the use of data which were averaged over several orbital cycles.

The first mapping experiment of this kind, an inversion of optical and soft X-ray light curves of ST LMi, was presented by Cropper & Horne [8]. They were facing all the difficulties mentioned but it was not possible to solve them at that time. They assumed a completely flat spot of emission for both types of radiation. ST LMi is not eclipsing, but the accretion region undergoes self-eclipses by the white dwarf. The emission spot found by them is arc-shaped with a core at one extremity in the X-ray map, while the optical map was closely coincident with some additional structure. The under-constrained problem of inverting the light curve leads to artifacts in the map, faint arms stretching north-east and north-west from the main accretion region. These follow the stellar limb as seen at the phases of ingress and egress of the accretion region. Also, an apparent (?) double spot emerges as a consequence of a shoulder in the light curve. Whether such a structure is real or due to internal absorption/extra emission from a raised mound [45] remains open as long as no extra constraints are applicable.

The inversion process is, at least in principle, much better constrained by a true eclipse. The unique system HU Aqr, shows both eclipses and self-eclipses. On the other hand, optical and X-ray light curves (see Figs. 11 and 14) in the high accretion state are strongly influenced by cyclotron beaming and intrinsic absorption which makes a successful mapping experiment almost impossible. Fig. 14 therefore compares only the results of a forward computation assuming a flat spot in one case and a raised spot in a second case. Eclipse egress, which is contrary to eclipse ingress not or only marginally affected by intrinsic absorption, lasts only 1.3 sec, which means that the accretion spot has an azimuthal extent smaller than $\sim 4°$. A fit to the light curve assuming a flat spot at a high northern co-latitude of only 9° gives a satisfactory fit to the wings of the bright phase, but could be finally excluded due to an imperfect phase match between center of bright phase and true phase zero (fit shown with dashed line). A spot with vertical extent of about $0.015\,R_{wd}$ at co-latitude 31° and same azimuth of 46° (solid line) gives the same fit to overall light curve than the flat spot model, but places the eclipse at the correct phase.

Most of the problems one encounters when optical and X-ray light curves are inverted can be overcome by inverting ultraviolet light curves. Ultraviolet radiation can reasonably be assumed to originate from a flat extended region, i.e. without vertical extent, it is not or only negligible affected by short-time

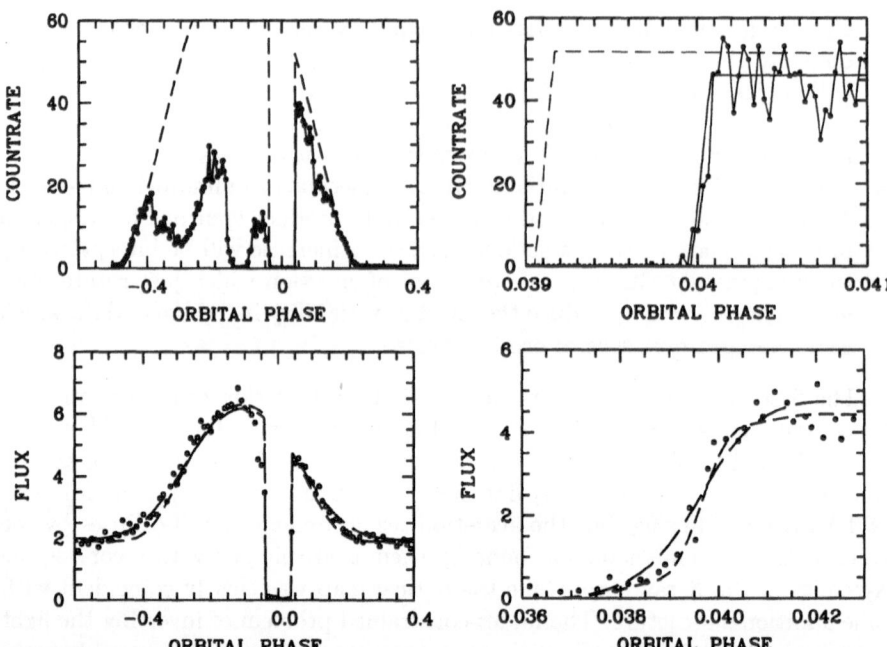

Fig. 14. High-state X-ray (ROSAT) and low-state ultraviolet (HST/FOS) light curves of HU Aqr together with synthetic light curve. Shown are light curves with full phase coverage and sections centered on eclipse egress.

variability, its spectral composition and angular characteristic is reasonably well understood by stellar atmosphere models. The problem then is to collect appropriate input data. There are two systems which received sufficient coverage by HST thus allowing a kind of mapping experiment to be performed, AM Her and HU Aqr [15] (Schwope et al., in preparation). AM Her was observed in both low and high states by IUE and HST [14,15]. The geometry of AM Her is similar to that of ST LMi with no eclipse, but a self-eclipse of the main accretion region. So far, HST-data were not mapped by an objective inversion method, but by a parameter fitting approach with optimization being achieved by a genetic algorithm [15].

The method of Gänsicke assumes a flat circular spot with some form of temperature decrease from a central peak down to the undisturbed photosphere. The parameters to be optimized are the orbital inclination, the location of the spot, the size of the spot, the distance to and the radius of the white dwarf, the extent of the spot, the photospheric and the central peak temperature. The spectral flux at given wavelength for a specific temperature is interpolated between data bank entries for hot, high-gravity, pure hydrogen model spectra.

The best fit using this model for AM Her (HU Aqr, see Fig. 14) predicts a rather large accretion spot of opening angle $\theta \simeq 70°$ with peak temperature $T_c = 47\,000\,\mathrm{K}$. However, without the extra constraint of an true eclipse it is

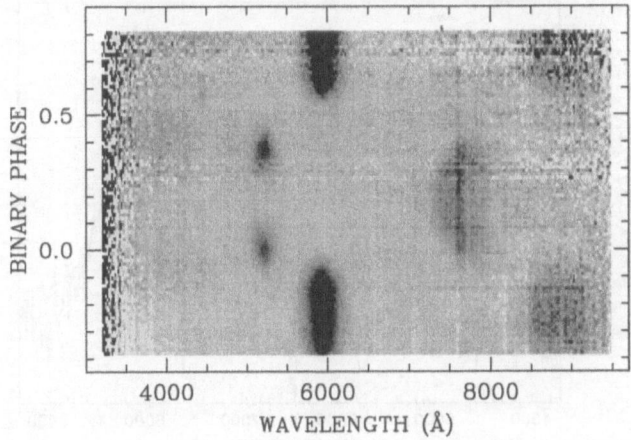

Fig. 15. Trailed low-resolution spectrogram of HS1023+39 showing cyclotron line emission from two accretion spots at ∼60 MG and ∼68 MG (the data were kindly provided by H.-J. Hagen [31]).

possible to trade the opening angle versus the peak temperature. A model with T_c as large as 200000 K with $\theta = 28°$ could not be ruled out. This means that the predictive power of the method applied to a non-eclipsing system is rather limited.

Some preliminary results of our modeling of HU Aqr are shown in Fig. 14 (bottom panels). As for AM Her, the best fit to the HST/FOS continuum data for the whole bright phase excluding the eclipse is a rather large spot with half opening angle $\theta \sim 50°$ and central temperature 34000 K (fit shown with long-dashed lines; the temperature of the undisturbed photosphere was 14000 K). This gives a smooth fit to the wings of the bright phase. It gives a bad fit to the eclipse data, which is much better fitted with peak temperature as high as 80000 K and an opening angle of 22° (the corresponding fit is shown with short-dashed lines, still a preliminary result).

The synoptic view allowed by the unique system HU Aqr illustrates the difficulties one encounters if one tries inversion of the light curve(s) in an objective manner: these are the limited number of photons, the short duration of the critical events, and the uncertain and wavelength-dependent geometry of the emitting region. On the X-ray side some progress will be made in the near future by the XMM-observations of eclipsing polars, on the ultraviolet side no significant observational progress can be expected in the foreseeable future unless some bright, eclipsing system will be newly discovered and thoroughly observed with HST.

Optical data might play a more important rule in the future if one makes full use of the directional characteristics of cyclotron radiation in the modeling process. A first such attempt has been undertaken by Potter et al. [27,28,29] with the concept of Stokes imaging (Potter, this volume). Although promising, this attempt suffers so far from the observational side from the use of broadband po-

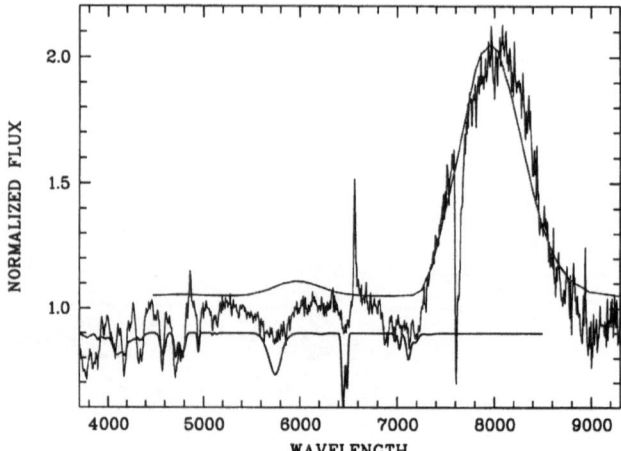

Fig. 16. Low-resolution normalized spectrum of RBS0206 (1RXS J012851.9–233931) showing a strong cyclotron line in the red ($B_c = 45 \pm 1\,\mathrm{MG}$) and a phostospheric Zeeman spectrum of a mean surface field $B_Z = 36 \pm 1\,\mathrm{MG}$ [41].

larimetry only, and from the modeling side from the neglection of an unpolarized background radiation and the use of homogeneous temperature models only. The active regions emerging from this inversion are rather large, contrary to the expectation that they should similar in size as the X-ray emitting regions.

The next logical step is the use of high-time resolution data with spectral and polarimetric resolution. This kind of data will allow proper modeling of the cyclotron spectrum, which inherently carries temperature and angular information about the emitting plasma and allows, under lucky circumstances, proper treatment of background radiation components. Some early examples of the necessary type of data and modeling were presented for MR Ser [37] and UZ For [32]. These systems showed at low accretion rate in their optical spectra deeply modulated cyclotron harmonics. Meanwhile other systems were discovered with similar properties, a trailed spectrum of one particularly beautiful example, HS1023+3900 [31], is reproduced in Fig. 15. Twin systems with similar cyclotron lines are RBS0206 [41] and HS0922+1333 [30]. These systems are distinguished from most of the others by completely un-blended cyclotron lines which appear and disappear as the cyclotron emitting plasma becomes self-eclipsed by the white dwarf. Similarly to the X-ray light curve of HU Aqr shown above in Fig. 14, a complete disentangling of the geometry is almost impossible without a genuine eclipse by the secondary star, but it seems to be only a matter of time until such a system will be found in the sky. It seems reasonable then to think of physical parameter eclipse mapping (see Vrielmann's contribution to this volume) of such a system with temperature, field geometry and field strength as physical parameters to be mapped.

7   The Structure of the White Dwarfs Magnetic Field

The first Zeeman spectra of AM Her were obtained in a low accretion state and published some 20 years ago [25]. Since then, photospheric Zeeman lines were detected in about a dozen systems (for a recent review see [54]).

Phase-resolved Zeeman spectroscopy of AM Her [25,55], MR Ser [37], and BL Hyi [38] has provided unambiguous evidence for field distributions which differ from a centered dipole, the usual first approximation. As a second approximation, those field structures were modeled by dipoles which were offset from the center of the star by 0.1–$0.3\,R_{\mathrm{wd}}$ along the dipole axis. However, these *ad hoc* models bear ambiguities and are far away from yielding unique solutions. The more appropriate approach is a multi-pole expansion or indirect imaging of the field without any pre-defined underlying morphology. This makes phase-resolved Zeeman spectroscopy/spectropolarimetry necessary. The targeted system should have a largely uncontaminated Zeeman spectrum of the white dwarf. This is not easy to achieve, since accretion-induced radiation components are usually prevalent in polars. Even in low accretion states, cyclotron emission and recombination radiation from the stream are often not negligible. An important role for such a mapping experiment might play the systems with persistent low accretion rate (HS1023, HS0922, RBS0206) introduced in the previous section. One template spectrum of RBS0206 showing a rather clean separation of the photospheric Zeeman spectrum and the accretion-induced cyclotron spectrum is reproduced in Fig. 16.

By application of an indirect imaging technique to a series of phase-resolved spectra the underlying field structure can be uncovered. A corresponding feasibility study for magnetic white dwarf stars was presented by Donati et al. [13]. At that time suitable data for single white dwarfs were missing. The main problem with these stars is the often unknown rotational period or, if known, the inappropriate high field strength. High-field stars with $B > 50\,\mathrm{MG}$, are presently inappropriate, because current Zeeman models cannot properly reflect the continuum polarization. On the other hand, large fields provide the largest Zeeman split and are thus most sensitive to variations of the field strength.

At least as long as continuum polarization is not properly addressed in Zeeman model spectra, the magnetic white dwarfs in polars form an important sample for such an indirect imaging experiment. Most of the magnetic white dwarfs in polars have a field strength below $50\,\mathrm{MG}$ (see the recent compilation in [54]) and all have known rotational periods. The presence of contaminating cyclotron radiation poses an extra difficulty to the method. On the other hand, the presence of one or two cyclotron line systems provides us with a boundary condition for the field strength at one or two particular places on the surface of the white dwarf. An additional complication is accretion heating of the polar cap(s) which has to be taken into account in the inversion process. Hence, a grid in magnetic field strength, orientation, temperature, and perhaps abundance is the minimum prerequisite for a successful mapping experiment which is otherwise technically feasible.

8   Summary

I have reviewed the results of indirect imaging techniques applied to AM Herculis stars (polars). The techniques discussed here are Doppler tomography, accretion stream mapping ASM, accretion spot mapping, and Zeeman imaging of the accreting white dwarf. Further techniques described in this volume by Dhillon and Potter are Roche tomography and Stokes imaging. Although rather small in number and rather faint in brightness, the polars offer a wide field for the application of tomographic and mapping techniques due to their preferred geometry, their handy period, and the multitude of physical processes which lead to radiation of different kind, at different wavelengths, and with fundamentally different properties.

The sites of emission studied here are the accretion stream, the accretion curtain and the irradiated hemisphere of the donor star which become visible in the light of reprocessed atomic line radiation in Doppler tomograms. We have found quasi-ballistic accretion streams which are slightly influenced by the presence of the magnetic field of the white dwarf. We have found in addition in a few systems magnetically controlled streams, which allow the orientation of the magnetic field to be investigated. Our theoretical understanding of Doppler maps needs to be developed in detail taking into account the excitation by the X-ray source and the ionization structure in the stream. Extended emission structures neither resembling a disc or a simple accretion curtain as seen in intermediate polars were uncovered in asynchronously rotating polars. Blue- and red-shifted emission lines are seen simultaneously, which is reminiscent of an accretion disc, but a convincing geometry for these systems is missing.

The donor stars became visible, i.e. resolved, in Doppler maps of photospheric absorption lines and the dramatic influence of X-ray irradiation on the structure of their photospheres became obvious. Again, a detailed theoretical understanding of the underlying physics needs to be developed. I'm quite confident that soon the first starspots on a late-type ZAMS (or near-ZAMS) donor star can be made visible by straightforward application of Doppler tomography or by Roche tomography. Then the question of magnetic activity at the bottom of the main sequence can be addressed by watching through a new observational window.

Accretion Stream Mapping is found to be a useful technique complementing Doppler tomography in order to derive the brightness map along the accretion stream. Both methods have their short-comes and a substantial step forward can be reached by their combination, Eclipse Doppler Tomography. The necessary input data can be obtained in the era of the giant 8-10m class telescopes.

Mapping experiments of the accretion spot using only self-eclipses by the white dwarf lack uniqueness due to the uncertain geometry of the accretion region. The geometry can in principle be nailed down using true eclipses by the secondary star. Since these are extreme short-lasting events, one is limited by the high degree of variability of the radiation (in X-rays), and the limited number of photons (UV and X-ray range). Some development can be expected from high time-resolved cyclotron spectroscopy in the optical which will eventually allow a physical parameter eclipse mapping of the accretion spot.

Zeeman imaging of the magnetic white dwarf finally will almost definitely be possible in the near future as soon as codes for atmospheric models are fully developed and as an almost uncontaminated Zeeman spectrum is obtained. A small number of low-accretion rate polars suited for this investigation were discovered recently. More will follow, if the present census of polars, which is heavily biased by the optical identification of X-ray counterparts, i.e. biased towards high-accretion rate systems, is enriched by optically selected systems, which may emerge from e.g. variability surveys. Once successful, the Zeeman images of white dwarf stars in polars will shed new light on the field structure of the white dwarfs themselves as well as of their progenitors.

Acknowledgements

I would like to thank Sonja Vrielmann, Danny Steeghs, and Boris Gänsicke, who shared my interest and their codes for exploring the secrets of polars. Special thanks go to Keith Horne and Henk Spruit for providing their Doppler tomography codes. Gary Schmidt, Susanne Friedrich, Ralf Geckeler, Rüdiger Staubert, Gyula Szokoly, Robert Schwarz, and H.-J. Hagen kindly allowed me to present their data of AR UMa, V1432 Aql, EUVE J1429–38, BY Cam, and HS1023+3900, respectively, in this review. K.-H. Mantel operated the MCCP during three beautiful nights under a clear Calar Alto sky. Without the help of my colleagues at the AIP, Robert Schwarz and Andreas Staude, this study would not have been possible.

The support of Henri Boffin and Danny Steeghs, the organizers of the workshop on *Astro-Tomography*, is gratefully acknowledged.

I'm deeply indebted to Esther Heisse for sharing her thoughts and poetry with me at the time of writing this review.

This project was supported by the Bundesministerium für Bildung und Forschung through the Deutsches Zentrum für Luft- und Raumfahrt e.V. (DLR) under grant number 50 OR 9706 8. The responsibility for the content of the publication lies with the author.

References

1. J. Bailey, 1995, in: ASP Conf. Ser. 85, 10
2. J. Bailey, M. Cropper, 1991, MNRAS, 253, 27
3. A. Bobinger, 2000, A&A, 357, 1170
4. A. Bobinger, H. Barwig, H. Fiedler, K.-H. Mantel, D. Simic, S. Wolf, 1999, A&A, 348, 145
5. J.M. Brett, R.C. Smith, 1993, MNRAS, 264, 641
6. C.G. Campbell, A.D. Schwope, 1999, A&A, 343, 132
7. M.S. Catalán, A.D. Schwope, R.C. Smith, 1999, MNRAS, 310, 123
8. M. Cropper, K.D. Horne, 1994, MNRAS, 267, 481
9. S. Davey, R.C. Smith, 1992, MNRAS, 257, 476
10. S. Davey, R.C. Smith, 1996, MNRAS, 280, 481
11. M.P. Diaz, J.E. Steiner, 1994, A&A, 283, 508

12. M.P. Diaz, J.E. Steiner, 1994, ApJ, 425, 252
13. Donati, J.-F., N. Achilleos, J.M. Matthews, F. Wesemael, 1994, A&A, 285, 285
14. B.T. Gaensicke, K. Beuermann, D. de Martino, 1995, A&A, 303, 127
15. B.T. Gänsicke, D.W. Hoard, K. Beuermann, E.M. Sion, P. Szkody, 1998, A&A, 338, 933
16. P.J. Hakala, 1995, A&A, 296, 164
17. M.K. Harrop-Allin, P.J. Hakala, M. Cropper, 1999, MNRAS, 302, 362
18. M.K. Harrop-Allin, M. Cropper, P.J. Hakala, C. Hellier, T. Ramseyer, 1999, MN-RAS, 308, 807
19. C. Heerlein, K. Horne, A.D. Schwope, 1999, MNRAS, 304, 145
20. C. Hellier, 1999, ApJ, 519, 324
21. R.H. Kaitchuck, E.M. Schlegel, R.K. Honeycutt, et al., 1994, ApJS, 93, 519
22. A.R. King, 1989, MNRAS, 241, 365
23. J. Kube, B.T. Gänsicke, K. Beuermann, 2000, A&A, 356, 490
24. J. Kuijpers, J.E. Pringle, 1982, A&A, 114, L4
25. D.W. Latham, J. Liebert, J.E., 1981, ApJ, 246, 919
26. R. London, R. McCray, L.H. Auer, 1981, ApJ, 243, 970
27. S.B. Potter, P.J. Hakala, M. Cropper, 1998, MNRAS, 297, 1261
28. S.B. Potter, 2000, MNRAS, 314, 672
29. S.B. Potter, M. Cropper, P.J. Hakala, 2000, MNRAS, 315, 423
30. D. Reimers, H.-J. Hagen, 2000, A&A, 358, L45
31. D. Reimers, H.-J. Hagen, U. Hopp, 1999, A&A, 343, 157
32. T. Rousseau, A. Fischer, K. Beuermann, U. Woelk, 1996, A&A, 310, 526
33. G.D. Schmidt, P. Szkody, P.S. Smith, et al., 1996, ApJ 473, 483
34. G.D. Schmidt, D.W. Hoard, P. Szkody, F. Melia, R.K. Honeycutt, R.M. Wagner, 1999, ApJ, 525, 407
35. A.D. Schwope, K. Beuermann, S. Jordan, 1995, A&A, 301, 447
36. A.D. Schwope, K. Beuermann, H.-C. Thomas, 1990, A&A, 230, 120
37. A.D. Schwope, K. Beuermann, S. Jordan, H.-C. Thomas, 1993, A&A, 278, 487
38. A.D. Schwope, K. Beuermann, S. Jordan, 1995, A&A, 301, 447
39. A.D. Schwope, S. Mengel, K. Beuermann, 1997, A&A, 320, 181
40. A.D. Schwope, K.-H. Mantel, K. Horne, 1997, A&A, 319, 894
41. A.D. Schwope, R. Schwarz, J. Greiner, 1999, A&A, 348, 861
42. A.D. Schwope, R. Schwarz, A. Staude, C. Heerlein, K. Horne, D. Steeghs, 1999, ASP Conf. Ser. 157, 71
43. A.D. Schwope, M.S. Catalán, K. Beuermann, et al., 2000, MNRAS, 313, 533
44. A.W. Shafter, K. Reinsch, K. Beuermann, et al., 1995, ApJ, 443, 319
45. M.M. Sirk, S.B. Howell, 1998, ApJ, 506, 824
46. K.B. Sohl, G. Wynn, 1999, ASP Conf. Ser., 157, 87
47. K.A. Southwell, M.D. Still, R.C. Smith, J.S. Martin, 1995, A&A, 302, 90
48. H. Spruit, 1998, astro-ph/9806141
49. D. Steeghs, K. Horne, T.R. Marsh, J.F. Donati, 1996, MNRAS, 281, 626
50. K.G. Strassmeier, S. Lupinek, R.C. Dempsey, J.B. Rice, 1999, A&A, 347, 212
51. S. Vrielmann, A.D. Schwope, 2000, MNRAS, in press
52. F.K. Walter, S.J. Wolk, N.R. Adams, 1995, ApJ, 440, 834
53. J.K. Warren, M.M. Sirk, J.V. Vallerga, 1995, ApJ, 445, 909
54. D.T. Wickramasinghe, L. Ferrario, 2000, PASP, 112, 873
55. D.T. Wickramasinghe, B. Martin, 1985, MNRAS, 212, 353

Tomography of Magnetic Accretion Flows

G. Wynn

Department of Physics and Astronomy, University of Leicester, Leicester LE1 7RH, UK

Abstract. In this review I highlight the importance of Doppler tomography to the analysis of the dynamics of accretion flows in the presence of magnetic fields. The focus of the review will be the magnetic cataclysmic variable stars, which harbour a large variety of magnetic phenomena. Doppler tomography along with observations of the spin state of the accreting star have led to the discovery of new forms of magnetic accretion flow within these binaries. I present a number of examples of these discoveries.

1 Introduction

Accretion flows are prevalent throughout the Universe. They are thought to play a pivotal role in star and planet formation, and drive the huge luminosities of active galactic nuclei. However many fundamental aspects of accretion physics remain unknown. In particular, we lack a firm understanding of accretion flows in the presence of magnetic fields. The process of accretion is perhaps best understood in the context of interacting binary stars. These systems are useful laboratories for the study of accretion processes since they reveal more about themselves than other astronomical objects. Consequently the focus of this review will be the role of Doppler tomography as a diagnostic on the nature of magnetic accretion flows in binary star systems.

The magnetic cataclysmic variable stars (mCVs) are particularly useful in the study of the process of accretion in the presence of intense magnetic fields. These objects are interacting binary stars in which a magnetic white dwarf (WD) accretes mass from a late-type companion star. The white dwarfs within these systems have large magnetic moments ($\mu_1 \sim 10^{32} - 10^{35}$ G cm^3) and magnetic stresses have a pervasive influence on the accretion dynamics. In contrast, neutron stars in the pulsing X-ray binaries have much lower magnetic moments ($\mu_1 \sim 10^{26} - 10^{30}$ G cm^3), and magnetic stresses only become important close to the surface of the neutron star. In this case accretion is known to take place via an extended disc. The accretion flows within mCVs take on a much wider variety of forms. However, these systems are usually divided into two classes: the AM Herculis stars (or polars) and the intermediate polars (or DQ Herculis stars). The main properties of these systems are summarized in Table 1 (a comprehensive review may be found in [47]). It is clear that the two classes are distinct. In particular, the rotational periods of the WDs ($P_{\rm spin}$) in the AM Her stars are observed to be closely locked to the orbital period ($P_{\rm orb}$), whereas the WDs within the intermediate polars rotate more rapidly than the orbital period. This period-locking in the AM Her stars, which contain the most strongly magnetic WDs, is thought to come about because the interaction between the magnetic fields of the two stars in the binary is able to overcome the spin-up torque of the accreting matter (see e.g. [11]). This suggests that magnetic effects are important throughout these objects. Indeed, observations of the AM Her stars paint a very different picture of the gas flow when compared to the non-magnetic CVs or the pulsing X-ray binaries. Doppler tomograms, for instance, show no sign of an accretion disc, but clearly show the signature of a magnetized accretion stream (see the review by Schwope in this volume for some examples of this). The intermediate polars on the other hand, represent the various systems which fill the phase space between the strongly magnetic AM Her stars and the non-magnetic CVs. In this review I will discuss the role of Doppler tomography in determining the nature of the accretion flow in the mCVs, concentrating mainly on the asynchronous intermediate polars.

Table 1. The properties of the magnetic cataclysmic variable stars by class

	AM Her stars	intermediate Polars
orbital periods	$P_{orb} \lesssim 2$ hours in general	$P_{orb} \gtrsim 3$ hours in general
WD surface field strength (B_*)	$\gtrsim 10$ MG	$\lesssim 10$ MG
magnetic moment ($\mu_1 \sim B_* R_1^3$)	$\gtrsim 10^{34}$ G cm^3	$\lesssim 10^{34}$ G cm^3
spin-orbit synchronization	$P_{spin} = P_{orb}$	$P_{spin} < P_{orb}$
mode of accretion	magnetized accretion stream	various

2   Magnetic Accretion Flows

The accretion flow between the two stars in a mCV is ionized by collisions as well as by the strong X-ray flux which emanates from the flow very close to the WD surface. Consequently, the accretion flow is highly conducting and motion relative to the magnetic field will cause magnetic stresses to alter its dynamics. The equation of motion of the gas in the binary system can be written as

$$\frac{\partial \mathbf{v}}{\partial t} + (\mathbf{v}.\nabla)\mathbf{v} = -\nabla \Phi_R - 2\Omega \wedge \mathbf{v} - \frac{1}{\rho}\nabla P - \frac{1}{\rho}\nabla\left(\frac{B^2}{8\pi}\right) + \frac{1}{\rho R_c}\left(\frac{B^2}{4\pi}\right)\hat{\mathbf{n}} \quad (1)$$

where Ω is the orbital angular velocity, B is the local magnetic field strength, R_c is the local radius of curvature of the magnetic lines of force and Φ_R is known as the Roche potential and includes the effect of both gravitation and centrifugal force. The term $-2\Omega \wedge \mathbf{v}$ is the Coriolis force per unit mass. The magnetic field exerts an isotropic pressure $B^2/8\pi$, and carries a tension $B^2/4\pi$ along the magnetic lines of force. These magnetic stresses act as a barrier to the accretion flow, and impede the motion of plasma across field lines. The details of the flow of plasma through the magnetosphere are complex and non-linear. Its motion is dynamic and inherently three-dimensional, unlike the flow in non-magnetic systems which takes the form a planar disc. The shear velocities between the plasma and the magnetic field are highly supersonic and in some cases can be relativistic. These features, together with the plethora of magnetohydrodynamic instabilities which affect the flow, have made detailed modeling of the accretion process exceedingly difficult. However, we can gain a broad understanding of the process of accretion in the presence of strong magnetic fields by examining the following 3 timescales: the local dynamical timescale $t_{\text{dyn}} \sim (R^3/GM_1)^{1/2}$, the magnetic timescale t_{mag} and the viscous timescale t_{visc}. I will refer to these timescales throughout the rest of this paper.

Most treatments of magnetic accretion predict that, at some point, the plasma flow is broken up by the magnetic stresses and progressively stripped of gas (either by instabilities or reconnection events). The tenuous stripped gas is quickly magnetized and forced into field-aligned flow. This magnetized flow reaches the magnetic poles of the primary star at highly ballistic speeds, and passes through a strong shock before accreting on to the WD. The hot, post-shock gas is a source of intense X-ray emission, which is modulated on P_{spin}. Thus, mCVs allow direct observation of the spin rates of the WDs and offer a unique insight into the angular momentum distribution within the binaries.

We can gain an estimate of the extent of the region within which the magnetic field is expected to play a role in determining the dynamics of the accretion flow (ie. where $t_{\text{mag}} \lesssim t_{\text{dyn}}$) by comparing the ram pressure of accreting material (ρv^2) and the local magnetic pressure ($B^2/8\pi$). These two quantities are defined to be equal at the Alfvén radius (R_a). Assuming a spherically symmetric accretion flow without significant pressure support we can deduce the spherical Alfvén radius (e.g. [6])

$$R_a \sim 5.5 \times 10^8 \left(\frac{M_1}{M_\odot} \right)^{1/7} R_9^{-2/7} L_{33}^{-2/7} \mu_{30}^{4/7} \text{ cm} \qquad (2)$$

where R_9 is the primary radius (10^9 cm), L_{33} is the system luminosity (10^{33} erg s^{-1}), and μ_{30} is the primary dipole moment (10^{30} G cm^3). Typically we find that R_a is greater than the system separation (a) in the AM Her stars, and magnetic effects are expected to dominate throughout the binary, as observed.

In the non-magnetic CVs the accretion process is unimpeded by the presence of a magnetic field. The accretion flow leaves the secondary star in the form of a thin stream from its surface (see [47] for a review of mass transfer via Roche lobe overflow). The point at which this stream flows from the secondary is termed the inner Lagrangian point (the L_1 point). The accretion stream adopts a circular, Keplerian orbit at a radius R_{circ} from the WD (because of energy dissipation via shocks in the gas flow). An accretion disc forms from the viscous evolution of the stream from R_{circ} in a time t_{visc}. In the case of a mCV we must consider the effect of the magnetic field on the disc formation process. If $R_a > R_{\text{circ}}$ then we have $t_{mag}(R_{circ}) < t_{visc}(R_{circ})$, and magnetic stresses will quickly (in a time $< t_{visc}$) dissipate the stream angular momentum and no accretion disc will form. It is possible to estimate R_{circ} from the conservation of angular momentum: $2\pi b^2/P_{orb} \sim (GM_1 R_{circ})^{1/2}$, where b is the distance from the WD to the L_1 point and M_1 is the mass of the WD. This results in the relation

$$\frac{R_{circ}}{a} \simeq (1+q)\left(\frac{b}{a}\right)^4 \qquad (3)$$

which is a function of the mass ratio $q = M_2/M_1$ only, here M_2 is the mass of the donor star. Typically we find $R_{circ} \lesssim a/10$ in mCVs. As noted above, $R_a \gg R_{circ}$ is satisfied in the AM Her stars and no accretion disc can form. In this case we have the hierarchy of timescales $t_{mag} \lesssim t_{dyn} \ll t_{visc}$ at L_1. In the case of the pulsing X-ray binaries, in which a magnetic neutron star is surrounded by an extensive accretion disc, one finds that $R_a \ll R_{circ}$ and hence $t_{mag}(R_{circ}) \gg t_{visc}(R_{circ})$, confirming that accretion disc formation is unaffected by the magnetic field. However, the inward spread of the accretion disc will eventually be halted by the magnetic field at some radius R_{in}. The usual definition of R_{in} is the point at which the magnetic field removes angular momentum from the disc at a greater rate than viscous stresses (e.g. [2])

$$-B_\phi B_z R^2|_{R_{in}} = \dot{M}\frac{d}{dr}(\Omega_K R^2)|_{R_{in}} \qquad (4)$$

where B_ϕ, B_z are the toroidal and poloidal field components respectively, and Ω_K is the Keplerian angular velocity. The value of R_{in} has an important effect on the equilibrium value of P_{spin}, which in turn has an important effect on the dynamics of the accretion flow. I shall return to this point in a later section. In the case of the asynchronous mCVs we find that in many cases $R_a \gtrsim R_{circ}$, which greatly complicates the accretion dynamics in these systems. A number of models have been put forward in an attempt to explain these systems. The 'standard' model of the accretion flow within the intermediate polars assumes them to be the WD analogue of the pulsing X-ray binaries: accreting via an extended accretion disc which is disrupted inside R_{circ}. However it has become clear (e.g. [8]) that the intermediate polars represent a variety of different magnetic phenomena. This realization has been prompted by observational techniques such as Doppler tomography, along with new methods for the theoretical treatment of the gas flow within the intermediate polars. In the following sections I will concentrate on these new results.

3   The Importance of Spin

The spin and orbital periods of the mCVs are shown in Fig. 1. The AM Her systems can be seen to follow the synchronization line along which $P_{spin} = P_{orb}$. The asynchronous systems occupy a wide range of parameter space between $10^{-3} \lesssim P_{spin}/P_{orb} \lesssim 1$. This fact alone hints at the variety of magnetic flows which are present within these systems.

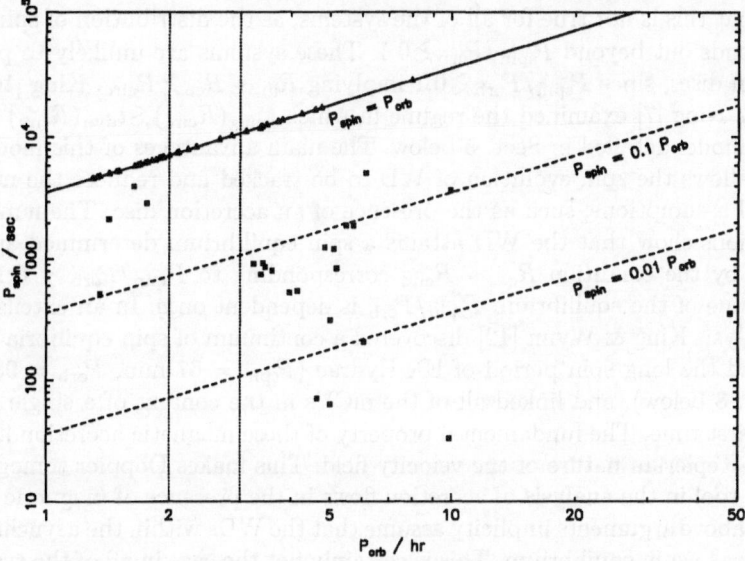

Fig. 1. Orbital and spin periods of the asynchronous mCVs: the AM Her stars are represented by triangles and the intermediate polars by squares. The vertical dotted lines represent the boundaries of the CV orbital period gap.

The spin periods of asynchronous mCVs allow us a unique insight into the angular momentum balance with the binary systems, and are a crucial diagnostic tool when comparing theoretical models. The spin rate of a magnetic WD accreting via a disc reaches an equilibrium when the rate at which angular momentum is accreted by the white dwarf is balanced by the braking effect of the magnetic torque on the disc close to R_{in}. Because of the complex nature of the disc-magnetosphere interaction, most models assume that the system is axisymmetric and in steady state. Many of these models find that the point at which the accretion disc is disrupted (R_{in}) is very close to the WD co-rotation radius (R_{co}). The radius R_{co} is defined to be the radius at which the magnetic field rotates at the same rate as the local Keplerian frequency, and hence $R_{co} = (GM_1 P_{spin}^2/4\pi^2)^{1/3}$ (see e.g. [13]). The relation $R_{in} \sim R_{co}$ implies that a small magnetosphere (low μ_1) results in fast equilibrium rotation and vice versa. Specifically, for a magnetic system with a truncated accretion disc we expect $R_{in} \sim R_{co} \ll R_{circ}$, which translates into the relation $P_{spin}/P_{orb} \ll 0.1$. This is certainly satisfied by the pulsing X-ray binaries where $P_{spin}/P_{orb} \lesssim 10^{-6}$. Figure 1 clearly shows that some of the asynchronous mCVs also follow this relation. Notably, in GK Per ($P_{spin} = 381$ s, $P_{orb} = 48$ hr) P_{orb} is so long that $R_a \ll R_{circ}$, and in DQ Her ($P_{spin} = 71$ s, $P_{orb} = 4.65$ hr) the very short P_{spin} presumably indicates a rather low magnetic field, leading to the same conclusion. It would seem then that some mCVs do indeed conform to the standard model and resemble the pulsing X-ray binaries in accreting via an extended Keplerian disc, which is disrupted close to the WD surface. However it is also clear from the figure that this is not true for all of the systems, as the distribution of spin periods extends out beyond $P_{\rm spin}/P_{\rm orb} \gtrsim 0.1$. These systems are unlikely to possess accretion discs, since $P_{\rm spin}/P_{\rm orb} \gtrsim 0.1$ implying $R_{\rm in} \sim R_{\rm co} \gtrsim R_{\rm circ}$. King [10] and Wynn & King [7] examined the regime in which $t_{\rm mag}(R_{\rm circ}) \lesssim t_{\rm dyn}(R_{\rm circ})$ utilizing the model detailed in Sect. 5 below. The main advantages of this model are that it allows the spin evolution of WD to be tracked and reduces the number of initial assumptions, such as the presence of an accretion disc. The numerical calculations show that the WD attains a spin equilibrium determined approximately by the condition $R_{\rm co} \sim R_{\rm circ}$ corresponding to $P_{\rm spin}/P_{\rm orb} \sim 0.1$. The exact value of the equilibrium $P_{\rm spin}/P_{\rm orb}$ is dependent on q. In an extension of this analysis King & Wynn [12] discovered a continuum of spin equilibria which explained the long spin period of EX Hydrae ($P_{\rm spin} = 67$ min, $P_{\rm orb} = 98$ min; see Sect. 8 below), and linked all of the mCVs in the context of a single model for the first time. The fundamental property of these magnetic accretion flows is the non-Keplerian nature of the velocity field. This makes Doppler tomography an ideal tool in the analysis of accretion flows in the presence of magnetic fields.

The above arguments implicity assume that the WDs within the asynchronous mCVs are in spin equilibrium. This is certainly not the case in all of the systems: very short spin period of AE Aqr ($P_{\rm spin} = 33$ s, $P_{\rm orb} = 9.88$ hr) does not imply the presence of a Keplerian accretion disc as the WD is currently a long way from spin equilibrum (see Sect. 7 below). This once again highlights the importance of relaxing the assumptions implicit in theoretical models of magnetic systems.

In the following sections I will highlight the role of Doppler tomography in the analysis of magnetic CVs. This technique, along with a generalized theoretical model has led to the discovery of new forms of magnetic accretion flow. After a summary of tomography as applied to magnetic systems, and a brief description of the theoretical models, I will discuss the cases of AE Aqr, EX Hya and RX1238. All of these systems have highlighted the power of Doppler tomography, along with information on the spin state of the WD, in unraveling the form of magnetic accretion flows. In Sect. 6 I highlight the importance of the AM Her stars in determining the details of the theoretical model.

4   Doppler Tomography of Magnetic CVs

Many of the attempts to examine the form of the complex accretion flows within the asynchronous mCVs have concentrated on the observed optical and X-ray power spectra of their light curves [9,16,17,19]. The systems show significant power at a number of frequencies including the spin (ω), orbital (Ω) and beat ($\omega - \Omega$) frequencies as well as a number of combinations of these. The presence of the beat frequency in many of the observed X-ray power spectra is important as it indicates that accretion rate on to the WD is controlled by some structure which is fixed in the orbital frame of the binary. However, the analysis of the power spectra failed to resolve the uncertainty about the basic accretion kinematics within the binaries. The main reason was that any theoretical model of the accretion flow which includes some sort of asymmetry about the white dwarf (which is static in the orbital frame) is able to account for the observed frequencies. This degeneracy is highlighted by the various models which have been put forward to explain the observations: e.g. spiral arms in a truncated accretion disc [15], an accretion stream which overflows the surface of a disc [1] and an accretion stream which 'switches' between the magnetic poles of the WD as it rotates relative to the secondary star [19].

In the face of this basic uncertainty and the large diversity in the possible models any further information on the kinematics of the accretion flow is crucial. In this respect the insight into the velocity field of the accretion flow offered via Doppler tomography is an essential component of the analysis of magnetic systems. To date, published tomograms exist for 12 of the AM Her systems and 12 of the intermediate polars. Axel Swchope's review within this volume includes a number of tomograms of the AM Her systems. These data show a marked contrast to the tomograms of disc accreting binaries (see e.g. the review by Danny Steeghs in this volume) and confirm the idea that these systems accrete via a stream and do not contain accretion discs. However, magnetic systems pose a problem in that they violate one of the 'axioms' of Doppler tomography presented by Tom Marsh in his review in this volume: that all motion is confined to the orbital plane of the binary system. This makes the interpretation of the tomograms of magnetic systems additionally complex. In his review Marsh points out that the best method of interpreting observed tomograms is to use a theoretical dataset to distinguish between models of accretion flows. In this respect theoretical models can be used to predict trailed spectograms, which in turn can be used to produce tomograms for direct comparison with observation. In this way the models can be used to determine the effects of motion out of the orbital plane. In order to facilitate this, it is important to develop a theoretical model which is flexible enough to produce tomograms from a variety of different flow regimes. I describe one such model in Sect. 5 below.

The fact that the observed line fluxes of the mCVs are modulated on the spin and, in some cases, beat frequencies allows a further extension of the Doppler tomography technique. To resolve the variations in the tomogram which occur as the white dwarf rotates, it is possible to compute Doppler maps as a function of the orientation of the white dwarf relative to the binary system. This was performed by Marsh and Duck [14] on observations of the intermediate polar FO Aqr. They termed this technique 'stroboscopic Doppler tomography', as it is analogous to the use of a stroboscope in freezing the motion of rotating objects. The results of this analysis were intriguing. The beat-resolved Doppler images showed a bright spot of emission from the region of the gas stream and secondary star, which varied in both position and brightness as the WD rotates. This motion could be traced back to complex features in the trailed spectra. The movement in the spot position is consistent with an azimuthally extended structure extending $\sim 120°$ around the WD, which is illuminated as the radiation from the magnetic poles sweeps through the system. The data show no evidence for an accretion disc. This technique revealed a lot of information which was encrypted within the trailed spectra, and otherwise difficult to access. In

Sect. 8 below I present new results using stroboscopic Doppler images of the intermediate polars EX Hya and RX1238.

5 A Model for Magnetic Accretion Flows

The equation of motion for gas flow within mCVs (Eq.(1)) includes magnetic pressure and tension terms. King [10] and Wynn & King [7] constructed a model of accretion flows in the mCVs by assuming that material moving through the magnetosphere interacts with the local magnetic field via a shear velocity-dependent acceleration. This is analogous to the assumption that the magnetic stresses are dominated by the tension term in Eq. (1), giving a magnetic acceleration

$$\mathbf{a}_{mag} \simeq \frac{1}{R_c \rho} \left(\frac{B^2}{4\pi} \right) \hat{\mathbf{n}} \tag{5}$$

which can be written in the general form

$$\mathbf{a}_{mag} \simeq -k(r, t, \rho)[\mathbf{v} - \mathbf{v}_f]_\perp \tag{6}$$

where $\mathbf{v}$ and $\mathbf{v}_f$ are the velocities of the material and field lines, and the suffix $\perp$ refers to the velocity components perpendicular to the field lines. The coefficient k contains the details of the plasma-magnetic field interaction, and will be a function of position, time and gas density in general. This can found, at least in approximate form, for the limiting cases of diamagnetic and magnetized accretion flows [22]. Some of the pros and cons of this model are summarized in Table 2. The treatments of King [10] and Wynn & King [7] assume that the plasma flow is inhomogeneous, diamagnetic and interacts with the field via a surface drag force, characterized by the timescale

$$t_{mag} \sim c_A \rho_b l_b B^{-2} \frac{|\mathbf{v}_\perp|}{|\mathbf{v} - \mathbf{v}_f|_\perp} \tag{7}$$

where c_A is the Alfvén speed in the medium surrounding the plasma, ρ_b is the plasma density, and l_b is the typical length-scale over which field lines are distorted. Plasma will exchange orbital energy and angular momentum with the field on this timescale, which is dependent on P_{spin} since $|\mathbf{v}_f| \sim 2\pi R / P_{spin}$. Material at radii greater than R_{co} will experience a net gain of angular momentum and be ejected from the binary or captured by the secondary star. On the other hand, material inside R_{co} will lose angular momentum and be accreted by the WD. An equilibrium will result when these angular momentum flows balance. Hence, this simple prescription allows the spin evolution of the WD in mCVs to be followed in detail. Once spin equilibrium is attained the predicted flow pattern can be directly compared to observation by computing theoretical Doppler tomograms. In the next sections I will describe the results of this process in a number of important cases.

Table 2. Pros and cons of the magnetic model

Pros	Cons
simplifies complex MHD	some physics is lost
easy to add to hydro codes	k is model dependent
spin evolution of WD	poor resolution close to WD

6   The AM Herculis Stars

The AM Herculis stars are unique among interacting binary star systems in that
magnetic fields dominate the motion of gas throughout the system. There is no
observational evidence for the presence of accretion discs within these systems.
The generally accepted model is that the accretion flow takes the form of a
magnetized stream. This stream flows from the L_1 point on the secondary star
down to the magnetic poles of the white dwarf. Close to the L_1 point the flow
is not necessarily along the magnetic field lines. This is commonly referred to
as the ballistic stream. However, in this region magnetic stresses may still affect
the velocity field of the flow, causing it to differ from the case of 'free' flow.
There is an abundance of observational evidence which supports this model.
In particular Doppler tomography shows clear evidence for the presence of a
magnetized accretion stream. Axel Schwope presents a number of examples of
these Doppler images in his review elsewhere in this volume. In this section I
will briefly outline the importance of Doppler tomography of the AM Her stars
as a probe of the plasma-magnetic field interaction.

Figure 2 shows the results of a calculation using the magnetic model de-
scribed in Sect. 5, in the regime in which $t_{\mathrm{mag}} \lesssim t_{\mathrm{dyn}} \ll t_{\mathrm{visc}}$ at L_1. The Doppler
images which result from this flow are also shown in the figure. The tomograms
are simply weighted by mass density, as is the case for all theoretical Doppler
images presented in this paper. These results are based on the parameters of
HU Aqr. The AM Her stars are observed to suffer phases of high and low mass
accretion rate (termed the high and low states). During the high and low states
the mass flow, and hence the density, of the accretion stream is altered. A lower
stream density implies (cf. equation 7) that magnetic effects are more important, and the resultant change in the trajectory of the accretion stream is seen
in the observational tomograms in the various accretion states (see the review
by Schwope). The theoretical calculations presented in Fig. 2 reflect this change
in the accretion rate ($\dot{M}$) by a change in t_{mag} (which is reduced by a factor 5
in the low state for the means of this example). The tomograms reproduce the
gross properties of the observational examples presented by Schwope, including
the change in velocity space trajectory between the high and low states. Moreover, the bright spot close to the position of the L_1 point is also reproduced
in the theoretical images, even though the secondary star is not included in the
calculation. The bright spot is caused by a stagnation point in the magnetic flow
close to the L_1 point. A similar analysis was performed by Heerlein et al. [7] on

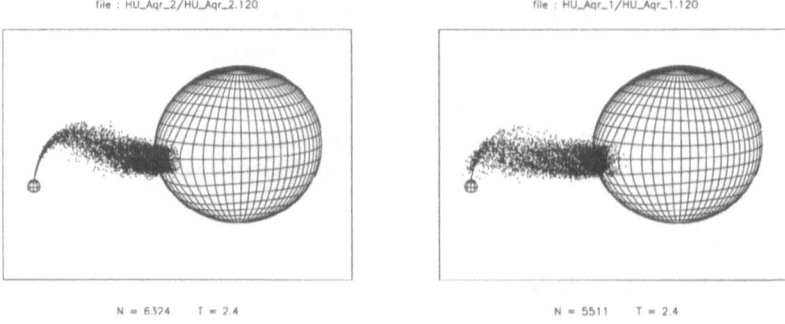

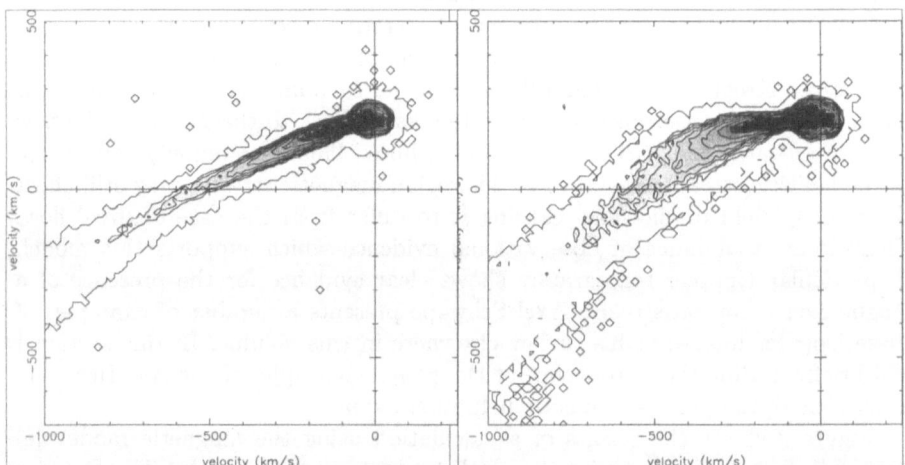

Fig. 2. Theoretical calculations of the gas flow within the AM Her stars (top), and the resultant Doppler tomograms (bottom). The results for low and high accretion states are shown in the left and right hand set of panels respectively.

HU Aqr using a different model, which involved gas stripping from the accretion stream, with encouraging results.

The tomograms of the AM Her stars offer a unique opportunity to test the magnetic model described above. This is because the trajectory of accretion stream in the velocity space of the Doppler images is very sensitive to the inclination angle of the magnetic dipole of the WD and the magnetic k parameter only. In some cases (such as HU Aqr) we have observational estimates of the dipole inclination. This allows the form of k parameter to be tested directly via theoretical tomograms such as those in Fig. 2. The high and low accretion states then enable the dependence of k on mass accretion rate to be determined. In this respect Doppler tomography of the AM Her stars offers a method of calibrating theoretical models of magnetic accretion flow. This work is already underway.

7   Case Study 1: AE Aquarii

AE Aquarii is an unusual cataclysmic variable star. It has an orbital period of 9.88 hours, amongst the longest known for a CV, and the WD rotational period is 33s, amongst the shortest known in a CV. The rapid rotation of the WD implies that it has a relatively weak magnetic moment (see Sect. 3). This fact, along with the long P_{orb} (and hence large R_{circ}), led to the assumption that AE Aqr was a CV analogue of the pulsing X-ray binaries. In this picture the WD accretes mass via a truncated accretion disc (as detailed above). However, a number of properties of AE Aqr remain unexplained by this model. The rotational period of the WD in AE Aqr was discovered to be increasing at a rate $\dot{P}_{spin} = 5.64 \times 10^{-14}$ s s^{-1} by de Jager et al. [3]. This spin down of the WD implies a spindown power $L_{spin} = -I\omega\dot{\omega} \simeq 6 \times 10^{33}$ erg s^{-1} (where I is the moment of inertia of the WD), which is a factor ~ 100 larger than the observed X-ray luminosity [4], and hence any possible accretion torque. Moreover if the X-ray luminosity is interpreted as accretion luminosity, the estimated rate of accretion on to the WD is a factor $\sim 10^3$ lower than expected from a 10 hour binary. These properties are puzzling and AE Aqr has many more unusual features, but perhaps the most problematical was the H_α Doppler tomogram presented in [21]. AE Aqr shows strong, highly variable, single-peaked Balmer emission lines and the resultant tomograms show no evidence of an accretion disc. The observed H_α tomogram is reproduced in Fig. 3, and shows that the emission is primarily in the lower left quadrant (v_x and v_y both negative), which forms the centre of a broad, asymmetric peak. It is normally very difficult to account for a tomogram of this form, because there is no obvious part of the binary system which moves with this velocity vector.

Wynn, King and Horne [21] and Eracleous & Horne [5] show that the spin-down of the WD in AE Aqr and its unusual Doppler tomogram could be explained by the rapidly spinning WD acting as a magnetic propeller. Most ($\sim$ 99%) of the transferred mass is centrifugally ejected from the binary on a time-scale $\lesssim t_{dyn}$. The required condition for this is $t_{mag}(R_{circ}) \lesssim t_{dyn}(R_{circ})$ and $R_{circ} \gg R_{co}$. An example of the flow pattern is shown in Fig. 4, and its Doppler image is presented in the lower panel of Fig. 3. What is noticeable is that the magnetic propeller model is successful in reproducing the features of the observed tomogram: i.e. broad, single-peaked emission lines centered at low velocities in the lower left hand quadrant. In the theoretical tomogram the most densely populated areas are located near the L_1 point, and where the gas is decelerating as it leaves the binary. Moreover the trajectory of the gas stream, and hence the form of the computed tomogram, was found to be very sensitive to the magnetic coefficient k. With k thus constrained by the Doppler images, the mass transfer rate was constrained by the spin down power: the rotational energy of the WD is carried of via the kinetic energy of the ejected gas. The mass transfer rate required to explain the spin down power is close to that expected for a 10 hour binary system. The low X-ray luminosity is explained because of the low rate of accretion on to the WD.

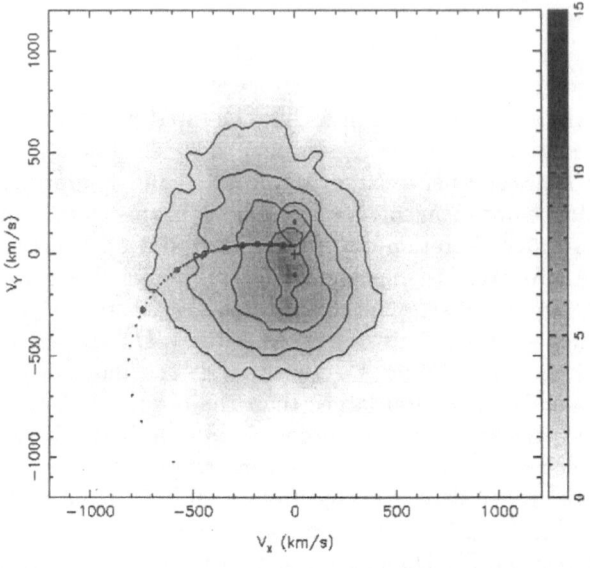

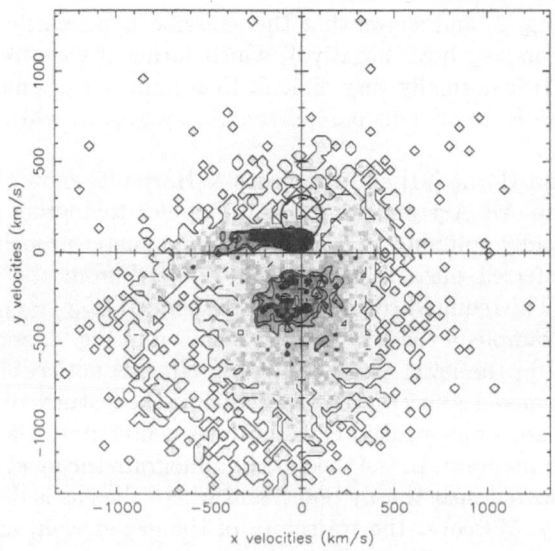

Fig. 3. Top: the observed H_α tomogram of AE Aquarii. Bottom: the theoretical, density-weighted tomogram of the flow depicted in Fig. 4

In the case of AE Aqr observations of the spin state of the WD along with tomography led to the discovery of a new form of magnetic gas flow in mCVs. In order to explain these observations it was important to employ a theoretical

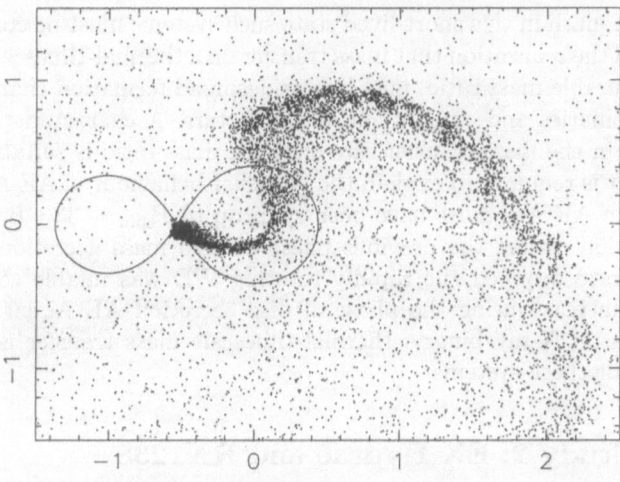

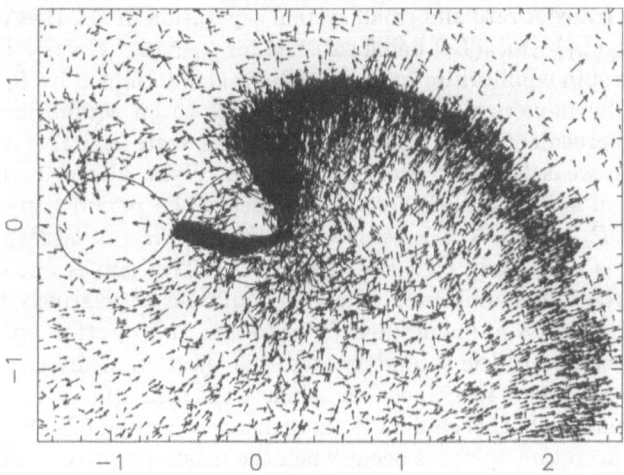

Fig. 4. The magnetic propeller in AE Aqr. The panels show a snapshot of the mass distribution (top) and velocity vectors (bottom) of the flow. The magnetic WD ejects ∼ 99% of the mass transfer stream. The scaling is units of the binary separation.

model which required few implicit assumptions. In particular the requirements that an accretion disc be present in the system and that the WD is in spin equilibrium were relaxed. Further results are still emerging from this discovery. We know that the WD in AE Aqr is spinning down on a timescale which is relatively short relative to the orbital evolution timescale of the binary. Since we

observe AE Aquarii in this short-lived state such systems must be common. This
has prompted the realization that mass transfer on a thermal-timescale, provoked
by an unfavourable mass ratio, may be a widespread formation channel for low-
mass X-ray binaries and cataclysmic variable stars. A characteristic feature is
a sharp drop in the mass transfer rate once the mass ratio is suitably reversed,
precisely what is required to explain the propeller behaviour of AE Aquarii. The
white dwarf in AE Aqr must have been spun up to $P_{\mathrm{spin}} \sim 33$ s by an episode
of disc accretion during the previous phase of high mass accretion rate. Once
the transfer rate dropped the rapidly rotating WD was unable to accept the
transferred matter. During the phase of disc accretion AE Aquarii may have
been a super-soft X-ray binary: thermal timescale mass transfer is a favoured
explanation for these systems.

8   Case Study 2: EX Hydrae and RX1238

The two systems EX Hya ($P_{\mathrm{spin}} = 67$ min, $P_{\mathrm{orb}} = 98$ min) and RX1238
($P_{\mathrm{spin}} = 36$ min, $P_{\mathrm{orb}} = 90$ min) are unlikely to contain accretion discs since their
equilibrium $P_{\mathrm{spin}}/P_{\mathrm{orb}}$ implies $R_{\mathrm{co}} \gg R_{\mathrm{circ}}$. The fact that both of these systems
lie below the orbital period gap (2 - 3 h), lead King & Wynn [12] to suggest that
the low mass transfer rate and small system separation in EX Hya results in the
condition $t_{\mathrm{mag}}(L_1) \lesssim t_{\mathrm{dyn}}(L_1)$ being satisfied for $\mu_1 \gtrsim 10^{33}$ G cm^3. This then implies that the spin equilibrium in these systems is determined by $R_{\mathrm{co}} \sim b$, which
is confirmed by numerical experiment and leads to an equilibrium $P_{\mathrm{spin}}/P_{\mathrm{orb}}$
in excellent agreement with observation. In this state EX Hya resembles an
asynchronous, weak-field AM Her star. King & Wynn also show that there is
a continuum of spin equilibria for systems below the period gap, between the
$R_{\mathrm{co}} \sim R_{\mathrm{circ}}$ ($P_{\mathrm{spin}}/P_{\mathrm{orb}} \sim 0.1$, see Sect. 3) and the $R_{\mathrm{co}} \sim b$ ($P_{\mathrm{spin}}/P_{\mathrm{orb}} \lesssim 0.7$)
equilibria. It is very likely that the system RX1238 occupies one of the states
close to the $R_{\mathrm{co}} \sim b$ equilibrium. The WD equilibria at relatively long P_{spin} in
EX Hya and RX1238 is maintained by accretion and ejection episodes as the
WD rotates relative to the secondary star. The ejection of material close to L_1
with high specific angular momentum balances the spin up torque caused by the
accretion of matter. In this case much more mass is accreted ($\gtrsim 90\%$) rather than
ejected. The accretion episodes occur when the magnetic poles point toward the
secondary star, and the ejection episodes occur as the poles pass over the limb
of the WD (with respect to the secondary). This results in two instances of each
episode per beat cycle. The form of the gas flow between the stars in EX Hya
and RX1238 is depicted in Fig. 7.

Figures 5 and 6 present, as yet unpublished, H_α Doppler tomograms of
EX Hya (Wheatley and Maxted, private communication) and RX1238 (Hellier,
private communication). Each of these sets of images have been computed as a
function of the orientation of the WD, as in the stroboscopic Doppler tomography technique described above, and are filtered back-projections. Both sets
of images are remarkably similar in that they show a hook-like emission region
stretching anti-clockwise from the position of the L_1 point. This emission region

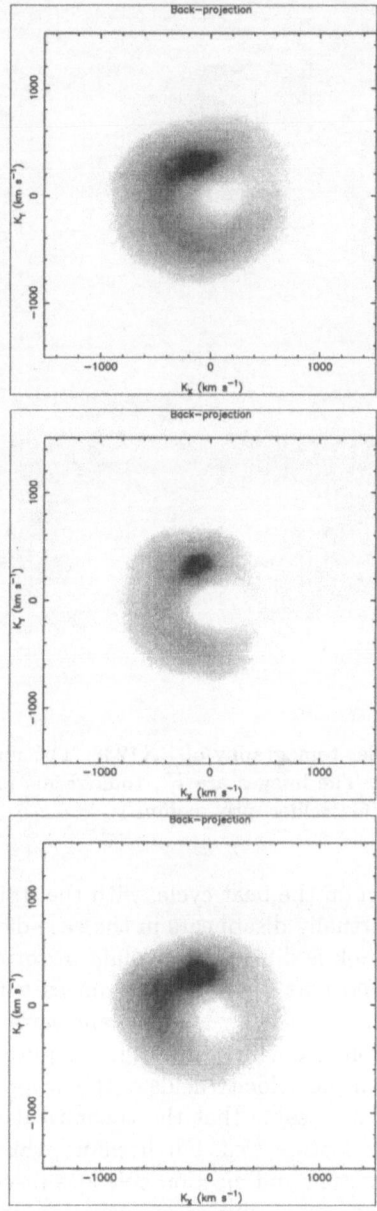

Fig. 5. Stroboscopic Doppler tomography of EX Hya. The images are H_α tomograms produced by the filtered back-projection technique (P. Wheatley and P. Maxted, priv. comm.)

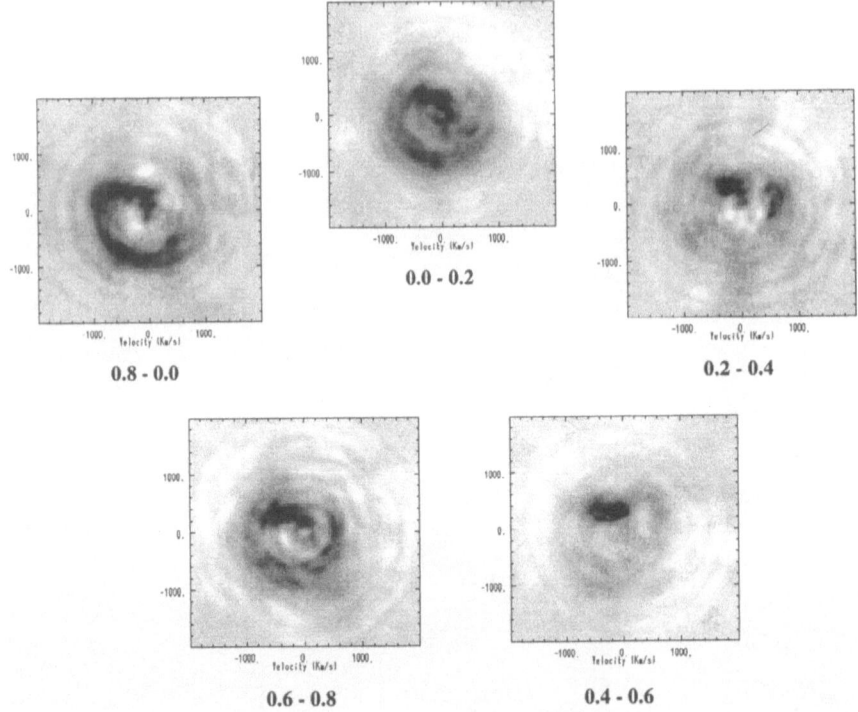

Fig. 6. Stroboscopic Doppler tomography of RX1238. The number refer to the beat phase range of each image. The images are H_α tomograms produced by the filtered back-projection technique (C. Hellier, priv. comm.)

shows significant variation on the beat cycle: with the 'tail' of the hook becoming less pronounced (it virtually disappears in the case of RX1238) through the cycle. The low velocity hook is difficult to explain in terms of the conventional disc accreting model, as no part of the binary moves with this set of velocity vectors. The fact the tomograms show a significant variation on the beat cycle suggests that this variation is magnetic in origin. Figure 8 shows the Doppler tomograms computed from the velocity fields of the accretion flows presented in Fig. 7. It is immediately noticeable that the theoretical tomograms reproduce the low-velocity, hook-like feature well. Furthermore, this feature varies on the beat cycle due to the accretion and ejection episodes discussed above. A tomogram constructed from the system during an accretion and an ejection episode is presented in Fig. 7. During the ejection episode the 'tail' of the hook is less pronounced. This agrees remarkably well with the observed data, and we could identify the phases during which the hook-like emission region is weaker as the ejection episodes which are required to keep the system in equilibrium. Figure 9 shows how this occurs in terms of the trajectories of gas elements through velocity space. Despite this success one problem remains: the observed tomograms

file : 1238.019

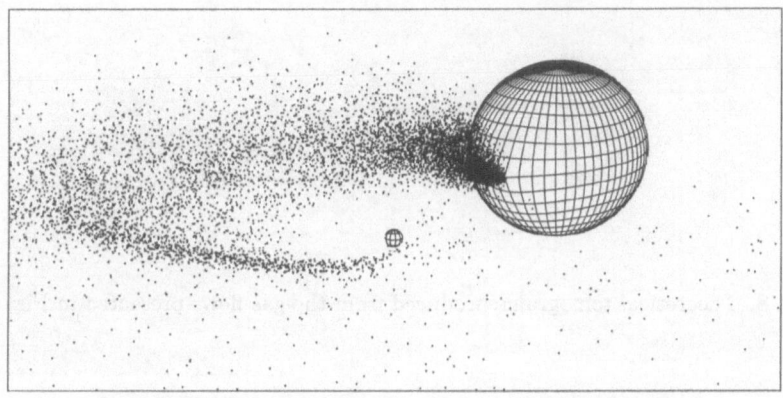

N = 9996   T = 1.9

file : 1238.021

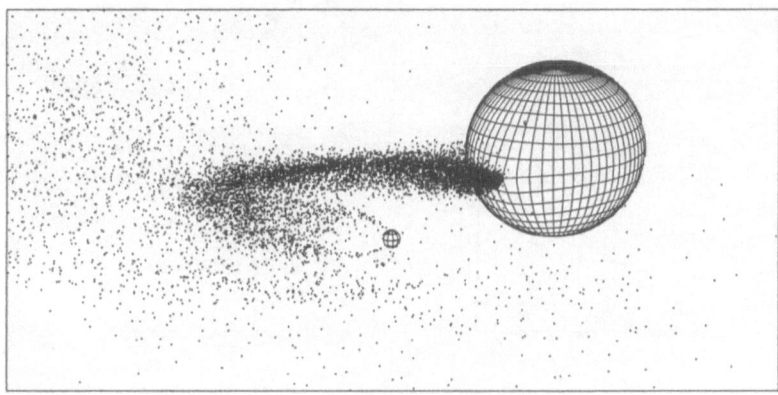

N = 10665   T = 2.1

Fig. 7. The expected gas flow pattern in the mCVs EX Hya and RX1238. The top panel shows an ejection event and lower panel shows the system during an accretion phase.

seem to indicate one ejection episode per beat cycle, whereas the theoretical data predict two. A possible explanation for this discrepancy is the neglect of illumination of the gas flow in the system. The line emission responsible for the observed tomograms could be driven in part by the illumination of the gas flow in the system by the X-ray emission from the white dwarf. This is also modulated on the beat cycle, and could be responsible for 'hiding' particular regions of the gas flow at certain phases.

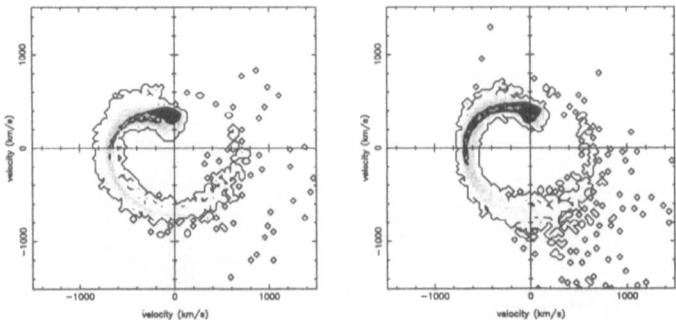

Fig. 8. Theoretical tomograms produced from the gas flows presented in Fig. 7.

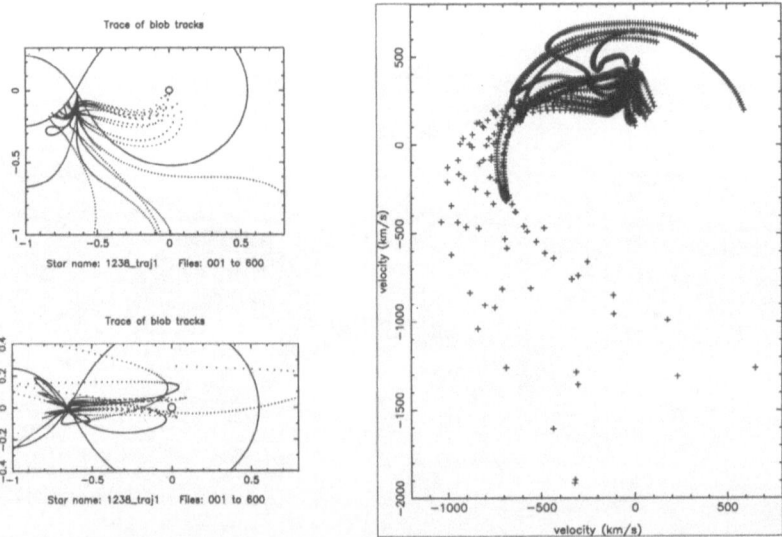

Fig. 9. Gas particle trajectories in real and velocity space. The panels on the left show the binary system from the top down with the orbit of the secondary being anti-clockwise (top left), and viewed from the leading side of the secondary's orbit at an inclination of 90° (bottom left). The particle trajectories are shown in the frame of the binary orbit. The right hand panel shows the trajectories of the particles through velocity space. This clearly shows the origin of the hook-like features in the H_α tomograms of EX Hya and RX1238.

9 Discussion

In this review I have attempted to highlight the importance of Doppler tomography, along with observations of the spin state of the accreting star, to the analysis of the dynamics of accretion flows in the presence of magnetic fields. We have reached the stage at which theoretical models can attempt to, simultaneously, predict both of these observational data sets. In doing so it is necessary to relax many assumptions about the form of the flow. This has led to the discovery of new forms of magnetic accretion flow such as the accretion and ejection events in EX Hya and RX1238, and the magnetic propeller in AE Aqr. Doubtless other forms of magnetic flow are yet to be uncovered. These discoveries have prompted further insights into the nature of the mCVs, and binary stars in general. In particular, the mass transfer history of AE Aquarii may prove to be very important to the study of the evolution of many types of interacting binary stars.

Doppler tomography has proved to be especially important in the analysis of gas flow around magnetic stars as it allows relatively easy access to information which is hidden in the trailed spectra. This is highlighted by the examples of stroboscopic Doppler tomography, which gives us the ability to examine the velocity field of the gas as a function of the orientation of the accreting star. This places tight constraints on theoretical models. In the case of the AM Her stars, the tomograms of many systems with different primary magnetic moments in the high and low accretion states offer us a unique opportunity to test and develop models of magnetic accretion flows. These models, which have been developed in an attempt to explain the gas flow within the mCVs, should find applications in other astrophysical systems in which accretion flow is affected by a strong magnetic field.

References

1. Allan A., Hellier C., Buckley D.A.H., 1999, In: Annapolis Workshop on Magnetic Cataclysmic Variables, ASP Conference Series, Volume 157, Coel Hellier & Koji Mukai (eds.), p. 57
2. Bath G.T., Evans W.D., Pringle J.E., 1974, MNRAS, 166, 113
3. de Jager O.C., Meintjes P.J., O'Donoghue D., Robinson E.L., 1994, MNRAS, 267, 577
4. Eracleous M., Halpern J., Patterson J., 1991, ApJ, 382, 290
5. Eracleous M., Horne K., 1996, ApJ, 471, 427
6. Frank J., King A.R., Raine D.J., 1992, Accretion Power in Astrophysics, 2nd edn., Cambridge University Press, Cambridge
7. Heerlein C., Horne K., Schwope A.D., 1999, MNRAS, 304, 145
8. Hellier C., 1996, In: Cataclysmic Variables and Related Objects, A.Evans and Janet H.Wood (eds.), Kluwer Academic Publishers, p. 143
9. Hellier C., 1992, MNRAS, 258, 578
10. King A.R., 1993, MNRAS, 261, 144
11. King A.R., Frank J., Whithurst R., 1991, MNRAS, 250, 152
12. King A.R., Wynn G.A., 1999, MNRAS, 310, 203
13. Li J., 1999, In: Annapolis Workshop on Magnetic Cataclysmic Variables, ASP Conference Series, Volume 157, Coel Hellier & Koji Mukai (eds.), p. 235
14. Marsh T.R., Duck S.R., 1996, New Astronomy, 1, 97
15. Murray J.R., Armitage P.J., Ferrario L., Wickramasinghe D.T., 1999, MNRAS, 302, 189
16. Norton A., 1993, MNRAS, 265, 316
17. Warner B., 1986, MNRAS, 219, 347
18. Warner B., 1995, *Cataclysmic Variable Stars*, Cambridge University Press

19. Wynn G.A., King A.R., 1992, MNRAS, 255, 83
20. Wynn G.A., King A.R., 1995, MNRAS, 275, 9
21. Wynn G.A., King A.R., Horne K.D., 1997, MNRAS, 286, 436
22. Wynn G.A., Leach R., King A.R., 2000, MNRAS, submitted

The Geometrical Configuration of Polars and Possible Reconstruction Artefacts of Eclipse Mapping Methods

J. Kube

Universitäts-Sternwarte Göttingen,
Geismar Landstraße 11, D-37073 Göttingen, Germany

Abstract. The eclipse mapping method allows one to reconstruct the brightness distribution along the accretion stream of polars. Since the geometry of accretion streams is more complicated than that of accretion disks in non-magnetic systems, it is essential to gain knowledge about the geometrical configuration of such a system. Here, I give a summary of the parameters with which the geometry can be described. I present some reconstruction artefacs that can be the results of wrong geometrical parameters.

1 Introduction

Polars are cataclysmic variable stars with a synchronously rotating magnetized white dwarf. Thus, the accreted matter does not form an accretion disk, but falls ballistically down until the magnetic pressure exceeds the ram pressure of the infalling material. From this stagnation region S on, the matter follows the magnetic field line passing through S. For simplicity, all models consider a dipole geometry of the magnetic field, although the real field configuration may be very different from that simple assumption.

The eclipse mapping method has been proven to be a successful tool for the reconstruction of the brightness distribution on accretion disks [18] (Baptista, this volume). Recently, full 3D-models of the more complicated accretion geometry of polars have been developed, allowing the application of the eclipse-mapping to this family of objects [2–5]. Besides this, analysis of the physical properties of the accretion stream imply slight changes in the simple ballistic-magnetic trajectory of the stream [6,7] if one considers the magnetic drag force. To compare these results with observations, it is neccessary to know the limitations and sensitivity of the reconstruction methods.

2 Geometrical Parameters

2.1 Model Geometry of Polars

The usual approximation for the geometry of the accretion stream of a polar is as follows: (1) Matter overflowing the secondary star's surface at the L_1-point moves on a ballistic trajectory towards the white dwarf. The stream is assumed to have a constant circular cross section [3,5] or to be a set of emitting points

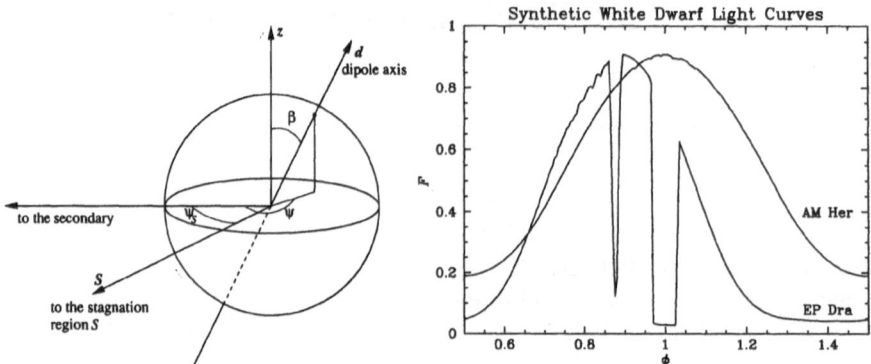

Fig. 1. Left: The three parameters which describe the position of the centered magnetic dipole and the field line on which the matter is accreted: tilt β, azimuth Ψ, azimuth of the stagnation region Ψ_S. **Right:** The position of the hot spot defines the difference between orbital and photometric phase (compare AM Her with EP Dra). In eclipsing systems (EP Dra), the eclipse width restricts (q, i), and the position of the dip gives hold to the position of the stagnation region. Synthetic light curves generated with the CVMOD code [5].

[4]. (2) At some point where the magnetic pressure equals the ram pressure the stream couples onto a single field line (stagnation region) and falls onto the white dwarf. The cross section is kept constant [3] or decreases according to the magnetic field geometry [5]. The field line is assumed to originate from a tilted and rotated dipole field. Thus, we have a "ballistic stream" and a "magnetic stream".

The position of the ballistic stream is given by the equations of motion [5], the configuration of the magnetic stream has to be described by three parameters (Fig. 1, left). The dipole tilt (β) and the angle between the line connecting the two stars, and the projection of the dipole axis to the orbital plane, (azimuth, Ψ) give the configuration of the magnetic dipole inside the white dwarf. Still, there is an infinite number of possible magnetic streams, from which we select the one that lies on the field line which intersects with the orbital plane at the stagnation region. The angle between the stagnation region and the center of the secondary star as seen from the center of the white dwarf we call Ψ_S, the azimuth of the stagnation region.

From all the binary parameters (period, mass etc.), we need the pair (i, q), which are inclination and mass ratio. All other parameters are only necessary for absolute calibrations of the emissivity of the resulting maps.

2.2 Interpretation of Observations

The determination of (i, q) is done by measuring the relative length of the eclipse of the white dwarf (Fig. 2). Having additional information about the mass ratio q from high resolution spectroscopy, one can derive the inclination. For well observed objects, one can have an accuracy of $\pm 2°$ for i and $\pm 20\%$ for q.

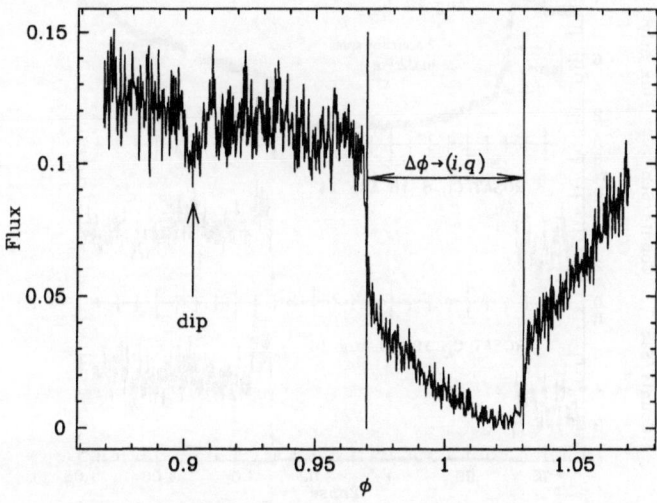

Fig. 2. The eclipse width constrains the pair (i, q) [18]. The position of the dip gives Ψ_S, since *beta* and Ψ are known from the light curve in Fig. 3. Data: HST-FOS UV light curve of UZ For.

The point where the magnetic stream hits the surface of the white dwarf is heated up to several 10^6 K. This makes this hot spot a UV and X-Ray source. Hence, X-Ray and UV light curves show a strong orbital variation which is due to the different viewing angles to the hot spot (Fig. 1, right). In eclipsing systems, one can use the timing of the ingress and the egress of the hot spots to determine their positions on the surface of the white dwarf. As long as accretion occurs only on one side of the orbital plane, one would expect to find a single hot spot. In some systems, e.g. in UZ For, two hot spots can clearly be seen, giving evidence to two-pole accretion (Fig. 3). Having the position of two hot spots, the position of the magnetic dipole is fixed (if the assumption of a centered dipole is correct).

Now we can take advantage from the fact that the accretion stream absorbs some of the radiation from the hot spot. This leads to a typical dip in most UV and X-ray eclipse light curves of polars. From analysis of the dip position (Fig. 2), the value for Ψ_S is fixed if β and Ψ are known. In those systems where a determination of all these three values has not yet been accomplished, one normally assumes $\Psi = \Psi_S$, which is a relatively poor approximation that astonishingly seems to work quite well with respect to eclipse mapping. An analysis of the long-time behaviour of the UV/X-ray-dip in UZ For can be found at [8].

Depending on the geometry, there are three different kinds of systems: (a) Those where the white dwarf is not eclipsed. Still most parts of the accretion stream could be eclipsed by the secondary and hence being mapped by the means of an eclipse mapping code. (b) Systems where the white dwarf is eclipsed (the classic description of "eclipsing"), but some parts of the accretion stream are not. This is probably the case for UZ For. Since no information about the

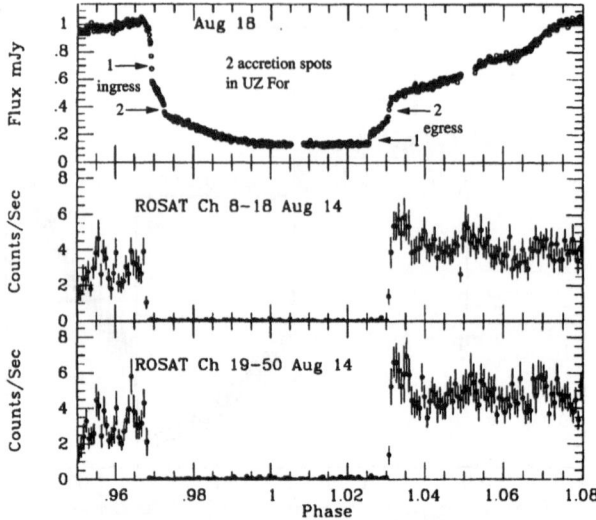

Fig. 3. UZ For shows the ingress and egress of two hot spots in the optical light curve (top). From the timing of these four events, one can derive the position of the accretion regions on the white dwarf. Original viewgraph taken from [9].

intensity distribution on the non-eclipsed parts of the stream is in the eclipse light curves, one cannot map details on these parts. (c) Systems where all the stream is eclipsed. HU Aqr is one of those systems. In these systems, the total eclipse helps in reconstructing the brightness distribution on the stream even if the geometry is not known too well.

3 The Eclipse Mapping Reconstruction Method and Its Limitations

The eclipse mapping methods is based on the fact that, in an orbiting system with high inclination, an object like the accretion stream is gradually eclipsed by another object, e.g. the secondary star. At a given time – written as orbital phase ϕ – one observes the flux $F(\phi)$. If V are all visible parts of emitting objects of the binary system, we can write

$$F(\phi) = \int_V I(r) f(n(r)) \, dS \tag{1}$$

where $I(r)$ is the intensity of an surface element at the position r and $n(r)$ is the normal vector on this surface element. $f(n(r))$ is the function describing the geometrical emission characteristics of the surface. For optically thick emission with no limb darkening, $f(n(r))$ becomes $\cos(\alpha)$, where α is the angle between $n(r)$ and the observer.

Since all mapping codes use grids of N finite surface elements, we can write the continuous formulation in (1) discretely as

$$F(\phi) = \sum_{i \in \mathcal{V}(\phi)} I_i A_i(\phi) \tag{2}$$

with I_i the intensity of the i-th surface element $(0 < i < N)$, $A_i(\phi)$ the projected surface of this surface element at orbital phase ϕ (assuming optically thick emission) and $\mathcal{V}(\phi)$ the set of visible surface elements at orbital phase ϕ.

Let $\Delta F_{1,2}$ be the flux difference between two different orbital phases ϕ_1 and ϕ_2. Using (2), $\Delta F_{1,2}$ reads like

$$\Delta F_{1,2} := F(\phi_1) - F(\phi_2) = \sum_{i \in \mathcal{V}(\phi_1)} I_i A_i(\phi_1) - \sum_{i \in \mathcal{V}(\phi_2)} I_i A_i(\phi_2) \tag{3}$$

For small phase differences, say $360° \cdot |\phi_1 - \phi_2| < 10°$, the projected surface areas don't differ much: $A_i(\phi_1) \approx A_i(\phi_2) =: A_i$. Let's assume for the following, that ϕ_1 and ϕ_2 are somewhere close to 0.95, which means that we are looking at the eclipse ingress of the accretion stream. With $\mathcal{I}_{1,2} := \mathcal{V}(\phi_1) \setminus \mathcal{V}(\phi_2)$, (3) becomes

$$\Delta F_{1,2} = \sum_{i \in \mathcal{I}_{1,2}} I_i A_i \tag{4}$$

For the egress, the respective formula is

$$\Delta F_{4,3} = \sum_{i \in \mathcal{E}_{3,4}} I_i A_i \tag{5}$$

From this equations, we can see that artefacts can be possible. We have to consider two cases: (i) If only the ingress of the accretion stream is used for eclipse mapping: The intensity decrease $\Delta F_{1,2}$ is attributed to all surface elements with $i \in \mathcal{I}_{1,2}$, so if $\mathcal{I}_{1,2}$ contains more than one element, one cannot distinguish between them. This is called *aliasing*. (ii) If ingress and egress data are available, one can have different sets of simultanously disappearing/reappearing surface elements. $\mathcal{E}_{i,j}$ denominates the surface elements reappearing between the phase steps ϕ_i and ϕ_j. If, say, elements $3, 4, 5 \in \mathcal{I}_{1,2}$ disappear at the same time, we can still distinguish their intensities, if $3 \in \mathcal{E}_{3,4}$, $4 \in \mathcal{E}_{4,5}$ and $5 \in \mathcal{E}_{5,6}$. In real polars, this is the case for the surface elements on the ballistic stream and on the magnetically funelled part of the accretion stream, if $\beta \neq 0$ (dipole tilt) and $i < 90°$ (inclination). Using this, aliasing effects can be reduced in many cases.

To avoid aliasing, one has to choose the size of the surface elements so that $|\mathcal{I}/\mathcal{E}_{1,2}| = 1$, that is, one surface elements appears or disappears at a time. How large is such a surface element? The size in direction of the accretion flow Δs is the interesting value. As we can see from Fig. 4,

$$\frac{\ell'}{\sin \xi} = \frac{\Delta s}{\sin 2\pi \Delta \phi} \tag{6}$$

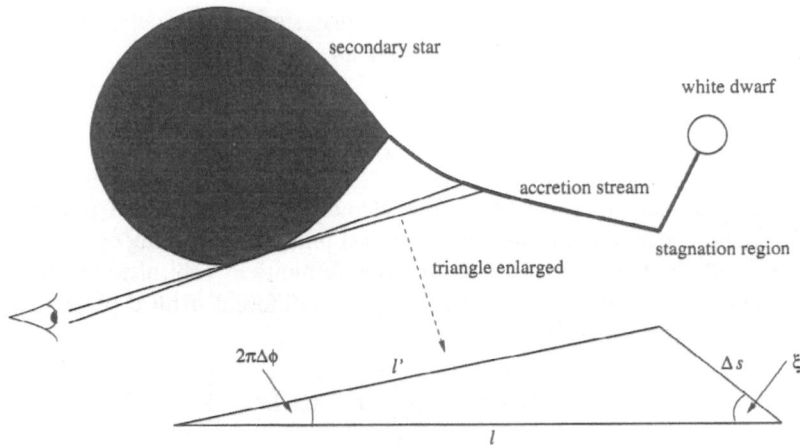

Fig. 4. Derivation of the smallest possible surface element's size, Δs. See (6).

with $\Delta \phi$ the orbital (time-) resolution. Since $\Delta s \ll \ell$, we can use $\ell' \approx \ell$. ℓ is also approximately the distance from the surface element to the center of the secondary star. Since we are interested in the absolute value of Δs, we then find

$$\frac{\Delta s}{\ell} \approx \frac{\sin 2\pi \Delta \phi}{|\sin \xi|} \qquad (7)$$

From (7), we learn that (a) the spatial resolution is highest perpendicular to the line of sight ($\xi = 90°$), (b) the spatial resolution decreases with distance from the secondary star and (c) the time resolution transforms nearly linearly into spatial resolution for small $\Delta \phi$, giving a useful resolution of $\approx 1/100 \cdot a$ for $\Delta \phi \leq 0.01$.

4 A Survey of Artefacts

To find out what typical artefacts look like and which geometry parameters are most sensitive to the developement of artefacts, we conducted some test calculations. For an imaginary polar, IM Sys, which is eclipsing but does not completely eclipse the accretion stream, we calculated three artificial light curves with different input intensity distributions: (i) constant brightness: the stream was given a constant brightness value for all surface elements. (ii) "point": The stream's intensity was set to 1 for all surface elements except all those near to the stagnation region which were set to 10. (iii) "ramp": The intensity increases from 1 at the L_1-point to 10 at the stagnation region, only the way to the nearer accretion spot is set to 10, the way to the further spot is set to 1. The phase coverage was choosen so that the ingress of all and the egress of only some of the surface elements was visible. The reconstruction of these light curves with noise-free data and a correct system geometry leads to a very good agreement between the original map and the reconstruction. We noisified the data with $S/N = 20$

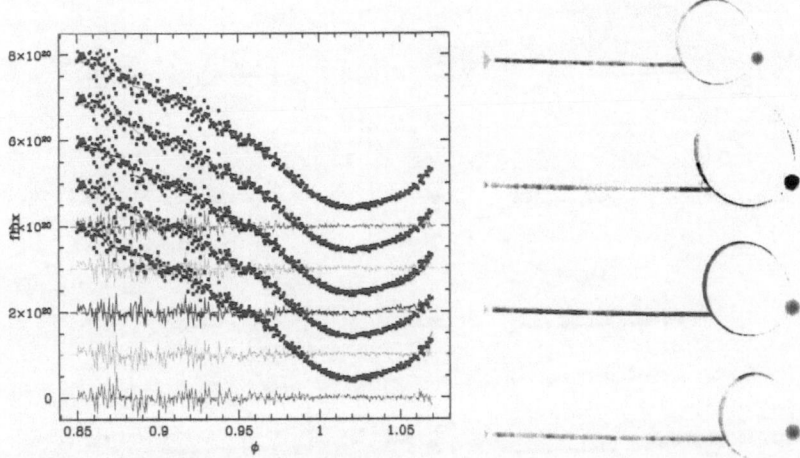

Fig. 5. Reconstrucion of the input map with equal brightness distribution. **Left:** light curves (points: data, lines: reconstruction and residuals), from top to bottom: correct geometry, errors (a) to (d). **Right:** intensity maps, from top to bottom errors (a) to (d).

and a constant noise of 1%, which still allows a near-perfect reconstruction if the geometry is known accurately.

Now we performed four test calculations with different errors in the geometry: (a) $(i, q) = (85°, 5.6)$ instead of $(i, q) = (82°, 5)$, (b) $\beta = 30°$ instead of $\beta = 20°$, (c) $\Psi = 45°$ instead of $\Psi = 35°$, and (d) $\Psi_S = 45°$ instead of $\Psi_S = 40°$. In Fig. 5, left, we show the light curves of the reconstructed intensity distributions. The upper light curve gives the reconstruction if the geometry was correct, then we have the four different wrong geometry parameters from top to bottom. The synthetic data is shown as points, the residuals between data and reconstruction are also shown for each light curve. The right hand panel in Fig. 5 and all of Fig. 6 show the reconstructed maps.

Analysis of the maps reveals the following: (I) all main features are reconstructed correctly. (II) The parameters most sensitive to artefacts are (i, q) and Ψ_S. (III) Typical artefacts are: • a slightly enhanced brightness on the northern magnetic stream (i/a, i/d, ii/a, ii/d), • wrong position of the brightness on the ballistic stream (i/a, ii/a, iii/a, iii/d).

Reasons for these two observed types of artefacts are: aliasing for the enhanced brightness on the non-eclipsed magnetic stream, a longer or shorter ballistic stream in the cases (a) and (d), giving different phasing for the eclipse timings of the stagnation region.

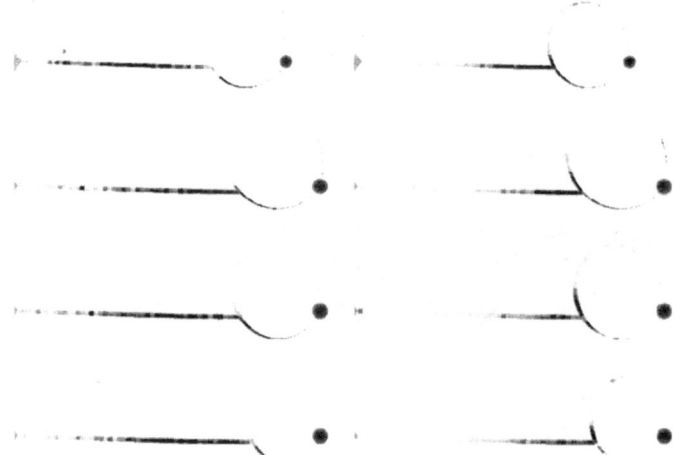

Fig. 6. Left: Reconstrucion of the input map with "point" brightness distribution. **Right:** Reconstrucion of the input map with "ramp" brightness distribution. Descriptions as in Fig. 5

5 Summary and Outlook

The eclipse mapping method has shown its capabilities also in non-disk systems like polars. Extensive test calculations show that artefact-free reconstructions are possible even for slightly wrong geometrical paramaters with medium to high-noise data. Nevertheless, great improvements of the reconstruction of the intensity distributions or even the tracking of physical parameters in AM Her systems will be made by a combination of eclipse mapping and doppler tomography.

References

1. K. Horne, 1985, MNRAS, 213, 129–141
2. P.J. Hakala, 1995, A&A, 296, 164–168
3. S. Vrielmann, A.D. Schwope, 1999, "Accretion stream mapping", in *Annapolis Workshop on Magnetic Cataclysmic Variables*, ed. by C. Hellier, K. Mukai (ASP, San Francisco), ASP Conference Series 157, pp. 93–98
4. M.K. Harrop-Allin, P.J. Hakala, M. Cropper, 1999, MNRAS, 302, 362–372
5. J. Kube, B.T. Gänsicke, K. Beuermann, 2000, A&A, 356, 490–500
6. C. Heerlein, K. Horne, A.D. Schwope, 1999, MNRAS, 304, 145–154
7. G.A. Wynn, A.R. King, 1995, MNRAS, 275, 9–21
8. M.M. Sirk, S.B. Howell, 1998, ApJ, 506, 824–841
9. J. Bailey, 1995, "Fundamental properties of polars", in *Cape Workshop on Magnetic Cataclysmic Variables*, ed. by B. Warner, D.A.H. Buckley (ASP, San Francisco, 1995), ASP Conference Series 85, pp. 10–20

Spot Mapping in Cool Stars

A. Collier Cameron

School of Physics and Astronomy, University of St Andrews,
North Haugh, St Andrews, Scotland KY16 9SS

Abstract. Active late-type stars display starspot activity with far greater area coverage, and at a wider range of latitudes, than is seen on the Sun. In this review I outline the signal enhancement and indirect imaging techniques available for mapping spots on stellar surfaces. I review recent developments in the use of starspots as tracers of surface differential rotation, and discuss the way in which starspot lifetimes appear to scale with stellar luminosity class.

1 Introduction

Starspot activity is a ubiquitous feature of rotating stars with outer convective zones. Even on the relatively inactive Sun, it is the most easily observed manifestation of dynamo activity, with an observational history dating back thousands of years. Sunspot activity has been monitored nearly continuously since the invention of the telescope; indeed, much of our knowledge of long-term fluctuations in the solar activity cycle has been derived from telescopic records maintained in the archives of the Observatoire de Paris and elsewhere since the early 17th century [33]. Sunspots serve as tracers of the solar differential rotation, and the oscillations in their surface coverage, latitude and magnetic polarity patterns [18] laid the foundations of our present picture of the solar dynamo as a self-sustaining phenomenon driven by convection and rotation.

Spot activity on stars with high rotation rates and deep convective zones is a great deal more vigorous than that seen on the Sun, to the extent that in very active systems such as the RS CVn binaries, starspots can modulate the flux received from the star by several tenths of a magnitude as the star rotates.

In seeking a fuller understanding of how solar and stellar dynamo activity depends on a star's interior structure and rotation rate, the challenge for observers is to obtain as complete as possible an observational picture of how the time-dependent behaviour of starspots varies with rotation, convective zone depth and the tidal and radiative effects of a close binary companion. We need to develop ways of determining what fraction of a given star's surface is occupied by spots at any given time, in order to trace the waxing and waning of starspot coverage over a stellar cycle. We need to determine whether starspots congregate in well-defined belts, and whether these belts show the same time-dependent latitude drifts as on the Sun, in order to learn about the geometry and temporal behaviour of the dynamo modes that may be present. We need to use starspots as flow tracers to measure the dependence of the surface rotation rate on stellar latitude, as a function of convective zone depth and rotation rate, in order

to learn how angular momentum conservation drives fluid circulation throughout the convective zone. We need to determine the polarities of the magnetic
fields that emerge from the stellar interior in these active regions, to discover
the stellar equivalents of Hale's polarity laws. And we have to be prepared for
surprises. We don't expect all stellar dynamos to behave in the same way. Even
the solar dynamo may well have changed its character completely as the young
Sun passed from being a rapidly rotating, fully convective protostar to a young,
fast-rotating main-sequence star, to its present comfortable middle age. More
changes are expected as it expands into red gianthood at the end of its life.
A full understanding of stellar dynamo activity requires us to explore a large
parameter space, encompassing all phases of stellar evolution where convection
and rotation are important, in both single and close binary stars.

All of these observations are relatively easy in the solar case, where the
stellar disk is resolvable on (sometimes embarrassingly) fine spatial scales. The
stars, however, present a tougher challenge. To date, all but a very few nearby
supergiant stars have stubbornly resisted all attempts at direct imaging of their
surfaces. Instead, we must turn to tomographic methods to attain the micro-
arcsecond resolution needed to study starspots in detail.

The purpose of this paper is partly to review the substantial advances that
have been made over the last decade, but principally to present the mathematical
foundations of Doppler imaging and related techniques, in sufficient detail to enable those interested in developing their own codes to implement the techniques
in their own software. In Section 2 I outline the basic principles of Doppler imaging of stellar surfaces. The types of data needed to achieve this are described in
Section 3 together with a recipe for combining the profiles of many spectral lines
via the method of least-squares deconvolution. Section 4 presents the "forward
problem" of converting an "image" of the stellar surface, described by a single
parameter, into a realistic set of synthetic data that can be compared directly
with observational data. Once such a model is established it can be used with
appropriate methods of statistical inference as described in Section 5, to derive
a map of the stellar surface. This has to be done in tandem with optimised
knowledge of other stellar parameters, as described in Section 6. Methods for
establishing the resolution and reliability of the resulting maps are summarised
in Section 7. The true power of stellar surface imaging, however, goes beyond
single "snapshots" of stellar surfaces, and begins to address time-dependent phenomena: spot lifetimes, differential rotation and activity cycles. Some examples
of recent successes in this area are described in Section 8.

2 How Doppler Imaging Works

The term "Doppler imaging" was coined by Vogt & Penrod [15], who demonstrated that travelling starspot bumps were observable in the line profiles of
HR 1099, and that an image of the stellar surface could be derived from them.

A photospheric absorption line in which rotation is the dominant broadening
mechanism displays time-variable irregularities if the visible surface of the star

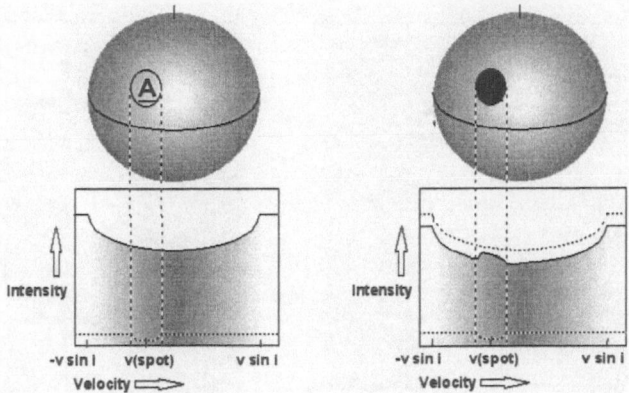

Fig. 1. Schematic illustration of the formation of a bright starspot "bump" in a rotationally-broadened line profile.

is mottled by dark spots. The effect of a cool, dark region on a rotationally-broadened line profile is illustrated in Fig. 1. The "missing" light of the spot consists of a continuum contribution that spans the line profile, plus a narrow line contribution that is Doppler shifted by an amount that depends on the projected distance of the spot from the stellar rotation axis. Removing this light causes an overall depression of the continuum, but less light is removed at the Doppler shift of the spot relative to the centre of the line. The observable signature of a dark spot on the stellar surface is therefore a bright bump in every photospheric absorption line in the star's spectrum. As the star rotates, the spots are carried across the stellar disc, causing the bumps to change their Doppler shifts in accordance with their projected distances from the star's rotation axis. Spots near the equator remain visible for half the stellar rotation cycle, tracing out a sinusoidal velocity variation with an amplitude equal to the stellar equatorial rotation velocity, $V \sin i$. Spots at progressively higher latitudes follow progressively lower-amplitude sinusoids. The fraction of the rotation cycle for which a spot remains visible depends on its latitude and the inclination of the stellar rotation axis to the line of sight. The times at which spot signatures cross the centre of the line profile thus reveal their longitudes, while the amplitudes of their sinusoids (or equivalently, their radial accelerations at line centre) tell us their latitudes. If high-latitude spots are present, the length of time for which they remain visible tells us the axial inclination of the star (Fig. 2).

3   Data Requirements

Doppler imaging can be used to map spots on any active, late-type star whose rotational Doppler profile is significantly broader than the intrinsic, unbroadened line profile emerging from any localised region on the stellar surface. Nature has been kind to us in this respect. Magnetically-active late-type stars with enough

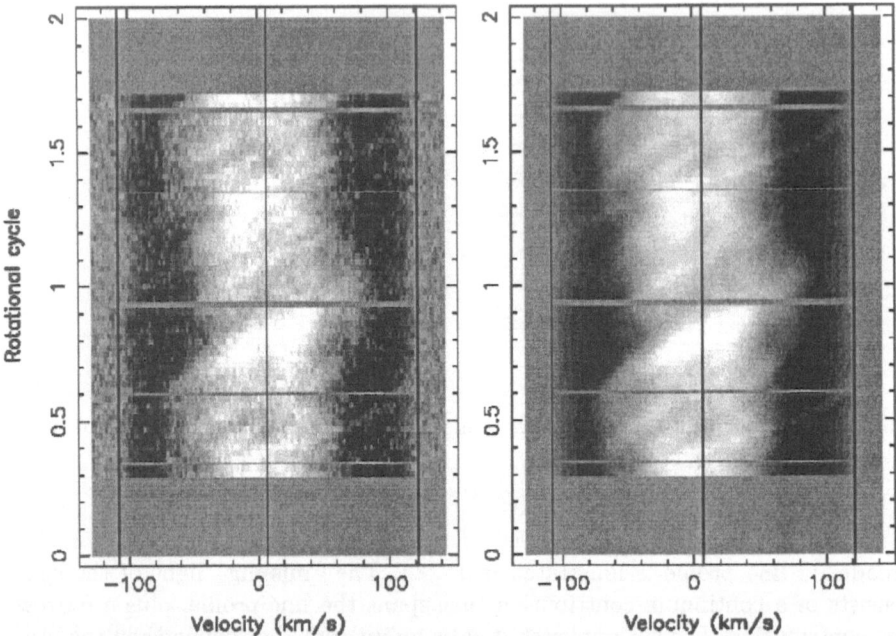

Fig. 2. A trailed spectrogram showing line profiles of the pre-main sequence G star RX J1508.6 -4423 [15] exhibits bright streaks due to starspots at a variety of latitudes crossing the stellar disc during the 0.31-day rotation cycle. Time increases upward. A lone spot close to the equator moves rapidly across the profile at phases 0.37 and 1.37. Higher-latitude substructures in an offcentred, dark polar cap have smaller velocity amplitudes and radial accelerations as they cross line centre. The left-hand panel shows the data, and the right-hand panel is the model derived from the reconstructed image. The theoretical profile of an unspotted star has been subtracted from all spectra to enhance contrast.

starspot coverage to qualify as likely targets for Doppler imaging tend to be rapid rotators almost by definition, rotation being a key ingredient in the dynamo process that generates the activity. Moreover, since an outer convective envelope is the other key ingredient, the stars concerned tend to be of spectral type mid-F or later. Their spectra therefore contain large numbers of metal absorption lines.

The need for a good supply of relatively unblended, intermediate-strength photospheric lines normally restricts Doppler imaging studies to the red part of the optical spectrum. The earliest Doppler imaging studies concentrated on the Fe I line at 6430Å and the Ca I line at 6439Å, and these remain firm favourites with practitioners of the art of Doppler imaging who employ single-order spectrographs with restricted wavelength coverage. The development of advanced line-stacking methods in recent years has relaxed somewhat the need to avoid blends, and has allowed Doppler imaging codes to utilise most of the optical spectrum as described in Section 3.1 below.

One weakness of the Doppler imaging method is that it is relatively insensitive to the temperatures of the dark features on the stellar surface. A bump in a line profile implies a measurable flux deficit, but the width of the bump cannot be less than that of the intrinsic line profile. A bump with this minimum width could be produced either by a warm, marginally-resolved spot, or a cooler, darker, unresolved one. Moreover, Doppler imaging is not sensitive to uniform coverage of small, unresolved spots in the "clean" photosphere. Other methods with greater temperature sensitivity are needed to resolve these ambiguities.

The simplest way to detect the presence of cool spots on a rotating, late-type star is through the use of broad-band photometry. Some idea of the fraction of the stellar surface occupied by spots can be gained from light-curves in two or more bands, sufficiently widely separated in wavelength that the contrast between the spots and the photosphere is measurably different. Vogt [49] showed that this information is sufficient to establish the relative temperatures of the spots and photosphere, and so to estimate the area coverage of spots needed to give the observed rotational modulation of the light-curve.

More recently, O'Neal, Saar & Neff [29] have used the strengths and ratios of near-IR TiO bands to establish the temperatures of spots on II Peg and other stars. Since TiO should be present in the spots but not in the photosphere, the overall TiO band strengths are a good diagnostic of total spot coverage. The II Peg results indicate that the surface coverage of spots on active stars can be 50% or more, considerably greater than the fractional spot areas that are usually found by Doppler imaging alone.

Ideally, Doppler imaging should be carried out over a wavelength range that includes the TiO bands used by O'Neal et al, and should be supported by simultaneous broad-band (usually VRI) photometry. This should suffice to establish the overall spot coverage at the time of observation, the relative temperatures of the cool spots and the bright photosphere, and the spatial distribution of spots over the stellar surface.

3.1 Combining Line Profiles

Most Doppler imaging studies are conducted using cross-dispersed echelle spectrographs on 2-m and 4-m class telescopes. The average echellogram records thousands of photospheric absorption lines of weak to intermediate strength, and a few stronger ones. Donati et al. [5] developed a method for utilizing the full information content of all the weak and intermediate-strength lines in a computationally efficient way, by combining them into a single composite profile via the method of least-squares deconvolution (LSD).

This method entails taking a list of lines with relative strengths appropriate to the type of star concerned, and computing via the method of least squares a "mean" line profile which, when convolved with the line pattern, gives an optimal match to the observed spectrum (Fig. 3). The deconvolved profile thus represents an average broadening function that is representative of all the lines recorded in the spectrum.

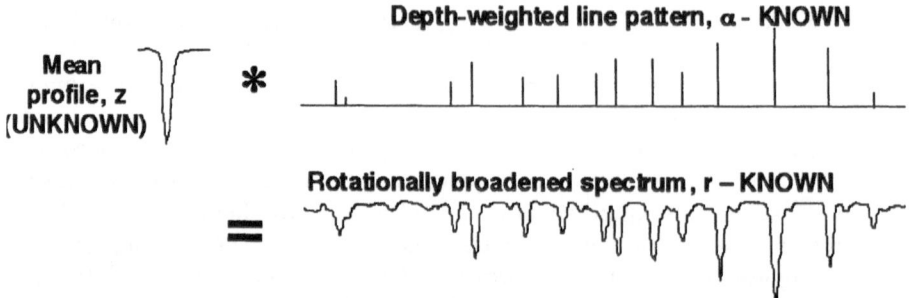

Fig. 3. Least-squares deconvolution treats the observed stellar spectrum as the convolution of a pattern of delta functions with the wavelengths and strengths of known absorption lines, with an unknown "average" line profile.

In terms of signal improvement, LSD is analogous to aligning and averaging (with appropriate weighting factors for line strength and local continuum signal strength on the recorded frame) the profiles of all the individual photospheric absorption lines recorded on each echellogram and included in the line list. If all lines were of equal strength and the continuum signal were constant over the whole frame, the signal-to-noise deconvolved profile would be roughly proportional to the square root of the number of line images used. An echellogram with 4900 line images should in principle yield a composite profile with S:N nearly 70 times greater than the signal-to-noise ratio of a typical single line in the original spectrum. In practice, the recorded continuum is not uniform, the lines have a wide variety of depths, and flat-fielding errors sometimes take their toll. Even so, the signal-to-noise ratio of the deconvolved profile is found to be some 30 times greater than that of a single line in the best-exposed parts of the original spectrum when 4000 or so lines are combined. The least-squares fitting procedure has the additional advantage that neighbouring, blended lines are treated simultaneously, thereby eliminating the sidelobes that would be produced by a simpler shift-and-add procedure or by cross-correlation.

Donati et al. [5] treat the observed residual spectrum r as the convolution of a "mean" line profile $z(v)$ with a set of weighted delta functions at the wavelengths of a comprehensive list of spectral lines. The profile is defined on a linear velocity scale, with a velocity increment chosen to be close to the average velocity increment per pixel in the extracted spectra, typically $\Delta v \sim 3$ km s^{-1} per bin. The elements A_{jk} of the convolution matrix A are computed by summing, over all spectral lines i, the fractional contribution of the element of the deconvolved profile at velocity v_k to the data pixel at wavelength λ_j when the centre of the deconvolved profile is shifted to the wavelength λ_i of each line in turn. Hence

$$A_{jk} = \sum_i w_i \Lambda[(v_k - c(\lambda_j - \lambda_i)/\lambda_i)/\Delta v]. \tag{1}$$

The triangular function Λ has the form $\Lambda(x) = 1+x$ for $-1 < x \leq 0$, $\Lambda(x) = 1-x$ for $0 < x < 1$, and is zero everywhere else. The line weights w_i, incorporated in the convolution matrix A, are proportional to the central depths of the lines as computed from a Kurucz model atmosphere for the appropriate spectral type.

The deconvolved profile $z(v)$ has the form of a line profile normalized to unit continuum intensity, but from which the continuum has been subtracted. Determination of $z(v)$ via least squares entails minimizing the magnitude of the misfit vector $|r - A.z|$,

$$\chi^2 = (r - A.z)^T.Q.(r - A.z), \tag{2}$$

weighted so as to make due allowance for the observational errors σ_j associated with the individual spectral bins. Here the inverse variances σ_j^2 associated with the N elements r_j of the residual spectrum r are incorporated via the diagonal matrix

$$Q = \text{Diag}[1/\sigma_1^2, \dots, 1/\sigma_N^2]. \tag{3}$$

The least-squares solution for z is found by solving the matrix equation

$$A^T.Q.A.z = A^T.Q.r. \tag{4}$$

Since the square matrix $A^T.Q.A$ is symmetric and positive-definite, the least-squares problem can be solved using efficient methods such as Cholesky decomposition [32].

The deconvolved profile z is expressed in units of the weighted mean continuum level. The deconvolution procedure compensates for local line blends, and so has the advantage over cross-correlation methods that outside the region occupied by the residual stellar profile, the deconvolved spectrum is flat. The formal errors on the M points of the deconvolved profile z are obtained in the usual way from the diagonal elements of the $M \times M$ covariance matrix

$$C = [A^T.Q.A]^{-1}. \tag{5}$$

Least-squares deconvolution has some disadvantages. Because it is based on the assumption that all lines have similar profiles, it loses any temperature-dependent information that could be gleaned from the relative strengths of the spot signatures in line with different excitation potentials. Future improvements to the method may allow subsets of lines belonging to different elements, or having different excitation potentials, to be selected, in order to recover some of this lost information. For the moment, however, it's worth noting that a multiplex gain of 30 in S:N is equivalent to the ratio of a 120-m telescope's light-gathering power to that of a 4-m telescope, and it's a lot cheaper.

4   The Forward Problem

4.1   Choice of Mapping Parameter

The first step in reconstructing an image of the stellar surface is to set up a grid of pixels on the stellar surface. Ideally these pixels should all be of roughly comparable area, as a rectangular grid of latitudes and longitudes is rather wasteful in the polar regions of the image. In most applications the latitude spacing of pixels is uniform, but the number of longitude bins at each latitude scales as the cosine of the latitude. Other approaches which are particularly useful on contact and semi-contact systems include the geodesic scheme of [21], which abandons a regular latitude-longitude grid in favour of a set of interlocking triangular surface elements.

The image consists of a vector f of image-parameter values $f_i, i = 1, \ldots, N$, which parametrise the brightness of the image on the surface grid. The image parameter should be both positive and additive if entropy-based arguments are to be used in computing the final image. This means that the total flux received from any set of distinct pixels of the stellar surface is linearly proportional to the area-weighted sum of the pixel values in the image. For example, bolometric surface brightness is additive, whereas temperature is not.

The image parameters used in the literature generally fall into two classes. The codes of Rice et al. [34], Jankov & Foing [25] and Berdyugina [4], for example, define the f_i to be bolometric surface brightnesses, effectively assigning a unique temperature to each pixel in the image. In this model, the form of the spectrum is a continuous function of f. There is no restriction on the mix of bright and dark features in the image. Operationally, it requires a grid of pre-computed synthetic spectra spanning the range of temperatures expected on the stellar surface. The main drawback of this method is that it rests on the implicit assumption that all temperature structure on the stellar surface is resolved by the imaging process. This may lead to problems in cases where the resolution of the image is poor, as in the case where $V \sin i$ is only a few times the intrinsic widths of the mapping lines. In this case, the bolometric surface brightness of a blurred spot lies somewhere between the surface brightnesses of the photosphere and the spots. If the form of the spectrum varies non-linearly with bolometric surface brightness, there is a danger that spurious features could result. Collier Cameron [9] found that this led to the appearance of spurious bright features in gaps between darker regions in entropy-regulated images.

Vogt et al. [50] apply a "thresholding" procedure to their images which alleviates this problem to some extent, at the cost of a clear definition of the image entropy.

The codes of Collier Cameron [9], Donati & Brown [11] and Hendry & Mochnacki [22] avoid this form of image instability by fixing the effective temperatures of the spots and photosphere, and defining the f_i to represent the fractional spot occupancies of the image pixels. The image pixel values all lie in the range $0 < f_i < 1$. In one sense this model is restrictive, in that it does not allow any other temperature components to be present. It is probably inappropriate for use in T Tauri stars (where accretion hotspots may be present as well as cool starspots) or on the inner faces of strongly irradiated stars in close binary systems. However, it is probably the safer option to use in situations where there is no other evidence for strong facular or other hotspots, and it copes well with unresolved dark spots. It also has the computational advantage that the local specific intensities only have to be calculated for two photospheric temperature components. A related technique was developed by Kürster [27] using a adaptation of the CLEAN algorithm.

4.2 The Local Line Profile

The local specific intensity emerging from any image pixel on the model star at a given wavelength in the observer's rest frame depends on the value of the image parameter, the angle between the outward normal and the line of sight, and (for spectroscopy) the Doppler shift of the pixel. The areas, outward normal vectors and Doppler shifts of all visible pixels must be computed at each rotation phase observed, using either a spherical model (for slowly rotating single stars) or a Roche model (for fast rotators and close binaries).

The computational overheads involved in computing synthetic spectra and photometry for all visible pixels at each phase would be prohibitive. The usual approach is to store specific intensities on a grid of wavelengths, temperatures and direction cosines $\mu_{ik} = \cos\theta_{ik}$, where θ_{ik} is the angle between the outward normal of pixel i at phase k and the line of sight. The accuracy of the modelling depends crucially on the oscillator strengths, damping parameters and turbulent broadening parameters of many weak or intermediate-strength lines whose tabulated values are not always reliable. The most convenient source of the relevant atomic information is the Vienna atomic-line database [26,31]. It is common practice to check the predicted flux spectrum against observations of a slowly-rotating star of similar spectral type to the Doppler-imaging target star. Once the elemental abundances, damping coefficients and turbulent broadening parameters have been adjusted to give a good fit to the strengths and shapes of a set of lines of the same element, the oscillator strengths can be adjusted to improve further the fits to individual lines. Any remaining small discrepancies between the spectra of the non-rotating template and the target star can then be dealt with by modifying the abundances.

The photometric intensities are computed over the effective wavelength ranges of the bands observed, and are stored on an interpolation grid of effective temperatures and direction cosines. These can be computed directly, or with the help of a grid of limb-darkening coefficients such as that of Wade & Rucinski [51]. The wavelength and temperature dependence of the limb darkening coefficient is very important if the spot covering fraction is high and the spots are warm enough to contribute a significant fraction of the photospheric flux.

4.3 Surface Integration

If we treat the image and the computed model of the data as vectors f and D respectively, a linear image-data transformation would be a matrix problem of the form $D = V.f$.

There are, however, two sources of nonlinearity in the imaging problem. First, the relation between the value of an image pixel and its contribution to the data is a non-linear function of both temperature and limb angle. Second, we cannot determine exactly the observed continuum flux in each spectrum, or the

photometric zero-point of each light-curve, even though the relative shapes of
the line profiles and light-curves are known to high precision. Instead we must
content ourselves with renormalising the individual line synthetic profiles at each
phase so that their inverse variance-weighted means match those of the observed
profiles at each iteration. Similarly, the synthetic light-curve in each photometric
band must be renormalized so that its inverse variance-weighted mean value
matches that of the observations. In both types of data the renormalization
process introduces a scale factor that is itself dependent on the computed fluxes.

Before rescaling, the computed flux D'_{jk} in spectral bin j of the line profile
at the kth phase is

$$D'_{jk} = \sum_i w_i \mu_{ik} \left[I(f_i, \lambda_j[1 - v_{ik}/c], \mu_{ik}) \right]. \tag{6}$$

Here w_i is the pixel area and μ_{ik} is set to the foreshortening cosine if the pixel is
visible, or zero if it is hidden. The convolution of the local line profile with the
rotation profile is effected through the radial velocity v_{ik} of pixel i at phase k.

The weighted mean flux in the computed profile at phase k is given by

$$\hat{D}_k = \frac{\sum_j D'_{jk}/\sigma_{jk}^2}{\sum_j 1/\sigma_{jk}^2} \tag{7}$$

where σ_{jk}^2 is the variance in the flux for the same observation.

Similarly, the mean flux in the observed line profile is

$$\hat{F}_k = \frac{\sum_j F_{jk}/\sigma_{jk}^2}{\sum_j 1/\sigma_{jk}^2} \tag{8}$$

where F_{jk} is the observed flux in spectral bin j at phase k.

The renormalised computed spectral data elements are

$$D_{jk} = \frac{\hat{F}_k}{\hat{D}_k} D'_{jk}. \tag{9}$$

A similar renormalisation is required for the photometry, since we know the
shape of the light-curve better than we know its zero-point. The unscaled flux
for the kth photometric observation in waveband j is

$$D'_{jk} = \sum_i w_i \mu_{ik} \left[I(f_i, \lambda_j, \mu_{ik}) \right]. \tag{10}$$

Here λ_j labels the effective wavelength of band j. The weighted mean flux of the
computed light-curve in band j is

$$\hat{D}_j = \frac{\sum_k D'_{jk}/\sigma_{jk}^2}{\sum_k 1/\sigma_{jk}^2}. \tag{11}$$

The mean flux in the corresponding observed light-curve is

$$\hat{F}_j = \frac{\sum_k F_{jk}/\sigma_{jk}^2}{\sum_k 1/\sigma_{jk}^2} \tag{12}$$

where F_{jk} is the observed flux in band j at phase k, and the renormalised computed photometric values are

$$D_{jk} = \frac{\hat{F}_j}{\hat{D}_j} D'_{jk}. \tag{13}$$

The matrix containing the marginal responses of all data elements D_{jk} to all image values f_i is needed for calculating the gradients of the discrepancy function and the regularising function in whatever iterative multi-dimensional search algorithm is used for the reconstruction. The image-data transformations given above have the non-linear form $\boldsymbol{D} = \boldsymbol{V}(\boldsymbol{f})$, but the effect on the synthetic data of small changes in the image can be treated by the local linear approximation

$$\delta \boldsymbol{D} = \boldsymbol{V}.\delta \boldsymbol{f}. \tag{14}$$

The elements of the visibility matrix give the marginal dependence of each data element on each image pixel, $V_{ijk} = (\partial D_{jk}/\partial f_i)$. For the spectral fluxes these have the values

$$V_{ijk} = \frac{1}{\hat{D}_k}\left(\hat{F}_k w_i \mu_{ik} \frac{\partial I(f_i, \lambda_j[1 - v_{ik}/c], \mu_{ik})}{\partial f_i} - D_{jk}\frac{\partial \hat{D}_k}{\partial f_i}.\right) \tag{15}$$

where

$$\frac{\partial \hat{D}_k}{\partial f_i} = w_i \sum_m \frac{\mu_{ik}}{\sigma_{mk}^2} \frac{\partial I(f_i, \lambda_m[1 - v_{ik}/c], \mu_{ik})}{\partial f_i}. \tag{16}$$

The visibility matrix elements for the photometric data are

$$V_{ijk} = \frac{1}{\hat{D}_j}\left(\hat{F}_j w_i \mu_{ik} \frac{\partial I(f_i, \lambda_j, \mu_{ik})}{\partial f_i} - D_{jk}\frac{\partial \hat{D}_j}{\partial f_i}.\right) \tag{17}$$

where

$$\frac{\partial \hat{D}_j}{\partial f_i} = w_i \sum_l \frac{\mu_{il}}{\sigma_{jl}^2} \frac{\partial I(f_i, \lambda_j, \mu_{il})}{\partial f_i}. \tag{18}$$

Because of the need to renormalize the synthetic data to match the means of the observed profiles and light-curves, the marginal dependences are themselves functions of the current image. This means that the visibility matrix needs to be re-calculated once per iteration, or whenever a new stellar image is used. Note that in the filling-factor spot model,

$$\frac{\partial I(f_i, \lambda_j, \mu_{ik})}{\partial f_i} = I_{\text{spot}}(\lambda_j, \mu_{ik}) - I_{\text{phot}}(\lambda_j, \mu_{ik}) \tag{19}$$

and the only source of nonlinearity in the model is the renormalization.

The rescaling of the spectral-line profiles decribed above is strictly only appropriate for the case where we are dealing with a flux spectrum possessing a well-defined continuum, so the equivalent widths of the lines are well defined. An alternative approach may be better suited to least-squares deconvolved profiles in situations where background light due to an early-type companion, scattered light or an accretion disc has been subtracted prior to deconvolution.

In this case, the observed and computed profiles should be orthogonalised and an optimal scale factor A_k computed for each spectrum in the series as follows:

$$A_k = \frac{\sum_j (F_{jk} - \hat{F}_k)(D'_{jk} - \hat{D}_k)/\sigma_{jk}^2}{\sum_j (D'_{jk} - \hat{D}_k)^2/\sigma_{jk}^2} \qquad (20)$$

and the synthetic data should be rescaled to give:

$$D_{jk} = A_k(D'_{jk} - \hat{D}_k). \qquad (21)$$

This minimises the χ^2 misfit to each observed line profile, expressed in the orthogonalised form $(F_{jk} - \hat{F}_k)$. In this case the expression for the linearised marginal dependences V_{ijk} is a little more complicated than the expressions given above.

5   The Inverse Problem

5.1   Regularised Least-Squares Strategies

Most authors have adopted some form of least-squares strategy in seeking the simplest image that can fit the data satisfactorily. Although in principle this could be achieved by least-squares methods, it it commonly the case that the number of image pixels exceeds the number of data points, making the problem ill-conditioned. The presence of random noise in the data combines with gaps in phase coverage to compound the problem. As a result, there is no unique image that can fit the data at a given value of the χ^2 statistic, which is given in matrix form by

$$\chi^2(\boldsymbol{f}) = (\boldsymbol{F} - \boldsymbol{D}(\boldsymbol{f}))^T.\text{Diag}[\sigma^2].(\boldsymbol{F} - \boldsymbol{D}(\boldsymbol{f})) \qquad (22)$$

In order to overcome this lack of uniqueness, a regularising function is imposed on the image. The first application of this technique to the starspot imaging problem was by Vogt, Penrod & Hatzes [49], who used the Shannon-Jaynes image entropy as the regularising function. For images where the mapping parameter f_i is bolometric surface brightness and so can take any positive value, the entropy takes the form:

$$S = \sum_i (f_i - m_i - f_i \ln \frac{f_i}{m_i}). \qquad (23)$$

If a filling-factor model is adopted, however, a restricted form of the entropy is used, which combines the entropy of the spot image f_i and the photospheric image $(1 - f_i)$:

$$S = \sum_i (-f_i \ln \frac{f_i}{m_i} - (1 - fi) \ln \frac{(1 - f_i)}{(1 - m_i)}). \tag{24}$$

In both cases the default model m_i is the value that a pixel will assume in the absence of any other constraint imposed by the data.

The other form of regularisation that has been used frequently in spot-mapping applications (e.g.[30,34] was developed by Tikhonov [44], and first applied to the Ap-star imaging problem by Goncharsky et al. [16,17]. Tikhonov regularisation seeks to minimise gradients in the image via the functional:

$$T = -\sum_i \|\nabla f_i\|^2. \tag{25}$$

The use of entropy as the regularising function minimizes spurious correlations between image pixels, whereas Tikhonov regularisation maximises the smoothness of the solution.

In either case, the image reconstruction problem becomes a constrained optimisation exercise, in which the f_i are adjusted iteratively to maximise the functional

$$Q = S - \lambda \chi^2 \tag{26}$$

This is tantamount to maximising S (or in the Tikhonov approach, T) over the surface of the hyper-ellipsoid in image space that is bounded by the constraint surface at some fixed value of χ^2. The Lagrange multiplier λ is set to a value such that the final solution lies on a surface with $\chi^2 \simeq M$, M being the number of measurements in the data set.

5.2   The Occamian Approach

In practice, there is little difference between Doppler images produced with Tikhonov and MaxEnt regularisation, when the images are derived from data sets that have high S:N and good phase coverage around the rotation cycle. At first this might seem surprising, since MaxEnt discourages correlations between neighbouring pixels in the image while Tikhonov encourages them. The reason for this lies in the smoothing that is naturally built into the image-data transformation by the finite width of the local line profile.

To understand this, it is helpful to think about the Hessian matrix of the image-data transformation. At each step, the linearised approximation to each element of the Hessian matrix is derived from the marginal dependences V_{ijk} of the image-data transformation:

$$H_{lm} = \frac{1}{2} \frac{\partial^2 \chi^2}{\partial f_l \partial f_m} = \sum_{jk} \frac{V_{ljk} V_{mjk}}{\sigma_{jk}^2}. \tag{27}$$

The eigenvectors of this matrix define the principal axes of the error ellipsoid in image space, and the associated eigenvalues give their associated dimensions. A small eigenvalue implies that the ellipsoid is extended in the corresponding direction, and that the data therefore do not constrain the linear combination of image pixels concerned. Conversely, a principal axis with a large eigenvalue represents a linear combination of pixels that are well-constrained by some part of the data. This illustrates well that noise, gaps in the data and the finite surface resolution imposed by the mapping line conspire to ensure that many of the image dimensions are not linearly independent. This "Occamian" approach was first applied to astrophysical imaging problems by Terebizh & Biryukov [43], and Berdyugina [4] adapted it for stellar surface imaging. Berdyugina showed that by using singular-value decomposition to determine the principal axes of the error ellipsoid and reject those eigenvectors with very small eigenvalues that simply added noise to the solution, it was possible to reconstruct a stellar surface image of comparable quality to those obtained by regularised least-squares.

The orthonormal basis vectors that define the image in Berdyugina's approach to surface imaging thus consist of linear combinations of (mostly neighbouring) pixels in the image map. Images that are well-constrained by the χ^2 criterion alone thus tend to be intrinsically smooth. When a regularising function is used, its effects should only become apparent when fitting data of very poor quality. Overfitting noisy data with MaxEnt causes the map to break up into structures smaller than can reasonably be resolved by the image-data transformation. Tikhonov regularisation, on the other hand, keeps the image smooth even when it is poorly constrained by the data.

5.3   Heterogeneous Datasets

Simultaneous photometry and spectroscopy of an active star provide important complementary information of the state of the stellar surface. Multicolour photometry is very sensitive to the total coverage of different temperature components on the stellar surface, but provides only crude insight into the spatial distribution of features on the stellar surface. The time-varying shapes of Doppler-broadened spectral-line profiles, on the other hand, provide a powerful probe of the fine spatial detail in the mottling of the stellar surface, but have little sensitivity to uniform coverage of unresolved spots.

Trying to fit photometry and spectroscopy simultaneously with historic MaxEnt or similar approaches is not simply a matter of adding together the χ^2 contributions from the two different types of data. The initial value of χ^2 when a flat light-curve from a blank image is compared with the data, is many times greater than that obtained for the spectroscopy where the spot "bumps" are typically only a few times greater than the observational errors. The photometry therefore dominates, and the image reconstruction process grinds to a halt trying to fit noise in the photometry before it even begins to take the spectroscopy into account.

The solution is to provide separate constraints for the photometry and the spectroscopy. Several authors have adopted related solutions to this problem.

Collier Cameron et al. [8] introduced separate Lagrange multipliers for the badness-of-fit parameters, seeking a maximum-entropy solution along the intersection of the image-space error ellipsoids for the spectroscopy and photometry:

$$Q(\boldsymbol{f}) = S(\boldsymbol{f}) - \mu\chi_s^2(\boldsymbol{f}) - \nu\chi_p^2(\boldsymbol{f}). \tag{28}$$

Equivalently, the χ^2 statistics for the two data sets can be combined with an appropriate weighting factor β to give

$$\chi_t^2 = \beta\chi_s^2 + (1-\beta)\chi_p^2, \tag{29}$$

allowing us to solve the problem in the conventional way:

$$Q(\boldsymbol{f}) = S(\boldsymbol{f}) - \lambda\chi_t^2(\boldsymbol{f}). \tag{30}$$

with $\mu = \lambda\beta$ and $\nu = \lambda(1-\beta)$.

This approach was used by Unruh et al. [46] to combine photometric and spectroscopic observations of AB Dor, while Barnes et al. [2] used the same method to combine spectral observations of PZ Tel from data sets of different spectral resolution and S:N secured at two different sites. Hendry & Mochnacki [22] also used independently-adjustable Lagrange multipliers for their spectroscopy and photometry of VW Cep. They pointed out that when the final solution is attained, the gradient of S in image space should lie in the plane spanned by the gradients of χ_s^2 and χ_p^2. This is automatically satisfied by the weighting-factor method, where

$$\nabla S = \lambda\beta\nabla\chi_s^2 + \lambda(1-\beta)\nabla\chi_p^2 \tag{31}$$

at the solution point where $\chi_s^2 = M_s$ and $\chi_p^2 = M_p$.

6   Dealing with Nuisance Parameters

In all stellar surface imaging problems, the image-data transformation is influenced by a number of additional physical parameters. In even the simplest case of a single, spherical star the spectral flux received at Earth depends on its surface brightness distribution, its angular diameter, the inclination i of its rotation axis to the line of sight, the degree of rotational broadening of the spectrum, and the stellar radial velocity V_r. In the case of a binary, the visibilities of the different parts of the stellar surface are governed by additional parameters describing the binary orbit and the equipotential surface configurations of the two stars.

In many cases these additional "nuisance" parameters can be estimated independently of the imaging process. For instance, $v\sin i$ and V_r can be determined from analysis of line profiles averaged around the stellar rotation cycle. It is not, however, sufficient simply to estimate these parameters in advance and calculate a surface image. Many authors [9,30,34,45,49] have explored the artefacts that appear in stellar images as a direct result of systematic errors in the nuisance parameters. These typically take the form of bright or dark axisymmetric

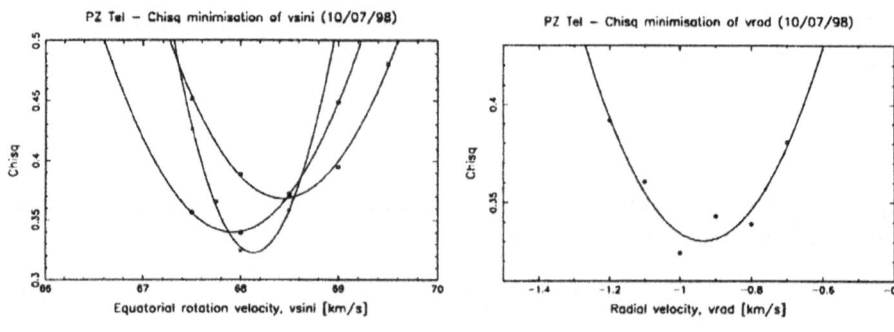

Fig. 4. Determination of radial velocity, average line EW and equatorial rotation speed for PZ Tel, using minimum attainable χ^2 for images computed with different parameter values [2]

structures at the poles and/or equator due to errors in line equivalent width or $V \sin i$, and spurious north-south streaks in images that are poorly sampled in phase with erroneous radial velocities.

In general, such errors produce additional structure in the image as the reconstruction algorithm struggles to fit the data. If the reconstruction converges at all, it generally yields a poor fit to the data. As computing speeds have increased over the last decade it has become feasible to make multiple reconstructions for a range of values of each of the nuisance parameters. The combination of parameters that yields either the lowest total spot coverage[10], or the best attainable fit to the data [1,34,40], is then selected as being the optimal set.

Rice & Strassmeier [36] pointed out in a recent study of the young K dwarf LQ Hya that a more self-consistent treatment of the stellar parameters is possible. The values of $V_e \sin i$ and i were derived by optimising the fits to the data. Combined with the period, these yield the radius of the star. By using the photometric zero points in three bandpasses as an additional constraint on the image, they were then able to ensure that the image yielded the correct total flux given the star's known parallax. Although their use of bolometric surface brightness as the mapping parameter makes it difficult to define the fraction of the stellar surface covered in spots, the resulting images showed spot activity to affect most parts of the image.

A more sophisticated treatment of the same idea has been adopted by Hendry & Mochnacki [22], in mapping the magnetically-active contact binary VW Cephei. In this approach, several of the nuisance parameters governing the binary orbit and stellar surface geometry were adjusted simultaneously with the image vector at each iteration in the image reconstruction, effectively incorporating the relevant nuisance parameters as additional image pixels. This required additional elements of the visibility matrix to be calculated, to give the marginal dependence of the synthetic data D_{jk} on the individual nuisance parameter values g_i. These derivatives generally had to be calculated numerically, since the nuisance parameters g_i affect the geometric kernel itself.

The resulting solution yielded an image and a set of system parameters that yielded an optimal fit to the data while maximising the entropy of the image. Most importantly, Hendry and Mochnacki [22] did not satisfy themselves merely by fitting the shapes of the line-profiles and light-curves in their data. They also included the independently-measured parallax of the system with the stellar surface flux to predict and match the standard V magnitude of the binary at each epoch. The results confirmed the long-standing suspicion that Doppler imaging without an absolute flux constraint under-estimates the total spot coverage on active stars. This should not come as too much of a surprise: the whole purpose of regularisation is to fit the data with the smallest possible number of spots in the image. Studies of TiO band strengths in active stars [29] point independently to the same conclusion.

While nobody has yet attempted a full Bayesian treatment of the nuisance parameters in Doppler imaging, the possible strategies that might be adopted for the single-star problem are easy enough to envisage. In the absence of any other information about the inclination of the stellar rotation axis to the line of sight, the prior probability for the inclination is $Pr(i) = \sin i$. The prior probability distribution for the distance d can be approximated as a gaussian based on independent parallax measurements. The priors for the projected rotational speed $V \sin i$, and radial velocity V_r can be approximated as gaussians based on least-squares estimates derived from rotationally-averaged spectra. Similarly the uncertainty in the average magnitudes $\hat{m}_V$, $\hat{m}_R$, etc. of the star at the epoch of observation yield gaussian priors. Even in the absence of a contemporaneous light-curve, the historical photometric record can be used to set priors for the magnitudes, albeit less well-constrained.

The flux received at Earth in the V band is related to the stellar image and the nuisance parameters by

$$f_V = \left(\frac{2\pi}{P_{rot}} \frac{V \sin i}{d. \sin i} \right)^2 \int \int I_V(M) \cos \theta(M) dM \tag{32}$$

This flux can then be converted to an apparent V magnitude using an appropriate zero-point and averaged over the appropriate times of observation. If a filling-factor model is used for the image, the average fraction of spots on the disc at any time can be determined from the V-band specific intensities and limb-darkening coefficients in the photosphere and spots. In practice, this requirement to match the mean light-curve level could be implemented by adopting this mean occupancy (or that required to match the maximum level of the light-curve) as the default value m for the image vector.

The search for the most probable solution should then take into account not just the fit to the data and the image entropy, but also the prior probabilities assigned to the final values of the nuisance parameters.

7   Surface Resolution and Noise

Although the final reconstructed image is usually displayed as a vector containing the most probable value for each pixel, a *bona fide* Bayesian treatment would

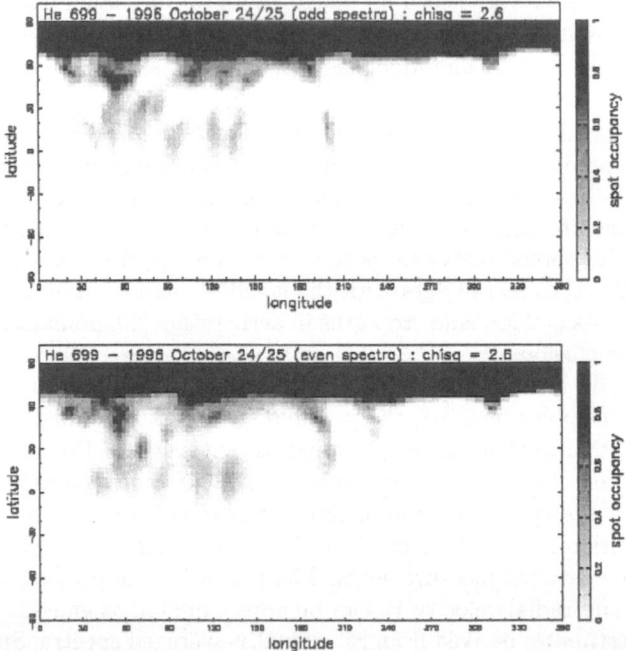

Fig. 5. Images of the rapidly rotating G dwarf He 699 in the α Persei cluster, derived from (top) odd-numbered spectra, (bottom) even-numbered spectra secured at the WHT on 1996 October 25/26 [1].

give the full posterior probability distribution for each pixel. At the very least, it should be possible to determine error bars on the image values. In doing so, we must appreciate that neighbouring pixel values are in general strongly correlated for the reasons outlined in Section 5.2 above. This means that it is not straightforward to establish the reliability of quantities derived from operations carried out in image space, such as cross-correlations of pairs of images taken at different times. Nonetheless, correlated errors are better than no error estimates at all.

Most efforts to determine the reliability of features seen in stellar surface images have taken the form of empirical consistency tests. Strassmeier et al. [37] compared the structure seen in images of EI Eri obtained from the same dataset with three independent codes. Collier Cameron & Unruh [7] compared images of AB Dor obtained from independent data in three different spectral lines. This approach has also been adopted by other authors, notably the long-running series of imaging papers by Strassmeier, Rice and collaborators in *A& A*.

A more quantitative approach was developed by Barnes et al. [1] for stellar images derived from densely-sampled series of line profiles obtained by least-squares deconvolution. The temporal sampling of these data was dense enough, and the S:N high enough, that images of comparable resolution could be derived from independent subsets of the data consisting of the odd-numbered and even-numbered exposures (Fig.5). Latitude-by-latitude cross-correlation of the resulting images yielded a strong peak at most latitudes. The width of this peak gives a good indication of the surface resolution attainable on the stellar surface. This method is analogous to the bootstrap method described by Marsh elsewhere in this volume for assessing the reliability of Doppler tomograms of accretion disks.

Berdyugina's Occamian method [4] has the useful feature that, once a solution is attained, the diagonal elements of the Hessian matrix serve as estimates of the inverse variances of the corresponding pixels in the image. While this approach does not yield so much information about the degree of blurring of the image (or equivalently, the scale of the smallest resolvable features), it does give an indication of which features in the image are significant enough to be believable.

8   Applications of Time-Series Doppler Imaging

In its present state of development, Doppler imaging is very successful in determining the locations of cool spots on stellar surfaces. It has long been known that the rotationally-modulated photometric light-curves of active dwarf stars show significant changes in morphology on timescales of a week or two. The active subgiants in RS CVn binaries appear to have more stable spot distributions which only show significant changes on timescales of a few months.

The solar analogy provides two likely explanations for the changing shapes of stellar light-curves. Individual sunspots have finite lifetimes, usually of little more than a month or so. On even shorter timescales, however, the relative positions of sunspot groups at different latitudes change, because the solar surface rotation rate decreases from the equator toward the poles.

8.1   Starspot Lifetimes

Doppler imaging has the ability to disentangle the effects of finite spot lifetimes from those of differential rotation. Several groups have carried out long-duration studies of individual stars. Berdyugina et al. [5,6] have mapped the spot distribution on II Peg at roughly three-monthly intervals since 1992. They find that maps taken a month or so apart (e.g. their 1998 October and November images) show almost identical spot structure. On timescales of order 3 months, some major spot groups persist but others emerge and disappear. In general, the lifetimes of individual active regions on this star appear to be of order 3 to 4 months. There is, however, a strong body of photometric evidence that the major spot groups tend to emerge at a pair of diametrically-opposed longitudes, which rotate at a stable rate. Similar results have been found for the RS CVn systems , σ Gem, and HR7275 [3].

Similar active-region lifetimes were deduced by Strassmeier & Bartus [42] in a record-breaking run of images of HR 1099 secured on 57 consecutive nights in 1996 November and December. In the "running mean" movie of the stellar surface reconstructed from these data, one persistent high-latitude spot dominates the map. Other satelite spots and appendages form and dissolve on timescales of 1 to 4 weeks, in agreement with Hall's [19] conjecture that smaller spots tend to have shorter lifetimes than the largest active regions. This is also, of course, true of the Sun.

A further intriguing possibility has been raised by a longer-term study of HR 1099 [50], incorporating a sequence of images running from 1981 to 1993. Like the authors cited above, they found major changes in the spot distribution on timescales of order 2 months or more. Some of the apparently longer-lived high-latitude spots appeared to rotate faster than the equatorial regions and to migrate gradually towards the pole, although the sparse sampling of the stellar surface distribution raises some doubts about the reality of this result. Strassmeier & Bartus [42] claimed to see some evidence of this poleward drift in some of the weaker spots in their densely-sampled sequence, but as the evidence is somewhat subjective and the required drift velocities are super-Alfvenic, it would be desirable to confirm these observations at the higher S:N ratios afforded by least-squares deconvolution methods.

Photometric evidence for similarly long-lived spot formation sites at diametrically-opposed longitudes has also been found on the rapidly rotating active dwarfs AB Dor and PZ Tel [23,24].

8.2   Starspots as Flow Tracers

Hall [19] studied a large sample of RS CVn light-curves, and came to the conclusion that the relative rate of differential rotation $\Delta\Omega/\Omega$ appears to decrease linearly with increasing rotation rate. This effectively means that the beat frequency $\Delta\Omega$ between the rotational frequencies at high and low latitudes is approximately independent of the stellar rotation rate.

High-resolution images based on densely-sampled series of least-squares deconvolved line profiles have been shown by Donati et al [12,14,15] and Barnes et al. [1,2] to achieve longitude resolution of order $3°$ at low and intermediate latitudes. This is demonstrated by cross-correlation of images derived from odd- and even-numbered spectra in a given time-series [1].

Donati & Collier Cameron [12] and Donati et al. [14] measured the latitude dependence of the rotation rate at the surface of AB Dor, by cross-correlating images taken four to six nights apart (Fig. 3). In these datasets, there was a substantial overlap in phase between most reliably mapped parts of the stellar surface on the nights concerned. The form of the differential rotation was found to be very similar to that of the Sun, and was fitted well with a solar-like differential rotation law of the form $\Omega(l) = \Omega(0) - \Delta\Omega \sin^2 l$.

Similar cross-correlation studies have since yielded unambiguous differential rotation results for the K dwarf PZ Tel [2], the pre-main sequence G star RX J1508.6 -4423 [15] and the M dwarf RE J1816 +541 (Barnes, this volume). It

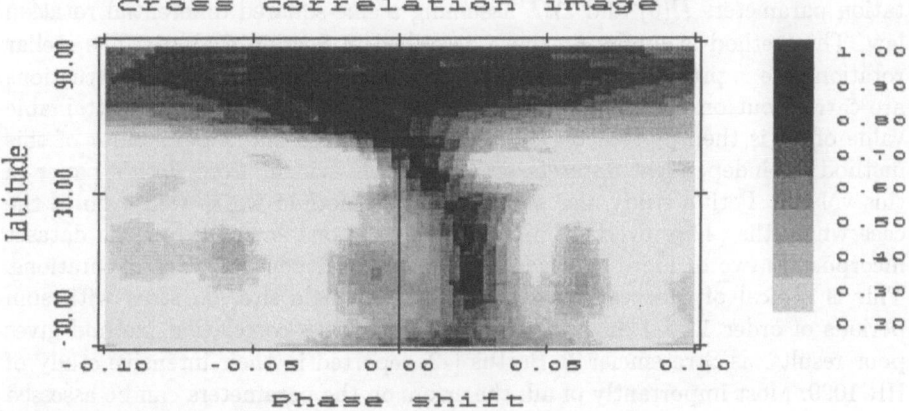

Fig. 6. Latitude-by-latitude cross-correlation function showing relative shifts in rotation phase for surface features on AB Dor between 1995 December 07 and 11 (from [12]).

has also yielded an upper limit for LO Peg [28]. Two images of the young main-sequence G dwarf He 699 in the α Per cluster, secured a month apart, failed to produce an identifiable cross-correlation ridge despite good phase overlap. The cross-correlation functions of images derived from the odd and even spectra on each night showed strong correlation ridges at zero shift over a wide range of latitudes, indicating that reliable flow tracers were present in the images. The most plausible explanation is that the lifetimes of the small-scale spots needed for a successful cross-correlation study have lifetimes less than a month. Similar studies of the G8 giant HD 51066 [39] and the FK Com star HD 199178 [41] also failed to produce identifiable cross-correlation ridges, probably for the same reason.

Solar-like differential rotation patterns have also been found on the weak-line T Tauri star V410 Tau [35] and the K0 giant IL Hya [52]. Sequences of Doppler images of a few other stars have been claimed to show that the polar regions rotate faster than the equatorial zones, notably HR 1099 [50], and HU Vir [38,20]. Confirmation of these results with higher S:N data and denser temporal sampling would provide a major challenge to our understanding of fluid circulation in stellar convective zones. However, we note that a subsequent study of HR 1099 by Petit (this volume) does not confirm the results of Vogt et al. [50], indicating a small but significant amount of solar-like differential rotation.

While the cross-correlation method has the advantage that it maps the differential rotation in a model-independent way, it is not straightforward to fit a differential rotation law to the results, as the errors on the cross-correlation function are difficult to quantify reliably. Indeed, this is true of any attempt to derive measurements from the reconstructed maps. A better-quantified way of determining differential rotation parameters has been developed by Donati et al [15], using the goodness of fit to the data to measure the differential rotation parameters $\Omega(0)$ and $\Delta\Omega$, assuming a sine-squared differential rotation law. The method is similar to those described in Section 6 above. The stellar rotation rate is prescribed as a function of latitude, and image reconstructions are carried out on a grid of values of $\Omega(0)$ and $\Delta\Omega$. The minimum attainable value of χ^2 is then plotted on this grid. The success and repeatability of this method for independent datasets on HR 1099 is evident from Petit's paper in this volume. Petit's study also shows that this method works well even in the case where the phase overlap on successive rotations is small, but the dataset incorporates two or more observations at each longitude over several rotations. This is typical of observations secured from a single site, on stars with spin periods of order 1 to 4 days. In such cases the cross-correlation method gives poor results, as Strassmeier & Bartus [42] reported in their intensive study of HR 1099. Most importantly of all, the errors on the parameters can be assessed directly from the χ^2 contours, which are derived directly from the data whose errors are independent and well-understood.

9   Conclusions

Tomographic imaging of stellar surface brightness distributions is now a mature branch of stellar astrophysics, capable of addressing fundamental questions about fluid circulation and magnetic-field generation in the outer convective zones of rapidly rotating stars. The availability of cheap, fast computers brings stellar surface imaging within the grasp of anyone with access to a high-resolution spectrographs on telescopes with apertures of 2 metres or more. While some improvements in imaging techniques are still possible and desirable, most of the existing codes currently in use are producing images of comparable quality showing similar features.

This places us in a good position to concentrate on the science rather than the nuts and bolts of debugging our codes. Differential rotation rates and starspot lifetimes need to be measured in representative samples of stars throughout the HR diagram. This will be time consuming, and may require large amounts of time on 4m-class (or bigger) telescopes for the fainter objects such as M dwarfs and T Tauri stars. The 8-metre telescope era may present us with the opportunities we need to secure these observations, as some observatories seek to run 4m-class telescopes more cheaply, allocating longer observing runs with fewer instrument changes. With careful target selection and efficient temporal sampling strategies, it should be possible for consortia of imaging groups to secure the necessary telescope time to track differential rotation on single and binary T Tauri stars, young main-sequence stars and evolved subgiants spanning a wide range of rotation periods and convective-zone depths. The results of existing long-term monitoring programmes, such as HR 1099 and AB Dor, need to be reassessed in order to determine whether different sampling strategies (perhaps securing images at intervals of a month or so rather than days and years) might yield new or better physical insights into their behaviour.

References

1. Barnes J. R., Collier Cameron A., Unruh Y. C., Donati J.-F., Hussain G. A. J., 1998, MNRAS, 299, 904
2. Barnes J. R., Collier Cameron A., James D. J., Donati J.-F., 2000, MNRAS, 314, 162
3. Berdyugina S. V., Tuominen I., 1998, A&A, 336, L25
4. Berdyugina S. V., 1998, A&A, 338, 97
5. Berdyugina S. V., Berdyugin A. V., Ilyin I., Tuominen I., 1998, A&A, 340, 437
6. Berdyugina S. V., Ilyin I., Tuominen I., 1999, A&A, 349, 863
7. Collier Cameron A., Unruh Y. C., 1994, MNRAS, 269, 814
8. Collier Cameron A., Jeffery C. S., Unruh Y. C., 1992, in Jeffery C. S., Griffin R. E. M., eds, Stellar Chromospheres, Coronae and Winds. Institute of Astronomy, Cambridge, p. 81
9. Collier Cameron A., 1992, in Byrne P. B., Mullan D. J., eds, Surface Inhomogeneities on Late-type Stars. Springer-Verlag, Berlin, p. 33
10. Collier Cameron A., 1995, MNRAS, 275, 534
11. Donati J.-F., Brown S. F., 1997, A&A, 326, 1135
12. Donati J.-F., Collier Cameron A., 1997, MNRAS, 291, 1
13. Donati J.-F., Semel M., Carter B., Rees D. E., Collier Cameron A., 1997, MNRAS, 291, 658
14. Donati J.-F., Collier Cameron A., Hussain G. A. J., Semel M., 1998, MNRAS, 302, 437
15. Donati J.-F., Mengel M., Carter B. D., Marsden S., Collier Cameron A., Wichmann R., 2000, MNRAS, 316, 699
16. Goncharsky A. V., Stepanov V. V., Khokhlova V. L., Yagola A. G., 1977, Soviet Astron. Lett., 3, 147
17. Goncharsky A. V., Stepanov V. V., Khokhlova V. L., Yagola A. G., 1982, SvA, 26, 690
18. Hale G. E., Nicholson S. B., 1938, Technical Report
19. Hall D. S., 1996, in Strassmeier K. G., Linsky J. L., eds, IAU Symposium 176: Stellar Surface Structure. Kluwer, Dordrecht, p. 217
20. Hatzes A. P., 1998, A&A, 330, 541
21. Hendry P., Mochnacki S., Collier Cameron A., 1992, ApJ, 399, 246
22. Hendry P. D., Mochnacki S. W., 2000, ApJ, 531, 467
23. Innis J. L., Thompson K., Coates D. W., Lloyd Evans T., 1988, MNRAS, 235, 1411
24. Innis J. L., Coates D. W., Evans T. L., 1990, MNRAS, 242, 306
25. Jankov S., Foing B. H., 1992, A&A, 256, 533
26. Kupka F., Piskunov N., Ryabchikova T. A., Stempels H. C., Weiss W. W., 1999, A&AS, 138, 119
27. Kürster M., 1993, A&A, 274, 851
28. Lister T. A., Collier Cameron A., Bartus J., 1999, MNRAS, 307, 685
29. O'Neal D., Saar S. H., Neff J. E., 1996, ApJ, 463, 766
30. Piskunov N. E., Tuominen I., Vilhu O., 1990, A&A, 230, 363
31. Piskunov N. E., Kupka F., Ryabchikova T. A., Weiss W. W., Jeffery C. S., 1995, A&AS, 112, 525
32. Press W. H., Flannery B. P., Teukolsky S. A., Vetterling W. T., 1992, Numerical Recipes: The Art of Scientific Computing (2nd edition). Cambridge University Press, Cambridge

33. Ribes J. C., Nesme-Ribes E., 1993, A&A, 276, 549
34. Rice J. B., Wehlau W. H., Khokhlova V. L., 1989, A&A, 208, 179
35. Rice J. B., Strassmeier K. G., 1996, A&A, 316, 164
36. Rice J. B., Strassmeier K. G., 1998, A&A, 336, 972
37. Strassmeier K. G. et al., 1991, A&A, 247, 130
38. Strassmeier K. G., 1994, A&A, 281, 395
39. Strassmeier K. G., Bartus J., Kovari Z., Weber M., Washuettl A., 1998, A&A, 336, 587
40. Strassmeier K. G., Rice J. B., 1998, A&A, 339, 497
41. Strassmeier K. G., Lupinek S., Dempsey R. C., Rice J. B., 1999, A&A, 347, 212
42. Strassmeier K. G., Bartus J., 2000, A&A, 354, 537
43. Terebizh V. Yu., Biryukov V. V., 1994, Ap&SS, 218, 65
44. Tikhonov A. N., 1963, Soviet Math. Dokl., 4, 1624
45. Unruh Y. C., Collier Cameron A., 1995, MNRAS, 273, 1
46. Unruh Y. C., Collier Cameron A., Cutispoto G., 1995, MNRAS, 277, 1145
47. Vogt S. S., 1981, ApJ, 250, 327
48. Vogt S. S., Penrod G. D., 1983, PASP, 95, 565
49. Vogt S. S., Penrod G. D., Hatzes A. P., 1987, ApJ, 321, 496
50. Vogt S. S., Hatzes A. P., Misch A. A., Kürster M., 1999, ApJS, 121, 547
51. Wade R. A., Rucinski S. M., 1985, A&AS, 60, 471
52. Weber M., Strassmeier K. G., 1998, A&A, 330, 1029

Imaging the Magnetic Topologies of Cool Active Stars

J.-F. Donati

Observatoire Midi-Pyrénées, 14 Avenue E. Belin, F-31400 Toulouse, France

Abstract. In the last decade, new indirect imaging techniques have been proposed to unveil the surface magnetic topologies of rapidly rotating active stars, which are found to be vastly different from that of the Sun. After a quick recall of how direct magnetic field (i.e. spectropolarimetric) signatures can be detected in such stars, I describe in detail to which extent surface magnetic field topologies can be recovered from extensive sets of time resolved, high signal to noise Zeeman signatures. I outline in particular how specific assumptions on the field structure (e.g. potential field, linear force-free field) can help improving the modelling, not only of the surface field map, but also of the large scale topology of the coronal field. I finally present the results obtained to date on how dynamo processes seem to operate in the atmospheres of cool active stars and how they differ from those of the Sun.

1 Introduction

Magnetic field is very often a crucial parameter in astrophysical problems. It indeed plays a key role basically everywhere in the universe, at all time and spatial scales, and in particular in stars, during their formation, their evolution and the last stages of their lives.

In the particular case of stars, the first concern is to understand the origin of large-scale magnetic fields, whether they are remnants (i.e. fossil fields) from older evolutionary stages (like those of some chemically peculiar stars), or whether they are self-generated (i.e. dynamo) fields resulting from turbulent motions in a partly ionised plasma within the stellar envelope (such as those of the Sun and other active late-type stars). Studying large-scale magnetic structures can also inform us on the impact of these fields on various transport processes operating within or around stars (such as convection, turbulence, diffusion, large-scale circulation, internal rotation, accretion or mass loss) and therefore on stellar evolution as well. For cool stars and the Sun in particular, a detailed and self-consistent picture of how large-scale fields are produced, evolve on both short and long timescales and affect the Sun and stars as well as their direct circumsolar/stellar environment is still crucially lacking at the moment, even though of obvious interest for us.

In order to progress significantly on these tracks, the best solution is probably to obtain maps of stellar magnetic topologies and to study how they vary with time, in very much the same way as done for the Sun but for a much wider range of stellar fundamental parameters (and in particular rotation rates and convective depths). Thanks to the new indirect stellar surface imaging techniques (see Cameron, this volume), it is now possible to resolve the surface of rapid rotators, and to recover the detailed structure of stellar magnetic fields. In this paper, I detail the various steps of this process, mainly the detection of Zeeman signatures and the reconstruction of magnetic field topologies from sets of rotationally modulated Zeeman signatures. I will explain in particular to which extent the orientation of field lines can be recovered, and how the modelling can be improved through specific assumptions on the field structure. I finally present the most recent results on how dynamo processes seem to operate in the atmospheres of cool active stars, and conclude by listing some very promising research directions in this field.

2   Detecting Zeeman Signatures of Cool Active Stars

The most unambiguous evidence for magnetic fields in stellar photospheres is undoubtedly the presence of circular and linear polarisation signals (called Zeeman signatures) in the profiles of spectral lines. As first observed by Zeeman at the end of the nineteenth century, these signatures strongly depend, not only on the field strength and atomic parameters of the spectral lines involved, but also on the orientation of the magnetic field vector with respect to the line of sight (see e.g. [19] for a detailed description of the atomic physics underlying the Zeeman effect); while a field oriented along the line of sight (i.e. longitudinal field) produces *circular* (or Stokes V) line profile polarisation only, a field perpendicular to the line of sight (i.e. transverse field) generates nothing but *linear* (or Stokes Q and U) line profile polarisation. The very fact that polarised Zeeman signatures are sensitive to the *vector* properties of the magnetic field mostly explains the four decades of unsuccessful attempts at detecting such signals in cool active stars (e.g.[2,3]) in these objects, the magnetic structure is indeed sufficiently complex for the stellar visible hemisphere to feature at any given time several field regions of opposite polarities, whose individual polarisation signatures thus mutually cancel out.

About a decade ago, Semel [22] proposed to measure Zeeman signatures in the particular case of *rapidly rotating* late-type stars. For such objects, regions of opposite magnetic polarities located at different radial velocities with respect to the observer contribute to the global Zeeman signature at different wavelengths and therefore no longer mutually cancel their effects in circular (or linear) polarisation (as they do on slow rotators). Following this idea, Donati et al. [16] were the first to detect successfully direct (i.e. polarisation) Zeeman signatures from a cool star other than the Sun (the K1 primary of the RS CVn system HR 1099). From time resolved spectropolarimetric observations of rapid rotators, one can thus obtain sets of rotationally modulated Zeeman signatures that can in principle be inverted back into a magnetic map of the surface field structure (with the help of a stellar imaging package, see Sect. 4), a method usually referred to as Zeeman-Doppler imaging.

Optimally, one would like to measure Zeeman signatures in individual line profiles, as one can do for solar magnetic regions for instance. However, in the particular case of late-type stars where Stokes V signatures rarely exceed 0.3% in relative peak-to-peak amplitude (due to dilution with non magnetic regions and cancellation with opposite polarity regions located within the same isovelocity strip), one has to extract the desired polarisation information from thousands of spectral lines simultaneously, with the help of cross-correlation tools such as 'Least-Squares Deconvolution' [7]. Using this technique, magnetic fields have been directly detected in about 20 objects of various evolutionary stages ranging from pre main sequence to red giant branch ([7,8] and mapped in six of them ([6,9]. Stokes Q and U Zeeman signatures are expected to be typically more than an order of magnitude smaller and have never been detected yet in cool stars other than the Sun. As an illustration, Fig. 1 presents the rotationally modulated Stokes V signature (extracted from about 1,500 spectral lines with Least-Squares Deconvolution) for the rapidly rotating young K0 dwarf AB Doradus [10]. In this graph, Zeeman signatures can be very clearly traced throughout their transit from the blue wing to the red wing of the mean line profile, as the corresponding individual magnetic regions on the stellar surface are carried across the visible disc by rotation. From the complex shape of the detected Zeeman signatures at any given rotation phase, and in particular the numerous zero crossings that the polarised signals feature throughout the line profile, one can already conclude that the field distribution is very complex, with many regions of opposite polarities simultaneously present on the visible stellar hemisphere.

3   Sensitivity of Dynamic Spectra to Field Orientation

Before straightforwardly attempting to invert sets of rotationally modulated Stokes signatures similar to that of Fig. 1, it is wise to explore how such data sets respond to different magnetic structures and in particular to different field orientations, by examining synthetic dynamic spectra corresponding to very simple field configurations. The elementary field distributions considered in this purpose all feature zero magnetic field over the stellar surface except in one small circular region (covering 5% of the total stellar area and located at rotation phase 0.5) in which the field strength is 1 kG and the field orientation is uniform. The only parameters that vary between the various elementary field distributions considered here are the spot latitude (set to either 20° or 70°) and the field orientation within the spot (assumed either purely radial – i.e. perpendicular to the stellar surface, meridional – i.e. directed along meridians, or azimuthal – i.e. lying along parallels). Altogether, six different elementary field distributions are used, whose corresponding synthetic Stokes V, Q and U dynamic spectra are shown on Fig. 2, Fig. 3 and Fig. 4 in the particular case of a star with an inclination angle i of 30° between the rotation axis and line of sight, and a line of sight projected equatorial rotation velocity $v \sin i$ of 40 km s^{-1}. For this computation, the intrinsic line profile is assumed to be a Gaussian with a relative central depth of 45%, a full width at half maximum of 10 km s^{-1} (including small and

AB Dor, 1996 Dec. 23 & 25, Stokes V

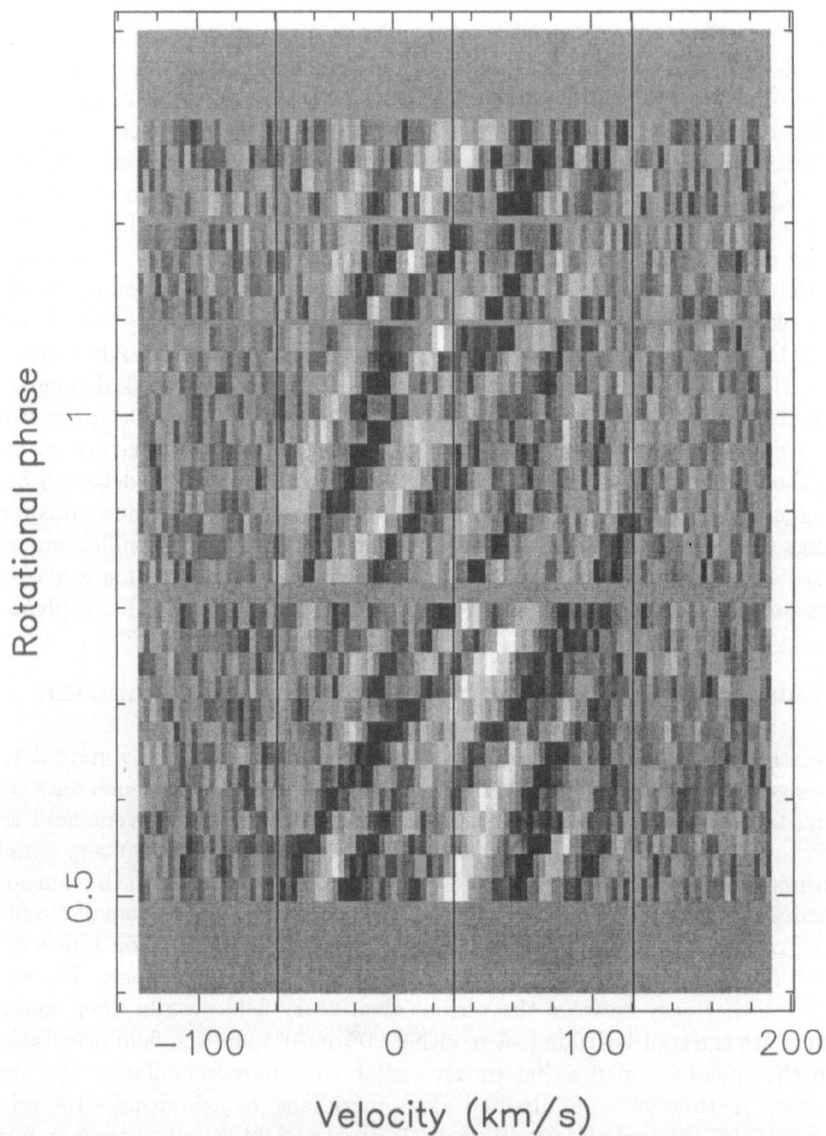

Fig. 1. Rotationally modulated Least-Squares Deconvolved Stokes V signature of AB Dor in 1996 December. Black and white code relative circular polarisation levels of −0.06 and 0.06% respectively. The central and side vertical lines depict the radial and rotational velocities of AB Dor.

large scale turbulence and instrumental broadening) and a Landé factor of 1.2. A linear continuum centre-to-limb darkening constant of 0.75 was also used. This is typical to what is observed for cool stars [6,9]. We also assume, following [9], that the weak magnetic field approximation (e.g. [25]) holds.

Unsurprisingly, we first find that the location and amplitude of the Zeeman signature trajectories in the dynamic spectra can be used to recover the longitude and latitude of the corresponding magnetic region, in exactly the same way as in conventional brightness Doppler imaging (see Cameron, this volume); low latitude features produce large velocity amplitude waves that are only visible for a fraction of the rotation cycle, while high latitude ones generate almost uninterrupted small velocity amplitude tracks.

To retrieve information on the orientation of field lines within the magnetic region, one has to look carefully on how the local Zeeman signature gets distorted during its transit throughout the line profile. For instance, the Stokes V Zeeman signature from a high latitude magnetic region will switch sign twice per rotation cycle if the field is either meridional (at phase 0.4 and 0.6 approximately) or azimuthal (at phase 0.0 and 0.5), while it remains constant in sign for a radial field. More generally, we observe that each Stokes V dynamic spectrum is related to a specific spot location and field orientation, except those for low latitude radial and meridional field regions which are found to be very similar. We can thus conclude that Stokes V dynamic spectra contain enough spatial information on the parent magnetic field structure to extract, not only the location of magnetic regions on the stellar surface, but also the orientation of field lines within these regions except for low latitude radial and meridional field regions expected to suffer some degeneracy (see [7] for more details).

Adding up Stokes Q and U dynamic spectra to the data set should in principle be sufficient to remove this degeneracy; as one can see on Fig. 3 and Fig. 4, dynamic spectra corresponding to low latitude meridional and azimuthal field regions are no longer similar. Time resolved Stokes Q, U and V observations of cool stars should thus be adequate, whenever available, to reconstruct unambiguously complex stellar surface magnetic field structures.

4   Recovering a Magnetic Topology

To validate the results of the previous section, the adequate test consists in trying to recover any given magnetic field structure (i.e. the magnetic surface topology of all three magnetic field components) from sets of synthetic dynamic spectra with the help of an automated stellar surface imaging package. Comparing the reconstructed image with the original one used to produce the synthetic data set informs us on how successful the process is.

The first package we use is that of Brown et al.[3], based on Skilling & Bryan's original algorithm for maximum entropy (MaxEnt) image reconstruction [23]; images reconstructed with this software are referred to as MaxEnt maps in the following. In this case, the quantities we reconstruct are the local magnetic fluxes for each field component on a 5,000 point grid at the surface

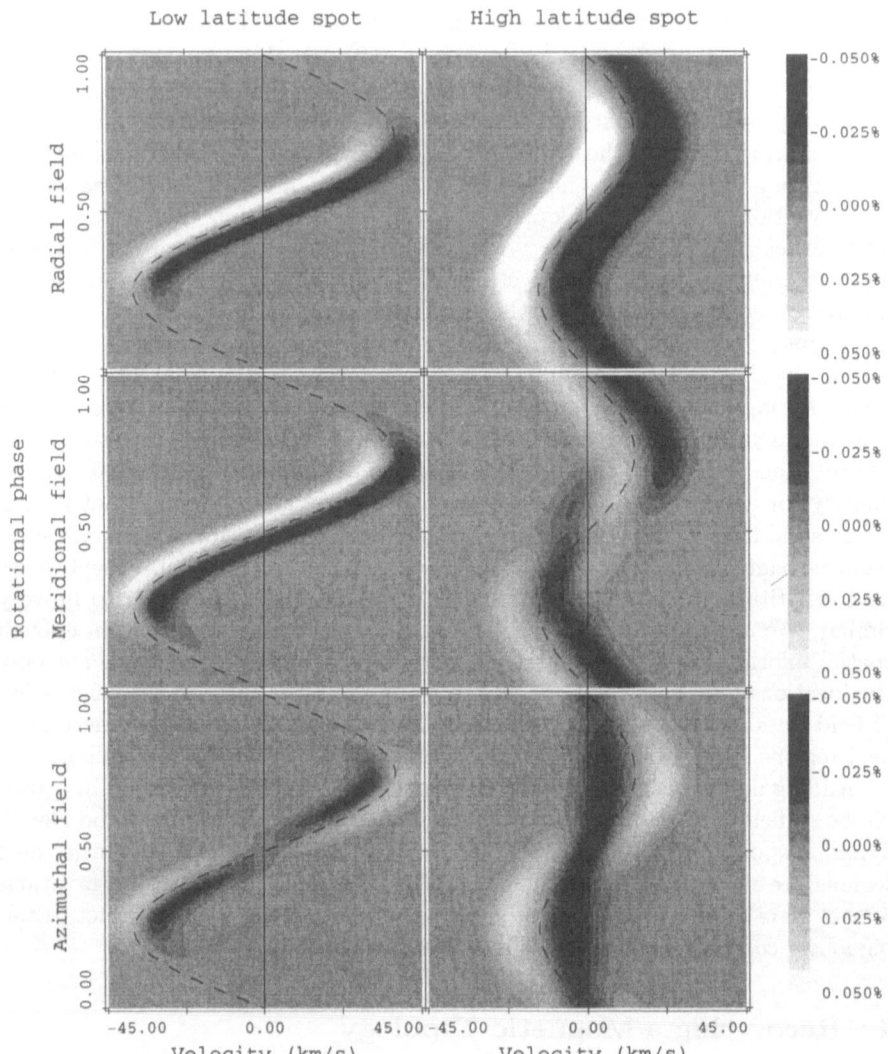

Fig. 2. Synthetic Stokes V dynamic spectra for 6 elementary magnetic field structures, consisting of a single magnetic region with uniformly oriented 1 kG magnetic field. The field orientation is either radial (top panels), meridional (middle panels) or azimuthal (bottom panels), while the spot latitude is either 20° (left panels) or 70° (right panels). The vertical line in the middle of each panel depicts the radial velocity of the synthetic star (set to zero), while the dashed line illustrates the radial velocity of the spot centre as the star rotates.

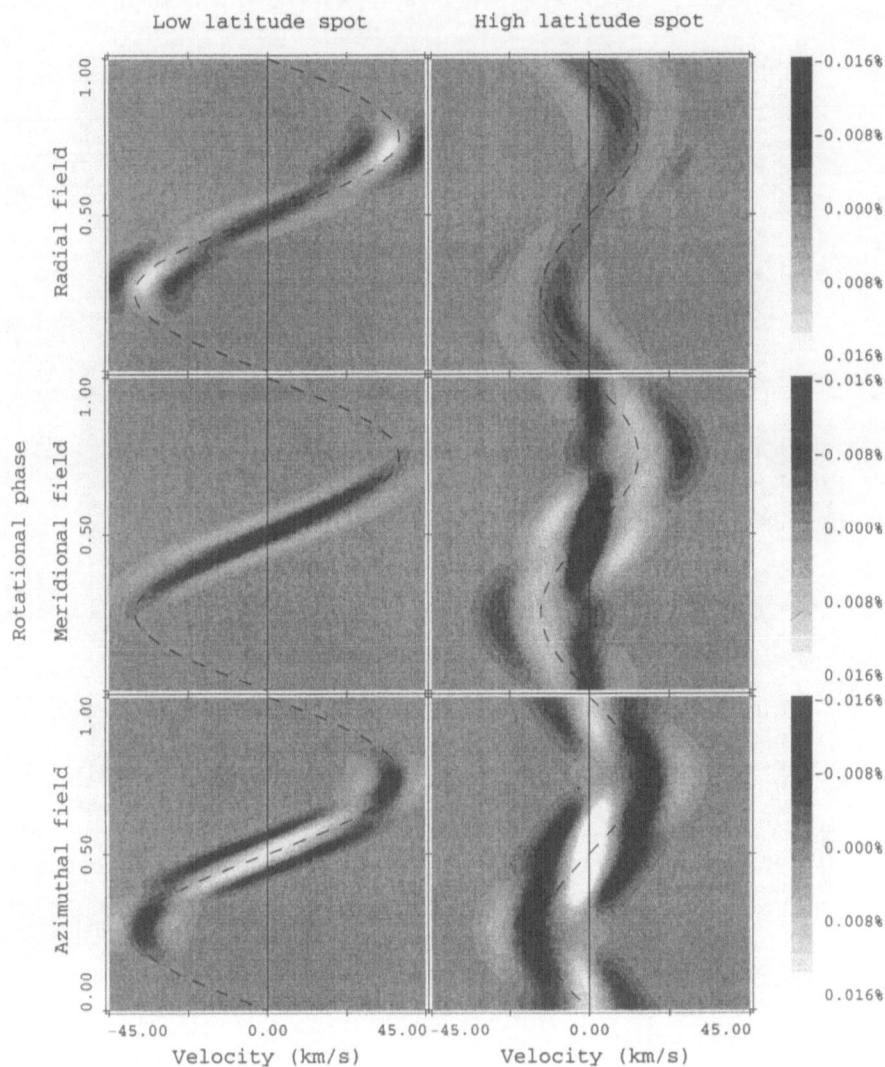

Fig. 3. Same as Fig. 2, but for Stokes Q dynamic spectra. Note that these signatures were expanded by a factor of 10.

of the star (corresponding to a maximum spatial resolution of about 4° at the stellar equator). We also developed a second method, based on spherical harmonics decomposition (SHaDe) of magnetic field components (with ℓ orders and m nodes of up to 40) similar to (though more powerful and versatile than) the preliminary attempt presented by Hussain et al. [17]. Once again, we use maximum entropy image reconstruction, though the quantities we now reconstruct are the

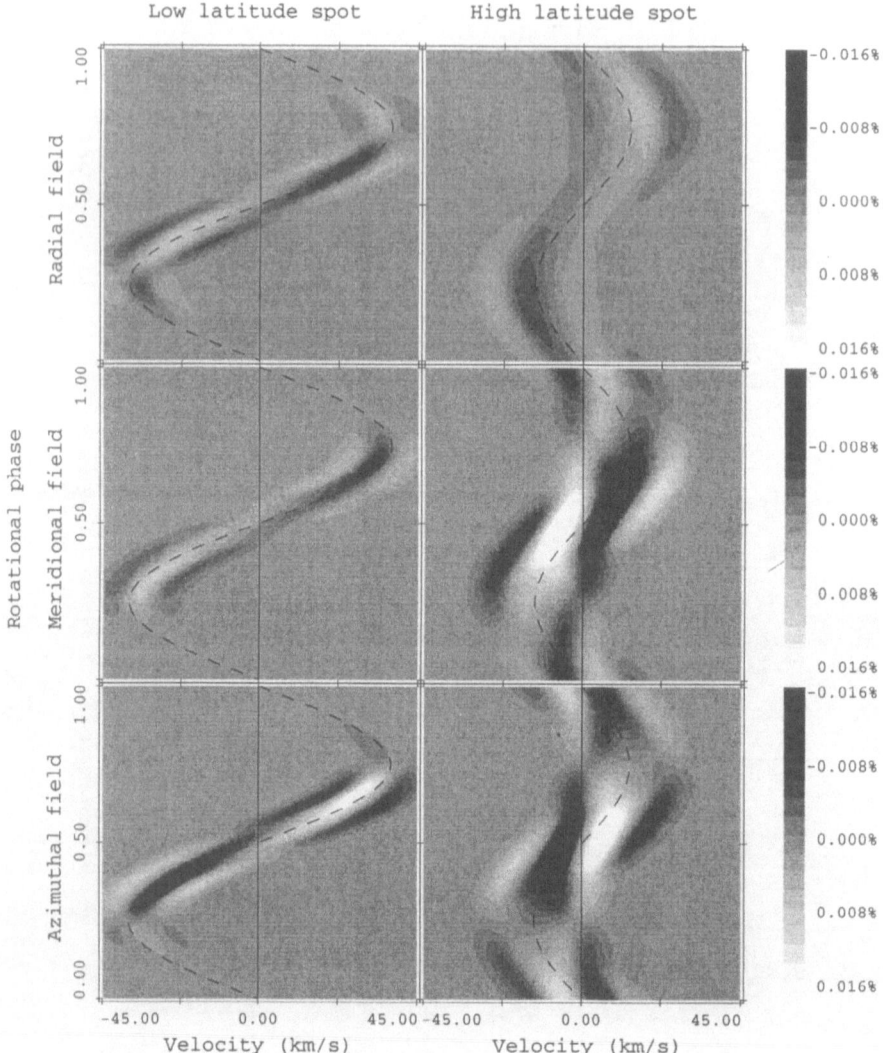

Fig. 4. Same as Fig. 2, but for Stokes U dynamic spectra. Note that these signatures were expanded by a factor of 10.

(complex) coefficients of the various spherical harmonics modes of each field component. Images reconstructed with this software are referred to as SHaDe maps in the following.

Even though the number of free parameters in the inversion process is about three times smaller for the second method, the computing time is considerably longer. The best way we found to compute how data points are affected by a change in the spherical harmonics mode coefficients (i.e. the response matrix) is to apply the response matrix of the MaxEnt process (indicating how data points are effected by a change in the local field components) onto another matrix whose columns contains the normalised surface maps of the spherical harmonics modes of interest; as both matrices are rather big (each containing about 10^7 points), this transformation requires of the order of 10^{11} floating point operations and can easily take a few CPU hours on a new generation workstation. Despite this drawback, SHaDe is especially interesting for several aspects. The main advantage is that we can dictate a priori relations between the spherical harmonics coefficients and impose specific constraints on the large scale field structure, e.g. that the field is potential (i.e. no currents) or force-free (i.e. currents lying along field lines). It ensures at the same time that the reconstructed field is realistic and obeys Maxwell's equations, a constraint that the MaxEnt inversion process cannot easily integrate. Other interesting advantages of SHaDe (compared to MaxEnt) are that the field can, in some cases, be recovered in cool regions (although they emit very few photons), or extrapolated rather easily up to the corona.

The best way of quantifying the reconstruction capabilities of Zeeman Doppler imaging as well as the specific advantages of both MaxEnt and SHaDe inversion processes is to try them on various test cases. Two such cases are presented below. The first one corresponds to a magnetic distribution for which the field is everywhere zero except in three regions (two circular spot and one ring), each with constant field strength (1 kG) and orientation (see Fig. 5, left column). This distribution is supposed to represent a typical (though admittedly oversimplified) magnetic distribution of a cool active star. For the second test, we use a dipole field structure with a 1 kG polar field strength and an axis of symmetry tilted by 70° with respect to the rotation axis (see Fig. 5, right column). Although dipole fields are not expected to resemble the complex magnetic structures of cool active stars, they are nevertheless very useful for us to check the general behaviour of the imaging process. For both tests, the stellar inclination angle i and projected rotation velocity $v \sin i$ are assumed to be 30° and 40 km s^{-1} respectively, and the intrinsic line profile model is the same as that described in Sect. 3.

For field structures like those of active stars, a quick glance at the reconstructed images demonstrates that MaxEnt reconstructions are fairly reliable (see Fig. 6). A more detailed look reveals that, as expected from Sect. 3 and discussed at length in [7], magnetic imaging from Stokes V data alone is very efficient at distinguishing azimuthal field from radial/meridional field regions, but suffers significant crosstalk between low latitude azimuthal and meridional field regions. We find that this crosstalk is significantly reduced when both circular and linear polarisation profiles are used in the reconstruction. The SHaDe reconstruction from Stokes V data alone (see Fig. 7) assuming either an unconstrained field structure or a linear combination of two linear force-free fields are

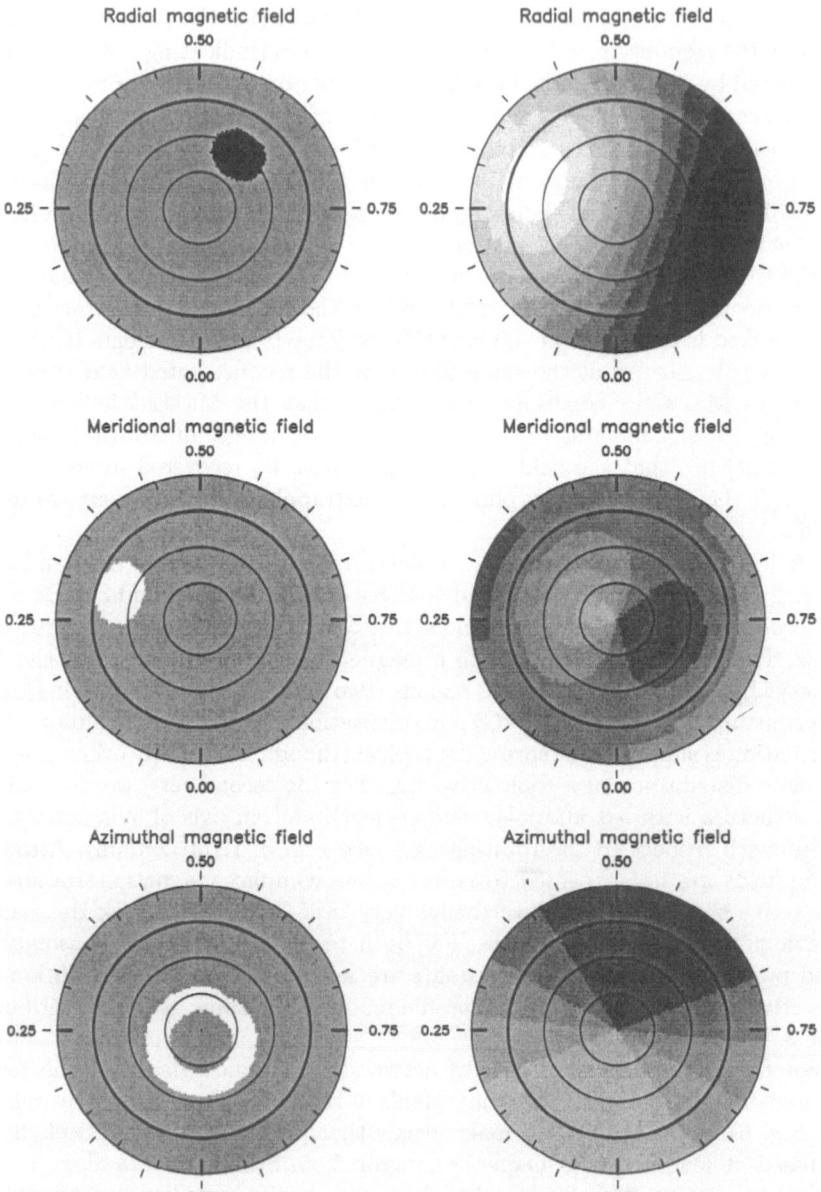

Fig. 5. Synthetic magnetic field structures used to evaluate the performance of Zeeman-Doppler imaging. Each column presents a single image through its magnetic field components in spherical coordinates. The star is shown in a flattened polar view with parallels drawn as concentric circles every 30° down to a latitude of −30°, the bold circle and the central dot denoting the equator and visible pole. Black and white code field intensities of −1 and 1 kG respectively. The radial ticks around each plot depict the rotational phases used in the reconstruction.

also acceptable[1]. The unconstrained SHaDe image only differs significantly from the first MaxEnt reconstruction in regions below the equator, i.e. where visibility is low and limb darkening strong. The second SHaDe image is also reasonably similar, although slightly less clean than the previous MaxEnt images; however, one should keep in mind that this SHaDe solution has the uppermost advantage of being much more realistic than the three others (as it automatically verifies Maxwell's equations), and even more realistic in particular than the input image. Note that crosstalk is still detected between low latitude radial and meridional field features in the two SHaDe images, as expected from the fact that Stokes V data alone were used in the reconstruction.

Our second series of tests unsurprisingly confirm Brown et al.'s earlier result [4] that dipolar magnetic structures cannot be properly recovered through a MaxEnt process when using Stokes V data alone (see Fig. 8, left column). Note in particular that both poles are reconstructed close to the subobserver latitude (60°) and not at all at their original location (20° and –20° for the positive and negative poles respectively). Even more disappointing is the findings that adding up linear polarisation profiles (i.e. Stokes Q and U) to the data set only *very marginally* improves the situation (see Fig. 8, right column), conversely to what Piskunov (1998) recently claimed. While the unconstrained SHaDe image is only slightly better (see Fig. 9, left column), the assumption that the field structure is a linear combination of two linear force-free fields (with α values of 0.001 and 1000 respectively) changes the situation completely, the solution being now very similar to the input image (see Fig. 9, right column). A SHaDe reconstruction assuming a potential field (not shown) does an even better job, producing a magnetic structure virtually indistinguishable from the original one.

Another interesting advantage of SHaDe reconstructions is that it can, in some cases, reconstruct magnetic flux hidden from the observer's view in very cool regions. As demonstrated in Fig. 10, the dipolar structure of the previous test case can still be properly reconstructed (assuming a dual linear force free field structure), even if both magnetic poles host large, very cool spots. This peculiarity should be very useful to extract accurate field structures in cases where brightness or abundance distributions are so contrasted that we get no information on the outshined parts of the star. This is of course a feature that MaxEnt reconstruction cannot provide, being intrinsically designed to produce the image with the least amount of information among those that fit the data to a given accuracy level (the missing flux being then simply not reconstructed on the image, see Fig. 10, left column)

Very similar results are obtained when using higher stellar inclination angles (say 60°) or higher projected rotation velocities (say 90 km s^{-1}).

We thus conclude that in all cases, the SHaDe imaging process performs as well as (for complex field structures), or significantly better than (in case of low

---

[1] Unsurprisingly, SHaDe reconstructions assuming either a potential field or a single linear force free field do not yield adequate fits to the data. Azimuthal field rings are indeed only compatible with high α linear force free topologies, whose almost pure toroidal structures are then incompatible with non zero radial field components.

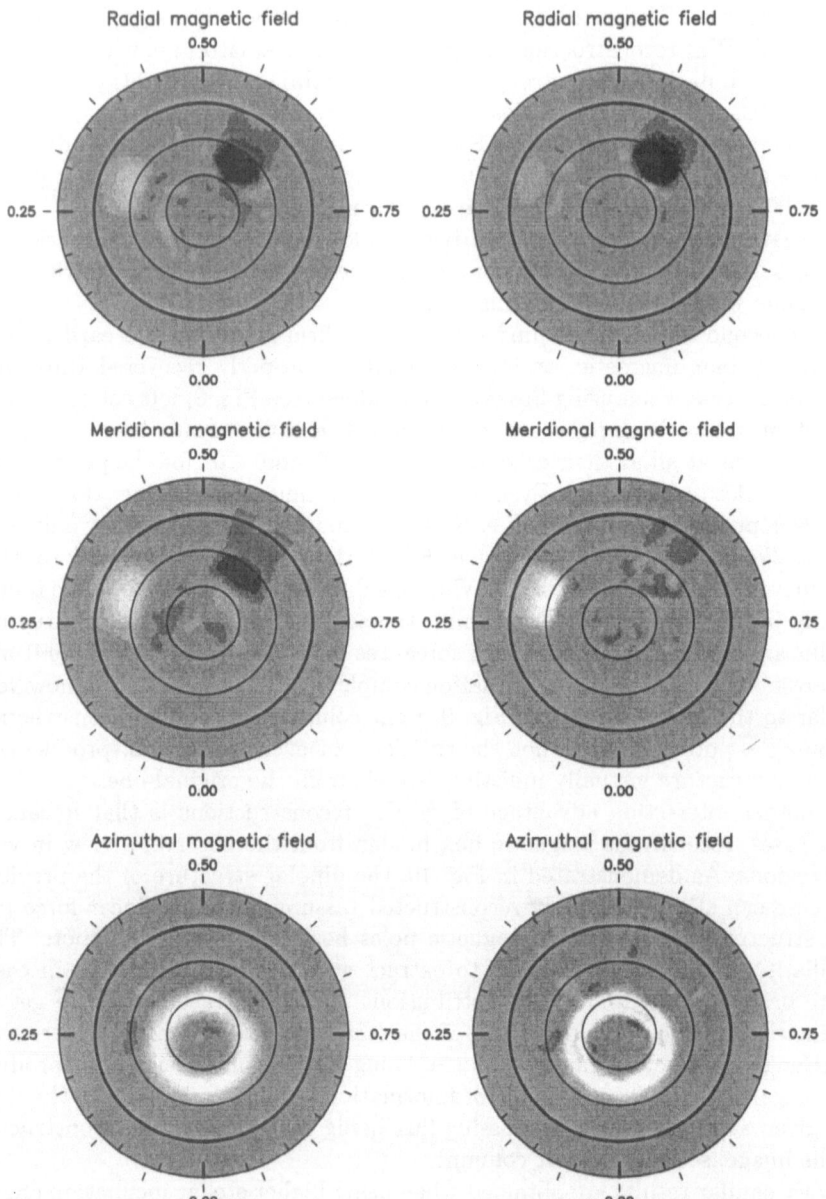

Fig. 6. MaxEnt reconstructions of the synthetic field distribution shown in the left column of Fig. 5, from Stokes V data only (left column) and from Stokes Q, U and V data (right column).

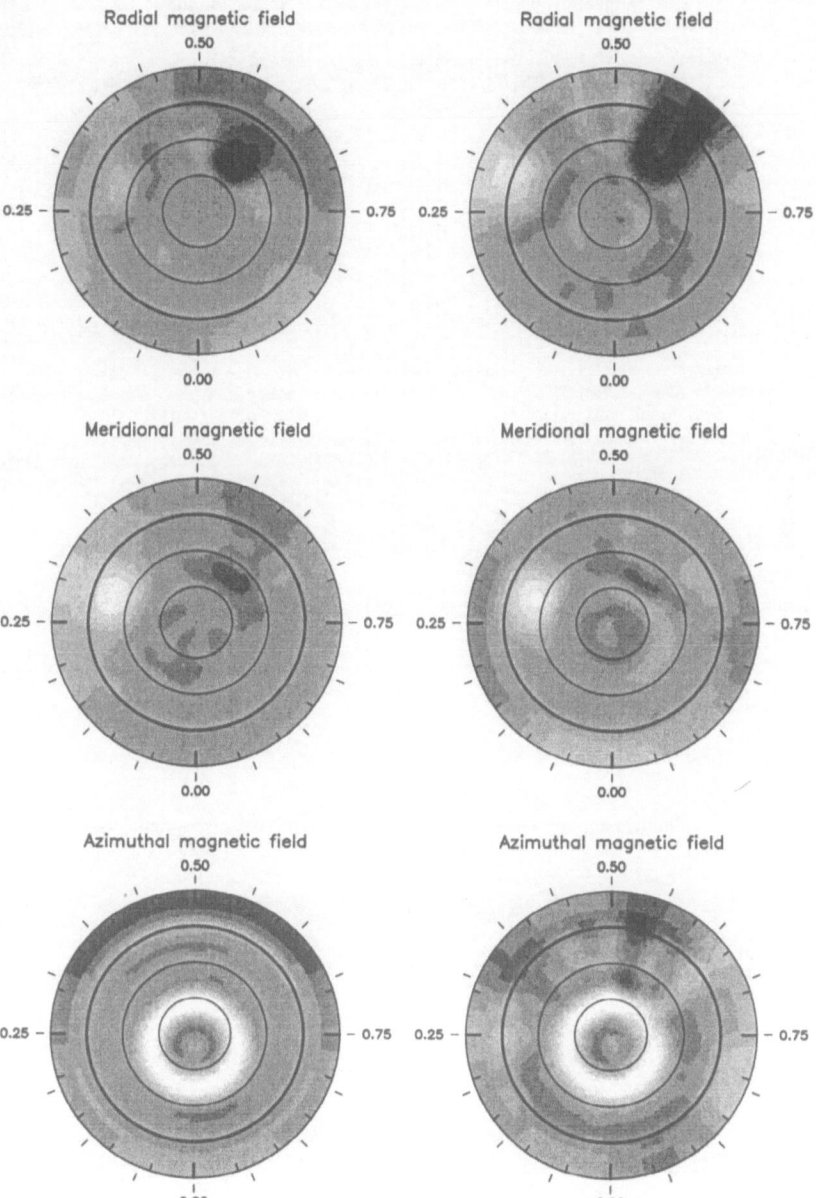

Fig. 7. SHaDe reconstructions of the synthetic field distribution shown in the left column of Fig. 5 from Stokes V data only, assuming either an unconstrained field structure (left column), or a linear combination of two linear force-free fields with α values of 0.001 and 1000 respectively (right column).

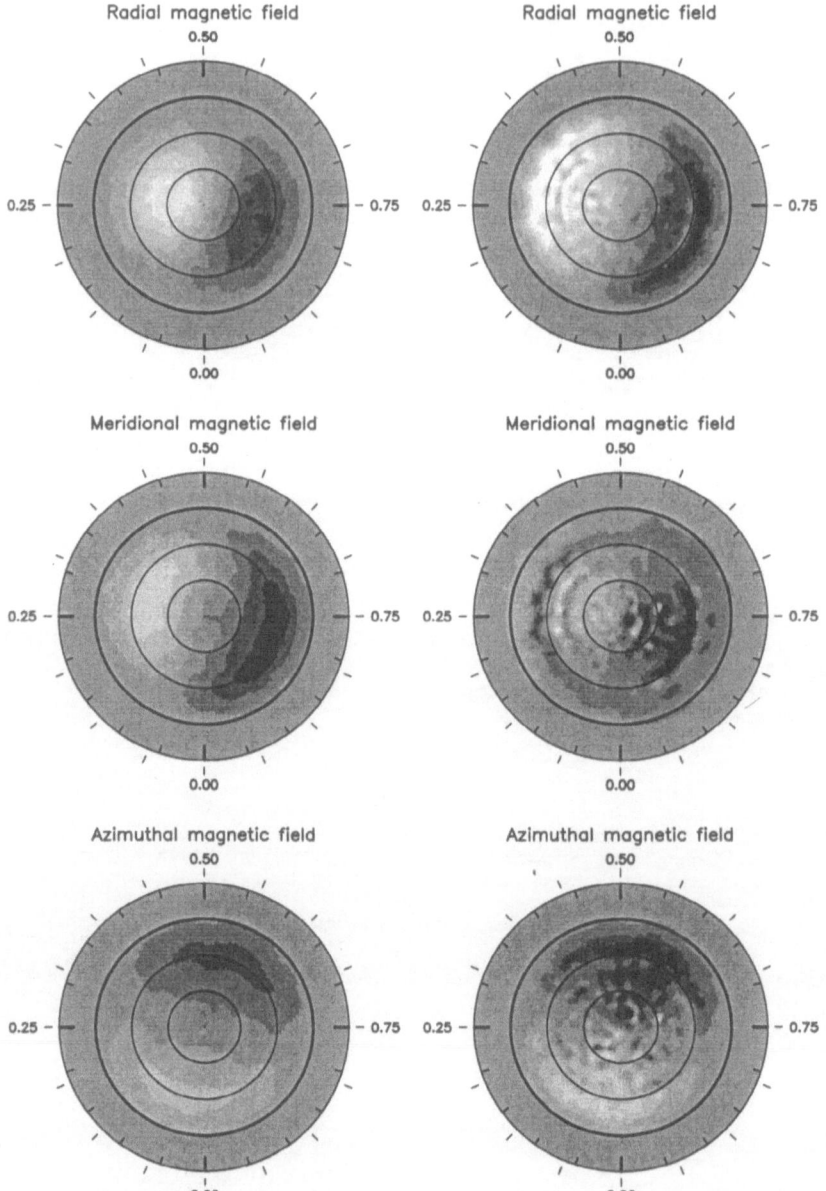

Fig. 8. Same as Fig. 6 for the synthetic field distribution shown in the right column of Fig. 5.

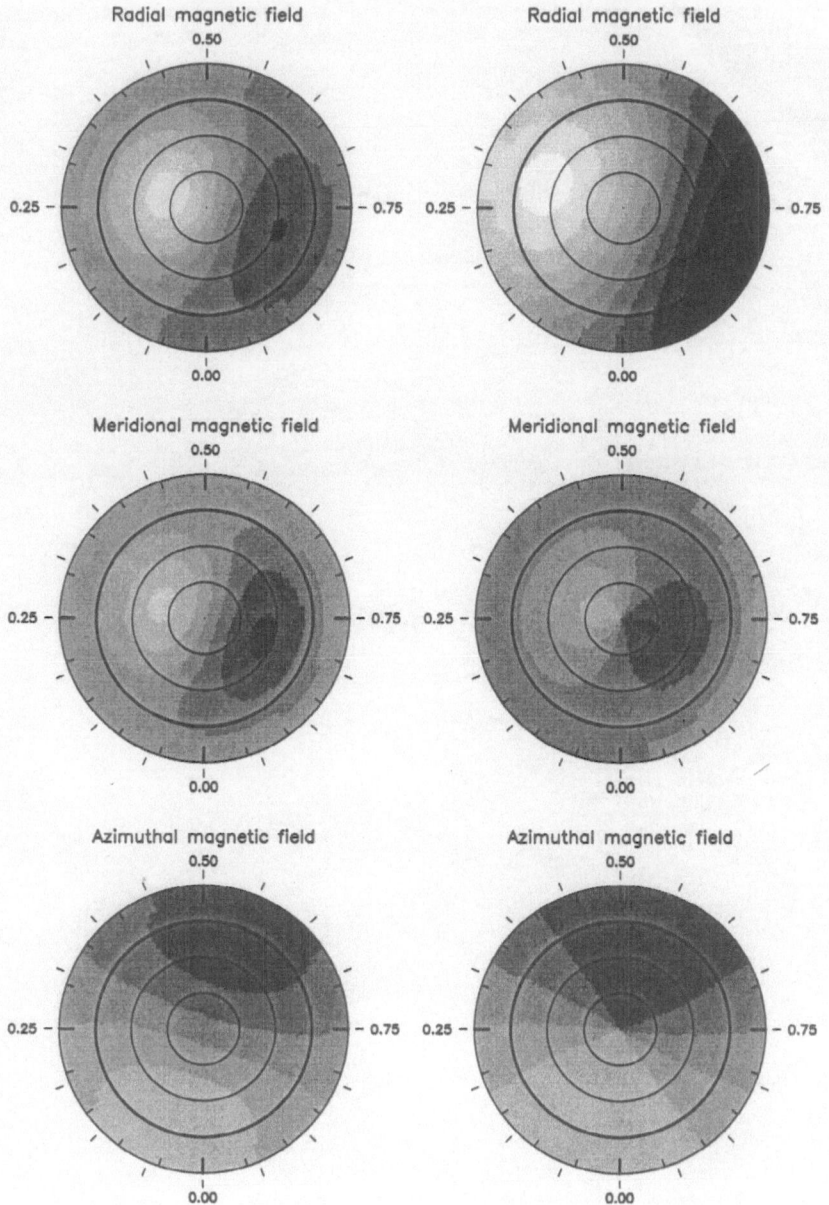

Fig. 9. Same as Fig. 7 for the synthetic field distribution shown in the right column of Fig. 5.

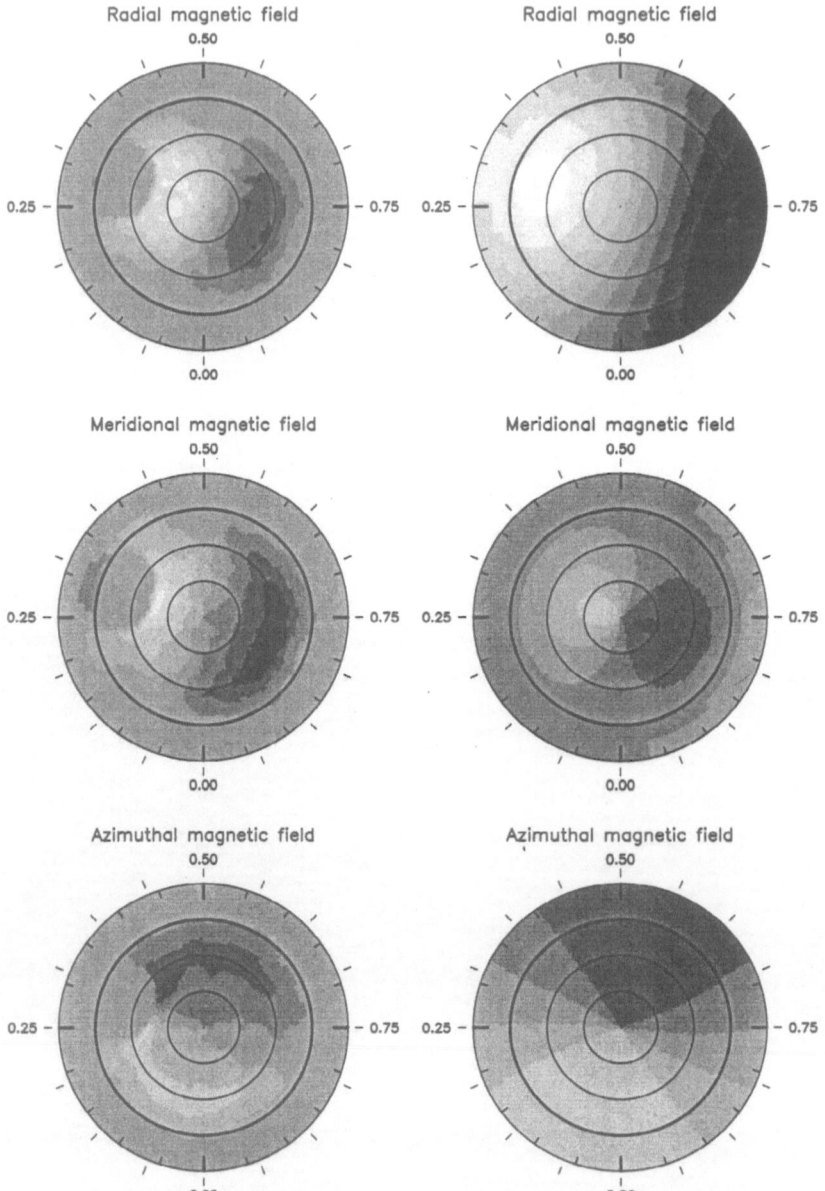

Fig. 10. MaxEnt (left column) and dual linear force free field SHaDe (right column) reconstructions of the synthetic field distribution shown in the right column of Fig. 5, from Stokes V data alone. We now assume that each magnetic pole host a very cool circular spot (each covering 5% of the total stellar area) and take into account the reconstructed brightness image in the inversion process.

order magnetic multipoles), the more conventional MaxEnt one. The assumption that the field structure is a linear combination of two linear force-free fields (with small and large α values respectively) is found to produce satisfactory results for all types of field structures. This is not altogether very surprising given the fact that this decomposition shares much similarity with the universal potential plus toroidal field decomposition.

5   Application to the K1 Subgiant of HR 1099

The ultimate test for illustrating how Zeeman-Doppler imaging performs and what we can learn on stellar magnetic topologies is to apply the afore mentioned inversion processes to a real data set. The one selected here focusses on the K1 primary star of the RS CVn system HR 1099, one of the most active star of the whole sky (with a rotation rate of 2.22 rad/d, i.e. about 10 times solar, and a relative convective depth of 85%). This data set was collected between 1998 Feb. 5 and 20 with the MuSiCoS high resolution spectropolarimeter [12] and the 2 m telescope Bernard Lyot atop Pic du Midi in France, and consists of 16 evenly phase spaced, high signal to noise ratio, Stokes V observations of the target star (S/N ranging from 4,700 to 12,200). The corresponding dynamic spectrum is shown in Fig. 11. For all reconstruction presented below, the stellar inclination angle and projected rotation velocity were respectively set to 38° and 40.5 km s^{-1} [6]. The intrinsic mean line profile is the same as that of Sect. 3 and 4.

The reconstructed field topologies are shown in Figs. 12 and 13. The MaxEnt image (see Fig. 12, left column) is very similar to other maps of the same star already published in the literature and derived from data sets collected at other epochs with a different instrumentation [6,8]. In particular, the most conspicuous and intriguing feature, the azimuthal field rings of positive and negative polarities encircling the star at low and high latitude respectively, are again clearly visible on this new image. Note that, due to significant offcentring towards phase 0.55, the high latitude (incomplete) azimuthal field ring includes the rotation pole and therefore shows up in spherical coordinates as two orthogonal bipolar groups centred on the rotation pole (one in the azimuthal field map and the other in the meridional field map). Various additional regions of radial field are also present at all latitudes at the surface of the star (the strongest ones being mostly of positive polarity). Low latitude meridional field regions coincide rather well with low latitude radial field regions and are thus likely attributable to the crosstalk effect discussed in Sect. 3.

The unconstrained SHaDe image (Fig. 12, right column) is perfectly compatible with the MaxEnt map and includes in particular all features mentioned above (i.e. the two azimuthal field rings plus several radial field spots) with very similar shapes and locations. The only noticeable difference is that the SHaDe field structure is slightly less spatially resolved than the MaxEnt one; this is again in agreement with the basic principles of maximum entropy image reconstruction whose goal is to produce the sharpest map compatible with the data.

HR 1099, 1998 Feb. 05–20, Stokes V

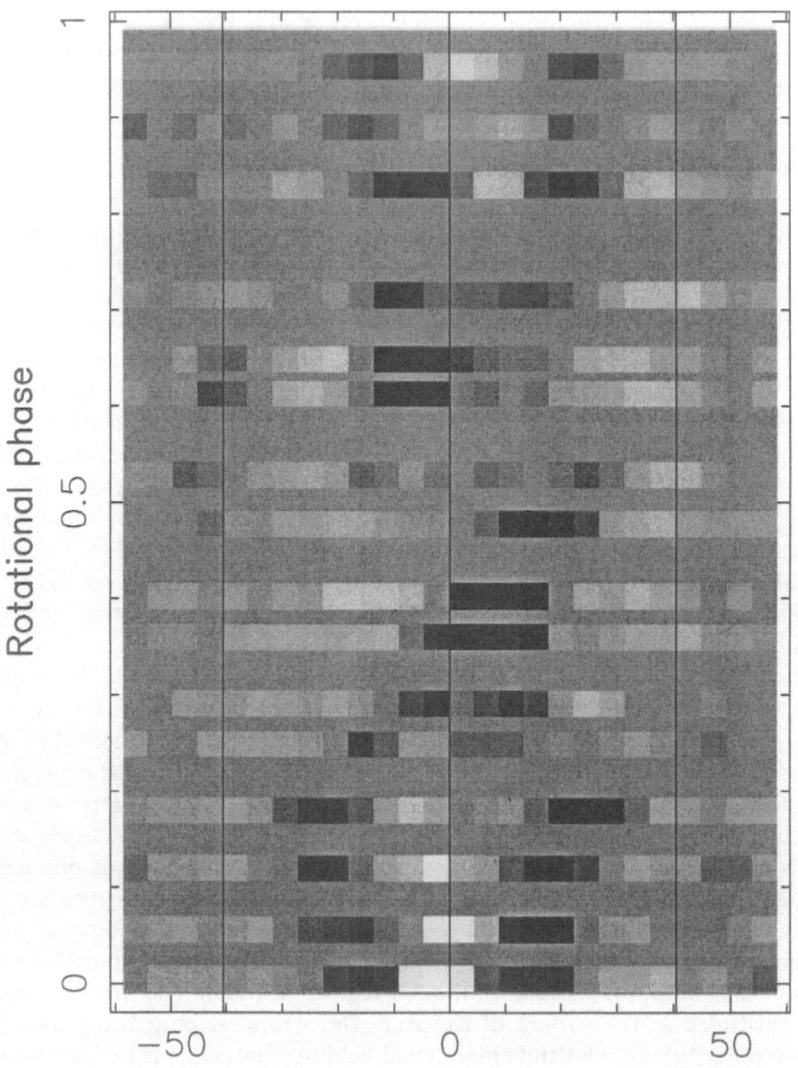

Fig. 11. Stokes V dynamic spectrum of the K1 subgiant of HR 1099, collected between 1998 Feb. 05 and 20. Black and white respectively code relative circular polarisation levels of –0.1% and 0.1%. Note that the velocity scale refers to the stellar rest frame. The two vertical lines on each side depict the rotational broadening of the mean line profile, while the middle one illustrates the line centre.

Assuming a dual linear force free field structure such as those used in Sect. 4, the SHaDe inversion process is able to converge very easily and produce a noise limited fit to the data. The corresponding field structure (Fig. 13, left column) is once again very similar to the two previous ones. The only noticeable differences are found to occur below the equator, a region on the stellar surface that contributes very little to Stokes V data as already explained in Sect. 4. Note that taking into account the brightness map reconstructed from Stokes I profiles only marginally changes the SHaDe image. From these three different images, we can already be rather confident on the reliability of the various map features; by taking pairs of images and computing the standard deviation over the differential map, we derive a typical error bar of order 120 G indicating that most of the afore mentioned features (whose strength ranges between 0.5 and 1 kG) are clearly detected.

To obtain still further evidence on the reality of the reconstructed surface features, we finally tried to run SHaDe reconstructions assuming successively a potential field structure and a single linear force free field with high α. While the former structure is incompatible with azimuthal field rings encircling the star (which would generate currents), the latter is close to a purely toroidal structure and thus forbids the presence of strong radial field regions. We obtain that the data cannot be fitted (i.e. the reduced χ^2 cannot decrease below 4.3) if the magnetic structure is forced to a potential field, strongly suggesting once again that the azimuthal field rings detected on HR 1099 are indeed real. Moreover, although the data cannot completely rule out a high α linear force free field, reaching a unit reduced χ^2 requires the SHaDe inversion process to perform an incredibly large number of iterations (in excess of 300, i.e. 5 to 40 times more than in the three previous cases). Unsurprisingly, the corresponding reconstructed map (Fig. 13, right column) exhibits much stronger contrasts and information content than the previous ones, thereby indicating that a single high α linear force free field is less likely to provide an adequate representation of the magnetic structure of HR 1099 than, say, the dual linear force free field model. Equivalently, we can conclude that the radial field regions discussed above are most likely real.

6 Dynamo Processes in Cool Active Stars

Among all magnetic features mapped at the surface of the K1 subgiant of HR 1099 (or other similarly active rapidly rotating cool stars, e.g. [6,8–10], the most intriguing ones are undoubtedly the regions in which the field is mostly azimuthal. The new SHaDe analysis presented above not only confirms the reality of these features, but also indicates that they represent the toroidal component of the large scale field structure. Similarly, the radial field spots we also reconstruct are also very likely real, and correspond to the poloidal component of the large scale field structure. We also find that the *polarity* of these features is essentially longitude independent at a given latitude; in certain cases, these regions even show up as complete azimuthal field rings encircling the star, as for

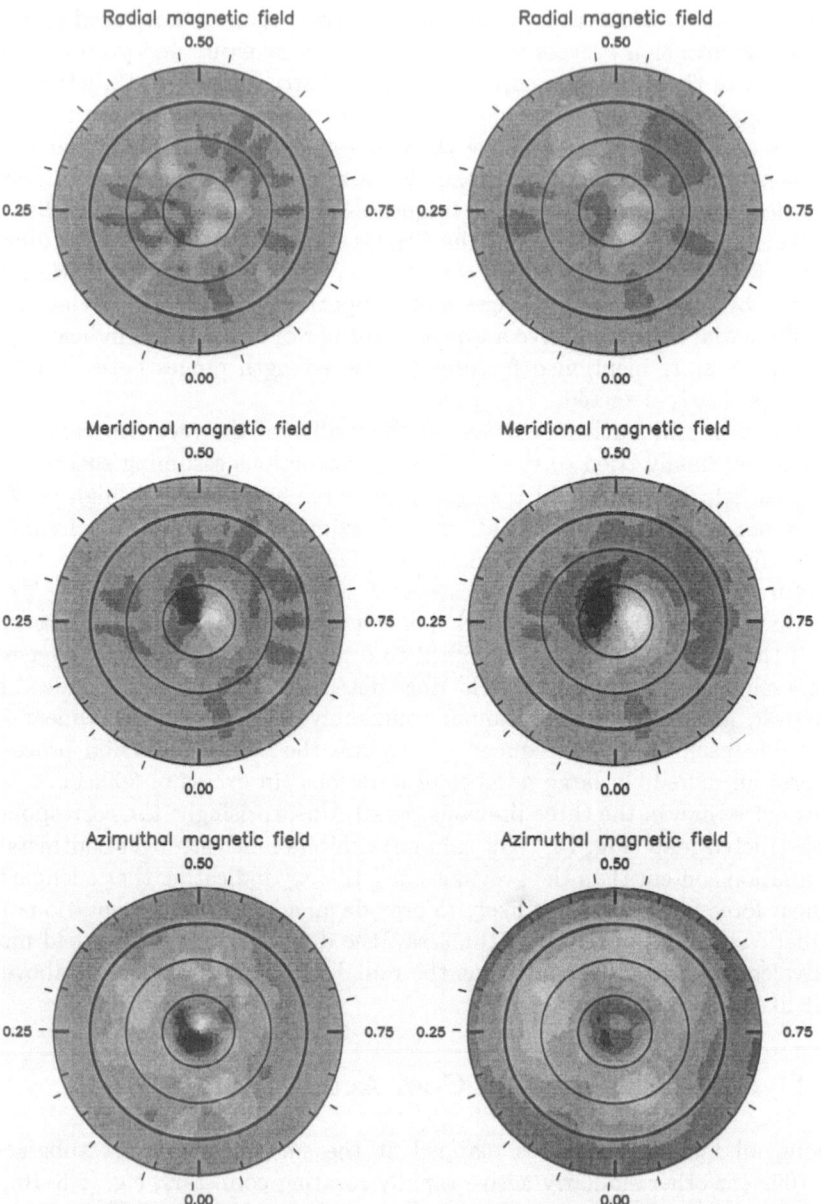

Fig. 12. MaxEnt (left column) and unconstrained SHaDe (right column) magnetic images of HR 1099. Both maps correspond to unit reduced χ^2 fits to the Stokes V data set shown in Fig. 11. The graphics conventions are the same as in Fig. 5.

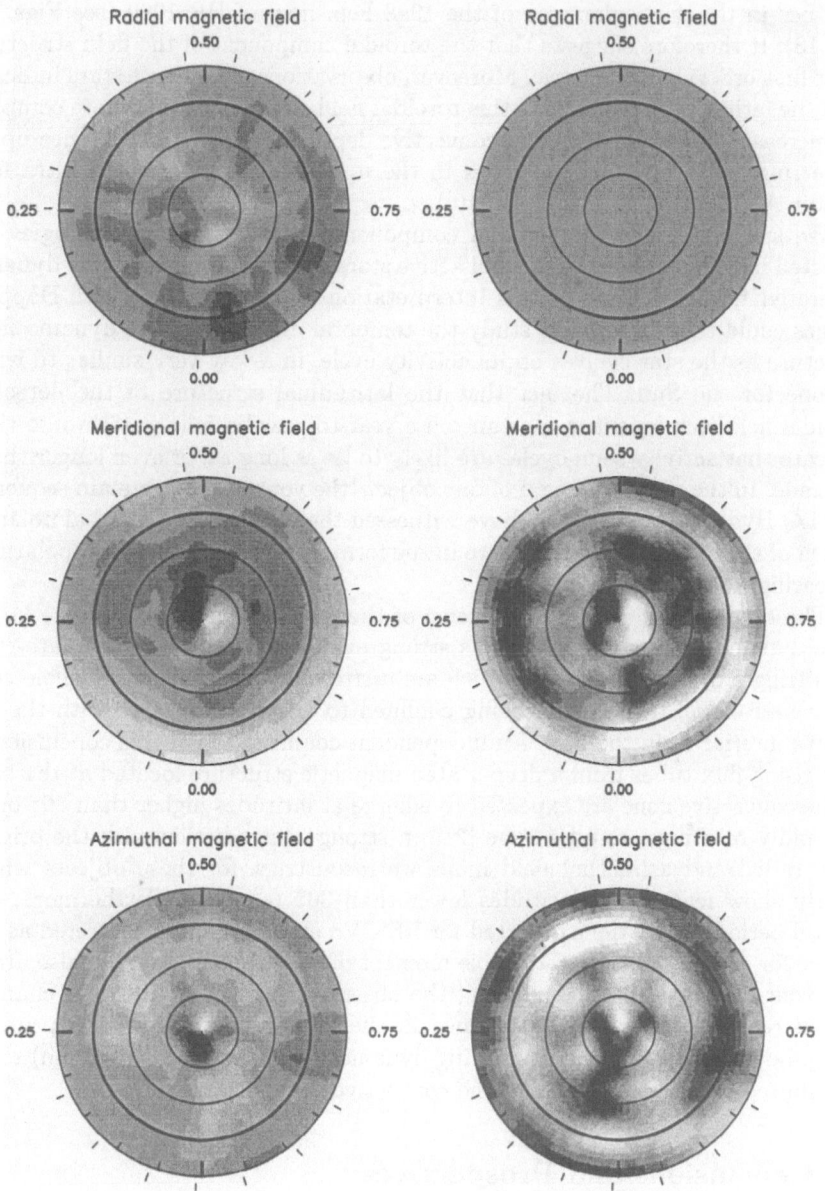

Fig. 13. SHaDe magnetic images of HR 1099, now assuming either a linear combination of two linear force free fields with $\alpha = 0.001$ and $\alpha = 1000$ (left column), or a single linear force free field with $\alpha = 1000$ (right column). Once again, both maps correspond to unit reduced χ^2 fits to the Stokes V data set.

instance in the particular case of the 1998 Feb. map of HR 1099 (see Figs. 12 and 13). It therefore suggests that the toroidal component of the field structure is, at first order, axisymmetric. Moreover, observations of different stars indicate that the latitudinal structure of this toroidal field component gets more complex for increasing rotation rates and convective depths, featuring for instance up to three rings of alternating polarities in the upper hemisphere of the ultra fast rotator AB Doradus (see Fig. 14, [10]).

We speculate that the toroidal component of the magnetic topologies we detected in the photospheres of cool active stars is that of the large scale dynamo generated field structure. If this interpretation is confirmed, Zeeman-Doppler images could then be used to study the temporal evolution of the dynamo field structure as the star evolves on its activity cycle, in a way very similar to what is done for the Sun. The fact that the latitudinal structure of the detected toroidal field component is constant on a year-to-year basis (e.g. [6] would thus indicate that activity semi-cycles are likely to be as long as, or even longer than, a decade. In the particular case of one object (the young zero age main sequence star LQ Hydrae), we may even have witnessed the beginning of a global polarity switch of the dynamo field structure in the form of new azimuthal field polarities appearing at high latitudes [6,13].

The detection at photospheric level of the toroidal component of the large-scale dynamo field is interpreted as strong evidence that very active late-type stars trigger dynamo processes which are *distributed* throughout the whole convective envelope, rather than being confined to an interface layer with the radiative interior as in the Sun. An independent confirmation of this conclusion is that rising flux tubes from a deep-seated magnetic structure located at the base of the convective zone are expected to emerge at latitudes higher than 50° or so in rapidly rotating late-type stars [21], in strong contradiction with the brightness, radial and azimuthal field maps we reconstruct for these objects which clearly show features at latitudes lower than 30° (e.g. [6]). Furthermore, the orbital period fluctuations detected for RS CVn systems (and interpreted as evidence for changes in the quadrupole moment of the primary star, [1]) also argue in favour of this picture; generating the observed quadrupole moment changes indeed requires the associated change in the internal rotation of the primary star (and therefore also the underlying dynamo processes producing them) to be distributed in a large fraction of the convective zone [6,18].

7   Conclusions and Prospectives

With Zeeman-Doppler imaging, one can now detect the complex magnetic field structures of rapidly rotating cool stars. From sets of rotationally modulated spectropolarimetric observations, one can in principle even faithfully reconstruct the magnetic field topology at the surface of the star. In that respect, the new SHaDe inversion process presented in this paper seems to be particularly robust and well suited for the reconstruction of all kinds of field structures. Present results suggest that dynamo processes of active stars are distributed within the

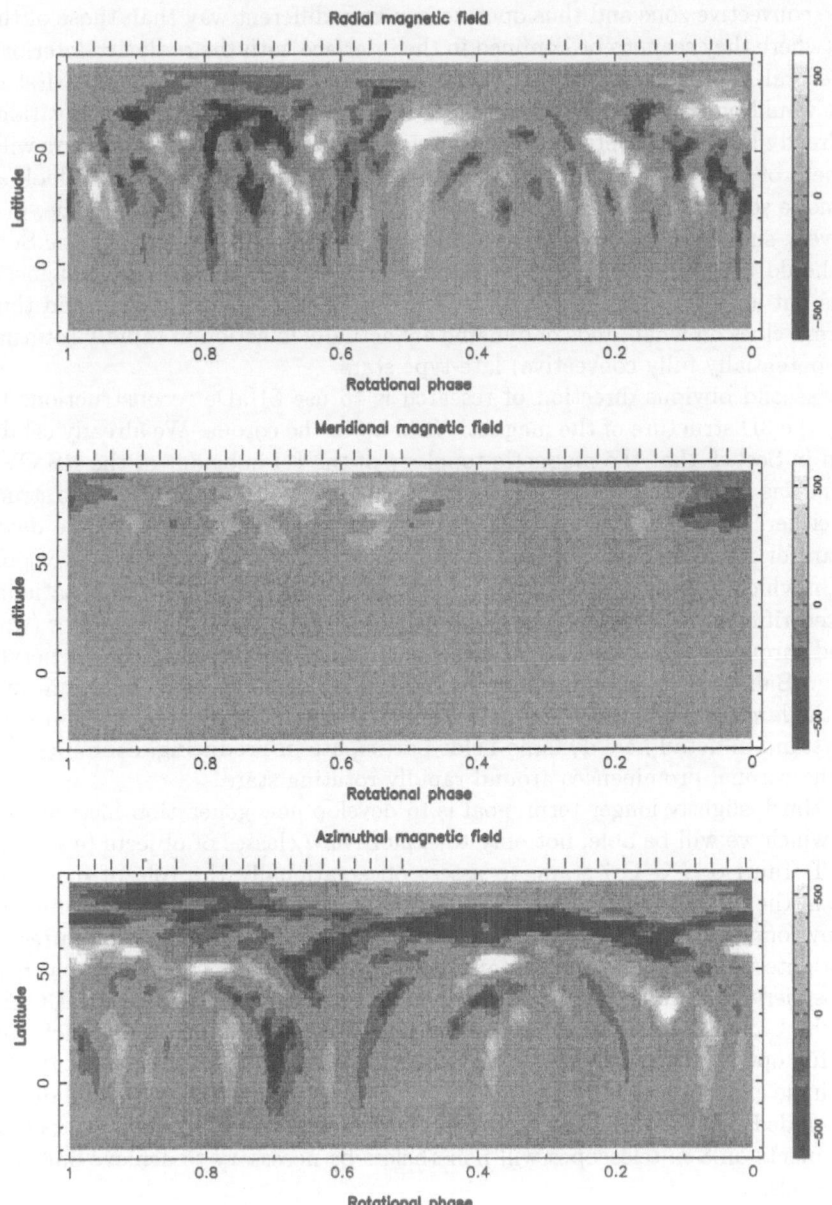

Fig. 14. MaxEnt image of the young ultra fast rotator AB Doradus, derived from the Stokes V data set of Fig. 1. The polarity of the azimuthal field component is found to be roughly longitude independent, switching from negative to positive and back to negative when going from the pole to the equator. The star is now shown in rectangular projection. Rotational phases of observations are depicted with vertical ticks above each panel, while black and white code field intensities of –600 and 600 G respectively.

whole convective zone and thus operate in a very different way than those of the Sun (where they seem to be confined in the interface with the radiative interior).

Several improvements in this field are expected in the near future. First of all, it would be very instructive to reanalyse the long term series of observations we already have (and continue to collect on a regular basis) for several stars with the help of the new SHaDe inversion process. The idea is of course to look at the mode structure of the magnetic field maps and its evolution with time, in a way very similar to that used by Stenflo [24] in the particular case of the Sun. We should thus eventually be able to derive not only the periods of magnetic cycles but also the spatial structure of the associated dynamo modes, and thus get fresh clues on what kinds of dynamo are actually in action in rapidly rotating (and potentially fully convective) late-type stars.

A second obvious direction of research is to use SHaDe reconstructions to study the 3D structure of the magnetic field up to the corona. We already established in Sect. 5 that the magnetic topology of the K1 subgiant of the RS CVn system was not a potential structure, and very likely not a linear force free structure either. Using the magnetic structures that yield the best fits to the data, we can for instance look at how the stable sites of the large scale field topology (in which ionised prominences can be trapped and resist both gravitational and centrifugal forces) match the locations of observed stellar prominences (witnessed through the fast moving absorption or emission transients they generate in, e.g., Balmer lines, [5]). Prominence data may thus help us to constrain further the large scale magnetic structure of active stars. Ultimately, the aim is to understand in detail how dynamo fields participate in producing, confining and ejecting coronal prominences around rapidly rotating stars.

A third, slightly longer term, goal is to develop new generation instruments with which we will be able, not only to explore new classes of objects (e.g. classical T Tauri or FU Ori stars, to assess observationally the role of magnetic fields in the interaction of pre-main sequence stars with their direct circumstellar environment), but also to detect linear polarisation Zeeman signatures of the brightest targets (and thus obtain the strongest possible constraints on their photospheric field structure). The new spectropolarimeter (named ESPaDOnS) that the Canada-France-Hawaii Telescope has started building, and which will yield full optical coverage (i.e. 370 to 1,000 nm) at a spectral resolution of 70,000 in a single exposure with 20% peak efficiency (atmosphere, telescope and detector included, [11]) is clearly an important step in this direction. New generation instruments on 8 m telescopes will nonetheless be necessary to achieve our ultimate goals.

All these results should improve our knowledge of how magnetic fields are produced and react on some of the most crucial transport processes operating within late type stars, and eventually bring us to the point where existing theories (on e.g. dynamo processes, prominence dynamics, coronal heating, accretion/ejection phenomena) that were specifically built to explain solar observations (i.e. the only available data at that time) will need to be reassessed and most likely revised in the broader context of both solar and stellar observations.

Ultimately, we should be able to get a better insight on how magnetic fields influence the evolution of late-type stars.

Acknowledgements

We warmly thank Moira Jardine and Andrew Collier Cameron for numerous fruitful discussions about various topics discussed in this paper.

References

1. Applegate J.H., 1992, ApJ, 385, 621
2. Borra E.F., Edwards G., Mayor M., 1984, ApJ, 284, 211
3. Brown D.N., Landstreet J.D., 1981, ApJ, 246, 899
4. Brown S.F., Donati J.-F., Rees D.E., Semel M., 1991, A&A, 250, 463
5. Cameron A.C., Robinson R.D., 1989, MNRAS, 236, 57
6. Donati J.-F., 1999, MNRAS, 302, 457
7. Donati J.-F., Brown S.F., 1997, A&A, 326, 1135
8. Donati J.-F., Brown S.F., Semel M., et al., 1992, A&A, 265, 682
9. Donati J.-F., Cameron A.C., 1997, MNRAS, 291, 1
10. Donati J.-F., Cameron A.C., Hussain G.A.J., Semel M., 1999, MNRAS, 302, 437
11. Donati J.-F., Catala C., Landstreet J.D., 1998, in it Proceedings of the "fifth CFHT users' meeting", Martin P., Rucinski S. (eds.), p. 50
12. Donati J.-F., Catala C., Wade G.A., et al., 1999, A&AS, 134, 149
13. Donati J.-F., Hussain A.C., Cameron A.C., et al., 2000, MNRAS (submitted)
14. Donati J.-F., Semel M., Carter B., Rees D., Cameron A.C., 1997, MNRAS, 291, 658
15. Donati J.-F., Semel M., Rees D.E., 1992, A&A, 265, 669
16. Donati J.-F., Semel M., Rees D.E., Taylor K., Robinson R.D., 1990, A&A, 232, L1
17. Hussain G.A.J., Jardine M.M., Cameron A.C., 2000, MNRAS (submitted)
18. Lanza A.F., Rodonò M., Rosner R., 1998, MNRAS, 296, 893
19. Mathys G., 1989, Fund. Cos. Phys. 13, 143
20. Piskunov N.E., 1998, CoSka, 27, 470
21. Schüssler M., Caligari P., Ferriz-Mas A., Solanki S.K., Stix M., 1996, A&A 314, 503
22. Semel M., 1989, A&A, 225, 456
23. Skilling J., Bryan R.K., 1984, MNRAS, 211, 111
24. Stenflo J.O., 1992, in *IAU Coll. 130, The Sun and Cool Stars: Activity, Magnetism, Dynamos*, Tuominen I., Moss D., Rüdiger G. (eds.), Springer, Berlin, p. 193
25. Stenflo J.O., 1994, in *Solar Magnetic Fields, polarised radiation diagnostics*, Ap&SS Lib. 189, p. 20

Differential Rotation of Close Binary Stars: Application to HR 1099

P. Petit[1], J.-F. Donati[1], G.A. Wade[2], J.D. Landstreet[3], J.M. Oliveira[4], S.L.S. Shorlin[3], T.A.A. Sigut[3], and A.C. Cameron[5]

[1] Laboratoire d'Astrophysique, Observatoire Midi-Pyrénées, 14 avenue Edouard Belin, 31400 Toulouse, France
[2] Département de Physique, Université de Montréal, C.P.6128, Succ. Centre Ville, Montréal H3C 3J7, Canada
[3] Physics and Astronomy Department, The University of Western Ontario, London, Ontario, Canada N6A 3K7
[4] ESA Space Science Department, SCI-SO/ESTEC, PB 299, 2200 AG Noordwijk, The Netherlands
[5] School of Physics and Astronomy, University of St Andrews, Scotland KY16 9SS

Abstract. We propose a new method for estimating differential rotation in binary stars, for which only moderate to poor phase coverage can be obtained (rotation period of order of a few days), preventing the use of conventional cross-correlation methods. Assuming a solar-like differential rotation law with two independent parameters (equatorial rotation rate Ω_{eq} and rotational shear between pole and equator dΩ), we reconstruct Doppler images for different values of the two parameters, and derive the optimal Ω_{eq}, dΩ and associated error bars from the corresponding χ^2 map. Simulations show that Ω_{eq} and dΩ can be recovered with good accuracy, even if the phase coverage per rotation cycle is poor, provided the total data set is long enough. From observations of the HR 1099 K1 subgiant secured in 1998, 1999 and 2000, we obtain that the equator rotates faster than the pole with a rotational shear about 3 times smaller than solar.

1 Introduction

Differential rotation is supposed to play a crucial role in amplifying and transforming the initial poloidal magnetic field of the sun into a toroidal field through dynamo processes. Many of the details of this general principle are still poorly understood, and a major aim of stellar differential rotation measurements is to evaluate the influence of stellar physical parameters on the rotation pattern to provide further constraints for hydrodynamic simulations.

Although it has been noticed for a long time that the rotation rate of the dark spots at the surface of the Sun decreases with increasing absolute latitude, a map of the rotation rate within the solar interior is only a recent result of the helioseismic techniques [11].

Doppler and Zeeman-Doppler imaging of fast rotators give access to the time evolution of stellar surface features, allowing the estimation of differential rotation parameters. We focus in this paper on the particular case of close binaries, to get some insight into the impact of tidal forces on the rotational shear of rapidly rotating stars.

2   Observations

For this study, we concentrate on the RS CVn system HR 1099 (K1 IV + G5 V), whose subgiant primary is one of the most active stars in the whole sky. Magnetic fields have been repeatedly detected and mapped on this star [7]. We observed HR 1099 during three seasons (January 1998 to March 2000) with the MuSiCoS spectropolarimeter [6] on the 2 m Télescope Bernard Lyot, Observatoire du Pic du Midi, France, and with the UCL Echelle Spectrograph [4] on the 3.9 m Anglo-Australian Telescope, Australia. All data were reduced with ESpRIT [4], and mean Stokes I and V profiles were derived by means of Least-Squares Deconvolution [4], using between 2600 (for TBL data) to 4500 (for AAT data) spectral features. The average signal to noise ratio (SNR) of the mean profiles is of order of 950. The complete observing log is listed in Table 1.

All imaging parameters (system orbital parameters, inclination i of the rotation axis to line of sight, and line of sight projected rotation velocity $v \sin i$ of the primary star) were adjusted by choosing the values which minimize the information content of the reconstructed image. The optimal values were consistent with previous findings of Donati [7].

Table 1. Description of HR 1099 data. For each observing run, we report (from left to right) the dates of first and last exposure, the telescope with which the spectra were secured, the number of spectra obtained in each sequence and the corresponding range of SNR per velocity bin of 4.5 and 1.4 km.s^{-1} for TBL and AAT data respectively.

Date	Tel.	n exp.	SNR
10 Jan 1998 - 15 Jan 98	AAT	28	930 - 1091
04 Feb 98 - 20 Feb 98	TBL	67	826 - 992
26 Feb 98 - 05 Mar 98	TBL	20	527 - 964
05 Dec 98 - 08 Dec 98	TBL	28	818 - 984
27 Dec 98 - 03 Jan 99	AAT	59	607 - 1098
13 Jan 99 - 31 Jan 99	TBL	20	439 - 860
19 Dec 99 - 29 Dec 99	AAT	28	976 - 1155
04 Feb 00 - 08 Mar 00	TBL	65	291 - 971

3   Cross Correlation Method

Differential rotation has already been measured on a few pre-main sequence (PMS) stars using cross-correlation techniques [2,3,6,8], by looking for longitudinal shifts of surface features between two Doppler images recorded about one week apart. The PMS stars observed so far have a rotation period of order of (or less than) one day, which allows one to achieve a dense and extended phase coverage of the star in one single night, and thus a good phase overlap between the two images. Furthermore, the timescale between both images is smaller than the spot lifetime, while the short data collection timespan for each image ensures that differential rotation itself does not blur the reconstructed spot pattern.

In the case of HR 1099 (and many other RS CVn systems), however, the phase coverage one can secure during one night can hardly exceed 0.1 rotation cycle. Several observing nights are therefore necessary to obtain one Doppler image of the primary star. Using the 1998 TBL data subset (see Sect. 2), we reconstructed a time series of images of HR 1099 by performing reconstructions from sliding selections of all spectra recorded on 6 consecutive nights[1]. Although the global spot distribution seems roughly constant through the sequence, the position of each individual spot clearly varies from one image to the other, some spots even splitting into two distinct features. It is however obvious that this apparent spot evolution is mostly due to the poor phase sampling[2] and is much larger than the rotational shear we are looking for; it is therefore impossible to look for differential rotation by cross-correlating two of these images.

4   Least Squares Determination
of the Differential Rotation Parameters

In this new method, we take the whole series of spectra secured during one season as a single data set to reconstruct one Doppler image, assuming a solar rotation law in the inversion process itself:

$$\Omega(l) = \Omega_{\mathrm{eq}} - \mathrm{d}\Omega \cdot \sin^2 l \tag{1}$$

where $\Omega(l)$ is the rotation rate at the latitude l, Ω_{eq} the rotation rate of the equator and $\mathrm{d}\Omega$ the rotational shear between pole and equator. Note that this reconstruction process allows us to use a data set secured over several weeks, provided the total observing time interval is smaller than the spot lifetime.

We run many reconstructions for different values of the two differential rotation parameters and choose the pair which minimizes the χ^2 of the reconstructed spectra at constant image entropy.

---

[1] the corresponding animation is available on:
http://webast.ast.obs-mip.fr/people/petit/movies.html

[2] A second animation created from a synthetic star with no differential rotation (available at the same URL), clearly shows how spurious spot motion is generated by poor phase sampling.

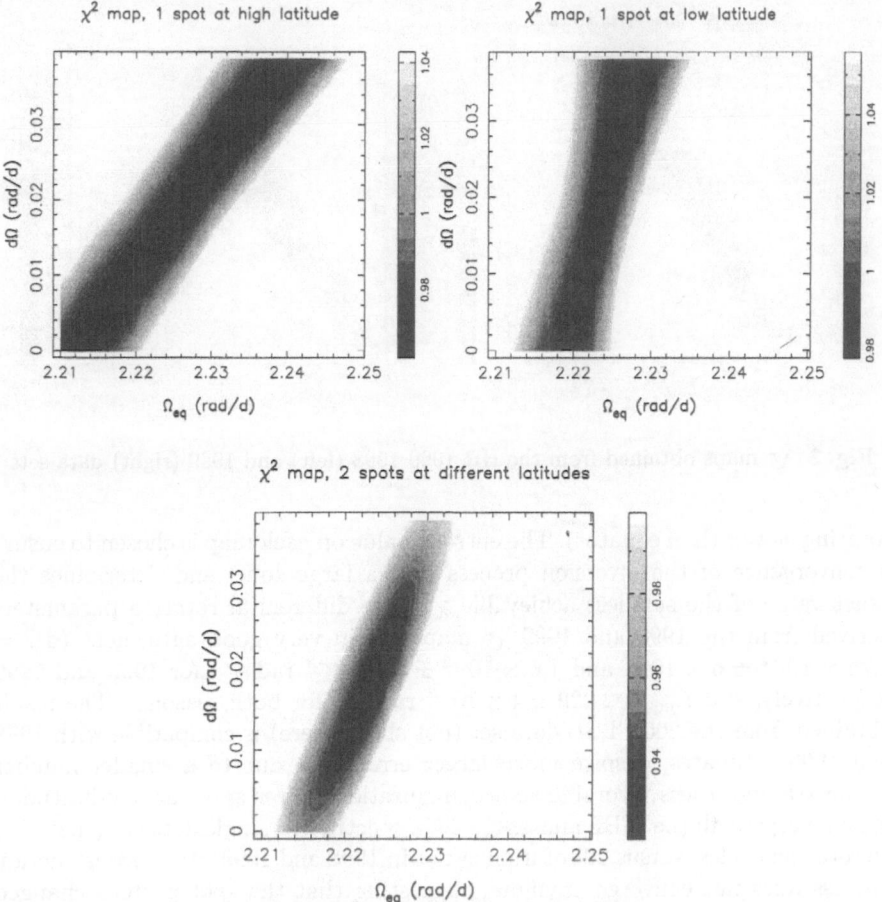

Fig. 1. χ^2 maps derived from simulations with the three synthetic stars discussed in text. The edge of the grey-scale plot corresponds to a 1% false alarm probability.

This method was tested on three synthetic differentially rotating stars. The first two stars host only one surface cool spot (located at latitude 60° and 30° respectively), while the third one includes two spots at different latitudes. The assumed differential rotation parameters are $d\Omega = 1.5 \times 10^{-2}$ rad.s^{-1} and $\Omega_{eq} = 2.222$ rad.s^{-1}. The phase sampling of the synthetic spectra corresponds to our 1998 HR 1099 data set (56 profiles spread over 19 rotation cycles). Figure 1 shows the resulting χ^2 maps; unsurprisingly, the differential rotation parameters are not constrained (though strongly correlated) when cool spots concentrate at a single latitude (upper frames). On the other hand, there is a clear χ^2 minimum, located close to the input differential rotation parameters, when spots cover a large range of latitudes (lower frame).

When applied to our HR 1099 data sets (Fig. 2), we obtain that the subgiant of the system is differentially rotating, with a solar-like rotational shear (pole

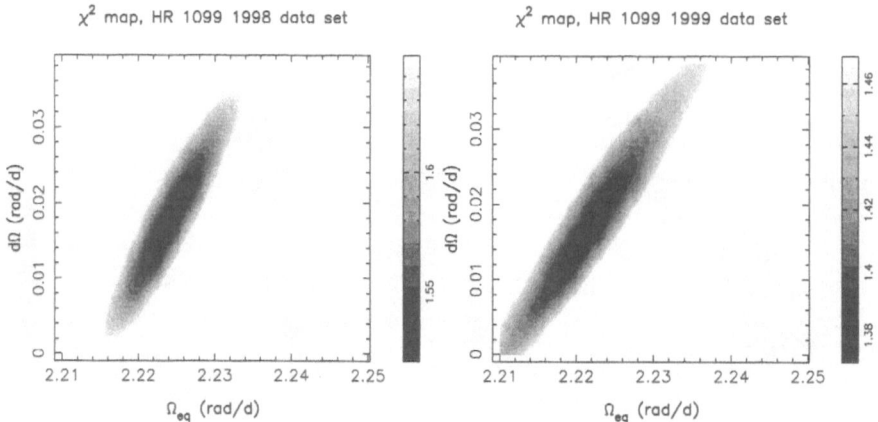

Fig. 2. χ^2 maps obtained from the HR 1099 1998 (left) and 1999 (right) data sets

rotating slower than equator). The entropy value on each map is chosen to ensure a convergence of the inversion process over a large zone, and determines the exact value of the smallest achievable χ^2. The differential rotation parameters derived from the 1998 and 1999 χ^2 maps are in very good agreement ($d\Omega = 1.75 \times 10^{-2} \pm 6 \times 10^{-3}$ and $1.6 \times 10^{-2} \pm 8 \times 10^{-3}$ rad.s^{-1} for 1998 and 1999 respectively, and $\Omega_{eq} = 2.222 \pm 4 \times 10^{-3}$ rad.s^{-1} for both seasons). The result obtained from the 2000 TBL data set (not shown here) is compatible with 1998 and 1999 estimates, despite much larger error bars due to a smaller number of spectra and a less favorable spot configuration (fewer spots at low latitude). When merging all (i.e. TBL and AAT) 2000 spectra in a single data set (spanning 28 rotation cycles, versus 20 rotation cycles in 1998 and 1999), the reconstruction process does not converge anymore, suggesting that the spot pattern changed beyond recognition between December 1999 and February 2000.

5 Discussion

Our estimate of the surface rotational shear of HR 1099 ($d\Omega = 1.7 \times 10^{-2} \pm 5 \times 10^{-3}$ rad.s^{-1} and $\Omega_{eq} = 2.222 \pm 3 \times 10^{-3}$ rad.s^{-1}) does not confirm previous measurements based on individual spot monitoring [12,13] concluding that the rotational shear was antisolar (pole rotating faster than equator). We think that our estimate is much more reliable than the published ones for the following reasons. First, the poor phase coverage inherent to all single site observations of HR 1099 makes it very difficult to determine accurately the location and migration rates of individual spots. Moreover, we find that such studies must be carried out on timescales smaller than about 2 months, much shorter than the interval between successive images in Vogt et al. [13]. Finally, the fact that two of our data sets, though fully independent, yield very similar results gives us further confidence in the reality of our detection. Note that the polar and equatorial periods we obtain ($P_{equator} = 2.8277 \pm 4 \times 10^{-3}$ d and $P_{pole} = 2.85 \pm 4 \times 10^{-3}$ d) encompass (within the error bars) all periods derived from photometry (ranging from $2.8245 \pm 1.2 \times 10^{-3}$ d to $2.8412 \pm 2 \times 10^{-4}$ d [9])

Assuming our estimate is correct, the lap time of HR 1099 (around 370 d) is 3 to 9 times greater than those of the Sun and PMS fast rotators. This result may suggest that the strong tidal forces operating in the convective zone of HR 1099 tend to synchronize rotation at all latitudes, thus strongly decreasing the rotational shear. This would be consistent with the only published theoretical study on differential rotation in binary stars [10], showing that differential rotation should be significantly weakened, though not totally suppressed, in the particular case of RS CVn systems.

More observations are still needed to compare in particular differential rotation parameters of single and double stars of similar evolutionary stage. Our study is also being carried out on other RS CVn systems (UX Ari, σ Gem). Moreover, a long-term monitoring of the rotational shear of HR 1099 is being undertaken to search for a potential correlation with the orbital period fluctuations recently detected on this system [7] and test the scenario of an exchange between magnetic and kinetic energy in the convective zone of the primary star [1].

References

1. Applegate J.H., 1992, ApJ, 385, 621
2. Barnes J.R., Cameron A.C., James D.J., Donati J.-F., 2000, MNRAS, 314, 162
3. Donati J.-F., Cameron A.C., 1997, MNRAS, 291, 1
4. Donati J.-F., Semel M., Carter B.D., Rees D.E., Cameron A.C., 1997, MNRAS, 291, 658
5. Donati J.-F., Cameron A.C., Hussain G.A.J., Semel M., 1999, MNRAS, 302, 437
6. Donati J.-F., Catala C., Wade G.A. *et al.*, 1999, A&AS, 134, 149
7. Donati J.-F., 1999, MNRAS, 302, 457
8. Donati J.-F., Mengel M., Carter B.D., Cameron A.C., Wichmann R., 2000, MNRAS, 316, 699
9. Henry G.W., Eaton J.A., Hamer J., Hall D.S., 1995, ApJS, 97, 513
10. Scharlemann E., 1982, ApJ, 253, 298
11. Schou J., Antia H.M., Basu S. *et al.*, 1998, ApJ, 505, 390
12. Strassmeier K.G., Bartus J., 2000, A&A, 354, 537
13. Vogt S.S., Hatzes A.P., Misch A.A., Kürster M., 1999, ApJS, 121, 547

Magnetic Doppler Imaging
of Chemically Peculiar Stars

N. Piskunov and O. Kochukhov

Uppsala Astronomical Observatory, Box 515, S-75120 Uppsala, Sweden

Abstract. We present the results of numerical experiments and real applications of our new Magnetic Doppler Imaging code INVERS10. The code is capable of simultaneously reconstructing the surface distribution of magnetic field vectors and one chemical element from a time sequence of line profiles measured in four Stokes parameters. The application of the code to a well studied magnetic Ap star α^2CVn recovered a field distribution, resembling an oblique dipole with the strength and orientation consistent with those derived with other techniques.

1 Introduction

Doppler Imaging (DI) is a well understood numerical technique that allows the reconstruction of stellar surface structures from the time variation of spectral line profiles. DI uses the inverse problem method to search for the optimal solution (map), while the multitude of solutions is restricted by means of the regularization technique (e.g. Maximum Entropy or Tikhonov) to ensure the convergence to a unique map. DI is a very computationally intensive technique due to the substantial amount of observational data that needs to be simulated (the number of spectral points × the number of rotational phases) and the complex physics that is required for accurate spectral synthesis (solution of radiative transfer equation (RTE) with complex blending). Despite of those difficulties, the DI was successfully applied to chemical composition imaging of CP stars and to active solar-type stars. Currently, several groups are continuing research in this areas using different versions of DI codes.

In case of Magnetic Doppler Imaging (MDI) the problem becomes even more difficult as we have to model the polarization information and simultaneously reconstruct multiple (at least three) maps. Piskunov [4] showed with numerical experiments that an incomplete set of data (less than 4 Stokes parameters) is fundamentally insufficient for successful inversion. Different approaches were suggested to deal with this problem. One is to restrict the magnetic field to a combination of multipolar components (e.g. [4]). This approach has the advantage of simplicity and stability imposed *a priori*, but it may be unable to tell us how close the model to the real field is. Another approach is to try a direct recovery of the distribution of magnetic vectors on the stellar surface and then compare it with the multipolar model. Such method is much more demanding computationally and requires extensive tests of stability and convergence, but it has the advantage of been model independent.

Our new MDI code INVERS10 follows the second approach. In the following we briefly discuss the method and the implementation of the code, present numerical experiments used to validate the code and conclude with magnetic map derived of a well studied magnetic Ap star α^2CVn.

2   Mathematical Formulation

Mathematically MDI is reduced to the problem of minimizing the functional:

$$\Phi(m) = \sum_{\phi,\lambda} \omega_{\phi\lambda}^{I} \left[I_{\lambda\phi}^{\text{calc}}(m) - I_{\lambda\phi}^{\text{obs}}(m) \right]^2 +$$

$$\sum_{\phi,\lambda} \omega_{\phi\lambda}^{Q} \left[Q_{\lambda\phi}^{\text{calc}}(m) - Q_{\lambda\phi}^{\text{obs}}(m) \right]^2 +$$

$$\sum_{\phi,\lambda} \omega_{\phi\lambda}^{U} \left[U_{\lambda\phi}^{\text{calc}}(m) - U_{\lambda\phi}^{\text{obs}}(m) \right]^2 +$$

$$\sum_{\phi,\lambda} \omega_{\phi\lambda}^{V} \left[V_{\lambda\phi}^{\text{calc}}(m) - V_{\lambda\phi}^{\text{obs}}(m) \right]^2 + \mathbf{R}(m)$$

where the summation over rotational phases ϕ and wavelengths λ estimates the discrepancy between synthetic and observed Stokes profiles I, Q, U, and V. $\omega_{\phi\lambda}$ are the weights reflecting the quality of the data and $\mathbf{R}$ is the regularization functional. The unknown is a vector function m defined on the stellar surface. For MDI at each surface element m includes the radial, meridional and longitudinal components of the field vector. Piskunov [4] has shown that in CP stars an inhomogeneous distribution of chemical elements may cause significant distortions in all Stokes profiles. In magnetically active late-type stars a similar effect may be caused by the temperature variations. Therefore, we included one more variable to the vector m: in the case of INVERS10 it is the local abundance of one chemical element.

Each Stokes parameter for a given rotational phase ϕ is computed by integrating the local Stokes parameter over the visible stellar hemisphere and convolving the result with the instrumental profile of the spectropolarimeter. The local Stokes parameters are computed by solving the magnetic RTE, which is a very time-consuming part of the calculations. Conventional DI codes often precompute the local profiles and store them in tables, so that during the inversion the solution of RTE is replaced by interpolation [6]. Unfortunately, in the case of MDI the required size of such a table and the time for computing it become prohibitively large and the magnetic RTE must be solved "on the fly". We have implemented two versions of INVERS10 based on Feautrier [1] and DELO RTE solvers [7]. The detailed comparison of the two methods is given in [3].

Due to the large amount of computations required for performing the inversion INVERS10 is designed to use parallel processing via Message Passing Interface (MPI). The radiative transfer and derivative calculations for different surface elements are distributed between different processors and computed in

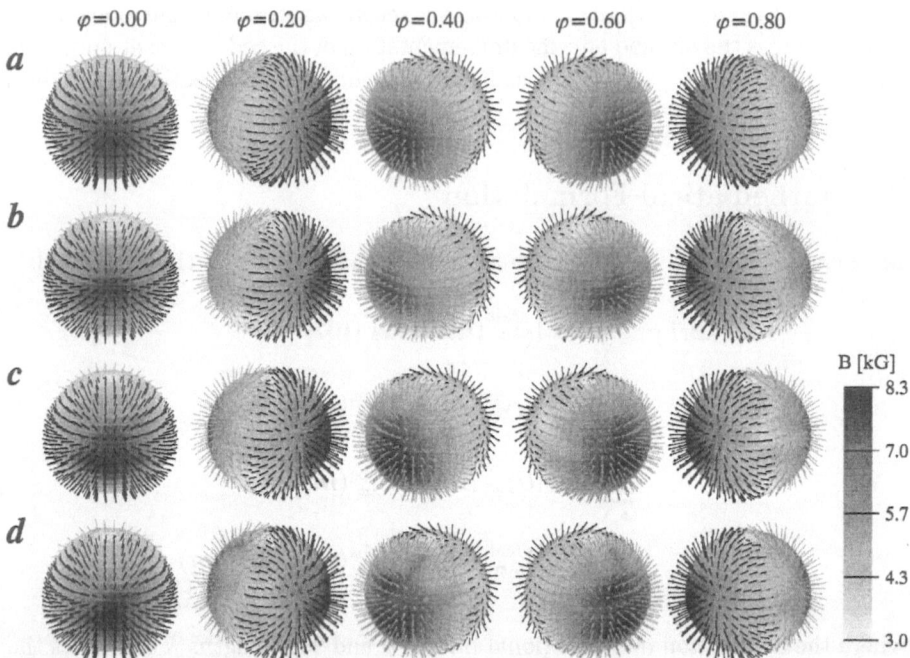

Fig. 1. Magnetic Doppler reconstruction of dipolar field for different stellar rotational velocities. Panel **a** is the true map, while panels **b**, **c**, and **d** show images, obtained for stars rotating with $v \sin i$ of 30, 10, and 5 km s^{-1}, respectively.

parallel. The surface elements are selected out of order leading to an automatic load balance and a linear scaling of speed with the number of CPUs. Tests with up to 64 CPUs confirmed these properties of the program.

3   Numerical Experiments

Numerical experiments allow us to study the properties of a DI code and, in particular, validate its stability and convergence. Each experiment includes the synthesis of fake observations, corresponding to a known stellar surface distribution, and their inversion.

Figure 1 shows the reconstruction of dipolar field for a model CP star with multiple spots of enhanced iron abundance. The reconstruction was done using all four Stokes parameters and the results (lower panels in Fig. 1) are very close to the initial image. The large amount of information contained in the Stokes parameters and their strong dependence on the orientation of the field vectors relative to the line of sight allows us to perform MDI even for very slow rotators ($v \sin i = 5$ km s^{-1}), which are out of reach for the conventional DI techniques based on intensity profiles only.

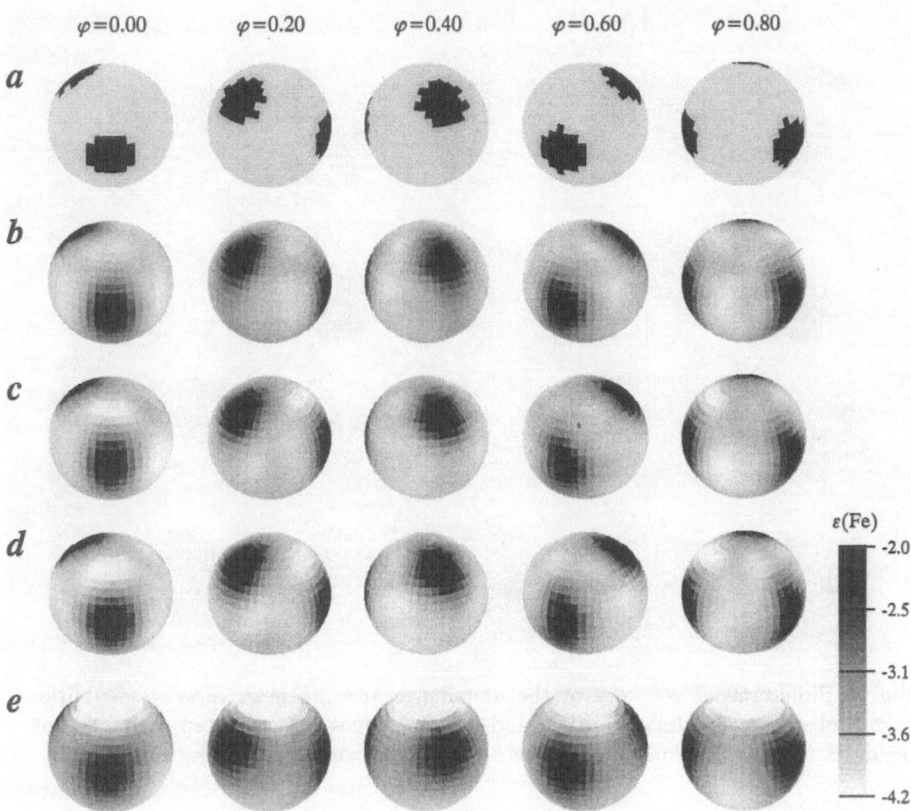

Fig. 2. Abundance distributions recovered simultaneously with field maps of Fig. 1. In addition to the true abundance map (panel **a**) and Doppler images for $v \sin i = 30, 10,$ and 5 km s^{-1} (panels **b**, **c**, and **d**, respectively), results of the reference non-magnetic reconstruction are shown on panel **e**.

Figure 2 further illustrates this particular advantage of MDI by comparing iron abundance distributions, recovered simultaneously with magnetic maps of Fig. 1. In the presence of a strong dipolar field abundance distributions were reconstructed fairly well for all rotational velocities, while an attempt to image the same iron map for a non-magnetic slowly rotating star (panel e in Fig. 2) failed to recover any latitude information about abundance spots and resulted in substantial smearing of the surface features.

Figure 3 shows the attempt to reconstruct the same field from the restricted data set (only I and V Stokes parameters). The results are strongly dependent on the initial guess and attempts to use Maximum Entropy or Tikhonov regularization drive the recovered field structure away from dipolar. This is not surprising, as the dipolar geometry does not minimize the functionals. Since there is a large amount of observational data for CP stars consisting of only two Stokes parameters, we have developed a specific regularization to use in MDI of these stars

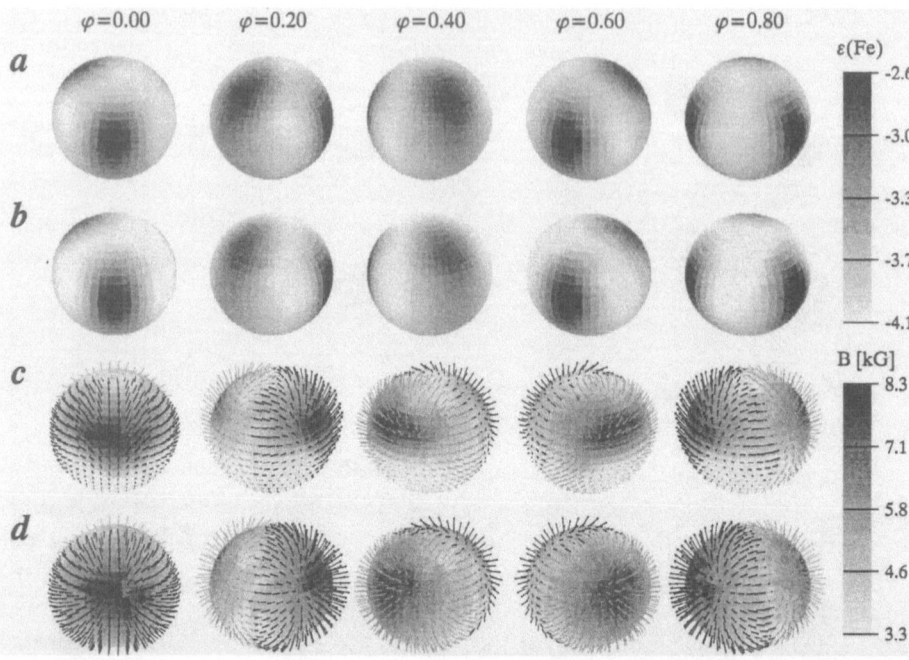

Fig. 3. Simultaneous recovery of the abundance and magnetic dipole distributions from Stokes I and V data. Panels **a** and **c** show images reconstructed with Tikhonov regularization, while **b** and **d** are images regularized with a multipolar functional.

[5]. It is based on determination of a best multipolar fit to the map obtained on each iteration. The difference between the current map and the multipolar fit is used as regularization. In this way the program is "encouraged" to reconstruct a field close to multipolar, but it is not restricted to multipoles. After adopting this regularization method we had no difficulties recovering dipolar fields from two Stokes parameters with good stability in respect to the initial guess (lower panel in Fig. 3).

4   MDI of Magnetic Ap Star α^2CVn

α^2CVn was one of the first discovered magnetic Ap stars and extensive work was dedicated to study the abundance distribution and magnetic field of this star. During numerical experiments with the INVERS10 we realized that magnetic geometries far from dipolar can still reproduce very smooth curves of effective magnetic field (disk average longitudinal field) indistinguishable from dipolar effective field curves. With MDI we expect to solve this ambiguity and verify if the field is truly dipolar. The reconstruction is based on 150 spectra in I and V Stokes parameters obtained with the SOFIN spectrometer at the Nordic Optical Telescope (La Palma). We used 3 Cr II lines for a simultaneous recovery of

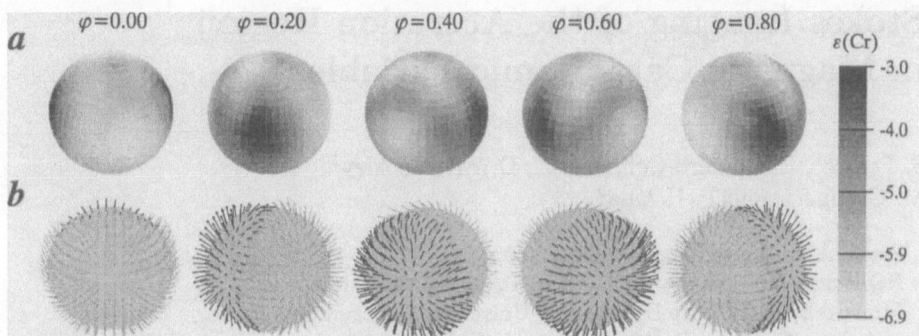

Fig. 4. MDI reconstruction of the magnetic field and Cr surface abundance distribution for α^2CVn.

the magnetic field and a chromium abundance map. The results are presented in Figure 4. Taking into account that not all four Stokes parameters were available, we see no significant difference from the dipolar field structure and field strength recovered in previous studies (e.g. [2]).

5 Conclusions

We have developed an MDI code capable of reconstructing the distribution of the field vectors and the abundance of one chemical element over the stellar surface. The code shows good numerical stability and convergence. Parallel implementation helps to deal with the large amount of computations involved in solving the magnetic RTE. A special regularization technique was introduced to deal with partial data sets when not all Stokes parameters are available. We found that the new code can be used successfully with very slow rotation velocities, typical for stars with strong fields. The application of the code to α^2CVn resulted in a preliminary confirmation of the dipolar field structure of this star.

References

1. Auer L.H., Heasley J.N., House L.L., 1977, ApJ, 216, 531
2. Glagolevskij Yu.V., Piskunov N., Khokhlova V.L., 1985, Sov. Astron. Lett., 11, 154
3. Piskunov N., 1999, in *Proceedings of the 2nd Workshop on Solar Polarization*, J. Stenflo and K.N. Nagendra (eds.), 1998, Bangalore, India, Kluwer Acad. Publ., ASSL 243, p. 515
4. Piskunov N., 1985, Sov. Astron. Lett., 11, 18
5. Piskunov N., Kochukhov O., 2001, A&A, in preparation
6. Piskunov N., Rice J.B., 1993, PASP, 105, 1415
7. Rees D.E., Murphy G.A., Durrant C.J., 1989, ApJ, 339, 1093

Stokes Imaging of the Accretion Region in Magnetic Cataclysmic Variables

S. Potter[1], E. Romero-Colmenero[1], D.A.H. Buckley[1],
M. Cropper[2], and P. Hakala[3]

[1] South African Astronomical Observatory,
   PO Box 9, Observatory 7935, Cape Town, South Africa.
[2] Mullard Space Science Laboratory, University College London,
   Holmbury St Mary, Dorking, Surrey, U.K.
[3] Observatory and Astrophysics Laboratory,
   FIN-00014, University of Helsinki, Finland.

Abstract. Modelling of polarized cyclotron emission from magnetic cataclysmic variables has been a pivotal technique for determining the structure and location of the accretion zones on the white dwarf. Stokes imaging of Potter et al. [8] is the first objective and analytical technique that robustly models the cyclotron emission and maps the accretion zones.

Polarisation modelling is discussed, followed by a summary of the Stokes imaging technique. We demonstrate this technique on new polarimetric observations of the magnetic cataclysmic variable V834 Cen.

1 Introduction

The usual picture of a cataclysmic variable is a binary system in which there is a Roche-lobe filling red dwarf (the secondary) and a white dwarf (the primary). The material that overflows the Roche-lobe of the secondary forms an accretion disc around the primary. The material then impacts on the surface of the white dwarf at the inner edge of the accretion disc. (See [15] for a review of Cataclysmic Variables).

In the magnetic Cataclysmic Variables (mCVs) subclass of polars, the white dwarf primary has a sufficiently strong magnetic field to lock the system into synchronous rotation and to actually prevent an accretion disc from forming. Instead, the material overflowing the Roche lobe of the secondary initially continues on a ballistic trajectory until, at some distance from the white dwarf, the magnetic pressure overwhelms the ram pressure of the ballistic stream. From this point onwards the accretion flow is confined to follow the magnetic field lines of the white dwarf (see [4],[15] and Schwope, this volume for a review of mCVs).

The supersonic free-falling material becomes subsonic at a shock which forms at some height above the white dwarf surface. The shock heated material reaches temperatures of $\sim$ 10-50 keV and is ionised. The hot plasma cools as it settles onto the white dwarf resulting in a stratified (in density and temperature) post-shock region. In the post-shock flow there are two cooling mechanisms: X-ray cooling in the form of bremsstrahlung radiation and, with sufficient magnetic fields, cyclotron cooling in the form of optical/IR cyclotron radiation.

2   Polarization Modelling

One of the most useful characteristics of the emission from the post shock flow in mCVs has been the polarised nature of the cyclotron emission. For example, when our line of sight is perpendicular to the post-shock flow, and therefore perpendicular to the magnetic field lines, we see an increase in the amount of linearly polarised emission. When the viewing angle changes to being parallel to the magnetic field lines that feed the post-shock, an increase in the amount of circularly polarised flux is observed. Therefore, as the white dwarf rotates, variations in the amount of circular and linear polarisation are observed repeating over the orbital period. The actual morphology of the polarisation variations will depend on the location, size and shape of the localised cyclotron emission region(s) on the surface of the white dwarf.

The polarisation observations can then be modelled by making use of cyclotron opacity calculations, for example those of Wickramasinghe & Wingett [16]. Their calculations considered a hemispherical blob of plasma with constant temperature and density immersed in a magnetic field. They then calculated the

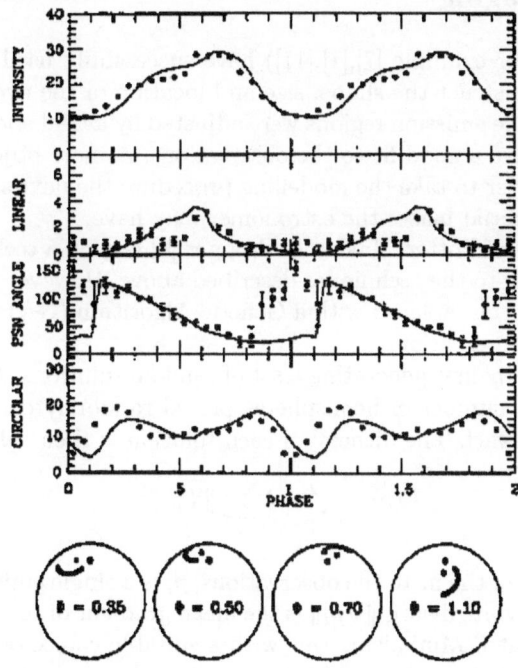

Fig. 1. Optical broad-band observations of V834 Cen [2] modelled with accretion arcs from [5]

polarised cyclotron emission, described by the Stokes parameters, one would expect from such a plasma as a function of viewing angle.

Polarimetric observations can then be modelled by assuming that the emission region(s) on the surface of the white dwarf are made up of several such hemispheres: one simply calculates the viewing angle (as a function of white dwarf rotational phase) to the artificially constructed emission region and then look up the corresponding values for the Stokes parameters as precalculated by, for example, [16]. The fit of the model light curves to the observations can then be improved by adjusting the shape, size and location of the artificial emission region.

Figure 1 shows an example of polarimetric observations of the mCV V834 Cen [2]. These observations show a bright and faint phase typical of polars. During the bright phase there are large amounts of linear and circular polarisation with clearly defined position angle variation. The solid curves represent the model light curves from [5], who found that an arc shaped emission region best reproduced the general morphology of the polarimetric observations. The linear flux arises when the emission region is on the limb of the white dwarf. The decrease in intensity and circular flux around phase $\phi \sim 0.0$ arises as a result of cyclotron self absorption by the post-shock flow. The viewing angle is such that we are looking directly down the accretion column at this phase.

3   Stokes Imaging

Several authors (for example [7],[1],[11]) have successfully used such modelling techniques to reconstruct the shape, size and location of the cyclotron emission regions. However the emission regions were adjusted by a trial and error approach until the model gave a good fit to the observations. A more objective technique was required in order to take the modelling procedure the next step forward and to eliminate any initial biases the astronomer may have.

Stokes imaging of Potter, Hakala & Cropper [8] is such a technique. It works in a similar manner to the techniques described above. However the trial and error approach has been replaced with a Genetic Algorithm (GA) which optimises the model to the observations.

The GA works by first generating a set of random solutions. A single solution consists of a large number of hemispheres placed randomly over the surface of the model white dwarf. The 'fitness' of each solution is then calculated using

$$F(p) = \chi^2 + \lambda \sum_i \|\nabla p_i\|^2 \qquad (1)$$

where χ^2 is the χ^2 of the fit to the observations, p_i is a single emission element on the surface of the white dwarf, $\|\nabla p_i\|$ is the mean gradient of the number of emission points at point i. Minimising this with a suitable choice of the Lagrangian multiplier λ will produce the maximum-entropy solution of the distribution of emission points. The GA then breeds all the solutions by using a type of natural selection procedure - the 'fittest' solutions have a higher probability of being selected for breeding. Eventually the improvement in fitness of the GA solutions starts to level out and a more analytical approach is used to improve the fit further. See [8] for details and test cases.

Stokes imaging has been applied to several mCVs, each with a different morphology and quality of observations. These include CP Tuc [12], ST Lmi [9], V347 Pav [10] and RXJ2115 [13]. In the case of RXJ2115, the white dwarf is not quite in synchronous rotation with the orbit. As a result, the accretion region moves along the surface of the white dwarf as different parts of the white dwarf are presented to the accretion stream. Even though the observations are of relatively poor quality, Stokes imaging shows how the emission region moves along the surface of the white dwarf over the beat period of about 7 days.

4   The mCV V834 Cen

Figure 2 shows new polarimetric observations of V834 Cen taken in March 2000 on the 1.0m at the SAAO using the UCT polarimeter [2]. The morphology of the light curves is very similar to the earlier observations by [3] displayed in figure 1. The most noticeable difference between them is the larger amount of linear flux in the new observations.

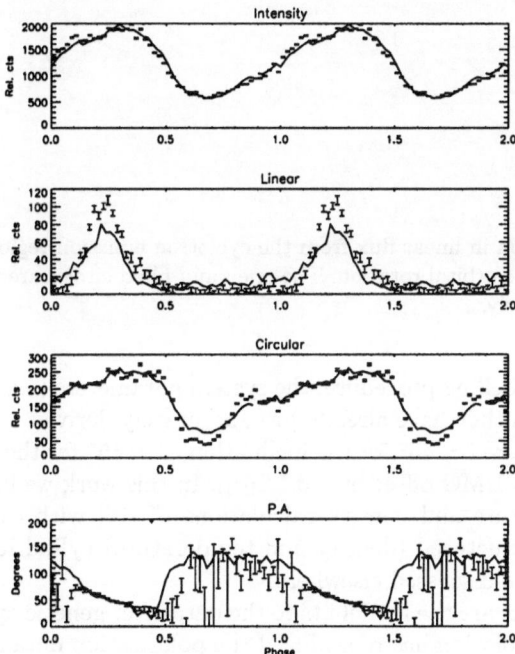

Fig. 2. New optical broad-band observations of V834 Cen. Solid curves show optimised model from Stokes imaging.

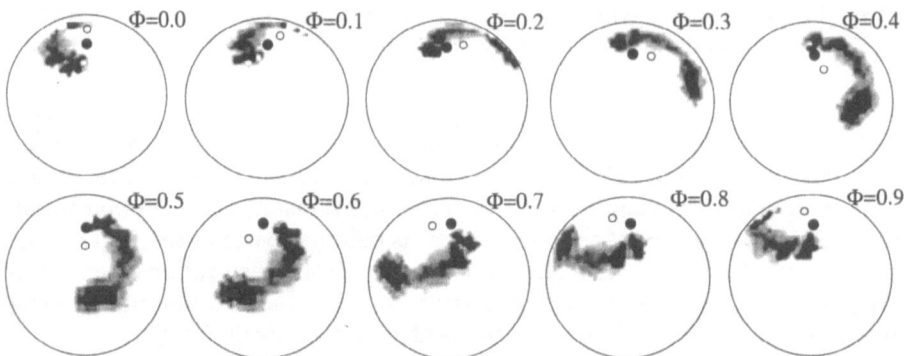

Fig. 3. The position of the cyclotron emission region as viewed from Earth for a complete orbital rotation. The open and filled circles are the magnetic and spin poles respectively.

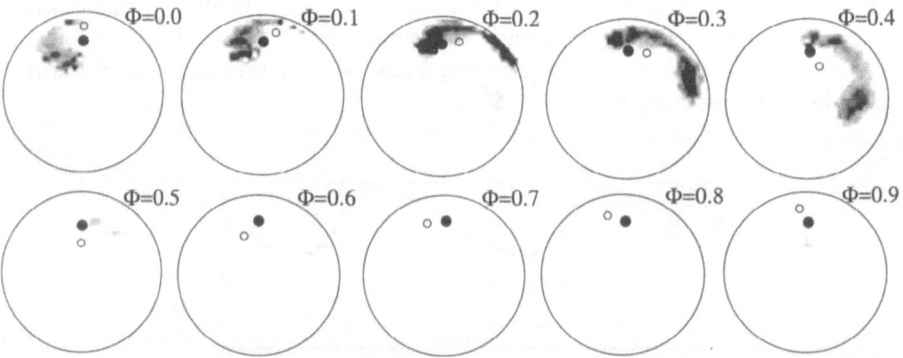

Fig. 4. The variation in linear flux from the cyclotron emission region as viewed from Earth for a complete orbital rotation. The open and filled circles are the magnetic and spin poles respectively.

During the modelling procedure, the system parameters (e.g. the inclination) were kept fixed, as they have already been accurately derived by [14],[5] and [6]. The values used were $i = 45°$ for the inclination, $\beta = 15°$ for the magnetic dipole inclination, $B_p = 31$ MG offset by $-0.1R_{WD}$. In this work we have replaced the constant temperature and density calculations of [16] with calculations using a more realistic structured (density and temperature) cylindrical shock. These calculations will be discussed elsewhere.

The solid curves are the model fits to the data after genetic optimisation. The optimised model solution has reproduced the polarisation data remarkably well. The gross morphology of all the light curves has been accurately predicted and in particular, the variation in position angle, the relative amounts of linear and circular flux, the peaks and dips in the circular polarisation and the intensity curve.

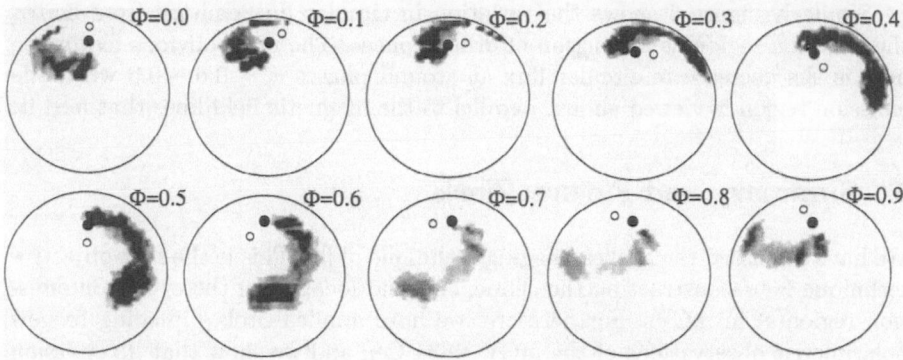

Fig. 5. The variation in circular flux from the cyclotron emission region as viewed from Earth for a complete orbital rotation. The open and filled circles are the magnetic and spin poles respectively.

Upon closer inspection, the model does not quite produce enough linear polarisation between phases $\phi \sim 0.1 - 0.3$. As explained with test data in [8], this arises because of the smoothing function used in equation (1) which has the effect of smoothing any sharp features that occur in the light curves. In addition the model has overestimated the amount of circular polarisation between phases $\phi \sim 0.6 - 0.7$. The dip at this phase has mostly arisen due to cyclotron self absorption when the accretion region is most face on. However, the model does not take into account the absorption of any remaining cyclotron emission by the accretion stream above the post-shock flow. Hence the overestimate of the circularly polarised flux.

Figure 3 shows the position of the cyclotron emission region on the surface of the white dwarf for a complete orbital rotation as viewed from Earth. The grey scale is an indication of the relative variation in electron number density across the emission region with the darker regions being more dense. As can be seen from figure 3, Stokes imaging has predicted a single emission region extended in an arc-like structure very similar to the prediction of [5]. They differ in that Stokes imaging actually models the density variations across the arc and in that Stokes imaging predicts a much broader arc. However, much of the broadness of the arc is in fact due to the resolution of the data and the smoothness term of equation (1). See [10] for a representation of the resolution of the technique.

Figure 3 shows the density distribution across the surface of the white dwarf. However, the observed intensity variations across the emission region depends on the viewing angle and hence on the orbital phase. For example, figure 4 shows the variation in linear flux emitted from across the emission region as a function of orbital phase. From figure 4 one can see how there is a distinct lack of linear flux between phases $\phi \sim 0.5 - 0.9$ and maximum linear emission at phases $\phi \sim 0.2 - 0.3$ when the emission region is viewed almost perpendicular to the magnetic field lines that feed it.

Similarly, figure 5 shows the variation in circular flux emitted from across the emission region as a function of orbital phase. The most obvious feature to note is the decrease in circular flux at around phases $\phi \sim 0.6 - 0.8$ when the emission region is viewed almost parallel to the magnetic field lines that feed it.

5 Summary and Future Work

We have discussed the Stokes imaging technique of [8]. This is the first objective technique for reconstructing the shape, size and location of the cyclotron emission region(s) in mCVs. Furthermore, we have applied Stokes imaging to new polarimetric observations of the mCV V834 Cen and we show that its emission region is consistent with an accretion arc in agreement with the modelling of [5].

Future enhancements to the model will include the addition of a further absorption coefficient above the post-shock flow as our modelling of V834 Cen has shown that cyclotron self absorption is insufficient in some circumstances.

It was pointed out during this workshop that the emission regions predicted by Stokes imaging are somewhat broader than expected. This is mainly due to the smoothness term of equation (1) and the resolution of the data. To investigate the effect of the smoothness term further we next plan to increase the resolution of the grid of emission points around the genetically optimised image and rerun Stokes imaging. This should eliminate the broadening factor introduced by the smoothness term. The broadening acts over the pixels in the image, therefore the smaller the pixels the less broad the emission region will become.

The next logical step will be to compare the results of Stokes imaging with that of other techniques such as Doppler mapping. To this end, we actually obtained simultaneous spectroscopy on V834 Cen from the 1.9m of SAAO with

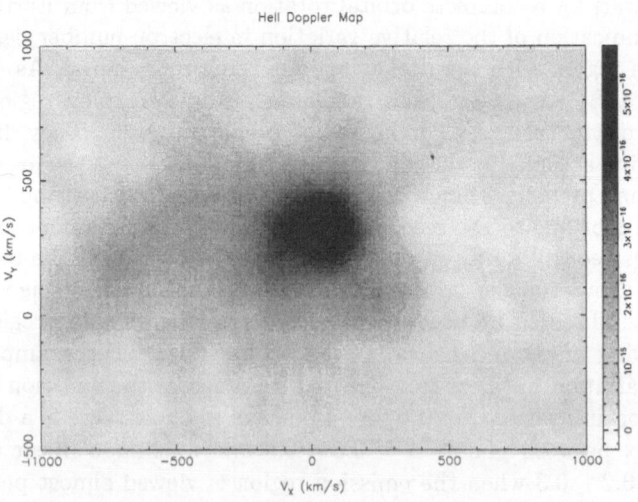

Fig. 6. The HeII Doppler map of V834 Cen.

the new polarimetric data presented above. Figure 6 shows the Doppler map of the HeII emission line of V834 Cen. The figure shows that there is emission at or around the secondary star and the ballistic part of the accretion stream is clearly evident. By fitting a model ballistic stream to the Doppler map we should be able to put constraints on where the stream becomes confined by the magnetic field of the white dwarf. By following the magnetic field lines from the optimised emission region back to the orbital plane we will then be able to compare the results of the two techniques.

References

1. Bailey J. A. et al. 1985, MNRAS, 215, 179
2. Cropper M., 1985, MNRAS, 212, 709
3. Cropper M., 1989, MNRAS, 236, 935
4. Cropper M., 1990, Space Sci. Rev., 54, 195
5. Ferrario L., & Wickramasinghe D. T., 1990, ApJ, 357, 582
6. Ferrario L., et al. 1992, MNRAS, 256, 252
7. Potter S. B., Cropper M., Mason K. O., Hough J. H. & Bailey J. A., 1997, MNRAS, 285, 82
8. Potter S. B., Hakala P. J. & Cropper M. 1998, 297, 1261
9. Potter S. B. 2000, MNRAS, 314, 672
10. Potter S. B., Cropper M. & Hakala P. J., 2000, MNRAS, 315, 423
11. Ramsay G., Cropper M., Wu K. & Potter S. B. 1996, MNRAS, 282, 726
12. Ramsay G., Potter S. B., Buckley D. A. H. & Wheatley P. J. 1999, 306, 809
13. Ramsay G., Potter S., Cropper M., Buckley D., A. H. & Harrop-Allin M. K., 2000, MNRAS, in press
14. Rosen S. R., Mason K. O. & Cordova F. A., 1987, MNRAS, 224, 987
15. Warner B., 1995, Cataclysmic Variable Stars, Cambridge Astrophysics Series 28. Cambridge Univ. Press, Cambridge
16. Wickramasinghe D. T. & Meggitt S. M. A., 1985, MNRAS, 214, 605

Doppler Images of the M Dwarf RE 1816 +541

J.R. Barnes[1] and A. Collier Cameron[2]

[1] Centro de Astrofisica da Universidade do Porto,
Rua das Estrelas, 4150 Porto, Portugal.
Guest investigator of the UK Astronomy Data Centre

[2] School of Physics and Astronomy, Univ. of St Andrews,
Scotland KY16 9SS

Abstract. M dwarfs are generally too faint to allow spectra with sufficient S/N ratios for Doppler imaging to be obtained in short enough exposure times. Using a large
number of photospheric absorption lines, we use a sophisticated cross-correlation technique to derive absorption profiles with S/N ratios of 350. The surface images of
RE 1816 +541 reveal a moderately spotted surface, with starspots appearing at all
latitudes. We attempt to measure the surface differential rotation differential rotation
rate and find a lower limit on the time it takes the equator to lap the poles of 80
+/- 9 days. This is in close agreement with other results and extrapolations of current
theoretical predictions.

1 Introduction

Indirect imaging is a powerful tool, allowing previously unimagined detail to be
revealed on the surfaces of rapidly rotating stars. Giants, RS CVn binaries, main
sequence dwarfs and T Tauri stars have all been the focus of studies in the past
17 years since the advent of Doppler imaging [15]. Yet, there still remain areas of
parameter space available to Doppler imaging techniques which have not been
studied due to the constraints of the technique. Conventionally, one or several
lines are used to invert a time series of spectra to obtain an image. We use the
technique of least squares deconvolution on each échelle spectrum in a time series
to obtain information from a large number of photospheric lines. This yields a
single stellar absorption profile with a high S/N ratio, thereby enabling images
of fainter objects to be derived. Even using this technique, there are only two
known single stars of sufficient brightness (feasible with 4m class telescopes),
with well known rotation periods. These are HK Aqr (Gl 890, SpT M0V) and
RE 1816 +541 (SpT M1-2V), with V = 11.83 and (B-V) = 1.45.

The ROSAT EUV Wide Field Camera source, RE 1816 +541 was detected at
the position RA 18h 16m 19.1s, Dec. +54 10 26 (J2000.0) with a positional error
of 63 arcsec and count rates of 6×10^{-3} s^{-1} [12]. This has been identified with the
object GSC 3904.00967, or EY Dra. The first detailed analysis of optical data was
carried out by Jeffries [9] who identified RE 1816 +541 as a dM1 or dM2 object
based upon the strength of its molecular MgH and TiO bands. Optical spectra
were secured on three occasions using the IDS spectrograph at the Isaac Newton
Telescope. These observations confirm chromospheric activity in the form of Ca
II H & K emission lines and Hα emission, a radial velocity (v_r) and projected

equatorial rotation velocity ($v\sin i$) of -21.9 ± 1.5km s^{-1} and 61 ± 1.5km s^{-1} respectively. An axial rotation period of 0.4589 d has been determined from V band photometric data [13] . More recently, the Hα line of RE 1816 +541 has been studied in detail [7] which indicates the presence of prominence clouds below the co-rotation radius. A very strong flare was also detected from pronounced effects in all emission lines, as well as in the photospheric absorption lines.

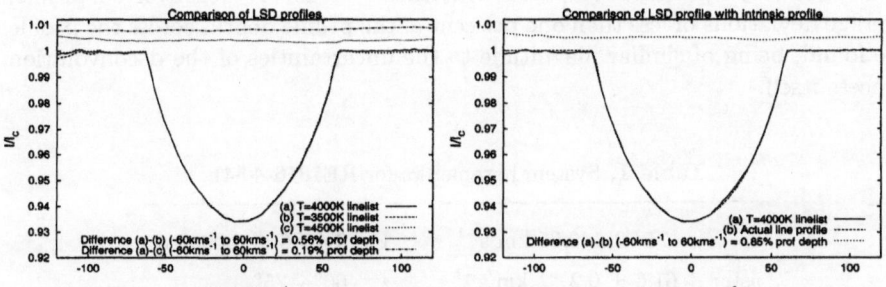

Fig. 1. Testing the validity of least squares deconvolution. As can be seen, the deconvolved profile shape is relatively independent of the line list used for $\Delta T = \pm 500$K). The mis-match of the deconvolved profile derived using the correct line list when compared with the intrinsic profile introduces a greater deviation, but of the same order of magnitude

2 Observations and Data Reduction

We have used the data set obtained by Eibe [7][1]. The spectra were taken at the 4.2 m William Herschel Telescope at the Observatorio del Roque de los Muchachos, La Palma on the nights of 1996 June 25-27, using the Utrecht Échelle spectrograph. The spectra cover the wavelength region 5350 - 9231 Å. The data were re-reduced using the STARLINK package, ECHOMOP.

3 Least Squares Deconvolution

The range of wavelengths in which are to be found the useful photospheric lines is 5350-7500Å. A theoretical line-list representing the positions and strengths (line depths) is required in order to act as a weighting mask for the least squares deconvolution technique [5]. A photospheric temperature of T = 4070K was determined using an appropriate (B-V)-temperature relation [6]. Line lists were obtained using the VALD [14] online database. At these temperatures, molecular bands begin to dominate stellar spectra, but unfortunately stellar atmospheric models do not include adequate molecular opacities. To this end, some

---

[1] extracted from the ING archive

preliminary tests seemed prudent to determine the validity of using the presently available line lists.

An artificial échelle spectrum was generated using a T = 4000K line list, an empirical rotation profile [8] and the wavelength calibration file obtained for the RE 1816 +541 data set. Least squares deconvolution was then applied using line lists with T = 3500K, T = 4000K and T = 4500K. The results are shown in Fig. 1 and reveal that use of an incorrect line list does not severely effect the deconvolved profile. Compared with the T = 4000K deconvolved profile, average deviations of less than one per cent of the profile depth, across the profile are found, being of similar magnitude to the uncertainties of the deconvolution process itself.

Table 1. System parameters for RE1816 +541

v_r	-21.72 ± 0.05 km s^{-1}	P	0.4586 ± 0.0002 d
$v\sin i$	61.6 ± 0.2 $^{-1}$ km s^{-1}	i	60° - 75°

Using a line-list with T = 4000K, we deconvolved the RE 1816 +541 spectral time series, carefully removing regions containing strong lines (e.g. Na I D and Hα), telluric lines and TiO bands. We use the technique described by Cameron [1] which makes use of the telluric lines to correct the radial velocities for shifts in the spectrograph, relative to the earth. The deconvolved time series are shown in Fig. 3. The mean S/N ratio of the input spectra was 32.9±3.3 and the multiplex gain 10.8±0.2, yielding a mean output S/N ratio of 354. This is significantly less than the maximum possible gain of (Number of lines)$^{1/2}$ = 23.5. We are confident that the line profile shape is recovered correctly, but it seems probable that the line depths used to create the mask in the deconvolution process are not very representative of the real spectrum. If this is the case, we are not maximising the information content available in each échelle spectrum, demonstrating the need for improved atmospheric models incorporating molecular opacities.

4   Stellar Parameters

Accurate stellar parameters have been determined from the least squares profiles. The whole data set was used, with spectra binned together into triplets. The results are summarised in Table 2, with v_r and $v\sin i$ errors estimated from the $\Delta\chi^2 = 1$ criterion. These parameters are in good agreement with those given by Jeffries [9]. The inclination of RE 1816 +541 is uncertain, and there is no parallax measurement to aid a determination. As discussed by [9], based upon the radial velocity alone (-21.72 ± 0.05) it seems likely that RE 1816 +541 is a member of the Local Association which predicts a value of -25.0 km s^{-1}. Given the assumption that the age of RE 1816 +541 is 50 Myr, a value of log L/L$_\odot$= -1.114 ± 0.039 is obtained from [3], which yields M$_v$ = 8.54 ± 0.12

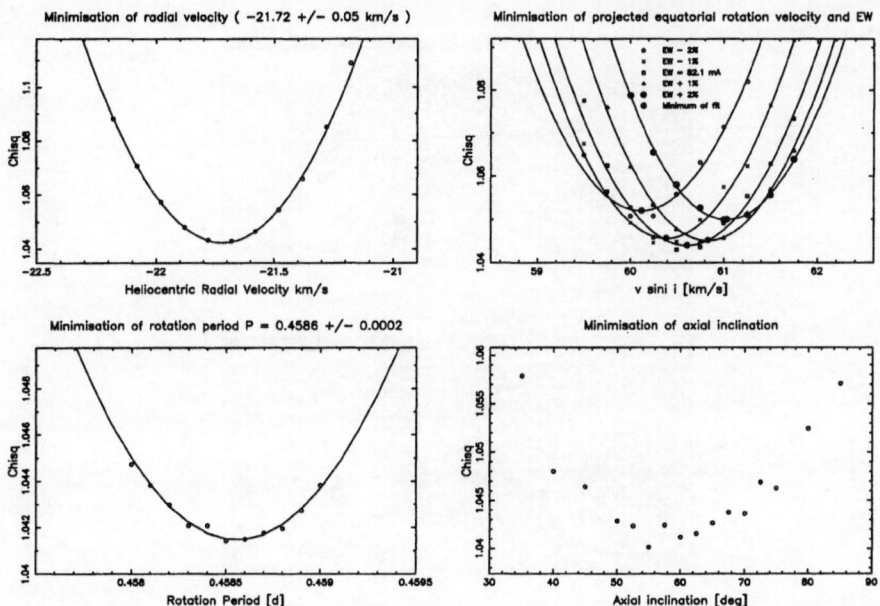

Fig. 2. Determination of RE 1816 +541 system parameters. A large number of maximum-entropy regulated fits are carried out for a series of combinations of parameters. The minimum χ^2 attained indicates the best-fit parameter

using the appropriate bolometric corrections [6]. Hence an estimate of D = 45.5 ± 2.1 pc is obtained. A value of $R = 0.601 \pm 0.034$ is determined and the theoretical temperature T = 3926 ± 100K, is in good agreement with that determined from the B-V colour (T = 4070K). Given the empirically determined $R\sin i = 0.549 \pm 0.002\,R_\odot$, an axial inclination of $i = 66°$ is determined (with $75° > i > 60°$). n the range 66° - 90°.

It is possible to determine the axial inclination from minimisation of the χ^2 statistic for inclinations below 60°. Above this value, a bias towards lower inclinations is favoured. The plot for minimisation of χ^2 vs axial inclination in Fig. 2 is minimised at around 55° - 60°. This indicates an inclination somewhat in excess of 60°, and for consistency with the above calculation, an inclination of $i = 70°$ is chosen for the image reconstructions.

5 The Images

The images are shown in Fig. 4 for each night of observations. It is apparent that there are similarities and differences, even within the regions of phase overlap, from one night to the next. This raises the question as to whether the features are evolving very rapidly, or whether the S/N ratio in the line profiles is too low to reliably constrain image features. To this end, synthetic data sets with the same parameters and the same S/N ratio were created from an "ideal", synthetic

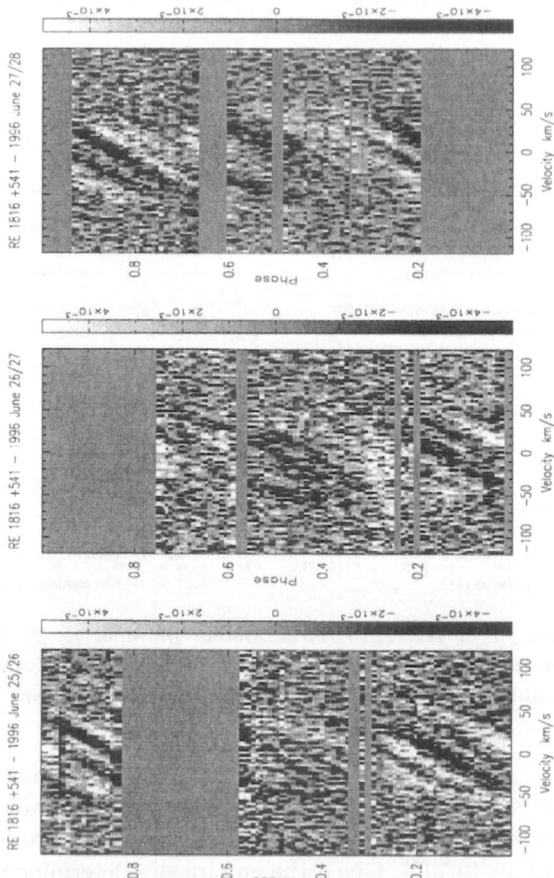

Fig. 3. Greyscale time series images of the deconvolved spectra for each night of observation. Phase is plotted against deconvolved velocity (zero corrected) bin.

image of RE 1816 +541. Images reconstructed from the data sets (which differ only in the random noise pattern) reveal that at a S/N ratio of 350, the features are not always consistently reproduced. Features above 40° latitude especially are less consistent, indicating that any potentially measurable profile distortions are strongly affected by noise. A S/N ratio of 1000 however yields much more reliable images, up to latitudes of 70°.

6   Differential Rotation

Using Doppler images, attempts have been made to measure the differential rotation for a number of stars [2]. The main sequence, single stars, all show surface shear rates which are comparable with the solar value. In addition to this apparent week dependency on rotation rate, there may also be only a weak

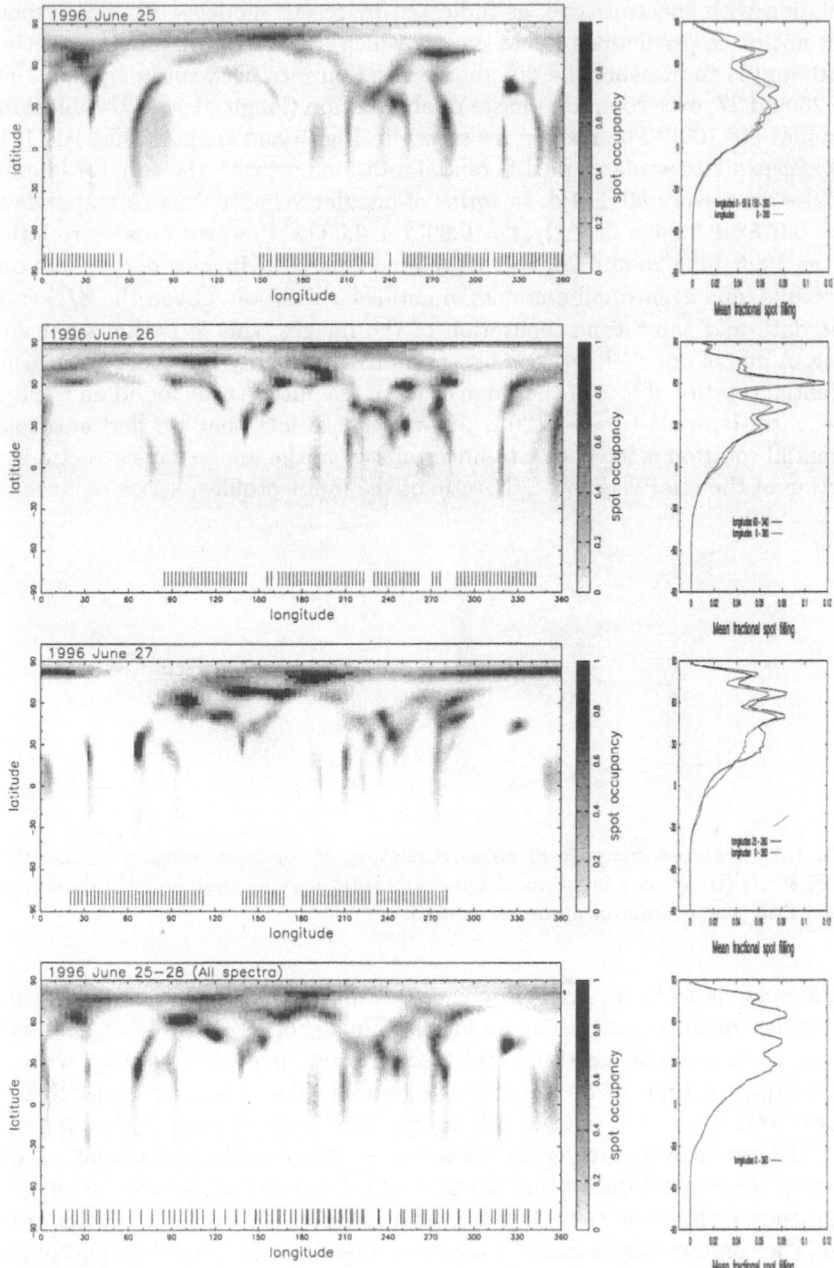

Fig. 4. RE1816 +541 Mercator maps for the nights of 1996 June 25, 26 and 27. The tick marks indicate the phases of observation. The plot to the right of each map is a representation of the mean spot filling as a function of latitude

correlation with spectral type, as indicated by recent models [10]. These models do not make predictions for M dwarfs which are mostly or fully convective. We attempted to measure the differential rotation rate between images on 1996 June 25 and 27, over common phases of observation (longitudes 0 - 270, blocking longitudes 50 - 160). The results are shown in Fig. 5 and suggest that RE 1816 +541 also exhibits a solar type differential rotation law, with the equator lapping the poles once every 80 ± 9 d. In terms of angular velocity, this corresponds to $\Delta\Omega = 0.0783 \pm 0.0088$ or $\Delta\Omega/\Omega = 0.0057 \pm 0.0006$. However cross-correlation between 1996 June 25 and 26, and 1996 June 26 and 27 images reveals less certain results, and even small amounts of anti-solar rotation. Given the S/N ratio of the data and short time separation of the images, this is perhaps not surprising. A gap of one day between observations is arguably too short to measure differential rotation if it is of the same order of magnitude as is found on the Sun (i.e. equator-lap-pole time $\sim$ 120 d). However, the fact that we find anti-solar differential rotation is likely due to uncertainties in the image reconstructions, a reflection of the relatively low S/N ratio of the input profiles.

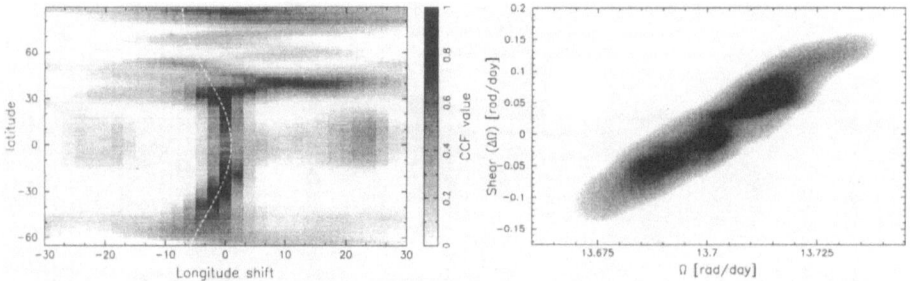

Fig. 5. (a) Greyscale image from cross-correlation of constant latitude strips (1996 June 25 & 27) (b) χ^2 as a function of equatorial rotation and photospheric shear rates Ω_{eq} and $\Delta\Omega$. The overall minimum is marked.

It is possible to incorporate a solar type differential rotation model [4] into the Doppler imaging code. Using an idealised image of RE 1816 +541, test data with the same observation times and a solar type differential rotation with an equator-lap-pole time of 50 days was generated. Two data sets with S/N = 350 and 1000 were created from which equator-lap-pole times of 29 ± 5 d and 49 ± 3 d were recovered using the empirical χ^2 minimisation technique, clearly indicating the need for data with high S/N ratios. Applied to the real data set, we find an equator-lap pole time of 102 ± 32 d, in good agreement with the result above. The contour plot in Fig. 5 however reveals other local minima and in fact the uncertainty associated with the 1-parameter 99% confidence interval for the photospheric shear rate is 133 d. Hence zero or anti-solar differential rotation cannot entirely be ruled out. The lack of correlation above 40° due to insufficient S/N ratio in the profiles is the limiting factor on constraining a reliable fit to the data, allowing only an upper limit on the rotational shear to be determined.

lation of constant latitude strips is perhaps less prone to systematic errors since the act of comparing images should cancel any systematics. Nevertheless, the equator-lap-pole time of 80 ± 9 d may only represent an upper limit on the differential rotation amplitude since artifacts in the maximum-entropy regularised images may bias the results. Further testing is required.

7   Discussion and Conclusion

There has been much interest in recent years over the form of the differential rotation in main sequence stars. Following the discovery of the region of strong radial shear at the base of the convection zone, models have been produced which can successfully reproduce the latitudinal emergence of magnetic flux on the Sun. Application of the model to rapid rotators suggested that the Coriolis force becomes a dominant factor in the latitude emergence of magnetic flux, and hence it has been predicted that we should see only starspots at intermediate latitudes, and not low latitudes, as on the Sun. Doppler imaging work by a number of authors has however shown that starspots usually occur at a range of latitudes, often in distinct latitude bands. Starspots are NOT however excluded from low latitudes. RE 1816 +541 also shows starspots at all latitudes, but it is unclear whether they are restricted to well defined latitude bands.

It is clear that a number of improvements need to be made before a widespread study of largely or fully convective M dwarfs becomes possible. Better model atmospheres with molecular opacities should greatly assist recovery of a higher S/N ratio profile. M dwarfs show a lower amplitude of modulation in V band lightcurves, often making period determinations difficult, though not impossible. The advent of 8 m class telescopes and more efficient spectrographs on 4 m class telescopes will hopefully greatly widen possibilities. Indeed, this analysis shows that the larger CCD chips now available are also crucial in order to obtain the S/N necessary to reliably constrain images and physical parameters such as differential rotation, if indeed it is significant at such late spectral types.

References

1. Collier Cameron A., 1999, in *IAU Colloquium 170, Precise stellar radial velocities*, Victoria BC Canada, eds J.B.Hearnshaw and C.D.Scarfe, ASP Conf. Ser., 185, 233
2. Collier Cameron A., Barnes J. R., Kitchatinov L., 1999, in 11th Cambridge Workshop on Cool Stars Stellar Systems and the Sun, ASP Conf. Ser. (In press)
3. D'Antona F., Mazzitelli I., 1994, ApJS, 90, 467
4. Donati J.-F., Mengel M., Carter B., Marsden S., Collier Cameron A., Wichmann R., 2000, MNRAS, 316, 699
5. Donati J.-F., Semel M., Carter B., Rees D. E., Collier Cameron A., 1997, MNRAS, 291, 658
6. Flower P., 1996, ApJ, 469, 355
7. Eibe M. T., 1998, A&A, 357, 757
8. Gray D. F., 1992, The observation and analysis of stellar photospheres. CUP, University of Cambridge

9. Jeffries R. D., 1993, MNRAS, 271, 476
10. Kitchatinov L. L., Rüdiger G., 1999, A&A, 344, 911
11. O'Dell M. A., Collier Cameron A., 1993, MNRAS, 262, 521
12. Pounds K. A. et al., 1993, MNRAS, 260, 77
13. Robb R. M., Cardinal R. D.,1995, IBVS, No. 4270
14. The Vienna Atomic Database project, Institut für Astronomie, Tuerkenschanzstrasse 17, A-1180 Wien, Austria, c/o W.W. Weiss, http://www.astro.univie.ac.at/~vald/
15. Vogt S. S., Penrod G. D., 1983, PASP, 95, 565

The Method of Spectra Disentangling and Its Links to Doppler Tomography

P. Hadrava

[1] Astronomical Institute of the Academy of Sciences of the Czech Republic, 251 65 Ondřejov, Czech Republic
[2] Physics Institute, NTNU, Trondheim, Norway

Abstract. The method of spectra disentangling fits time series of spectra of multiple stellar systems by a superposition of a priori unknown spectra of component stars with Doppler shifts due to orbital motion with orbital parameters not known in advance. The first part of this general problem – the decomposition of spectra using known radial velocities – was previously solved by the method of tomographic separation. The other part – the measurement of radial velocities for known component spectra – was best solved by the cross-correlation technique. The method of Fourier disentangling solves simultaneously both parts of this problem in a very effective way. It is shown here that its principle can also be used for Doppler tomography.

1 From Radial-Velocities Measurement to Disentangling and Beyond

The observations of binary stars in general, and their spectroscopy in particular, yield most of information about the stellar properties.

1.1 Radial-Velocity Curves

In the standard approach one measures wavelengths of some effective centers of spectral lines and calculates instantaneous radial velocities (RVs) of the binary components from them. By fitting the RV-curves obtained in this way one gets the orbital parameters of the system, or, by simultaneous fitting of light-curves, one gets also some other physical parameters of the system. This procedure is based on several implicit, but sometimes not well satisfied assumptions, especially:

- the lines originating in individual components can be well distinguished and measured, and
- the measured centers of lines correspond to the RVs of the centers of mass of the components.

The first assumption is violated whenever the spectral lines are highly blended. This is troublesome especially in the case of early-type stars, for which the line widths are often comparable with the RVs amplitudes.

The standard **method of cross-correlation** [10,16] or its two-dimensional generalization [20] can improve the measurement of RVs in such cases. However, it requires at least an approximate knowledge of spectra of individual component stars because these are used as the template spectra to be cross-correlated with the individual exposures of spectra of the binary.

On the other hand, the information about the component spectra, which is of great interest by itself, can be obtained by comparing the spectra of the binary observed in orbital phases with different RV-shifts. A simple method of separating the component spectra from two exposures with known RVs has been developed by Ferluga et al. [3,4]. Problems with stability of the solution caused by the noise are overcome by an alternative method of the so-called **tomographic separation** [1], which solves the overdetermined set of equations for the component spectra by fitting many exposures at once, thus diminishing their noise. This method is based on the analogy of the superposition of Doppler-shifted component spectra in different orbital phases with the tomographic projections in different angles of two parallel but displaced linear objects.

1.2 Spectra Disentangling

Applying iterative steps of cross-correlation and spectra separation, it would thus be, in principle, possible to fit the observed spectra of binaries by simultaneous solution of component spectra and the RVs, skipping the questionable step of recognizing individual spectral lines and measurement of their wavelengths. Alternatively, the spectra separation can be merged with the solution of RV-curves into a single procedure yielding directly the component spectra and the orbital parameters and skipping the determination of individual RVs as well. The problem formulated in this generality has been named 'disentangling' by Simon and Sturm [31] and this name will be used in this meaning hereafter.[1] From the mathematical point of view, the decomposition of the spectra of a binary requires the solution (or the least-square fit) of the set of $k \times N$ linear equations

$$\begin{pmatrix} \mathbf{M}_{A1} \ \mathbf{M}_{B1} \\ ... \\ \mathbf{M}_{Ak} \ \mathbf{M}_{Bk} \end{pmatrix} \begin{pmatrix} \mathbf{I}_A \\ \mathbf{I}_B \end{pmatrix} = \begin{pmatrix} \mathbf{I}(t_1) \\ ... \\ \mathbf{I}(t_k) \end{pmatrix} \tag{1}$$

---

[1] Simon and Sturm understood by disentangling the step of the decomposition (i.e. the separation) of the component spectra only, which is alternated with the optimization of orbital parameters to minimize χ^2, i.e. the residual noise in the observed spectra. However, compared to the standard procedure by means of RVs, the simultaneous solution of both these parts of the problem has an additional advantage because of the proper treatment of weights and errors of the information contained in the individual exposures. The method thus deserves its own name. Simon and Sturm argued that the problem of disentangling is different from that of tomography, the former being algebraic while the later is an analytical one. However, it will be shown here that the decomposition can be formulated analytically as well, while in the numerical solution, both methods must be represented algebraically.

for $2N$ unknowns. On the right-hand side of this equation there are k sub-vectors $\mathbf{I}(t_l)$ of the observed spectra (the number of exposures $k \geq 2$), each one of the dimension N corresponding to the number of pixels (typically of the order of 10^3). On the left-hand side, the submatrices $\mathbf{M}_{jl}$ are off-diagonal matrices with unit diagonals shifted according to the Doppler-shifts of the component stars A, B at times $t_l|_{l=1}^k$ of the exposures, the vectors $\mathbf{I}_A$ and $\mathbf{I}_B$ are the separated spectra of the components. To find the solution, Simon and Sturm used the method of singular value decomposition.

An alternative method of Fourier disentangling has been developed from the cross-correlation technique [5]. Equation (1) can be expressed in terms of convolution in logarithmic wavelength-variable $x = ln(\lambda/\lambda_0)$ of the intrinsic component spectra $I_j(x)$ with shifted δ-functions

$$\sum_{j=1}^{n} \delta(x - v_j(t_l)) * I_j(x) = I(x, t_l) , \tag{2}$$

where we have already admitted also a greater number n of the components with different RVs v_j. The Fourier transform (in variables $x \rightarrow y$) separates this equation into independent sets of equations

$$\sum_{j=1}^{n} \exp(iyv_j(t_l))\tilde{I}_j(y) = \tilde{I}(y, t_l) \tag{3}$$

for the individual Fourier modes. The original set of equations in $n \times N$-dimensional space (or $2N$-dimensional for a simple binary) thus splits into N systems of dimension n, what facilitates its numerical solution. To diminish the noise, it is again desirable to have the system overdetermined, i.e. the number of exposures $k > n$. The (linear) least-square fits of Eq. (3),

$$\frac{\partial S(y)}{\partial \tilde{I}_j(y)} = 0 , \tag{4}$$

where

$$S(y) = \sum_{l=1}^{k} \left| \tilde{I}(y, t_l) - \sum_{j=1}^{n} \exp(iyv_j(t_l))\tilde{I}_j(y) \right|^2 , \tag{5}$$

will correspond to the best fit of the observed spectra, provided the residual noise is white. The (non-linear) optimization of $S \equiv \int S(y)dy$ with respect to the orbital parameters p implicitly contained in RVs $v_j(t; p)$ (with substituted solutions of Eq. (4)) solves the general problem of spectra disentangling.

Already the first experience with this Fourier disentangling (implemented in the code named KOREL, see [7]) showed that it yields more accurate RVs in a less laborious way than the cross-correlation technique.[2] In addition, the disentangled spectra enable to determine temperatures, rotational broadening or chemical compositions of the stars, see e.g. [8,9,21].

---

[2] KOREL calculates RVs by cross-correlation of disentangled component spectra with the individual exposures, what explains its predominancy in comparison with the

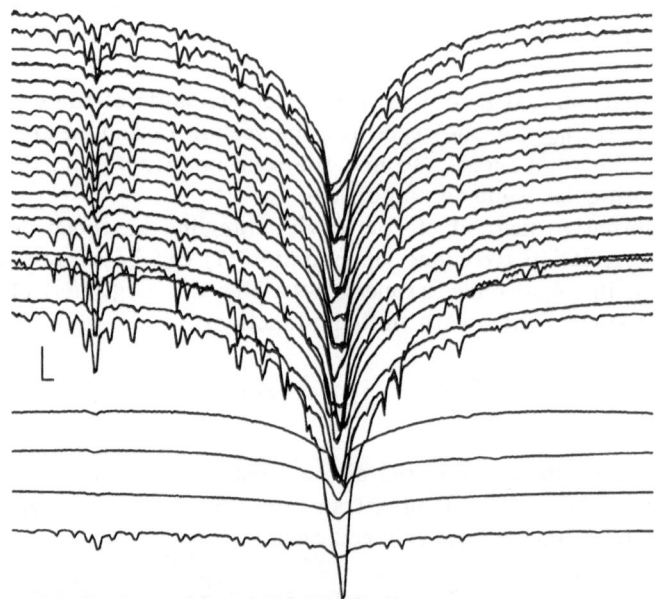

Fig. 1. Example of disentangled H_α line profiles of the triple star 55 UMa. Twenty source exposures with superimposed reconstructed spectra are at the top. The lower four spectra correspond to the separated components and the telluric lines

Regarding to the above mentioned analogy between the separation of binary spectra and the tomography of object of size $2 \times N$ pixels, it is obvious that the method of Fourier decomposition used in the disentangling of n components could be simplified for tomography of object $n \times N$ by a proper choice of shifts v_j of individual rows proportional to their distances (see Section 2.1). Generally, however, the RVs of each component of a multiple stellar system can follow different laws. This enabled to disentangle spectra of some triple stars (see e.g. [2] or Fig. 1). Another application could consist in separating the spectrum of a hot spot in interacting binaries or spectra of individual spots in a spotted accretion disk [13]. In the later case the spots can be either due to the impact of a gaseous stream in interacting binaries or due to some tidally driven hydrodynamic waves (i.e., the Doppler shifts need not to correspond to the velocities of propagation of the spots) or the spots can freely orbit around the central body like a plan-

standard cross-correlation using arbitrarily chosen template spectra (not to speak about the comparison with manual RV-measurements). However, the direct solution of orbital parameters by KOREL is even better than the solution of RV-curves yielded by KOREL (which can be advantageous in combination with published RVs) due to the proper handling of the errors.

etary system. The disentangling could also be used for verification of suspected extrasolar planetary systems and distinguishing faint companions.[3]

It would be desirable to have a method enabling to separate spectra of arbitrarily sampled parts of a smooth accretion disc or stellar surfaces. Unfortunately, this problem would require an inversion of mapping from 3-dimensional space (2-dimensional surface + frequency) into a 2-dimensional one (phase + frequency) and it does not have a unique solution. For example, an observed Gaussian line-profile can be a convolution either of a narrow intrinsic line-profile with rotational broadening of a cloud with Gaussian density distribution, or of a wide Maxwellian profile with a negligible rotational broadening. The standard Doppler tomography restricts the problem assuming the spectrum of local emission to be a δ-function [14], while the disentangling solves for arbitrary spectra of objects with geometry restricted to a finite or at maximum a 1-dimensional set of points. A mean way, based e.g. on the maximum entropy method [34] with a bias [12] corresponding to theoretical models in the additional dimension, could improve the solution. For example, the double-peaked emission of a Keplerian ring cannot be distinguished from a blended doublet of a point-like source by a pure mathematical spectra separation, but its confrontation with model spectra in a larger frequency region must solve the problem. Similarly, a deconvolution with a theoretical local line profile should enable to 'sharpen' spatial resolution of a tomographic mapping.

1.3   Relative Line Photometry and Line-Profile Variations

Another way of improvement and generalization of the disentangling method consists in abandoning the assumption on constancy of the line-profiles in the course of the whole orbital period. The first step has been developed [6] to account for changes of line strengths keeping the shapes of lines constant. Using this generalization included in the code KOREL, it is possible, e.g., to separate efficiently telluric lines from the observed spectra. However, the initial motivation to vary the line strengths was to explain and to remove scatter of RVs observed during eclipses. According to a simple consideration of the eclipse of stellar disk, one would expect that the change of line shape during the eclipse due to the rotational (i.e. the Schlesinger – Rossiter) effect must be at least comparable to the change (i.e. a decrease) of the line strengths. Nevertheless, the experience shows that the differences between the limb darkening in lines and the continuum must be taken into account to explain the appearance of not only the dim out of spectral lines of the eclipsed star, but also their enhancement at the initial and final phases of the eclipse. This effect decreases the influence of rotation, so that the change of shape of the line-profile can be neglected for not too fast rotators. Moreover, this effect yields a tool for an eclipse probing of the vertical structure of the stellar atmospheres.

---

[3] E.g., the lines of the companion, about 4 magnitudes fainter than the primary, were decomposed from the spectra of AR Cas [11].

The next development of the disentangling method should include the changes of the shape of the intrinsic line-profiles caused either by the rotational effect and tidal distortion including the effects of gravity darkening, or by nonradial pulsations or by stellar spots. This can be achieved replacing the simple shifted δ-functions in Eq. (2) by the Fourier transform of (shifted) broadening profiles. The investigation of some of these problems is in progress in cooperation with R.E. Wilson (see [19] for a review).

2 Application of Fourier Decomposition for Computed Tomography

2.1 Method of Oblique Projection

Let us suppose that the distribution of emissivity (per unit square in the velocity space) in a rotating source is given by a time-independent function $J(v_x, v_y)$. If there is no absorption of the emitted radiation[4] then the flux observed at the phase φ at radial velocity v_r is given by the Radon transform

$$F(v_r; \varphi) = \int J(v_x, v_y)\delta(v_r - v_x \cos\varphi - v_y \sin\varphi)dv_x dv_y = \quad (6a)$$

$$= \int J(\frac{v_r - v_y \sin\varphi}{\cos\varphi}, v_y)\frac{dv_y}{\cos\varphi} . \quad (6b)$$

The obliquely projected flux profile defined by

$$G(v_r; \varphi) \equiv \cos\varphi F(v_r \cos\varphi; \varphi) \quad (7)$$

then reads

$$G(v_r; \varphi) = \int J(v_r - v_y\tan\varphi, v_y)dv_y =$$

$$= \int J(v_r, v_y) \overset{v_r}{*} \delta(v_r - v_y\tan\varphi)dv_y . \quad (8)$$

Performing Fourier transform in variable $v_r \to r$, we get equation

$$\tilde{G}(r; \varphi) = \int \tilde{J}(r, v_y)e^{iv_y r\tan\varphi}dv_y . \quad (9)$$

For a chosen discretization in the v_y-space the integration can be replaced here by summation and for a given set $\{\varphi_l|_{l=1}^k\}$ of observations, this set of equations can be solved with respect to $\tilde{J}$ in the same way as Eq. (3) is solved by Eq. (4). Alternatively, one can substitute here $\varphi = \text{atan}\frac{z}{r}$ and to solve for $\tilde{J}(r, v_y)$ by inverse Fourier transform in z.

---

[4] I.e., either the source is optically thin, or J is given on the surface which does not change its inclination with respect to the line of sight.

Due to the loss of the observed information for $\varphi \to \frac{\pi}{2}$ in Eq. (7), $\tilde{J}(r, v_y)$ is undefined for $r = 0$, and $J(v_x, v_y)$ is thus defined up to an additive function of v_y. This inconvenience can be overcome either by fixing $\lim_{v_x \to \pm\infty} J(v_x, v_y) = 0$, if the emission profile is localized within the computed region, or by combining the results calculated for (nearly) perpendicular directions of the initial phase $\varphi = 0$.

2.2 Fourier Reconstruction Technique

The angular asymmetry of the above described method using the obliquely projected flux is avoided in the Fourier reconstruction technique[5] based on the so-called projection-slice theorem which can be obtained by the Fourier transform of Eq. (6a),

$$\tilde{F}(r, \varphi) = \int J(v_x, v_y) e^{ir(v_x \cos\varphi + v_y \sin\varphi)} dv_x dv_y =$$
$$= \tilde{J}(r \cos\varphi, r \sin\varphi) . \tag{10}$$

Consequently, the Fourier transform $\tilde{F}$ of perpendicularly projected flux F is equal to the cross-section in the corresponding direction of the two-dimensional Fourier transform $\tilde{J}$ of the emissivity distribution J. The emissivity J can thus be obtained by 2-dimensional inverse Fourier transform from $\tilde{J}$ interpolated from polar coordinates in which the radial sections $\tilde{F}$ are obtained from the observations.

Regarding the experience from the medical computed tomography [15], where the method of Fourier reconstruction has been found about 20-times faster than the widely used method of filtered back projection, the variants of Fourier methods seem to be worth of further investigation.

Acknowledgement

This study was supported by the grant A3003805 of the Grant Agency of the Academy of Sciences of the Czech Republic.

References

1. Bagnuolo W.G. Jr., Gies D.R., 1991, ApJ, 376, 266
2. Drechsel H., Weeber M., Lorenz R., Hadrava P., 1997, Astron. Ges. Abstr. Ser., 13, 207
3. Ferluga S., Floreano S., Mangiacapra D., 1991, La Lettre de l'OHP, 6, 3
4. Ferluga S., Floreano S., Bravar U., Bédalo C., 1997, A&AS, 121, 201
5. Hadrava P., 1995, A&AS, 114, 393
6. Hadrava P., 1997, A&AS, 122, 581

---

[5] However, the angular asymmetry of the oblique projection may turn into advantage if the observations do not cover sufficiently all phases.

7. Hadrava P., 2000, 'KOREL – the user's guide', http://sunk1.asu.cas.cz/~had/korel.html
8. Harmanec P., Hadrava P., Yang S., Holmgren D., North P., Koubský P., Kubát J., Poretti E., 1997, A&A, 319, 867
9. Hensberge H., Pavlovski K., Verschueren W., 2000, A&A, 358, 553
10. Hill G, 1993, In: *New Frontiers in Binary Stars Research*, ed. by K.-C. Leung, I.-S. Nhu (Calif. Astron. Soc. of the Pacific, San Francisco) p. 127
11. Holmgren D., Hadrava P., Harmanec P., Eenens P., Corral L.J., Yang S., Ak H., Božić H., 1999, A&A, 345, 855
12. Horne K., 1985, MNRAS, 213, 129
13. Karas V., Kraus P., 1996, PASJ, 48, 771
14. Marsh T.R., Horne K., 1988, MNRAS, 235, 269
15. Placidi G., Alecci M., Colaciechi G., Sotgiu A., 1998, J. Magn. Reson., 134, 280
16. Simkin S.M., 1973, A&A, 31, 129
17. Simon K.P., Sturm E., 1994, A&A, 281, 286
18. Skilling J., Bryan R.K., 1984, MNRAS, 211, 124
19. Wilson R.E., 1994 PASP, 106, 921
20. Zucker S., Mazeh T., 1994, ApJ, 420, 806
21. Zverko J., Žižňovský J., Khokhlova V.L., 1997, Contrib. Astron. Obs. Skalnaté Pleso 27, 41

Fourier Disentangling of Composite Spectra

S. Ilijić[1], H. Hensberge[2], and K. Pavlovski[3]

[1] Faculty of Geodesy, University of Zagreb, Kačićeva 26, 10000 Zagreb, Croatia
[2] Royal Observatory of Belgium, Ringlaan 3, B-1180 Brussel, Belgium
[3] Department of Physics, University of Zagreb, Bijenička 32, 10000 Zagreb, Croatia

Abstract. The Fourier disentangling algorithm can be applied on a time series of observed composite spectra to obtain orbital parameters and component spectra, in the assumption that the intrinsic spectra do not vary with time. Applications are shortly reviewed with the purpose to emphasize the power of the method. Thereafter, the progression of noise from the input data into the disentangled spectra and the orbital parameters is discussed. It is concluded that no bias is introduced by purely random noise when the necessary precautions are taken. Systematic noise is presently discussed from an empirical viewpoint.

1   Introduction

Attempts to separate the spectra of unresolved multiple systems go back to Wrigth [16], and result today in several algorithms that are designed to reconstruct the component spectra and determine their relative motion from optimization of *one* merit function (Fourier disentangling [3,4]; disentangling by singular value decomposition [13]) or that could do so (tomographic separation [1]). They start from a time-series of composite spectra, without using template spectra or a-priori assumptions about the shape of the spectra. The method has been mostly applied to well-detached binaries (in accord with the assumption that the intrinsic component spectra do not vary with time), but has been used also in the case of systems as complicated as β Lyr [2]. Most applications have concentrated on strong lines, due to hydrogen, helium or magnesium. The method yields precise radial velocities (RVs), even near conjunction and in the case that the spectra of the components show appreciable rotational broadening which leads to a composite spectrum of blended features. The improvement in the precision of the RV, relative to a cross-correlation technique, is documented e.g. in Fig. 7 of Harmanec et al. [5]. Especially the determination of mass ratios and orbit eccentricities profits from the improved RVs. Further profit comes from the inclusion of the light-curve in the case of eclipsing binaries, and an iterative application of the disentangling algorithm (providing RVs) and a light-curve + RV analysis code (putting e.g. tighter constraints on the light-ratio). Such a scheme is shown in Fig. 1 of Hensberge et al. [6].

In more difficult cases, the disentangling method was used for the *detection* itself of a faint spectral component [2,8]. The RV variation of the detected features then serves as decisive argument to attribute them to the previously unseen component or to matter moving with it.

Going one step further, the disentangled spectra may be analyzed using the methods applied for single stars. Temperatures and, if not known more accurately from the dynamical analysis, gravities can be estimated from spectroscopy, rather than using relations based on photometric colours. Rotational velocities and abundances can be estimated without interference of the other component. Atmospheric parameters of components of binaries were derived from disentangled spectra for the components of DH Cep [15], Y Cyg [14] and V578 Mon [6] (see Fig. 1). In the latter case, a differential abundance analysis shows that both very young main-sequence stars have the same chemical composition as the single stars in the same open cluster [10].

Precise stellar parameters obtained in this way have been used to derive stellar ages and distances. In the case of massive binaries, often found in open clusters or associations, it provides an independent track to the age of these groups. Another application concerns the detection of subtle spectral variability due to forced non-radial oscillations in binaries [5,7,8]. The reconstruction of an "average" spectrum for each of the components allows to search for travelling bumps or other characteristic profile variations by using difference spectra (observed spectra minus the composite model computed from the disentangled spectra).

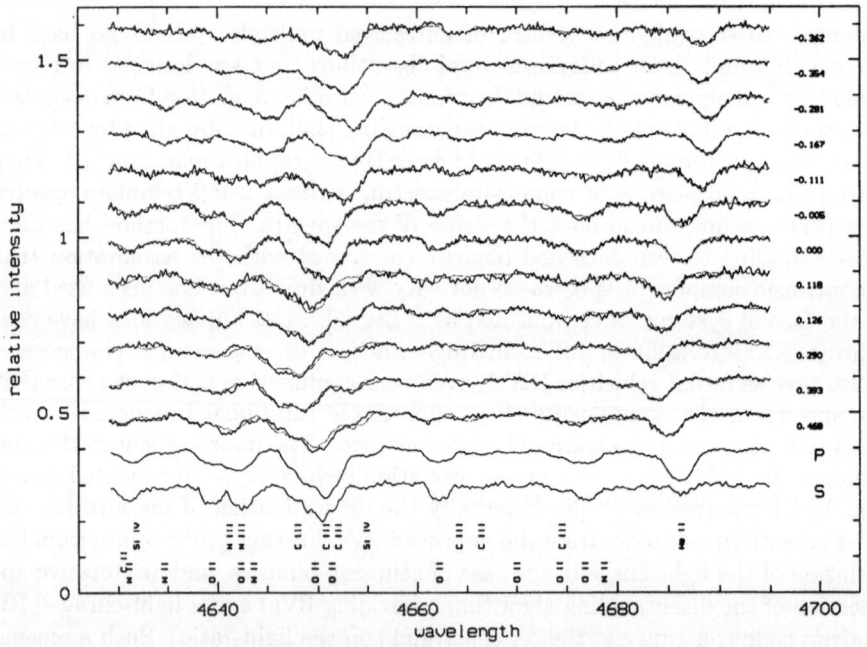

Fig. 1. Extract of the observed spectra (the orbital phase relative to the primary eclipse is indicated at the r.h.s.) and the disentangled spectra of the hotter (P) and the cooler component (S) of V578 Mon. The composition of P and S at the observed phases is overplotted. Shifts in relative intensity of 0.1 are applied for clarity

Disentangled spectra are significantly less noisy than each of the input spectra. Then, especially for the information present in the weaker spectral features, systematic noise becomes the limiting factor rather than random noise. Simon et al. [14] recognized that the most significant source of error lies in the rectification of the spectra. The progression of random and systematic noise to the orbital parameters and the disentangled spectra should thus be understood in detail in order to be able to put reliable error estimates on the final results.

2   Tests on Noise Progression

Simon & Sturm [13] have shown, for their singular value decomposition algorithm, that the spectrum of the primary of V453 Cyg, disentangled from 7 composite, $S/N \approx 50$, spectra taken out of eclipse matches the spectrum of the primary observed during total eclipse in the secondary minimum. Tests on synthetic spectra, with varying resolution and noise, and on composites of single star spectra have shown the intrinsic power of the method. The least-squares merit function depends sensitively on the assumed RVs and the intrinsic spectra were recovered even with a small number of input spectra (4) and spectra set up of broad hydrogen and diffuse helium lines only [12].

Hynes & Maxted [9] used synthethic OB spectra and present their results quantitatively, the main drawback being the limited accuracy of the standard deviation derived from only 10 independent realizations of Poisson noise. They caution against deriving uncertainties from the curvature of the χ^2 space near its minimum, since that was seen to underestimate the errors when the surface is locally not sufficiently smooth. They suggest that in real spectra ripples on that surface will be enhanced by short-scale auto-correlation in the noise of the input spectra, introduced by rebinning the observed data to a regular logarithmic wavelength grid. They note shortly that the rectifiation errors they introduced did not have a significant effect on the RV amplitudes.

We are presently performing tests using a Fourier disentangling algorithm. A 3-D grid of input datasets was prepared to perform Monte Carlo trials at each of its points, varying the number of composite input spectra (5, 10 or 20), their signal-to-noise ratio ($S/N = \infty$, 100, 50 or 25), and the light ratio of the components (1 or 3). The intrinsic synthetic component spectra ($T_{eff} = 30000$ K (27500 K), $\log g = 4.0$ (4.5) and $v \sin i = 120$ km s^{-1} (100 km s^{-1})), the orbital parameters (RV semi-amplitudes $K_1 = 100$ km s^{-1} and $K_2 = 200$ km s^{-1}, eccentricity $e = 0.1$ and periastron longitude $\omega = 60°$), the resolution of 10 km s^{-1} and spectral range (4630–4695 Å, almost identical to the one used by Simon & Sturm [12]) are kept unchanged. The spectral lines appear in both components with different relative intensity and combine into composite spectra with complicated blends of medium-intensity features as shown for the massive binary V578 Mon in Fig. 1. At each point of our 3-D grid of input datasets with finite S/N, disentangling was performed on 200 sample realizations of the dataset, such that standard deviations computed from a given sample are accurate within 10%.

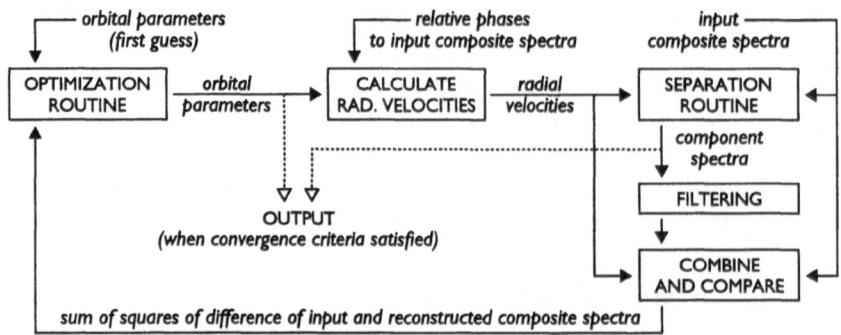

Fig. 2. The flowchart of the disentangling algorithm

The Fourier disentangling algorithm (Fig. 2) is based on a spectral separation routine implemented after Hadrava [3,4] (see also this volume), which in a non-iterative way calculates the component spectra given the composite spectra (data) and a set of RVs. The component spectra are computed from the requiring that, for the given RVs, they fit the observed spectra in the least-squares sense when mixed into reconstructed composite spectra. The success of this fit depends sensitively on the RVs. The optimization routine uses the downhill simplex minimization algorithm [11] to find the optimal set of RVs. The RVs are obtained through variation of K_1, K_2, e, ω, and an overall phase-shift $\Delta\phi$ with regard to the relative phases specified with the input spectra; $\Delta\phi$ correlates strongly with ω and only their sum is well-determined. For each sample dataset the routine was started ten times, with different first guesses for the 5 parameters, to check whether the same solution is found consistently.

The low pass filtering of component spectra that we introduced, prior to reconstructing the composite spectra for comparison to the data, improves the convergence properties of the algorithm by avoiding unphysical fine structure into the estimate of goodness of fit. The Savitzki–Golay smoothing filter [11] preserving 4th order curvature over a total width of 11 data pixels was checked to affect insignificantly the synthetic spectra of the individual components. The applied filtering is similar to the oversampling of the component spectra before the computation of the goodness of fit, as discussed by Simon et al. [14], and is a remedy against the ripples discussed by Hynes & Maxted [9].

3   Random Noise Tests: Conclusions

The conclusions we present are drawn from the analysis of the distribution of the estimates for the orbital parameters and of the disentangled spectra at each wavelength separately, obtained in 200 trials at each grid point:

- Bias in the disentangled spectra is introduced by the Fourier transformations applied on non-cyclic functions, mostly at the ends of the wavelength interval but strictly speaking at any wavelength. Caution in the choice of the end-points, apodizing and/or removal of the spectral ends in the disentangled spectra suffice to keep this bias below the desired level.

• The rms scatter at a wavelength point shows no dependence on wavelength or spectral structure and is well reproduced considering the progression of uncorrelated noise, $(S/N)_{1,2} \approx \langle (S/N)_{\text{input}} \rangle \, \ell_{1,2} \, \sqrt{m-2}$ where $\ell_{1,2}$ is the fraction of the total light due to the considered component (light factor), and m is the number of input spectra. The latter relation sets clearly the level at which one should bother about systematic errors in the input data.

• The orbital parameters are reproduced without measurable bias only if proper care is taken in the Fourier analysis (see above) and in the optimization procedure (Sect. 2).

• The rms scatter of the orbital parameters scales with the number of input spectra, the S/N and the intrinsic velocity information content of the component spectra as expected from the progression of uncorrelated random noise.

• The absolute level of this random scatter may be larger than predicted (less than a factor of 2); however, a better error progression theory should be developed (see below). If confirmed, the extra noise may be due to neglecting that the intensity on each pixel of the disentangled spectra was an unknown or that there exists correlation in the noise of the rebinned input spectra.

The theoretical estimate for the variance on $K_{1,2}$ was connected to the variances on the input velocities with a factor computed for the case of a single-lined binary in a circular orbit,

$$\text{Var}\, K_{1,2} = \frac{\sum_j \frac{\sin^2 \phi_j'}{\text{Var}\, v_{1,2}(j)}}{\sum_j \frac{\cos^2 \phi_j'}{\text{Var}\, v_{1,2}(j)} \sum_j \frac{\sin^2 \phi_j'}{\text{Var}\, v_{1,2}(j)} - \left(\sum_j \frac{\cos \phi_j' \sin \phi_j'}{\text{Var}\, v_{1,2}(j)} \right)^2} \tag{1}$$

wherein ϕ_j' is the phase of the j-th input spectrum as measured from the phase of maximum RV. Variances on the RV were computed after Hensberge et al. [6]:

$$\text{Var}\, v_{1,2}(j) = \frac{1}{\sum_i J_{1,2}^2(i) \, \text{Var}^{-1} I_j(i)} \times \frac{\ell_{1,2}(j)^{-2}}{1 - \text{Corr}^2(J_1, J_2)_j} . \tag{2}$$

Equation (2) was derived for the case of a spectrum with two components and takes into account: the variances of the input data; the intrinsic velocity information content of the component spectra, expressed by the gradients $J_{1,2}(i) = \mathrm{d}\, I_{1,2}(i)/\mathrm{d}\, i$ and the dilution factor $\ell_{1,2}^{-2}$; and the loss of precision due to correlation between the spectral gradients of the component spectra. However, it assumes perfectly disentangled spectra. For m observations of similar quality, fairly homogeneously distributed over the orbit, Eq. (1) simplifies to $\text{Var}\, K_{1,2} \approx \frac{2}{m} \times \langle \text{Var}\, v_{1,2} \rangle$.

4   Systematic Noise: Comments

We have verified the claim by Hynes & Maxted [9] about rectification errors (Sect. 2), from tests with continuum offsets on each input spectrum taken from a uniform distribution of offsets within 2%. However, in practice rectification errors are of a different nature (especially in echelle spectra). One might expect them to correlate with the underlying composite spectrum of weak lines and thus with wavelength and orbital phase.

An empirical indication of the effects to be expected can be gained from disentangling a piece of composite spectrum which contains a very broad and undeep (e.g. $\approx 1\%$) diffuse interstellar band, but not allowing for a third component in the disentangling. This shows that low-frequency errors are transferred to the output spectra and are *amplified* according to the light factor. In other words, such errors are most obvious in the spectrum of the fainter component when normalized to its intrinsic continuum. This sensitivity can be exploited to correct a-posteriori at least partially for rectification errors (noting that any change to the continuum of the disentangled spectrum of one component implies a related change to that of the other component). Unfortunately, this is a time-consuming, iterative and somehow subjective procedure.

Acknowledgements

The authors acknowledge support through 'IUAP P4/05', financed by the Belgian DWTC/SSTC, and through grant #007002 of the Croatian Ministry of Science and Technology. S.I. thanks the Director of the Royal Observatory of Belgium for support during his stay in June/July 2000. We are grateful to P. Hadrava for his cooperation.

References

1. Bagnuolo W.G. Jr., Gies D.R., 1991, ApJ, 376, 266
2. Bisikalo D.V. et al., 2000, A&A, 353, 1009
3. Hadrava P., 1995, A&AS, 114, 393
4. Hadrava P., 1997, A&AS, 122, 581
5. Harmanec P. et al., 1997, A&A, 319, 867
6. Hensberge H., Pavlovski K., Verschueren W., 2000, A&A, 358, 553
7. Holmgren D. et al., 1997, A&A, 322, 565
8. Holmgren D. et al., 1999, A&A, 345, 855
9. Hynes R.I., Maxted P.F.L., 1998, A&A, 331, 167
10. Pavlovski K., Hensberge H., 2000, In: *Poster Proc. IAU Symp. 200, Birth and Evolution of Binary Stars*, ed. by B. Reipurth, H. Zinnecker, p. 109
11. Press W.H., Teukolsky S.A., Vetterling W.T., Flannery B.P., 1992, *Numerical Recipes in FORTRAN*, 2nd edn. (Cambridge University Press, Cambridge)
12. Simon K.P., Sturm E., 1992, In: *Proc. of the 4th ESO/ST-ECF Data Analysis Workshop*, ed. by P.J. Grosbøl, R.C.E. de Ruijsscher (ESO), p. 91
13. Simon K.P., Sturm E., 1994, A&A, 281, 286

14. Simon K.P., Sturm E., Fiedler A., 1994, A&A, 292, 507
15. Sturm E., Simon K.P., 1994, A&A, 282, 93
16. Wright K.O., 1954, Publ. DAO Victoria (Canada), 10, 1

Doppler Tomography
of Eclipsing and Non-eclipsing Algols

M.T. Richards

Department of Astronomy, University of Virginia,
P.O. Box 3818, Charlottesville, VA 22903-0818, USA

Abstract. Since the advent of Doppler tomography, Hα tomograms of eleven eclipsing Algol-type binaries and an ultraviolet tomogram of U Sge have been made from difference profiles. Tomograms of CX Dra, a non-eclipsing Algol, were made from the observed and difference Hα, Hβ, He I λ 6678, and Si II λ 6371 profiles. Tomograms produced at multiple epochs demonstrate stability in systems like β Per and TX UMa but variability in others like U Sge, U CrB, and CX Dra. These tomograms show that the Algols contain a diverse range of circumstellar structures that include a gas stream, accretion annulus, transient accretion disk, and/or chromospheric emission source in a short-period Algol, or a classical accretion disk in a long-period system. Some systems, like U CrB and U Sge, alternate between stream-like and disk-like states. These results were obtained by extracting information that was formerly irretrievable from the observed spectra of these systems.

1 Introduction

The Algol-type binaries are close, semi-detached, interacting binary star systems which contain a cool F–K giant or subgiant secondary star that fills its Roche lobe and is losing mass to a hot B–A main sequence primary star (see Fig. 1). The term "Algol Paradox" arose after it was discovered that the less massive secondary star in an Algol-type binary was more evolved than its companion. This was contrary to the theory of stellar evolution in which the less massive star should evolve more slowly than its companion. The paradox led astronomers to realize that mass transfer was an active process in close binaries and had caused the originally more massive secondary star to lose a large portion of its mass to its companion, resulting in a reversal of the mass ratio in the system [6,20]. The Algols are still in the first phase of mass transfer by Roche lobe overflow, but they are in the late part of that first phase [48]. Many evolutionary models assume that the evolved star is still in the hydrogen burning phase (Case B), or has evolved beyond the onset of helium burning in the core or a shell (Case C). In addition, these models assume conservative evolution, i.e., no mass loss from the binary, although Tout & Eggleton [47] found evidence of mass loss from these systems through winds.

The classical Algols are typically in a slow stage of mass transfer with rates of $(10^{-11} - 10^{-7})$ $M_\odot$ yr^{-1}. However, systems like U Cep ($P \sim 2.49$ days) have variable orbital periods and mass transfer rates of $(10^{-6} - 10^{-5})$ $M_\odot$ yr^{-1} [30], and are termed "active Algols." Very active Algol-like systems like β Lyr ($P = 12.93$

days) also have variable orbital periods and are classified as "Serpentids" [8]. This review will focus on those Algols that are in the slow mass transfer phase. In this process, the gas free falls toward the primary under the influence of the Coriolis force in the rotating frame of the binary. In the long-period Algol systems ($P > 6$ days), where the primary is small relative to the binary separation, mass transfer results in a *classical accretion disk*. This disk is typically a large, permanent structure characterized by strong double-peaked emission lines broadened by the orbital motion of the gas in the disk. The blueshifted and redshifted emission peaks represent the approaching and receding lobes of the disk, respectively. A study of the Hα spectra of 9 long-period Algols by Olson & Etzel [27] showed that a steady state is never truly reached. Similar disks are found in the dwarf novae, low mass X-ray binaries, and some novae-like cataclysmic variables. However, the stars in a CV are faint relative to the accretion disk while the stars in an Algol binary are bright relative to any accretion regions.

The accretion process is more complicated in the short-period Algols ($P < 6$ days) because the primary is large relative to the binary separation and the gas flow from the L_1 point is not deflected sufficiently by the Coriolis force to avoid contact with the primary star. As a result, the high-velocity ($\sim 500 - 600$ km s^{-1}) gas stream is shocked by a grazing or direct impact with the slowly rotating primary ($v \sin i \leq 150$ km s^{-1}). This star-stream interaction leads to *direct impact accretion* [28] and results in a variety of accretion structures. The heating of the impact site results in a high temperature accretion region that can be observed at ultraviolet wavelengths in the form of absorption in several uv resonance lines (e.g., Si IV, C IV, S III, Al III [28]). Some of the gas from the impact region is released into orbits that repeatedly splash off the primary star until energy losses due to these collisions cause the gas to be accreted onto the mass gainer [44]. The temperature of the impact region does not increase beyond $\sim 10^6$K, and hence does not generate intense X-ray radiation. Our current understanding is that almost all of the X-ray radiation from Algol systems arises from magnetic activity on the late-type mass-losing companion (e.g., [46,51]), and not from any part of the accretion structure.

A quasi-stable disk-like accretion structure that surrounds the primary is found in the intermediate Algols which have orbital periods of (4.5 – 6) days. This structure is called a *transient accretion disk* [16]. It is variable and asymmetric, with a velocity field that is close to, but less than, the Keplerian (or circular) velocity. Typically, the emission from a transient disk is weak compared to the combined continuum flux of the system, whereas the emission from a classical disk is generally much stronger. In addition, the line profiles from a transient disk appear to be broadened by supersonic turbulence, while the dominant broadening mechanism in a classical disk is the near-Keplerian motion of the gas in the disk [15]. Moreover, the gas in a transient disk is highly variable and may disappear in as little as one orbit, while classical disks are considerably more stable.

Another structure called an *accretion annulus* [39] can be found in systems with $P < 4.5$ days. This structure is an asymmetric, sometimes clumpy, circum-

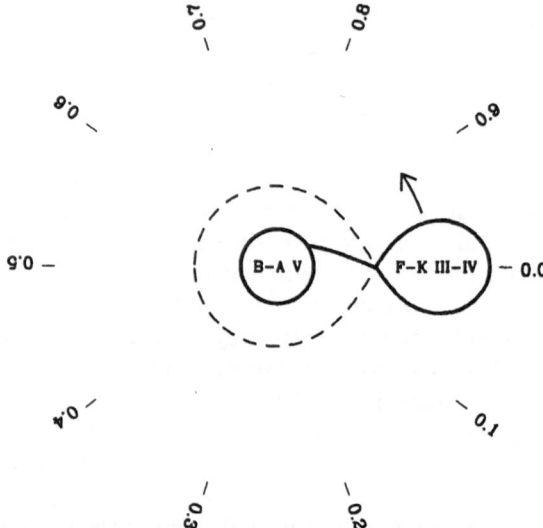

Fig. 1. Roche geometry of a typical short-period Algol-type binary system showing the predicted path of the gas stream, the range of spectral types, the spectral classes, and the lines of sight for different orbital phases.

primary distribution of gas with sub-Keplerian velocities, hence lower angular momentum than the quasi-stable disks. In these binaries, double-peaked $H\alpha$ emission is rarely detected in the observed spectra because the emission is weak relative to the combined continuum flux of the stars. In this case, the direct collision between the gas stream and the star produces a shock region on the star, but the gas eventually forms a region of gas around the star which might be clumpy because of the variable gas density in the region. This annulus is gravitationally bound to the star since it does not have sufficient angular momentum to form a stable accretion disk. The signature of this annulus is weak double-peaked emission in the $H\alpha$ difference profiles, with the peaks close to the rest wavelength of the line. Evidence of the annulus would not be detected readily in the observed profile.

The ballistic calculations of Lubow & Shu [22] were used by Peters & Polidan [28] and Kaitchuck, Honeycutt & Schlegel [17] to demonstrate why the division between short- and long-period systems occurs at about 5 days. For periods above 5–6 days, the stars are wide enough to form an accretion disk around the primary. However, for orbital periods less than 5 days, the primary is large relative to the orbital separation so that the trajectory of the gas stream crosses the locus of the photosphere of the primary; i.e., the gas stream makes direct impact with the surface of the star. These arguments are readily understood with the aid of the $r - q$ diagram, a plot of the radius of the mass gainer (r in units of binary separation) versus the mass ratio of the binary, $q = M_S/M_P$, where M_S is the mass of the secondary star and M_P is the mass of the primary star. Two reference curves ω_d and ω_{min} are drawn onto the diagram where ω_d

is the fractional radius of a classical accretion disk for different mass ratios, and the curve ω_{min} represents the distance of closest approach of the gas stream measured from the center of mass of the primary. If a system falls below the ω_{min} curve the gas stream will miss the primary star and will form a classical accretion disk fed by the gas stream. Alternatively, if a system is located above the ω_d curve the gas stream will strike the primary directly, resulting in an impact region and possibly a transient disk or an accretion annulus.

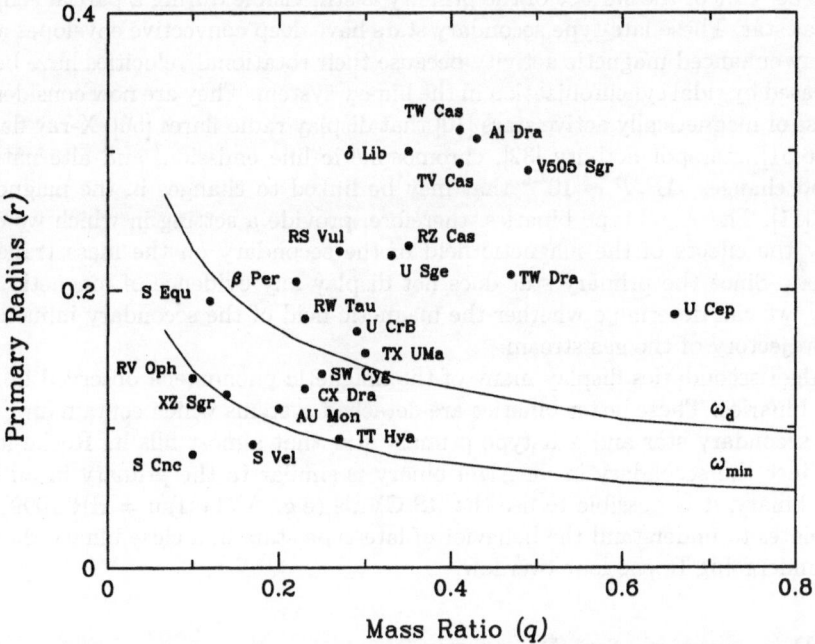

Fig. 2. The r–q diagram, where r is the radius of the primary star in units of the separation of the binary and q is the mass ratio. The two curves plotted ω_d and ω_{min} are the smallest radius of a stable accretion disk and the distance of closest approach of the gas stream, respectively. (Taken from [37].)

The observed properties of the circumstellar material produced by direct impact accretion in short-period Algols have been found from studies of both Balmer-line and ultraviolet spectra and also UBV photometry. The regions identified include the gas stream [38], a transient accretion disk [16], a denser region near impact site called a "localized region" [33,49], a "high rotational velocity region" [33,34,49], a "high temperature accretion region" near the impact site (UV spectra: [28]), and a circumstellar bulge which has been identified in UBV photometry [26] as a hot band around the equator of the primary. The circumstellar gas is distributed well beyond the orbital plane of the binary to heights comparable to the radius of the mass gaining star [33]. The circumstellar gas is also highly variable from one orbit to the next, with masses of $(10^{-12} - 10^{-10})$ M$_\odot$ and densities that are typically high: $N_e \sim (10^9 - 10^{11})$ cm^{-3} [7,16,23,26,28,30,33,34]. Some systems have extremely low circumstellar carbon abundance, $C \sim 10^{-3}C_{\odot}$, indicative of CNO processed material transferred from the secondary star [28].

Until recently, the secondary star was essentially ignored because it contributes only 5–10% of the total light at optical wavelengths. The result is that the primary dominates the light of the binary at all wavelengths from ultraviolet to infrared. The spectral types and luminosity classes of Algol secondaries are usually not known to high accuracy, except in totally eclipsing systems, because up to 30–40% of the surface of the primary is still visible during a partial eclipse of that star. These late-type secondary stars have deep convective envelopes and display enhanced magnetic activity because their rotational velocities have been increased by tidal synchronization in the binary system. They are now considered a class of magnetically active stars [10] that display radio flares [50], X-ray flares [45,46,51], starspot activity [32], chromospheric line emission, and alternating period changes $\Delta P/P \sim 10^{-5}$ that may be linked to changes in the magnetic field [11]. The Algol-type binaries, therefore, provide a setting in which we can study the effects of the magnetic field of the secondary on the mass transfer process. Since the primary star does not display any evidence of magnetic activity, we can determine whether the magnetic field of the secondary influences the trajectory of the gas stream.

Algol secondaries display many of the magnetic phenomena observed in RS CVn binaries. These latter binaries are detached systems which contain an F–G type secondary star and a K-type primary star that almost fills its Roche lobe [9]. Since the secondary in an Algol binary is similar to the primary in an RS CVn binary, it is possible to use the RS CVn's (e.g., V711 Tau = HR 1099) as templates to understand the behavior of late-type stars in a close binary that is not undergoing Roche lobe overflow.

2   Preparation for Tomography

The first Doppler maps of an Algol binary were presented by Jones & Richards [14]. Hα tomograms have now been made for 11 eclipsing Algols (RZ Cas, δ Lib, RW Tau, β Per, TX UMa, U Sge, S Equ, U CrB, RS Vul, SW Cyg, TT Hya) and 1 non-eclipsing Algol (CX Dra), while only one ultraviolet tomogram has been published (U Sge: [19]). Tomograms produced at multiple epochs for β Per, TX UMa, U Sge, U CrB, and CX Dra demonstrate stability in β Per and TX UMa but variability in U Sge, U CrB, and CX Dra. These tomograms show that the Algols contain a diverse range of circumstellar structures. The tomogram of a short-period Algol can display a gas stream, a chromospheric emission source, an accretion annulus, or a transient accretion disk, while the tomogram of a long-period Algol displays a classical accretion disk [3,36,38]. Some short-period systems display transient accretion disks whose intensity changes with epoch [3]. These results were obtained by extracting information that was formerly irretrievable from the observed spectra of these systems.

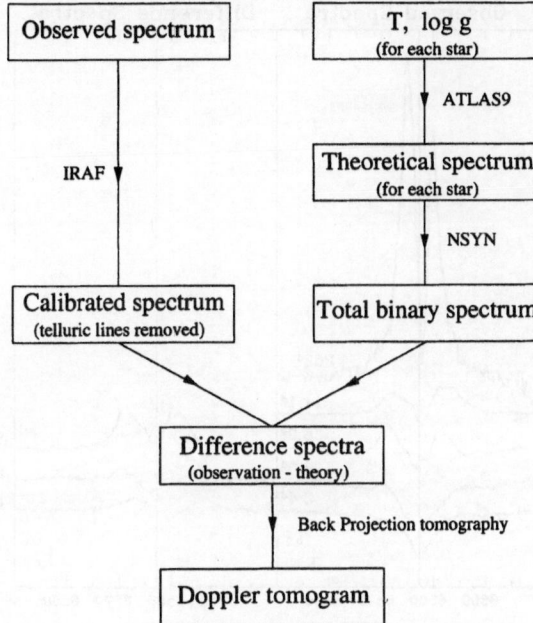

Fig. 3. Data reduction procedure employed by Richards & Albright (e.g., [34,37]).

The procedures used to create and interpret these tomograms can be summarized in a paragraph, but required a lot of effort from a single research group. It started with an intense observing campaign to collect mostly Hα spectra seen at phases around the entire orbit of 18 Algols. The observations were closely spaced to permit the detection of transitions in the profiles, and were obtained typically within 3 orbital cycles to reduce the influence of secular variations. Then these spectra were compared with theoretical spectra which represented the photospheric contributions of the stars, and difference profiles were calculated. The observed and difference profiles were studied and morphological groupings were identified. Doppler tomograms were generated from the spectra, and then hydrodynamic calculations were used to interpret these tomograms.

As stated in Section 1, the observed Hα spectra of the Algols display characteristic strong double-peaked line profiles if the orbital period is long (e.g., RZ Oph: P=262 days to TT Hya: P=6.95 days; Fig. 4). However, as the orbital period decreases below 6 days, the Hα profile looks increasingly like a distorted absorption profile. For a long time, these absorption profiles were thought to be entirely due to the orbital variations of the stellar line profiles. However, closer examination demonstrates that the changes detected in the observed profiles over an orbital cycle are more complex than expected from the stars alone. This was demonstrated in the case of TX UMa by Albright & Richards [1]. Moreover, within a given binary, there are significant changes in the observed profiles with time (e.g., U Sge [2]). These spectra show clear evidence of contributions from

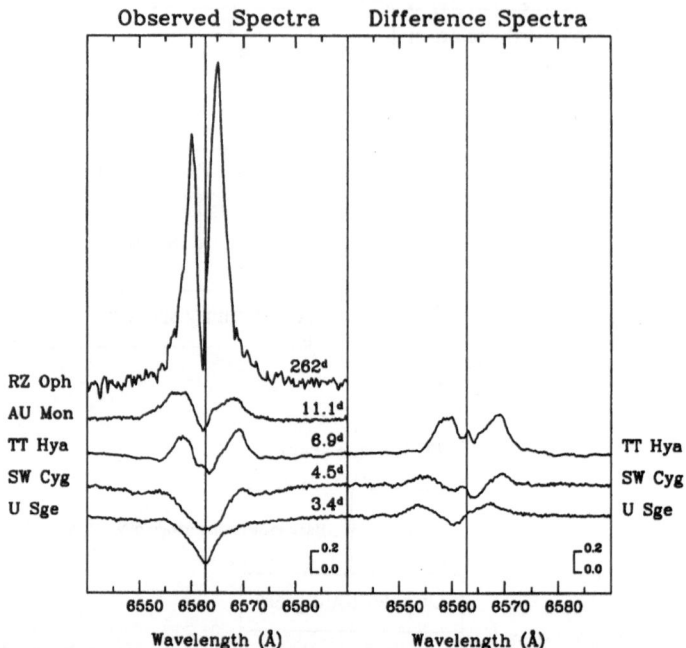

Fig. 4. Appearance of the observed Hα line with orbital period, P (left panel): RZ Oph ($P = 262$ days), AU Mon ($P = 11.1$ days), TT Hya ($P = 6.9$ days), SW Cyg ($P = 4.5$ days), and U Sge ($P = 3.4$ days). Difference spectra for TT Hya, SW Cyg, and U Sge are also shown in the right panel. Adapted from [36].

non-photospheric sources, so it is necessary in systems with short orbital periods to remove the photospheric line contributions.

The subtraction procedure enhances the weak emission and absorption contributions from circumstellar gas relative to the luminosity of the stars. While these contributions can be seen as distortions of the observed profiles, they cannot be studied in any detail unless the photospheric component is actually removed. Richards & Albright [37] calculated "difference profiles" by subtracting the combined theoretical LTE stellar photospheric Hα line profile from the observed line profile at each orbital phase. Theoretical calculations were used to represent the photospheric component because of the accuracy and flexibility offered by those calculations. Kurucz ATLAS9 model atmospheres [21] were used to produce the theoretical line profile of each star. Then, the combined line profile of the binary was generated within the NSYN line profile synthesis code which includes the effects of gravity darkening, limb darkening, the reflection effect, and non-synchronous rotation of the stars [1,33,34]. Richards [34] demonstrated that non-LTE effects are not significant for Algol primaries ($T_e < 20,000$ K), so the LTE ATLAS9 models are a good approximation to the photospheric line profiles. Moreover, since the gas stream does not penetrate down to the photospheric regions where the hydrogen lines are formed in the slow mass transfer

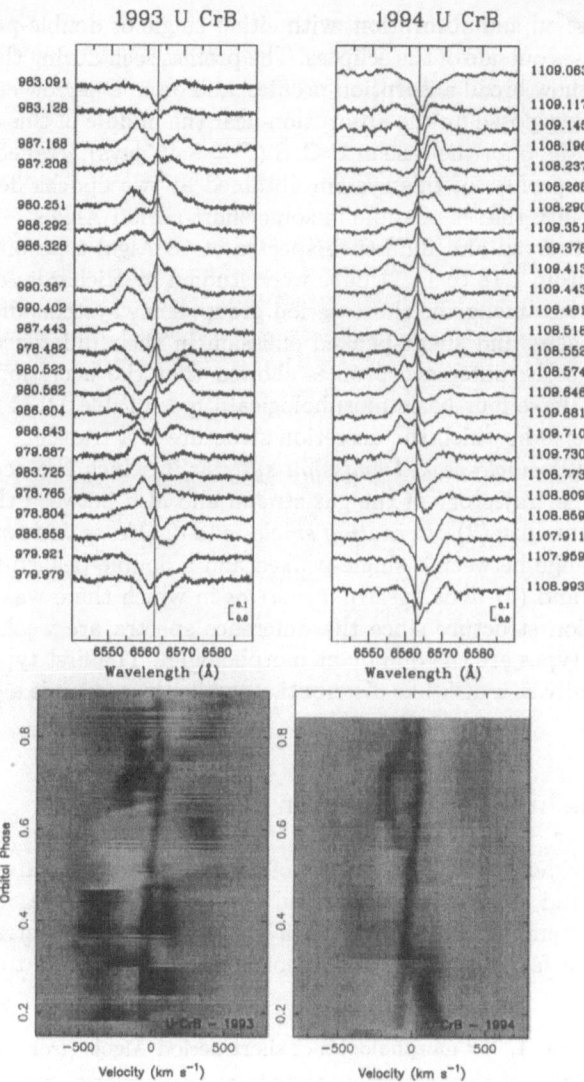

Fig. 5. Some Hα difference profiles and trailed spectrograms of U CrB for 1993 and 1994. This figure illustrates the kind of variability found in some short-period Algols. (Adapted from [37].)

Algols [33], the differencing procedure is valid for these systems. The subtraction procedure is justified only if it is assumed that the non-photospheric sources are optically thin, and this is a reasonable assumption, at least to first order.

These difference profiles represent the contributions of all non-photospheric gas flows in the binary, namely those produced by Roche lobe overflow or the chromosphere of the magnetically-active secondary star. They typically show blends of emission and absorption, with either single or double-peaked emission at orbital phases outside of the eclipses. The profiles seen during the eclipse of the primary star show broad absorption profiles, and often a narrow emission feature can be seen superposed on the absorption near the middle of this eclipse. This is illustrated in Fig. 5 for the case of U CrB ($P = 3.45$ days). Trailed spectrograms and difference profiles of this system obtained at two epochs demonstrate the kind of variability that is possible in some short-period Algols.

The Hα observed and difference spectra of 18 Algol-type binaries with orbital periods from 1.18 to 11.11 days were studied by Richards & Albright [37]. They found that among the short-period group, some systems display alternating double-peaked and single-peaked emission in their difference profiles. The analysis of the Hα difference profiles showed that the accretion structures in Algol binaries have four basic morphological types (Table 1): (1) *double-peaked emission systems* in which the accretion structure is a transient or classical accretion disk; (2) *single-peaked emission systems* in which the accreted gas was found along the trajectory of the gas stream and also between the two stars in an accretion annulus; (3) *alternating single- and double-peaked emission systems* which can change between a single-peaked and a double-peaked type within an orbital cycle; and (4) *weak spectrum systems* in which there was little evidence of any accretion structure since the difference spectra are weak at all phases. The first two types are the dominant morphologies. The first type can be interpreted physically as a disk-like distribution, while the second is a gas stream-like distribution.

3 The Technique of Doppler Tomography

We can find evidence of different types of circumstellar material through intense examination and study of the observed or difference spectra. However, there is always the temptation to produce a model of the gas distribution that is based on some favored or popular theory. The technique of tomography is a

Table 1. Hα morphologies of short-period Algols (from [37])

Type	P (days)	Morphology	Examples
1		Double-peaked emission	
	$P > 6$	permanent	CX Dra, TT Hya, AU Mon
	$4.5 < P < 6$	variable	SW Cyg
2	$2.7 < P < 4.5$	Single-peaked emission	RW Tau, β Per, TX UMa, S Equ, RS Vul
	$1.2 < P < 2.8$	(weak near $\phi \sim 0.5$)	RZ Cas, TW Dra
3	$2.7 < P < 4.5$	Alternating emission types	U Sge, U CrB, and possibly RW Tau, TX UMa, S Equ, RS Vul
4	$P < 2.4$	Weak spectrum systems	V505 Sgr, AI Dra, TW Cas, TV Cas, δ Lib

fairly unbiased way to produce an image. The more generalized technique of *computerized tomography* was formulated by the Austrian mathematician Johann Radon [12,31] and has been used in the field of medicine to reconstruct 3D images of parts of the human body from 2D pictures or "slices" (e.g. CAT scans). In astronomy, the technique of Doppler tomography was introduced by Marsh & Horne [24], and summarized by Robinson, Marsh & Smak ([43]: eqns. 9–15), Kaitchuck et al. [18] and by Marsh (this volume). Horne [13] introduced the technique of back-projection tomography as a simpler version of the MEMSYS code. Tomography was applied first to the classical accretion disks in cataclysmic variables (e.g. [18,25]), and later was used to locate the emission sources in the Algol-type binaries.

The images of Hα emission sources were obtained with a back-projection Doppler tomography code [13,18]. Although Hα spectra were obtained around the entire binary orbit, only the emission-dominated Hα difference profiles from orbital phases outside of primary eclipse were used in the code. The basic procedure is to convert the wavelength scale of each spectrum to a velocity scale relative to the rest wavelength of the line (e.g., Hα) in the rest frame of the binary. Then, within the tomography code, the profiles are passed through a fast-Fourier filter with a Gaussian of FWHM $= 3\sigma$, where σ is the velocity spacing of the pixels in the spectra. Every point in the binary is assigned a velocity (V_x, V_y, V_z), but the reconstruction is simplified by assuming that the z component of the velocity is constant. This avoids the difficulties involved in producing a three-dimensional image, so the tomograms display only the emission sources in the orbital plane of the binary.

The intensity at each point in velocity space (V_x, V_y) is calculated from the entire data set assuming an S-wave variation $(V_r = V_y \sin(2\pi\phi) - V_x \cos(2\pi\phi))$, with equal weights applied to each orbital phase. This procedure assumes that the emission line width is due mostly to Doppler broadening, but supersonic turbulence may be important in regions like the star-stream impact region. The data were sampled regularly to reduce the effect of unevenly spaced data, but larger gaps in the data resulted in linear streaks on the tomograms. The final result is a reconstructed image of the emission sources in the binary plotted in velocity coordinates. The tomograms were usually scaled individually to emphasize the strongest features in each system [3], but they were plotted on the same intensity scale to permit easy comparison in cases where multiple tomograms were generated for a given binary.

In a Doppler tomogram, the double-peaked emission line profiles from an accretion disk will map into a toroidal structure located between the two large circles centered on the mass gainer (Fig. 6). These circles represent the circular Keplerian velocities at the surface of the star (*solid outer circle*) and at the Roche lobe of the mass gainer (*dashed inner circle*). The image reconstruction preserves the angular positions of emission sources relative to the stars [43], but distances in the tomogram depend on the velocity field of a given structure. As a result, the inner edge of a Keplerian disk is found on the outer edge of the tomogram and *vice versa* so the sources far from the mass gainer in velocity coordinates are

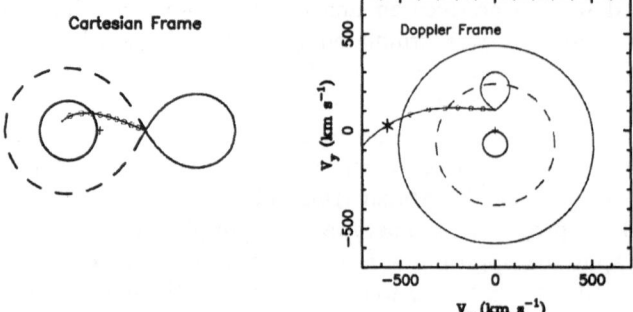

Fig. 6. Cartesian Roche geometry and gas stream trajectory for U Sge and the corresponding diagram in velocity space. The path of the gas stream is the curved trajectory marked at intervals of a tenth of the distance from the L_1 point to the distance of closest approach to the mass gainer. The large outer solid and dashed circles in the Doppler frame represent the circular (Keplerian) velocities at the photosphere and Roche surface of the mass gaining star, respectively. The asterisk marks the location where the gas stream makes impact with the stellar photosphere.

located close to that star in spatial coordinates. The gravitational path of the gas stream (*solid trajectory marked with small circles*), the Keplerian image of this path (*dot-dashed trajectory*), the location where the gas stream strikes the star (*asterisk*), and the surfaces of the stars, assuming synchronous rotation, are also plotted on the tomogram [38].

4   Tomograms of the Eclipsing Algols

The systematic collection of spectra required to produce a good tomogram could not have been accomplished without the availability of an observing facility, like the 0.9m Coudé Feed Telescope at Kitt Peak National Observatory, which allocated multi-night observing shifts to observe stars. Table 2 shows the number of spectra used to obtain the tomograms of each system. Doppler tomograms were made from the Hα difference spectra of eleven eclipsing Algols with $P = (1.20 - 6.95)$ days: RZ Cas, δ Lib, RW Tau, β Per, TX UMa, S Equ, U Sge, U CrB, RS Vul, SW Cyg, and TT Hya [3,36,38]. These tomograms were produced from the emission-dominated Hα difference profiles and illustrate several sources of Hα emission: (1) The gas stream, (2) transient or permanent accretion disks, (3) star-stream impact region, (4) disk-stream impact region, (5) circumprimary bulge and (6) the chromosphere of the secondary. In addition, an ultraviolet tomogram of U Sge was made from Si IV λ 1394 difference profiles, and a 23-year study of CX Dra spectra resulted in Doppler tomograms at four wavelengths and a comprehensive model for this binary.

Table 2. Source of Hα tomograms

Binary	P_{orb} (days)	Spectral type	Year	Telescope	No. data
RZ Cas	1.1953	A2V+G6III	1994	KPNO–Coudé Feed	28
δ Lib	2.3274	A0IV+G1IV	1994	KPNO–Coudé Feed	51
RW Tau	2.7688	B8V+K0III-IV	1994	KPNO–Coudé Feed	30
β Per	2.8673	B8V+K2IV	1976	DDO	59
			1992	NSO–McMath–Pierce	135
			1994	KPNO–Coudé Feed	36
TX UMa	3.0633	B8V+G0III-IV	1992	NSO–McMath–Pierce	81
			1993	NSO–McMath–Pierce	114
			1994	KPNO–Coudé Feed	27
U Sge	3.3806	B7.5V+G4III-IV	1993	KPNO–Coudé Feed	106
			1994	KPNO–Coudé Feed	48
S Equ	3.4361	B8V+F9 III-IV	1993	KPNO–Coudé Feed	57
U CrB	3.4522	B6V+G0III-IV	1993	NSO–McMath–Pierce	161
			1994	KPNO–Coudé Feed	47
RS Vul	4.4777	B5V+G1III-IV	1993	KPNO–Coudé Feed	81
SW Cyg	4.5730	A2V+K1IV	1994	KPNO–Coudé Feed	36
CX Dra	6.6960	B2.5V+F5III	1994	KPNO–Coudé Feed	54
TT Hya	6.9534	A3V+G6III	1994	KPNO–Coudé Feed	21
RS CVn: V711 Tau	2.8377	K1IV+G5V	1994	KPNO–Coudé Feed	27

4.1   Evidence of the Gas Stream

The Hα tomograms of RZ Cas (1994), δ Lib (1993), β Per (1976/77, 1992), U Sge (1994), S Equ (1993), U CrB (1994), and RS Vul (1993), all show elongated emission along the gravitational trajectory of the gas stream from the L_1 point toward the primary star ([38,39]; Fig. 7). It is fitting that the Algols provided the first convincing images of the gas stream in the entire class of interacting binaries. The collimation of the gas stream was most pronounced in the U CrB (1994) and U Sge (1994) tomograms, while other tomograms of these systems show that the gas stream emission does not dominate the tomogram all the time. This dominant stream, or stream-like state, occurs when the difference profiles follow a simpler S-wave variation than at other times (cf. Fig. 5 [37]). When the Hα difference spectra had stronger double-peaked emission, the appearance of the tomogram changed to include an accretion disk. Such changes can occur within an orbital cycle ([2]; Fig. 14).

4.2   Evidence of Accretion Disks

The tomograms of TT Hya (1994), SW Cyg (1994), U CrB (1993), and U Sge (1993) illustrate that the dominant emission source appears as a torus of gas about the primary star and within the expected locus of an accretion disk (Fig. 8 [3]). These images are reminiscent of the classical accretion disks seen in the tomograms of dwarf novae (e.g., U Gem) shown by Kaitchuck et al. [18]. The

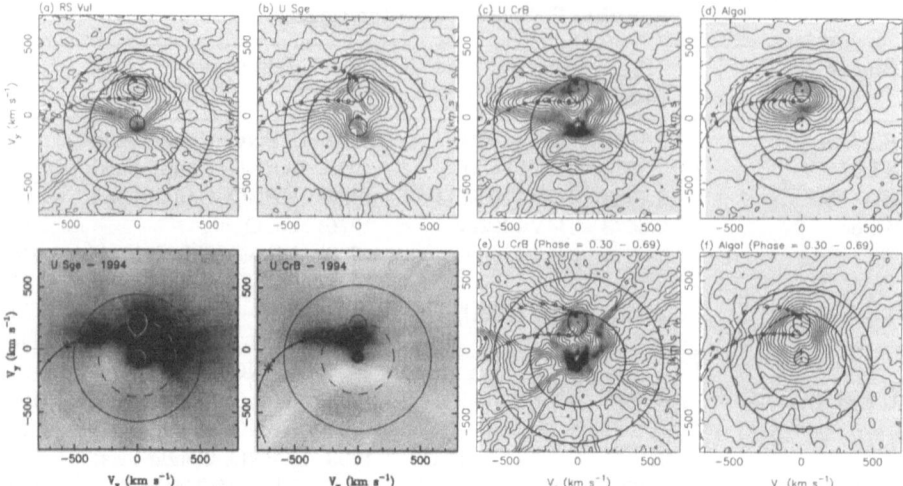

Fig. 7. Doppler tomograms of (a) RS Vul (1993), (b) U Sge (1994), (c) U CrB (1994) and (d) β Per (1976/77). The contour images all display distinct gas stream components that follow the free fall trajectory from the L_1 point in the rotating frame of the binary. Frames (e) and (f) show the tomograms for the phase interval $\phi = 0.30 - 0.69$ of U CrB and β Per, and demonstrate that the disk-like structure surrounding the primary star is enhanced when the gas stream is occulted by that star. (Adapted from [38].)

gas distribution seen in these four Algol-type systems depends on orbital period and their position in the $r - q$ diagram (Fig. 2). Systems with $P > 4.6$ days (e.g., SW Cyg: $P = 4.57$ days, and TT Hya: $P = 6.95$ days) display permanent disk emission in the tomogram with little evidence of any emission from a gas stream. These systems are wide enough to accommodate disk formation, and their disks were more symmetric than in U CrB, and U Sge. However, the structures seen in SW Cyg and TT Hya have different properties: The disk emission in SW Cyg represents an accretion structure that touches the surface of the primary star, while the disk in TT Hya does not touch the primary and extends to the Roche surface of that star.

The 1993 tomograms of U CrB ($P = 3.38$ days), and U Sge ($P = 3.45$ days) display emission within the locus of a Keplerian disk in addition to emission along the gas stream (Fig. 8). Both emission sources are always present but there was pronounced variability in the relative strengths of the emission components so that the gas stream dominates at some epochs (Fig. 7). Albright & Richards [3] found that the disks in U CrB and U Sge are twice as bright as the disk in TT Hya and almost ten times brighter than the disk in SW Cyg. In general, the tomograms of Algols with $P < 4.5$ days have a variety of emission sources that include a transient accretion disk at some epochs. Changes seen in the Hα difference profiles of RW Tau, TX UMa, S Equ, and RS Vul (Table 1) suggest that tomograms of systems with $2.7 < P < 4.5$ days should display transient

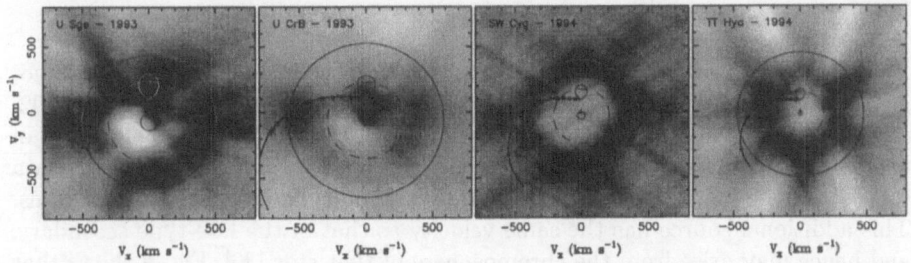

Fig. 8. Doppler tomograms of U Sge (1993), U CrB (1993), SW Cyg (1994) and TT Hya (1994) adapted from [3]. These show evidence of emission within the locus of an accretion disk, but the structures seen in U Sge and U CrB are transient disks.

accretion disks as well as gas stream emission, but no tomograms of these latter systems can be made in the disk-like state without additional observations.

4.3   Star-Stream and Disk-Stream Impact Regions

Another emission source can be seen in the tomograms of SW Cyg (1994), as well as in U Sge (1993) and U CrB (1993) when the disk emission is at least comparable to the gas stream. This source is concentrated close to the predicted location of the impact between the gas stream and the photosphere of the primary (asterisk: Fig. 8).

Another source of emission can be seen in the tomogram of U Sge (1994) to the right of the line of centers (Fig. 7). This source would also be found to the right of the line of centers in a Cartesian map since angular information is preserved in the transformation from velocity to linear dimensions. The emission may arise from a "localized region" like the one found in β Per [34] that is produced when the rotating disk strikes the gas stream after one orbit of the primary.

4.4   A Circumprimary Bulge

Another emission source is roughly centered on the synchronous velocity of the primary star (Fig. 7) and can be seen in the tomograms of U CrB (1993, 1994), S Equ (1993), RS Vul (1993), and SW Cyg (1994). It may arise from a circumprimary equatorial band or 'bulge' that is produced when the stellar photosphere is hit by a gas stream moving at speeds of over $5-10$ times the rotational velocity of the star [3]. Photometric evidence of this bulge suggests that it is formed during epochs of enhanced mass transfer [26]. A hot equatorial band is created by the transfer of angular momentum from the gas stream to the photosphere and causes the photosphere to rotate temporarily at super-synchronous rates. The bulge can be enhanced in the tomograms by using only those spectra from $\phi = 0.30 - 0.69$ where any gas stream emission would be occulted ([38]; Fig. 7). This emission source is strongest in U CrB (1994), which also displays the strongest gas stream emission of all the Algols.

4.5   Influence of Magnetic Activity

A prominent emission source associated with the secondary is seen in the tomograms of six Algols (Fig. 9, Fig. 14): δ Lib (1993), RW Tau (1994), β Per (1992, 1994), TX UMa (1992, 1993, 1994), S Equ (1993) and RS Vul (1993). Most of these systems are known to be magnetically active based on radio and X-ray data (cf. Section 1). The RW Tau (1994) tomogram looks like the β Per tomograms. This additional source has the same velocity as that of the late-type secondary, and hence may arise from the chromosphere of that star [36]. Fig. 9 shows that the strongest emission source in RS Vul is centered on the velocity of the cool G1 III–IV star, while in the remaining Algol systems the chromospheric source is weaker relative to other emission sources in the binary. The TX UMa (1993) tomogram shows the chromospheric emission source following the contour of the Roche lobe, with the gas stream in the predicted position. Tomograms obtained at multiple epochs for β Per and TX UMa suggest that the emission sources in these systems are more stable than those found in systems like U CrB or U Sge.

A tomogram of the RS CVn binary V711 Tau was made from observed Hα spectra to provide a check on the back-projection procedure [36]. As expected, the tomogram of V711 Tau displays no evidence of mass transfer via Roche lobe overflow, but contains a strong source centered on the more active K1 IV component, with weaker emission extending toward the less active, fainter G5 IV companion. A double chromospheric source is not obvious in this case because of the relative strengths of the K to G chromospheres. The stronger chromospheric Hα source in V711 Tau has the same velocity as the more active component, in agreement with the results for the Algols. This was a good first-order result since the differencing procedure applied to the Algols assumed that all the non-photospheric gas was optically thin, which is not correct in the optically thick chromospheric environment. When compared with the Hα emission from V711 Tau, the chromospheres of the Algols are $(0.2 - 6.2)$ times as powerful since they contribute $\sim (4 - 30)\%$ of the continuum flux at this wavelength. The chromospheric source in RS Vul was the strongest of all the systems studied, and was 6 times stronger than that in V711 Tau.

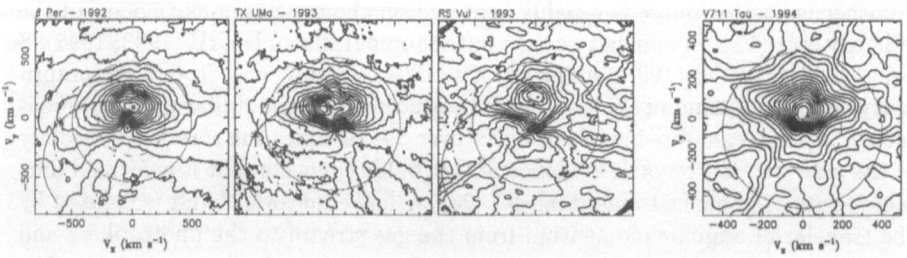

Fig. 9. Contour Doppler tomograms of magnetically active systems: β Per (1992), TX UMa (1993), RS Vul (1993), and V711 Tau (1994). (Taken from [36].)

4.6   The Absorption Zone

The Hα tomograms of U Sge (1993) and U CrB (1993) contain a region where emission is entirely absent (clear region in Fig. 8). This region was termed the "absorption zone" by Richards & Albright [3,37]. It is located on the side of the primary opposite the secondary star that would be visible from $\phi \simeq 0.2 - 0.8$. This region can be seen in other systems (e.g., S Equ and RS Vul) even when there was good orbital coverage. Fig. 10 shows that this zone is close to the star-stream impact site and overlaps a hotter region found in the ultraviolet tomogram of U Sge that was derived from Si IV λ1394 difference profiles [19].

4.7   Ultraviolet Tomogram of U Sge

Archival IUE spectra of U Sge obtained from 1983 to 1994 were used by Kempner & Richards [19] to study the distribution of hot gas (T~ 10^5 K) in the binary. The orbital variation detected in several uv resonance lines suggested the presence of circumstellar gas at these wavelengths. A Doppler tomogram was made from the Si IV λ 1394 line, which was the strongest resonance line in the SWP spectra of U Sge. In this case, difference profiles were calculated by subtracting the Si IV spectrum of a B8 V standard star from the observed Si IV spectrum of U Sge. The residual emission seen in the difference spectra was strongest between phases $\phi = 0.3$ and $\phi = 0.7$, with a strength of up to 0.2 of the continuum flux. The tomogram produced from 17 Si IV λ 1394 difference spectra shows a source of emission preferentially on the side of the mass gaining star away from the mass loser (between phases 0.3 to 0.7). The location of this UV region is consistent with the location of the Hα absorption zone seen in tomograms of U Sge and U CrB and with theoretical predictions of a high temperature accretion region [28].

4.8   CX Dra: A 23-Year Study of a Non-eclipsing Binary

The first non-eclipsing Algol to be studied in detail was CX Dra ($P = 6.696$ days, $q = 0.24$, $i = 53°$). Several hundred Hα, He I λ 6678, Hβ, and Si II λ 6371 spectra of CX Dra were obtained over 23 years from 1975 to 1998 at six observatories in five countries: Ondřejov Observatory (Czech Republic), McDonald Observatory (U.S.A.), Kitt Peak National Observatory (U.S.A.), Observatoire de Haute Provence (France), David Dunlap Observatory (Canada), and Okayama Observatory (Japan). The multiwavelength study led by P. Koubský resulted in the refinement of the orbital solution of CX Dra; equivalent width measurements which showed short-, medium-, and long-term behavior of the difference profiles; calculation of the Balmer decrement; velocity maps based on the velocity curves of the Hα and He I difference emission peaks; trailed spectrograms of the Hα, Hβ, He I, and Si II lines; and Doppler tomograms at these four wavelengths [40].

They found that: (1) the circumstellar environment in CX Dra changes in cycles of hundreds of days. The length of the cycles is variable. These cycles may be part of a super 4000-day cycle. (2) The equivalent width of the difference

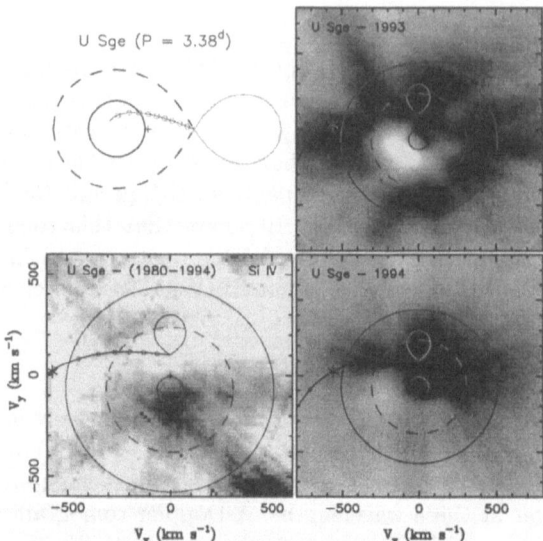

Fig. 10. Hα and Si IV λ 1394 tomograms of U Sge. A scale model of the binary is shown in the upper left corner of the figure.

Hα and He I λ 6678 are modulated with the orbital period of 6.696 days. The Hα, He I λ 6678, and Si II λ 6371 emission are always present in the spectra. (3) The radial velocities of the Hα emission peak follow an S-wave. The resulting velocity map shows that the source of the single-peaked emission lies close to the L_1 point, roughly between the primary and L_1 point. (4) Doppler tomograms constructed for Hα observed and difference profiles shows that the emission comes from a region of low velocity, a gas stream, and an accretion disk. The Hβ emission arises from a region that is cospatial with the Hα source. The Doppler tomograms for the He I λ 6678 and Si II λ 6371 lines suggest that emission originates also from a disk around the primary star. These tomograms all show remarkably consistent emission structures even though they were obtained from different observing sites at different wavelength dispersions (from 0.10 Angs./pixel at Haute Provence to 0.26 Angs./pixel at Ondrejov). (5) The model based on the equivalent widths of the difference profiles, the trailed spectrograms and Doppler tomograms of the Hα, He I λ 6678, Si II λ 6371, and Hβ lines suggest that the main source of the Hα emission is about halfway between the stars at a distance of 0.49a from the primary star; and the He I and Si II emission sources arise from a disk of radius $r_d = (2.0 - 3.5)R_P$ centered on the primary star (Fig. 11). The Hα tomogram of CX Dra derived from observed spectra, resembles those of U CrB (1993) and U Sge (1993) derived from difference spectra when these two eclipsing systems were in a disk-like state.

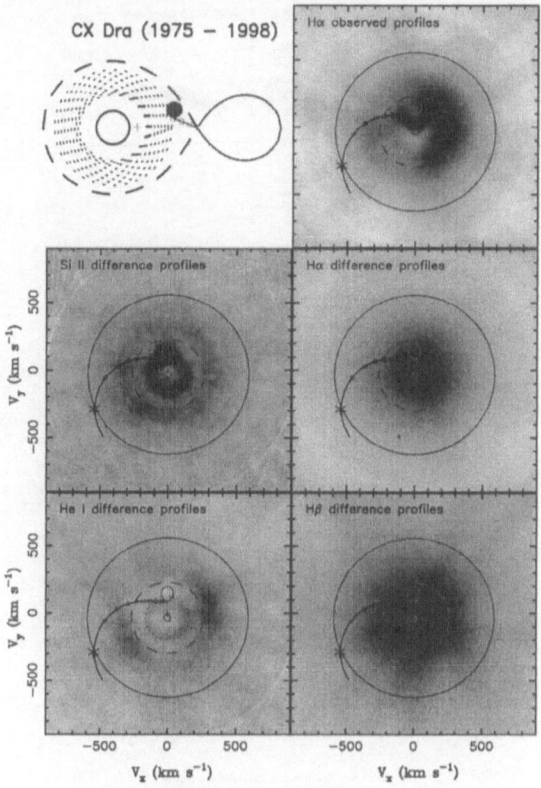

Fig. 11. Hα, Hβ, He I (λ 6678), and Si II (λ 6371) tomograms of CX Dra and the model derived from the tomograms. (Adapted from [40].)

5 Interpretation of Tomograms Using Hydrodynamic Simulations

The first hydrodynamic calculations to simulate the structures formed as a result of direct impact accretion in Algol binaries were made by Blondin, Richards, & Malinkowski [4]. They performed two-dimensional (2D) hydrodynamic simulations of β Per, and investigated the accretion flows and their dependence on various parameters such as the density and velocity of the gas stream. They used the 2D hydrodynamics code named Virginia Hydrodynamics-1 (or VH-1) which is based on the piecewise-parabolic method (PPM) developed by Colella & Woodward [5], and is formulated as a Lagrangian calculation followed by a remap onto a fixed Eulerian grid. The original code was set up in a Cartesian grid, and did not include gravitational or coordinate (i.e., Coriolis and centrifugal) forces. The code was modified to include a cylindrical polar grid so that the gas flow could be followed in concentric zones about the star. This was required for the study of the gas flow in the accretion region around the primary

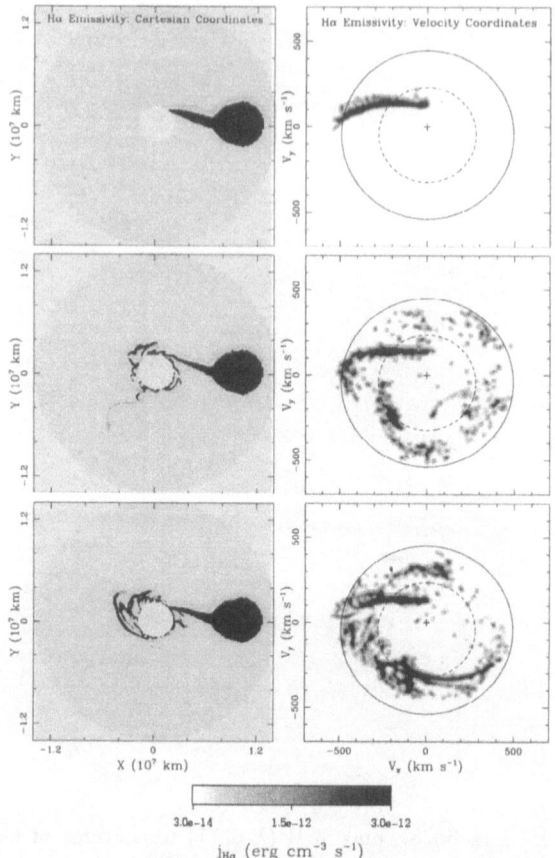

Fig. 12. The evolution of the Hα emissivity of the gas flow in β Per. From top to bottom the images show the system after time $t = 9.7^h = 0.14\ P_{orb}$, $t = 34.7^h = 0.5\ P_{orb}$, and $t = 68.9^h = 2.0\ P_{orb}$, respectively. The location of Hα emission in Cartesian coordinates is shown on the left while the distribution of Hα emission in velocity coordinates is shown on the right. (Taken from [41].)

star in the direct-impact binaries. Gravity and Coriolis forces and optically-thin radiative cooling were included.

The simulations of Blondin et al. [4] produced an extended accretion annulus whose structure was dependent upon the radiative cooling rate of the circumstellar gas around the primary. In particular, when the initial stream density was high ($n_s \sim 10^9 \text{cm}^{-3}$), the accreted gas was confined to a thin ring near the surface of the primary because of efficient cooling. However, when the stream density was decreased ($n_s \sim 10^8 \text{cm}^{-3}$), the gas expanded to fill most of the Roche lobe of the primary because inefficient cooling permitted the gas to heat and expand. These simulations illustrated the properties of the flow quite effectively in the Cartesian coordinate frame, but they did not display the velocity

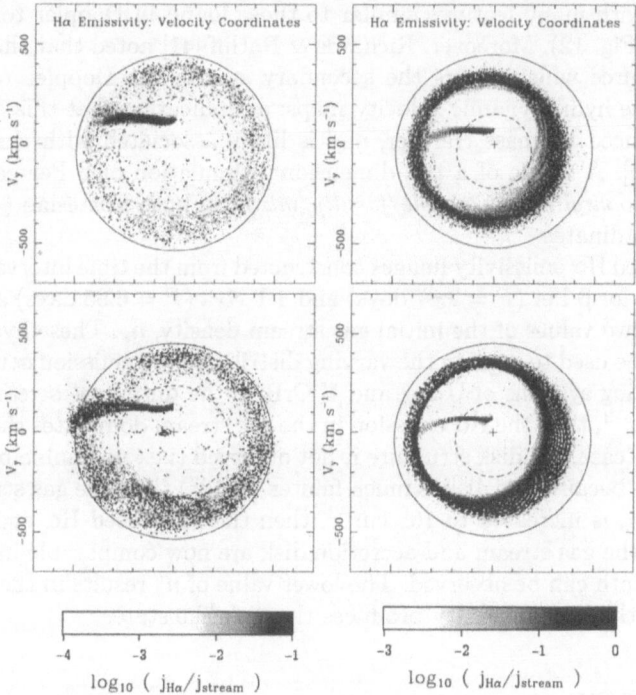

Fig. 13. Integrated Hα emissivity images constructed from the time interval $1.0\,P_{orb} < t < 2.0\,P_{orb}$ for β Per ($P = 2.87$ days) and TT Hya ($P = 6.95$ days). The emissivity is scaled to the mean emissivity in the gas stream for each simulation. The initial gas stream density in the simulations was $n_s = 10^8$ cm^{-3} (top), $n_s = 10^9$ cm^{-3} (bottom). (Adapted from [41].)

structure of the gas flows and did not include the emission line characteristics. As a result, the simulations of that initial work of Blondin et al. [4] could not be directly compared with the reconstructed Doppler images derived from spectroscopic observations of β Per.

Richards & Ratliff [41] extended the work of Blondin et al. [4] to address the need for hydrodynamic simulations which could be directly compared with observational results. In particular, they performed 2D simulations of β Per ($P = 2.87$ days) and TT Hya ($P = 6.95$ days). The former system was chosen to illustrate the case of direct-impact accretion where an accretion annulus should form, and the latter binary was chosen to illustrate the case where a classical accretion disk should form. The earlier work was also extended by producing maps of the Hα emissivity of the circumstellar gas in both Cartesian and velocity coordinates [41]. The velocity maps produced from the simulations were compared with the Doppler tomograms of the Algols [3,36,38,39] in order to correlate the emission features in the Doppler maps with corresponding structures in Cartesian space. The simulations produced asymmetric accretion structures with many features similar to those found in Doppler tomograms of the Algols (Fig. 12). Moreover, Richards & Ratliff [41] noted that the prominent emission source which covers the secondary star in the Doppler tomogram is absent in the hydrodynamic velocity maps; an indication that this extra source is not produced by mass transfer, and is likely associated with chromospheric emission [36]. A movie of a two-dimensional simulation of β Per can be found at *www.astro.virginia.edu/people/faculty/mtr8r*, in both Cartesian (spatial) and velocity coordinates.

Integrated Hα emissivity images constructed from the time interval $1.0\,P_{orb} < t < 2.0\,P_{orb}$ for β Per ($P = 2.87$ days) and TT Hya ($P = 6.95$ days) are shown in Fig. 12 for two values of the initial gas stream density, n_s. These hydrodynamic results can be used to explain the varying distribution of emission sources seen in the alternating systems of U Sge and U CrB. If the initial gas stream density is $n_s = 10^8$ cm^{-3}, then the Hα emission in the gas stream dominates the integrated map. In this case, the disk structure is not observed (i.e., no double peaked emission is seen) because the disk is much fainter ($\sim 10^2$) than the gas stream at this density. If n_s is increased to 10^9 cm^{-3}, then the integrated Hα emissivity plot shows that the gas stream and accretion disk are now comparable in brightness, and hence both can be observed. The lower value of n_s results in the stream-like state while the higher density produces the disk-like state.

6   Summary

Since the technique of Doppler tomography was introduced by Marsh & Horne [24], tomograms have been made for eleven eclipsing Algol-type binaries, one non-eclipsing Algol (CX Dra), and the RS CVn binary of V711 Tau. The tomograms of the eclipsing systems were produced from Hα difference profiles because the observed profiles were dominated by photospheric absorption. These tomograms are summarized in Fig. 14, which shows the Roche geometry and tomograms for ten systems with orbital periods of (1.20 – 6.95) days: RZ Cas, δ Lib, β Per, TT Hya, U Sge, U CrB, S Equ, RS Vul, SW Cyg, and TT Hya. RW Tau is omitted because its tomogram is similar to the β Per (1992) tomogram. These tomograms display several sources in the orbital plane: the gas stream along the predicted gravitational trajectory; transient or permanent accretion disks; a star-stream impact region where the gas stream strikes the stellar photosphere; a disk-stream impact region where the disk strikes the incoming gas stream; a circumprimary bulge produced by the impact of the high velocity gas stream onto the slowly rotating photosphere; the chromosphere of the secondary; and an absorption zone that overlaps with the locus of hotter gas seen in a uv tomogram. In addition, tomograms of CX Dra derived from a 23-year study show evidence of variability with epoch. The Hα and Hβ emission arises from a hot spot where the gas stream strikes the outer edge of an accretion disk, while the He I and Si II emission arise primarily from an accretion disk of radius $r_d = (2.0 - 3.5)R_P$.

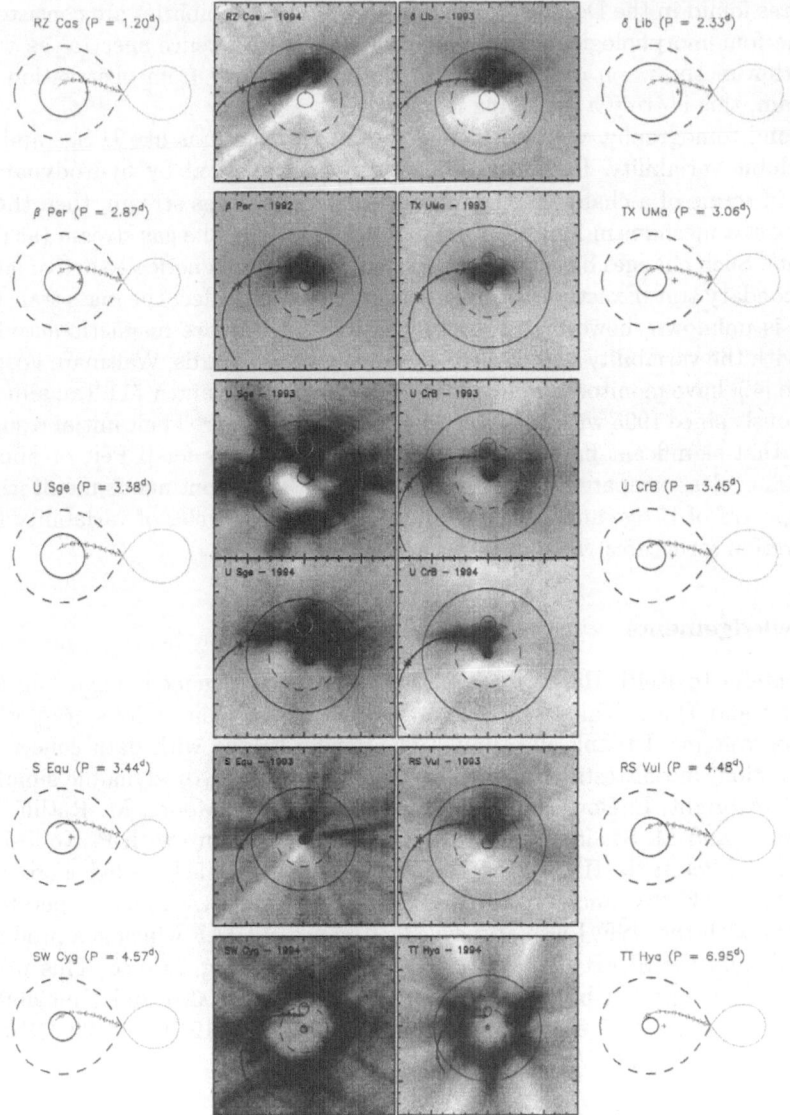

Fig. 14. Cartesian Roche geometry and Hα Doppler tomograms of the Algol-type binaries RZ Cas, δ Lib, β Per, TT Hya, U Sge, U CrB, S Equ, RS Vul, SW Cyg, and TT Hya. The tomograms are displayed in order of increasing orbital period from top to bottom.

These tomograms are consistent with the emission features found in the observed and difference profiles. Moreover, 2D hydrodynamic simulations can reproduce many of the features seen in the tomograms, with the exception of chromospheric emission. It is pleasing to note that the categories of accretion

structures found in the Doppler tomograms of Algol-type binaries are consistent
with the four morphological types based on the Hα difference spectra, as well
as location in the $r - q$ diagram. Given the long journey from observation to
tomogram, this is truly a satisfying result.

Beyond tomography, we need to understand why systems like U Sge and U
CrB exhibit variability. If this variability can be explained by hydrodynamic
models in terms of a change in the initial density of the gas stream, then there
has to exist a mechanism for altering the initial density of the gas stream (at the
L_1 point). Such changes have to be associated with the magnetically-active late-
type secondary star. Exactly how magnetic activity could affect the mass transfer
process is unknown, however it may be possible to compare magnetic activity
cycles with the variability found in U Sge and U CrB. Richards, Waltman, Foster
& Ghigo [42] have monitored radio flaring activity in β Per and V711 Tau almost
continuously since 1995 with the Green Bank Interferometer. Their initial results
suggest that significant flares occur roughly every 49 days for β Per, or about
17 orbital cycles, compared to ~120 days for V711 Tau. Continuous monitoring
of the spectra of U Sge and U CrB would provide exact cycles of variability for
the accretion structures. The journey has only just begun.

Acknowledgements

I am grateful to Keith Horne for his advice and encouragement regarding tomography and D.M. Whittle for suggesting the application of tomography to
the Algol systems. I thank all of my students who helped with data collection
and reduction, initial tests of the tomography code, and hydrodynamic simulations: G. Albright, L. Bowles, R. Jones, J. Kempner, L. Moore, M. Ratliff, E.
Rosolowsky, and M. Swain. I enjoyed the CX Dra adventure with P. Koubský,
V. Šimon, G. Peters, R. Hirata, P. Skoda, and S. Masuda. I acknowledge observing grants from KPNO and NSO (NOAO), operated by AURA under cooperative
agreement with the NSF. I also acknowledge the use of IRAF which is a product
of NOAO, and the SIMBAD database at CDS, Strasbourg, France. This work
was partially supported by AFOSR grants F49620–92–J–0024 and F49620-94-
1-0351; NSF grants AST 91114214, AST 9315108, INT–9512791, AST 9618461,
and AST 0074586; and NASA grant NAG 5-3056.

References

1. Albright G.E., Richards M.T., 1993, ApJ, 414, 830
2. Albright G.E., Richards M.T., 1995, ApJ, 441, 806
3. Albright G.E., Richards M.T., 1996, ApJL, 459, L99
4. Blondin J.M., Richards M.T., Malinkowski M., 1995, ApJ, 445, 939
5. Colella P., Woodward P.R., 1984, J. Comp. Phys., 54, 174
6. Crawford J.A., 1955, ApJ, 121, 71
7. Cugier H., Molaro P., 1984, A&A, 140, 105
8. Guinan E.F., 1989, Sp. Sci. Rev., 50, 35

9. Guinan E.F., Giménez A., 1994, in *The Realm of Interacting Binary Stars*, ed. by J. Sahade, G.E. McCluskey, Jr., Y. Kondo, (Kluwer, Dordrecht), pp. 51–110
10. Hall D.S., 1987, Pub. Astron. Inst. Czech., 70, 77
11. Hall D.S., 1989, Sp. Sci. Rev., 50, 219
12. Herman G.T., 1980, *Image Reconstruction From Projections, The Fundamentals of Computerized Tomography*, (Academic Press, New York)
13. Horne K., 1992, in *Fundamental Properties of Cataclysmic Variable Stars, Twelfth North American Workshop on Cataclysmic Variables and Low Mass X-Ray Binaries, San Diego, January 3–5, 1991*, ed. by A.W. Shafter (SDSU, San Diego), p. 23
14. Jones R.D., Richards M.T., 1992, BAAS, 24, 768
15. Kaitchuck R.H., 1989, Sp. Sci. Rev., 50, 51
16. Kaitchuck R.H., R.K. Honeycutt, 1982, ApJ, 258, 224
17. Kaitchuck R.H., R.K. Honeycutt, E.M. Schlegel, 1985, PASP, 97, 1178
18. Kaitchuck R.H., Schlegel E.M., Honeycutt R.K., Horne K., Marsh T.R., White J.C., Mansperger C.S., 1994, ApJS, 93, 519
19. Kempner J.C., Richards M.T., 1999, ApJ, 512, 345
20. Kopal Z., 1995, Annales d'Astrophys., 18, 379
21. Kurucz R.L., 1990, in *Stellar Atmospheres, Beyond Classical Models, Proceedings of the Advanced Research Workshop* Trieste, Italy, ed. by L. Crivellari et al., NATO ASI Series (Kluwer, Dordrecht), p. 441
22. Lubow S.H., Shu F.H., 1975, ApJ, 198, 383
23. McCluskey G.E.,McCluskey C.P.S., Kondo Y., 1991, ApJ, 378, 281
24. Marsh T.R., Horne K., 1988, MNRAS, 235, 269
25. Marsh T.R., Horne K., Schlegel E.M., Honeycutt R.K., Kaitchuck R.H., 1990, ApJ, 364, 637
26. Olson E.C., 1980, ApJ, 241, 257
27. Olson E.C., Etzel P.B., 1998, preprint
28. Peters G.J., Polidan R.S., 1984, ApJ, 283, 745
29. Peters G.J., 1989, Sp. Sci. Rev., 50, 9
30. Plavec M.J., 1983, ApJ, 275, 251
31. Radon J., 1917, reprinted in 1983, Proc. Symp. in Appl. Math. 27, 71
32. Richards M.T., 1990, ApJ, 350, 372
33. Richards M.T., 1992, ApJ, 387, 329
34. Richards M.T., 1993, ApJS, 86, 255
35. Richards M.T., Albright G.E., 1993, ApJS, 88, 199
36. Richards M.T., Albright G.E., 1996, in *Stellar Surface Structure, IAU Symposium 176*, Vienna, Austria, ed. by K. Strassmeier, J. Linsky (Kluwer, Dordrecht) pp. 493–500
37. Richards M.T., Albright G.E., 1999, ApJS, 123, 537
38. Richards M.T., Albright G.E., Bowles L.M., 1995, ApJL, 438, L103
39. Richards M.T., Jones R.D., Swain M.A., 1996, ApJ, 459, 249
40. Richards M.T., Koubský P., Šimon V., Peters G.J., Hirata R., Skoda P., Masuda S., 2000, ApJ, 531, 1003
41. Richards M.T., Ratliff M.A., 1998, ApJ, 493, 326
42. Richards M.T., Waltman E.B., Foster R.S., Ghigo F., 1998, in *Tenth Cambridge Workshop on Cool Stars, Stellar Systems, and the Sun*, ed. by R.A. Donahue, J.A. Bookbinder, ASP Conf. Ser., Vol. 154 (ASP, San Francisco), pp. 1546–1550
43. Robinson E.L., Marsh T.R., Smak J.I., 1993, in *Accretion Disks in Compact Stellar Systems*, ed. by J.C. Wheeler (World Scientific, Singapore), pp. 75–116

44. Smak J., 1989, Sp. Sci. Rev., 50, 107
45. Stern R.A., Uchida Y, Tsuneta S., Nagase F., 1992, ApJ, 400, 321
46. Stern R.A., Lemen J.R., Schmitt J.H.M.M., Pye J.P., 1995, ApJ, 444, L45
47. Tout C.A., Eggleton P.P., 1988, MNRAS, 231, 823
48. Trimble V., 1983, Nature 303, 137
49. Vesper D.N., Honeycutt R.K., 1993, PASP, 105, 731
50. Wade C.M., Hjellming R.M., 1972, Nature, 235, 270
51. White N.E., Culhane J.L., Parmer A.N., Kellet B.J., Kahn S., van den Oort G.H.J, Kuijpers J., 1986, ApJ, 301, 262

Tomography of the Atmosphere of Long-Period Variable Stars

A. Jorissen[1], R. Alvarez[1], B. Plez[2], D. Gillet[3], and A. Fokin[4]

[1] Institut d'Astronomie et d'Astrophysique, Université Libre de Bruxelles,
  C.P. 226, Boulevard du Triomphe, B-1050 Bruxelles, Belgium
[2] GRAAL, Université Montpellier II, cc072, 34095 Montpellier cedex 05, France
[3] Observatoire de Haute-Provence, 04870 Saint-Michel l'Observatoire, France
[4] Institute for Astronomy of the Russia Academy of Sciences,
  48 Pjatnitskaja, 109017 Moscow, Russia

Abstract. This paper presents a new tomographic technique to derive the velocity field across the atmosphere of long-period variable (LPV) stars. The method cross-correlates the optical spectrum with numerical masks constructed from synthetic spectra and probing layers of increasing depths. This technique reveals that the line doubling often observed in LPV stars around maximum light is the signature of the shock wave propagating in the atmosphere of these pulsating stars.

1 Introduction

Long-period variable stars (LPVs) are cool giant stars showing more or less periodic light variations with amplitudes of several magnitudes in the visual and with periods of several hundred days. It is known since long that the brightness variations of LPVs go along with spectral changes such as the doubling of the absorption lines around maximum light [7]. This line-doubling phenomenon is generally believed to reflect the differential bulk motions occurring in the large and tenuous atmosphere of LPVs and associated with its pulsation (and with the accompanying shock wave).

The tomographic[1] method described in Sect. 2 aims at deriving the velocity as a function of depth in the atmosphere. When applied to a temporal sequence of spectra, the method is able to reveal the outward propagation of the shock wave, as shown in Sect. 3.

2 Tomography of the Atmosphere

The study of the velocity field in the atmosphere of LPVs poses a special challenge, as the spectrum of these stars is extremely crowded, particularly in the optical domain. The cross-correlation technique provides a powerful tool to overcome this difficulty. The information relating to the line doubling (velocity shift

---

[1] The word *tomography* is used here in its etymological sense (*'cut display'*), which differs somewhat from the broader sense in use within the astronomical community (reconstruction of a structure using projections taken under different angles).

and line shape) is in fact distributed among a large number of spectral lines, and can be summed up into an average profile, or more precisely into a cross-correlation function (CCF). If the correlation of the stellar spectrum with a mask involves many lines, it is possible to extract the relevant information from very crowded and/or low signal-to-noise spectra.

The tomographic method rests on our ability to construct reliable synthetic spectra of late-type giant stars [10–12], and from those, to identify the depth of formation of any given spectral line. Rigorously, one should make use of the contribution function (CF) to the flux depression [1] to estimate that depth. However, it would be a formidable task to compute the CF for each line appearing in the optical spectrum of LPV stars. For the sake of simplicity, the 'depth function' $D = D(\lambda)$ is used instead, which provides the geometrical depth corresponding to monochromatical optical depth $\tau = 2/3$ at the considered wavelength λ. This function expresses the depth from which the emergent flux arises, and is not expected to differ much from the depth of line formation in the Eddington-Barbier approximation and for sufficiently strong lines [8].

Different masks M_i are then constructed from the collection of N lines $\lambda_{i,j}(1 \leq j \leq N)$ such that $D_i \leq D(\lambda_{i,j}) < D_{i+1} = D_i + \Delta D$, where ΔD is some constant optimized to keep enough lines in any given mask without losing too much resolution in terms of geometrical depth. Each mask M_i should thus probe lines forming at (geometrical) depths in the range $D_i, D_i + \Delta D$ in the atmosphere. These masks are then used as templates to correlate with the observed spectra of the LPV stars. This procedure should provide the velocity field as a function of depth in the atmosphere.

A more detailed description of the method can be found in [3].

3   Application to the Mira Variables RT Cyg and X Oph

A two-month-long monitoring of the two Mira-type stars RT Cyg ($P = 190$ d; $6.0 \leq V \leq 13.1$) and X Oph ($P = 329$ d; $5.9 \leq V \leq 9.2$) was performed around their maximum light with the fibre-fed echelle spectrograph ELODIE [4] in August-September 1999. Both stars were observed each night (weather-permitting), resulting in 32 spectra for both RT Cyg and X Oph. The spectrograph ELODIE is mounted on the 1.93-m telescope of the Observatoire de Haute Provence (France), and covers the full range from 3906 Å to 6811 Å in one exposure at a resolving power of 42 000.

In Fig. 1, the nightly sequence of CCFs obtained for RT Cyg and X Oph with a K0III mask is presented. It can be seen that the sequence of CCFs for RT Cyg follows the so-called Schwarzschild scenario sketched in Fig. 2: at phase ∼0.80, a single absorption peak is seen at −113 km s^{-1}; then, coming closer to maximum light, the CCF becomes more and more asymmetric (phases 0.80–0.90), until the blue component becomes clearly visible around phase 0.90. The blue peak continues to strenghten and becomes even stronger than the red peak after phase 0.96. The intensity of the red component, that remained almost constant until this phase, then starts to fade away.

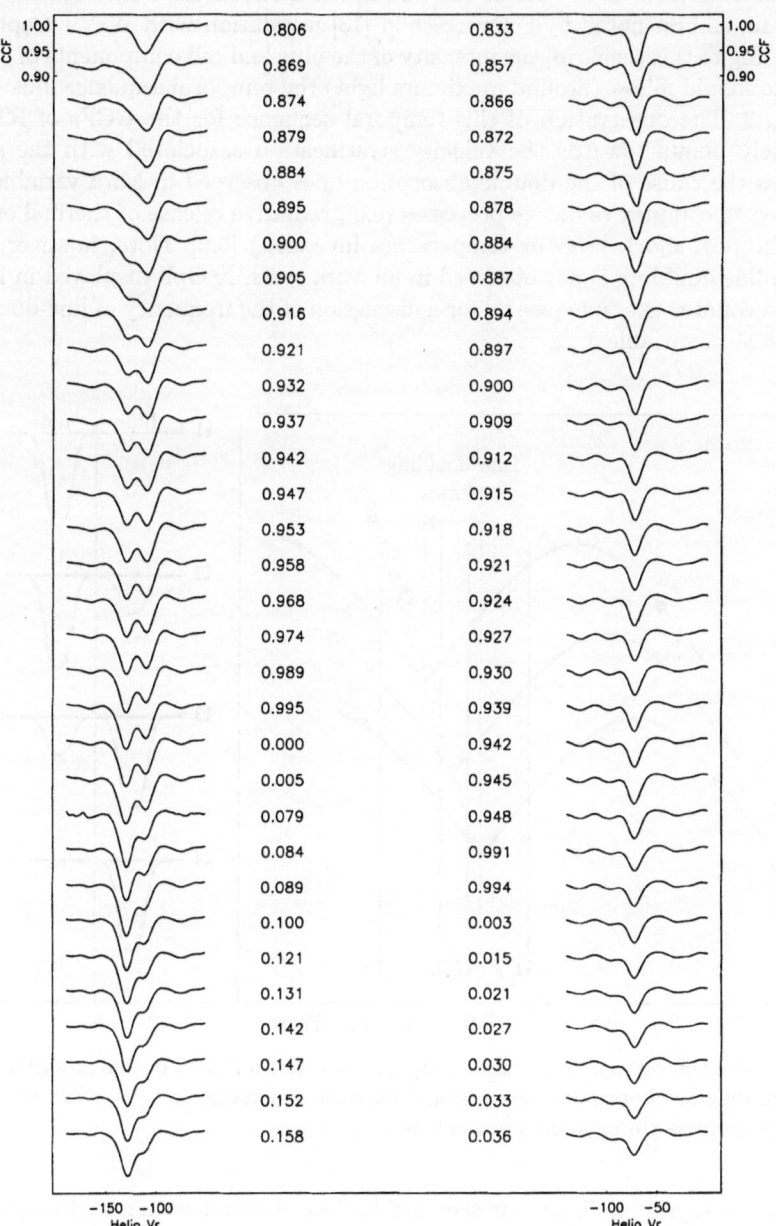

Fig. 1. Sequence of cross-correlation profiles of RT Cyg (left side) and X Oph (right side) obtained with a K0III mask in August-September, 1999. The labels beside each CCF refer to the corresponding visual phase based on the AAVSO ephemeris [9]

This behaviour is typical of the passage of a shock wave through the photosphere, as first noted by Schwarzschild [13] in relation with W Vir Cepheids. According to this scenario, the intensity of the blue and red components of a double line should follow (around maximum light) the temporal sequence illustrated in Fig. 2. The observation of this temporal sequence for the CCFs of RT Cyg definitely points towards the velocity stratification associated with the shock wave as the cause of the double absorption lines observed in Mira variables, as opposed to complex radiative processes (e.g., radiative release of thermal energy into the post-shock layer or temperature inversion) [5,6]. Note, however, that such a line doubling is not observed in all Mira stars, X Oph displayed in Fig. 1 being a counter-example (see [3] for a discussion of the frequency of line-doubling among Mira variables).

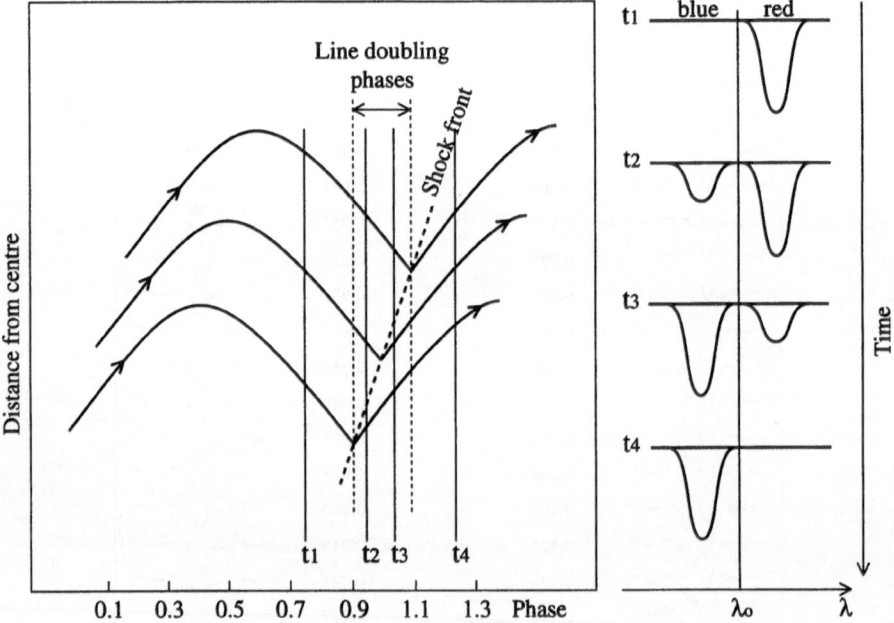

Fig. 2. The Schwarzschild scenario: temporal sequence followed by the intensity of the red and blue components of absorption lines close to maximum light, when the shock wave propagates through the photosphere

The tomographic technique sketched in Sect. 2 moreover allows to watch the outward propagation of the shock wave in RT Cyg.

Fig. 3 displays a sequence of CCFs probing increasing depths in the atmosphere of RT Cyg at four different phases around maximum light (see [2] for more details). At phase 0.905, it can clearly be seen that the line doubling occurs only for the 3 deepest masks. At a later phase (0.953), the line doubling involves 3 more masks further out, translating the upward motion of the shock.

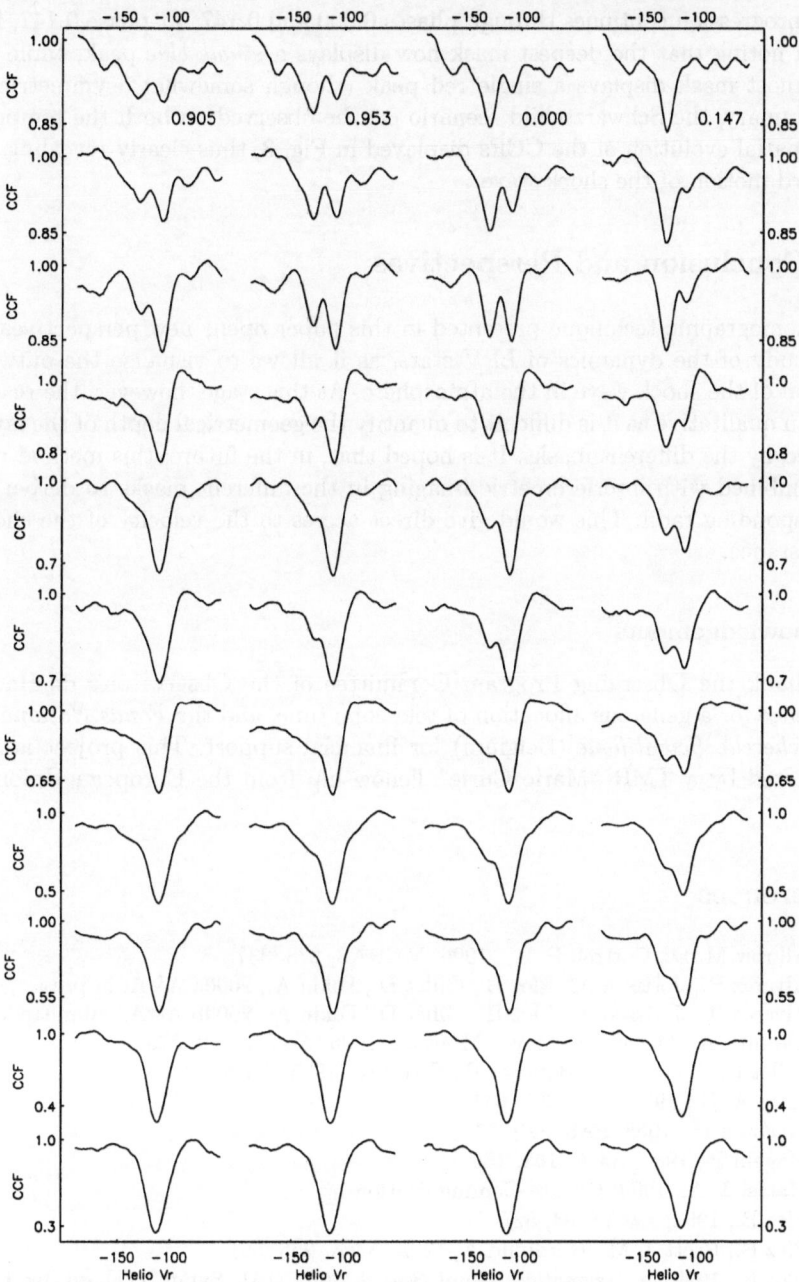

Fig. 3. Sequence of CCFs probing layers of increasing depth (from bottom to top) in the atmosphere of RT Cyg at 4 different phases (0.905, 0.953, 0.000 and 0.147)

This progression continues through phases 0.000 and 0.147. At phase 0.147, it is worth noting that the deepest mask now displays a *single blue* peak, while the outermost mask displays a single red peak (though somewhat asymmetrical). In summary, the Schwarzschild scenario can be observed on both the temporal *and* spatial evolution of the CCFs displayed in Fig. 3, thus clearly revealing the upward motion of the shock wave.

4   Conclusion and Perspectives

The tomographic technique presented in this paper opens new perspectives for the study of the dynamics of LPV stars, as it allows to visualize the outward motion of the shock wave in the atmosphere. At this stage, however, the results remain qualitative as it is difficult to quantify the geometrical depth of the layers probed by the different masks. It is hoped that, in the future, this method may be combined with interferometric imaging in the different masks to derive the corresponding radii. This would give direct access to the velocity of the shock, for instance.

Acknowledgements

We thank the Observing Program Committee of the Observatoire de Haute-Provence for a generous allocation of telescope time, and the *Fonds National de la Recherche Scientifique* (Belgium) for financial support. This project is also supported by a TMR "Marie Curie" Fellowship from the European Union to R.A.

References

1. Albrow M. D., Cottrell P. L. , 1996, MNRAS, 278, 337
2. Alvarez R., Jorissen A., Plez B., Gillet D., Fokin A., 2000a A&A, in press
3. Alvarez R., Jorissen A., Plez B., Gillet D., Fokin A., 2000b A&A, submitted
4. Baranne A., Queloz D., Mayor M. et al., 1996, A&AS, 119, 373
5. Gillet D., Maurice E., Bouchet P., Ferlet R., 1985, A&A, 148, 155
6. Karp A. H., 1975, ApJ, 201, 641
7. Maehara H., 1968, PASJ, 20, 77
8. Magain P., 1986, A&A, 163, 135
9. Mattei J. A., 1999, Private Communication
10. Plez B., 1992, A&AS, 94, 527
11. Plez B., Brett J. M., Nordlund Å, 1992, A&A, 256, 551
12. Plez B., 1999, in, *Asymptotic Giant Branch Stars* (IAU Symp. 191), ed. by T. Le Bertre, A. Lèbre, C. Waelkens, (Astron. Soc. Pacific 1999) pp. 75–83
13. Schwarzschild M., 1952, in: *Transactions of the IAU, Vol. VIII*, ed. by P.T. Oosterhoff (Cambridge University Press), pp. 811–812

Eclipse Mapping of Accretion Discs

R. Baptista

Departamento de Física, UFSC, Campus Trindade, 88040-900, Florianópolis, Brazil

Abstract. The eclipse mapping method is an inversion technique that makes use of the information contained in eclipse light curves to probe the structure, the spectrum and the time evolution of accretion discs. In this review I present the basics of the method and discuss its different implementations. I summarize the most important results obtained to date and discuss how they have helped to improve our understanding of accretion physics, from testing the theoretical radial brightness temperature distribution and measuring mass accretion rates to showing the evolution of the structure of a dwarf novae disc throughout its outburst cycle, from isolating the spectrum of a disc wind to revealing the geometry of disc spiral shocks. I end with an outline of the future prospects.

1 Introduction

Accretion discs are cosmic devices that allow matter to efficiently accrete over a compact source by removing its angular momentum via viscous stresses while transforming the liberated gravitational potential energy into heat and, thereafter, radiation [15]. They are widespread in astrophysical environments, from sheltering the birth of stars to providing the energetics of quasars and active galactic nuclei. It is, however, in mass-exchanging binaries such as non-magnetic Cataclysmic Variables (CVs) that the best environment for studies of accretion discs are possibly found. In these close binaries mass is fed to a white dwarf by a Roche lobe filling companion star (the secondary) via an accretion disc, which usually dominates the ultraviolet and optical light of the system [51].

Accretion discs in CVs cover a wide range of accretion rates, $\dot{M}$, and viscosity regimes. For example, the sub-class of dwarf novae comprises low-mass transfer CVs showing recurrent outbursts (of 2–5 magnitudes, on timescales of weeks to months) which reflect changes in the structure of the discs – from a cool, optically thin, low viscosity state to a hot, optically thick, high viscosity state – and which are usually parameterized as a large change in the mass accretion rate ($\dot{M}= 10^{-11}\ M_\odot\ yr^{-1} \mapsto 10^{-9}\ M_\odot\ yr^{-1}$) [35]. On the other hand, nova-like variables seem to be permanently in a high viscosity state, presumably as a result of the fact that the accretion rate is always large.

The temperatures in CV discs may vary from 5000 K in the outer regions to over 50000 K close to the disc centre, and the surface density may vary by equally significant amounts over the disc surface. Therefore, the spectrum emitted by different regions of the accretion disc may be very distinct. Additionally, the bright spot (formed by the impact of the gas stream from the inner Lagrangian point on the disc rim), the white dwarf at disc centre, and the secondary star may all contribute to the integrated light of the binary. Because what one directly observes is the combination of the spectra emitted from these diverse regions and sources, the interpretation of disc observations is usually plagued by the ambiguity associated with composite spectra. The most effective way to overcome these difficulties is with spatially resolved data.

Two complementary indirect imaging techniques were developed in the 1980's that provide spatially resolved observational constraints on accretion discs on angular scales of micro arcseconds – well beyond the capabilities of current direct imaging techniques. One is Doppler Tomography [28], which is treated in detail by T.R. Marsh is this volume. It uses the changes in line emission profile with orbital phase to probe the dynamics of accretion discs and is applicable to binaries over a large range of orbital inclinations, although it is restricted to emission line data.

The other is Eclipse Mapping [19]. It assembles the information contained in the shape of the eclipse into a map of the accretion disc surface brightness distribution. While its application is restricted to deeply eclipsing binaries, eclipse mapping can be used with continuum as well as line data. When applied to time-resolved spectroscopy through eclipses this technique delivers the spectrum of the disc at any position on its surface. Information on the radial dependence of the temperature and vertical temperature gradients (for optically thick regions), or temperature, surface density and optical depth (where the disc is optically thin) can be obtained by comparing such spectra with the predictions of models of the vertical disc structure. The spatial structure of the emission line regions over the disc can be similarly mapped from data of high spectral resolution. Moreover, an eclipse map yields a snapshot of an accretion disc at a given time. By eclipse mapping an accretion disc at different epochs it is possible to follow the secular changes of its radial brightness temperature distribution – for example, throughout the outburst cycle of a dwarf nova – allowing crucial tests of accretion disc instability and viscosity models.

After more than a decade of experiments, eclipse mapping has now become a mature and well established technique. There are already many good reviews on this topic in the literature [20,21,51,54]. The main aim of this review is therefore not to provide another description of the technique, but to make a summary of the results obtained so far giving emphasis on the impact they had on our understanding of the physics of accretion discs.

2   The Maximum Entropy Eclipse Mapping Method

2.1   The Principles

The three basic assumptions of the standard eclipse mapping method are:

- the surface of the secondary star is given by its Roche equipotential,
- the brightness distribution is constrained to the orbital plane, and
- the emitted radiation is independent of the orbital phase.

The first assumption seems reasonably robust. The others are simplifications that do not hold in all situations. A discussion on the departures from the two last assumptions is presented in Sect. 2.5.

A grid of intensities centred on the white dwarf, the eclipse map, is defined in the orbital plane. The eclipse geometry is specified by the inclination i, the binary mass ratio q ($=M_2/M_1$, where M_2 and M_1 are the masses of, respectively, the secondary star and the white dwarf) and the phase of inferior conjunction ϕ_0 [19,21]. Given the geometry, a model eclipse light curve can be calculated for any assumed brightness distribution in the eclipse map. A computer code then iteratively adjusts the intensities in the map (treated as independent parameters) to find the brightness distribution the model light curve of which fits the data eclipse light curve within the uncertainties. The quality of the fit is checked with a consistency statistics, usually the reduced χ^2. Because the one-dimensional data light curve cannot fully constrain a two-dimensional map, additional freedom remains to optimize some map property. A maximum entropy (MEM) procedure [44,45] is used to select, among all possible solutions, the one that maximizes the entropy of the eclipse map with respect to a smooth default map.

Figure 1 illustrates the simulation of the eclipse of a fitted brightness distribution while showing the comparison between the resulting model light curve and the data light curve. The geometry in this case is $q = 0.3$ and $i = 81°$. The left-hand panels show the data light curve (small dots) and the model light curve (solid line) as it is being drawn at five different orbital phases along the eclipse (indicated in the lower right corner). The right-hand panels depict the corresponding geometry of the binary for each orbital phase, in which the secondary star progressively occults the accretion disc as well as the white dwarf and the bright spot. The middle panels show the best-fit disc brightness distribution and how it is progressively covered by the dark shadow of the secondary star during the eclipse. At phase $\phi = -0.08$ only a small fraction of the outer, faint disc regions are eclipsed and there is only a small reduction in flux in the light curve. The eclipse of the bright spot at the edge of the disc and of the bright inner disc regions occur at about the same time (slightly after $\phi = -0.04$) and coincide with the steepest ingress in the light curve. The flux at phase $\phi = 0$ does not go to zero because a significant fraction of the disc remains visible at mid-eclipse. The asymmetry in the egress shoulder of the light curve maps into an enhanced brightness emission in the trailing side of the disc (the right hemisphere of the eclipse map in Fig. 1).

2.2 The Expressions

The expressions governing the eclipse mapping problem are as follows. One usually adopts the distance from the disc centre to the inner Lagrangian point, R_{L1}, as the length scale. With this definition the primary lobe has about the same size and shape for any reasonable value of the mass ratio q [19]. If the eclipse map is an N points flat, square grid of side λR_{L1}, each of its surface element (pixel) has an area $(\lambda R_{L1})^2/N$ and an associated intensity I_j. The solid angle

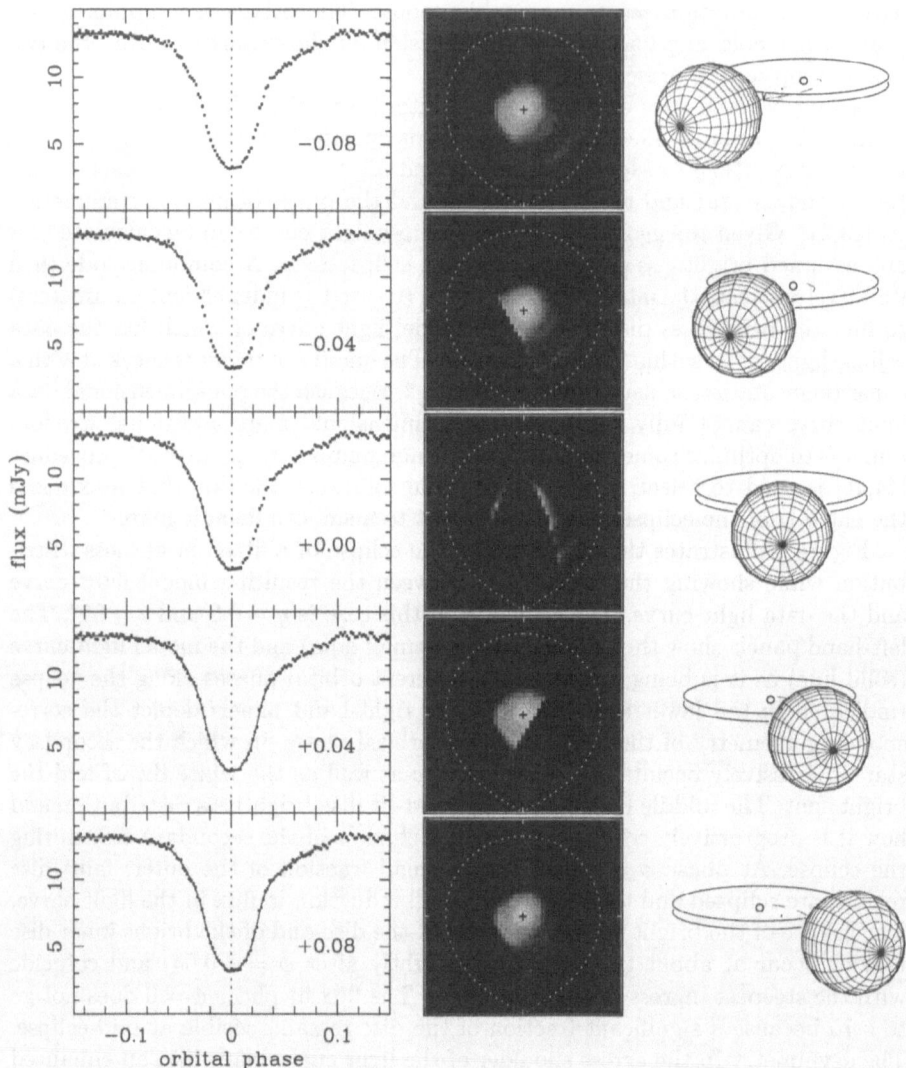

Fig. 1. Simulation of a disc eclipse ($q = 0.3, i = 78°$). Left-hand panels: data light curve (dots) and model light curve (solid line) for five different orbital phases (indicated in the lower right corner). Middle panels: eclipse maps in a false color blackbody logarithmic scale. Roche lobes for $q = 0.3$ are shown as dotted lines; crosses mark the centre of the disc. The secondary is below each panel and the stars rotate counter-clockwise. Right-hand panels: the corresponding geometry of the binary for each orbital phase.

comprised by each pixel as seen from the earth is then

$$\theta^2 = \left[\frac{(\lambda R_{L1})^2}{N} \frac{1}{d^2} \right] \cos i \,, \tag{1}$$

where d is the distance to the system. The value of λ defines the area of the eclipse map while the choice of N sets its spatial resolution.

The entropy of the eclipse map p with respect to the default map q is defined as

$$S = -\sum_{j=1}^{N} p_j \ln\left(\frac{p_j}{q_j}\right) , \qquad (2)$$

where p and q are written as

$$p_j = \frac{I_j}{\sum_k I_k} , \qquad q_j = \frac{D_j}{\sum_k D_k} . \qquad (3)$$

Other functional forms for the entropy appear in the literature [21,51]. These are equivalent to Eq. (2) when p and q are written in terms of proportions.

The default map D_j is generally defined as a weighted average of the intensities in the eclipse map,

$$D_j = \frac{\sum_k \omega_{jk} I_k}{\sum_k \omega_{jk}} , \qquad (4)$$

where the weight function ω_{jk} is specified by the user. A priori information about the disc (e.g. axi-symmetry) is included in the default map via ω_{jk}. Prescriptions for the weight function ω_{jk} are discussed in Sect. 2.3. In the absence of any constraints on I_j, the entropy has a maximum $S_{max} = 0$ when $p_j = q_j$, or when the eclipse map and the default map are identical.

The model eclipse light curve $m(\phi)$ is derived from the intensities in the eclipse map,

$$m(\phi) = \theta^2 \sum_{j=1}^{N} I_j V_j(\phi) , \qquad (5)$$

where ϕ is the orbital phase. The occultation function $V_j(\phi)$ specifies the fractional visibility of each pixel as a function of orbital phase and may include fore-shortening and limb darkening factors [21,39,54]. The fractional visibility of a given pixel may be obtained by dividing the pixel into smaller tiles and evaluating the Roche potential along the line of sight for each tile to see if the potential falls below the value of the equipotential that defines the Roche surface. If so, the tile is occulted. The fractional visibility of the pixel is then the sum of the visible tiles divided by the number of tiles.

The consistency of an eclipse map may be checked using the χ^2 as a constraint function,

$$\chi^2 = \frac{1}{M} \sum_{\phi=1}^{M} \left(\frac{m(\phi) - d(\phi)}{\sigma(\phi)}\right)^2 = \frac{1}{M} \sum_{\phi=1}^{M} r(\phi)^2 , \qquad (6)$$

where $d(\phi)$ is the data light curve, $\sigma(\phi)$ are the corresponding uncertainties, $r(\phi)$ is the residual at the orbital phase ϕ, and M is the number of data points in the light curve. Alternatively, the constraint function may be a combination of χ^2

and R-statistics [8],

$$R = \frac{1}{\sqrt{M-1}} \sum_{\phi=1}^{M-1} r(\phi)\, r(\phi+1)\,, \qquad (7)$$

to minimize the presence of correlated residuals in the model light curve [7]. For the case of uncorrelated normally distributed residuals, the R-statistics has a Gaussian probability distribution function with average zero and unity standard deviation. Requiring the code to achieve an $R = 0$, is equivalent to asking for a solution with uncorrelated residuals in the model light curve.

The final MEM solution is the eclipse map that is as close as possible to its default map as allowed by the constraint imposed by the light curve and its associated uncertainties [21,39]. In matematical terms, the problem is one of constrained maximization, where the function to maximize is the entropy and the constraint is a consistency statistics that measures the quality of the fitted model to the data light curve. Different codes exist to solve this problem. Many of the eclipse mapping codes are based on the commercial optimization package MEMSYS [44]. Alternative implementations using conjugate-gradients algorithms [7,8], CLEAN-like algorithms [47] and, more recently, genetic algorithms [13] are also being used.

2.3 Default Maps

A crucial aspect of eclipse mapping is the selection of the weight function for the default map, which allows the investigator to steer the MEM solution towards a determined type of disc map. A list of different prescriptions for ω_{jk} is given in Table 1.

Choosing $\omega_{jk} = 1$ (option A) results in a uniform default map and will lead to the *most uniform eclipse map* consistent with the data. This happens not to be a good choice for eclipse mapping because it results in a map severely distorted by criss-crossed artifacts [11,19,47]. This effect may be reduced by setting the weight function as a Gaussian profile of width Δ (option B), which results in the *smoothest map* that fits the data.

The third case (option C) sets D_j as an axi-symmetric average of the eclipse map and will lead to the *most nearly axi-symmetric map* that fits the data. It suppresses the azimuthal information in the default map while keeping the radial structure of I_j on scales greater than Δ_R. This seems a reasonable choice for accretion disc mapping because one expects the disc material to be roughly in Keplerian orbits, so that local departures from axi-symmetry will tend to be diminished by the strong shear. This is a commonly used option and is also known as the default map of full azimuthal smearing.

The full azimuthal smearing default results in rather distorted reproduction of asymmetric structures such as a bright spot at the disc rim. In this case, the reconstructed map exhibits a lower integrated flux in the asymmetric source region, the excess being redistributed as a concentric annulus about the same

Table 1. Prescriptions for default maps

weight functions		reference
A) most uniform map:	$\omega_{jk} = 1$	[19]
B) smoothest map:	$\omega_{jk} = \exp\left(-\frac{d_{jk}^2}{2\,\Delta^2}\right)$	[19]
C) most axisymmetric map: (full azimuthal smearing)	$\omega_{jk} = \exp\left[-\frac{(R_j-R_k)^2}{2\,\Delta^2}\right]$	[19]
D) limited azimuthal smearing: (constant angle θ)	$\omega_{jk} = \exp\left[-\frac{1}{2}\left(\frac{(R_j-R_k)^2}{\Delta_R^2} + \frac{\theta_{jk}^2}{\Delta_\theta^2}\right)\right]$	[40]
E) limited azimuthal smearing: (constant arc length s)	$\omega_{jk} = \exp\left[-\frac{1}{2}\left(\frac{(R_j-R_k)^2}{\Delta_R^2} + \frac{s_{jk}^2}{\Delta_s^2}\right)\right]$	[9]

NOTES: d_{jk} is the distance between pixels j and k; R_j and R_k are the distances from pixels j and k to the centre of the disc; θ_{jk} is the azimuthal angle between pixels j and k; and s_{jk} is the arc-length between pixels j and k.

radial distance. By limiting the amount of azimuthal smearing it is possible to alleviate this effect and to start recovering azimuthal information in the accretion disc. Two prescriptions in this regard were proposed. Rutten et al. [40] limited the amount of azimuthal smearing by averaging over a polar Gaussian weight function of *constant angles* along the map (option D) while Baptista, Steiner & Horne [9] have chosen to use a polar Gaussian function of *constant arc length* through the map (option E).

Figure 2 shows the effects of the three last weight functions when applied to an artificial map containing three Gaussian spots at different radial distances from its centre. The default with constant angles is more efficient to reproduce asymmetries in the inner disc regions (such as an accretion column, or the expected dipole pattern for velocity-resolved line emission mapping, or in presence of a bipolar wind emanating from the inner disc), while the default of constant arc length is more efficient in recovering asymmetries in the outer parts of the disc (such as a bright spot at disc rim, or in the case of an eccentric disc). The choice between these two default functions, in a given case, is defined by whether it is more important to have a better azimuthal resolution at the inner or outer disc regions. The radial profile is not affected by this choice.

Other possibilities concerning the default map were proposed in terms of the combination of different weight functions [11,47]. Particularly, the mix of the smoothest and the most axi-symmetric defaults is, in a sense, equivalent to the default of limited azimuthal smearing and leads to similar results. Another interesting proposal is that of a negative weight function, that may be used to

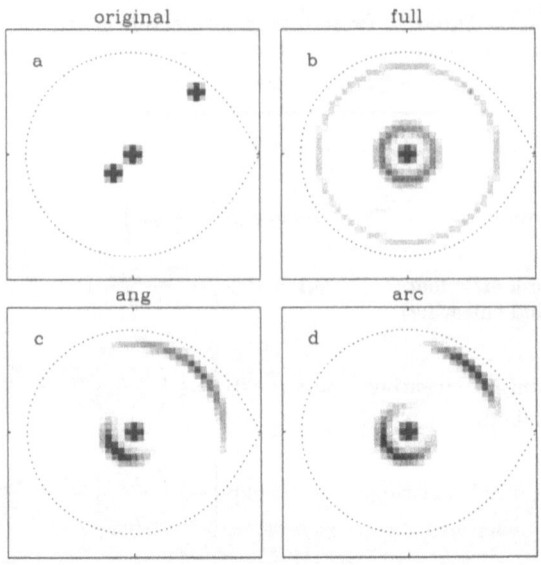

Fig. 2. Effects of the different weight functions for the default map. (a) The original map, with three Gaussian spots. The corresponding default map obtained using the default of (b) full azimuthal smearing; (c) constant angles; and (d) constant arc length. From [9]

avoid or minimize a certain map property (e.g. the presence of the undesired criss-crossed arcs) [47].

2.4 The Uneclipsed Component

Rutten, van Paradijs & Tinbergen [43] found that the entropy function can be a useful tool to signal and to isolate the fraction of the total light which is not coming from the accretion disc plane. They noted that when the light curve is contaminated by the presence of additional light (e.g. from the secondary star) the reconstructed map shows a spurious structure in the regions farthest away from the secondary star (the upper lune of the eclipse maps in Fig. 1, hereafter called the 'back' side of the disc). This is because the eclipse mapping method assumes that all the light is coming from the accretion disc, in which case the eclipse depth and width are correlated in the sense that a steeper shape corresponds to a deeper eclipse. The addition of an uneclipsed component in the light curve (i.e., light from a source other than the accretion disc) ruins this correlation. To account for the extra amount of light at mid-eclipse and to preserve the brightness distribution derived from the eclipse shape the algorithm inserts the additional light in the region of the map which is least affected by the eclipse, leading to a spurious front-back disc brightness asymmetry. Since the entropy measures the amount of structure in the map, the presence of these spurious structures is flagged with lower entropy values.

The correct offset level may be found by comparing a set of maps obtained with different offsets and selecting the one with highest entropy. Alternatively, the value of the zero-intensity level can be included in the mapping algorithm as an additional free parameter to be fitted along with the intensity map in the search for the MEM solution [4,41]. A detailed discussion on the reliability and consistency of the estimation of the uneclipsed component can be found in [9].

2.5 Beyond the Standard Assumptions

The standard eclipse mapping assumes a simple flat, geometrically thin disc model. Real discs may however violate this assumption in the limit of high mass accretion rate. Disc opening angles of $\alpha \gtrsim 4°$ are predicted for $\dot{M} \gtrsim 5 \times 10^{-9}\ M_\odot\ yr^{-1}$ [31,46]. At large inclinations ($i \gtrsim 80°$) this may lead to artificial front-back asymmetries in the eclipse map (similar to those discussed in Sect. 2.4) because of the different effective areas of surface elements in the front and back sides of a flared disc as seen by an observer on Earth. In extreme cases, this may lead to obscuration of the inner disc regions by the thick disc rim (e.g. [25]).

Motivated by the front-back asymmetry that appeared in the flat-disc map and by the difficulties in removing the asymmetry with the assumption of an uneclipsed component, Robinson et al. [36,37] introduced a flared disc in their eclipse mapping of ultraviolet light curves of the dwarf nova Z Cha at outburst maximum. They found that the asymmetry vanishes and the disc is mostly axisymmetric for a disc opening angle of $\alpha = 6°$.

Simulations [39] show that eclipse mapping reconstructions obtained with the flat-disc assumption result in good reproduction of the radial temperature distribution of flared accretion discs provided that the inner disc regions are not obscured by the disc rim (Fig. 3).

Wood [54] shows that it is usually impossible to distinguish between a flared disc and an uneclipsed component to the total light. Both effects lead to the appearance of spurious structures in the back regions of the disc, and eclipse maps obtained with either model may lead to equally good fits to the data light curve. Baptista & Catalán [2] pointed out that spectral eclipse mapping could help in evaluating the importance of each of these effects in a given case. If the uneclipsed component is caused by an optically-thin, vertically-extended disc wind, the uneclipsed spectrum shows a Balmer jump in emission plus strong emission lines, while in the case of a flared disc the spurious uneclipsed spectrum should reflect the difference between the disc spectrum of the back (deeper atmospheric layers seen at lower effective inclinations) and the front (upper atmospheric layers seen at grazing incidence) sides and should mainly consist of continuum emission filled with absorption lines.

Because of the assumption that the emitted radiation is independent of the orbital phase, in the standard eclipse mapping method all variations in the eclipse light curve are interpreted as being caused by the changing occultation of the emitting region by the secondary star. Thus, out-of-eclipse brightness changes (e.g. orbital modulation due to anisotropic emission from the bright spot) has to be removed before the light curves can be analyzed. The usual approach

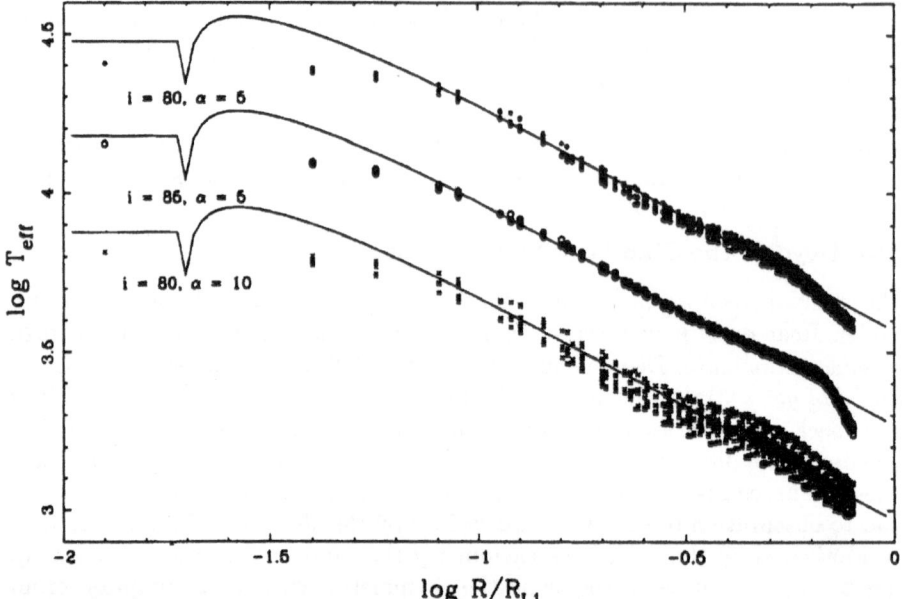

Fig. 3. Examples of reconstructed radial brightness temperature distributions with the flat-disc eclipse mapping for the case of flared discs. The inclinations i and disc opening angle α are indicated for each case. From [39]

is to interpolate the out-of-eclipse light curve across the eclipse phases [19,43]. An alternative approach is to apply a light curve decomposition technique to separate the contributions of the white dwarf, bright spot and accretion disc [56,57]. This technique however requires high signal-to-noise light curves and good knowledge of the contact phases of the white dwarf and bright spot, which limits its application to a few objects.

A step to overcome these limitations was done by Bobinger et al. [11] with the inclusion of a disc rim in the eclipse mapping method. The out-of-eclipse modulation is modeled as the fore-shortening of an azimuthally-dependent brightness distribution in the disc rim. This procedure allows to recover the azimuthal (phase) dependency of the bright spot emission. It however requires a good estimate of the outer disc radius.

The more advanced code of Rutten [39], including a flared disc, the disc rim, and the surface of the Roche-lobe filling secondary star, expanded the eclipse mapping method into a three-dimensional mapping technique. Nevertheless, it comes along with a significant increase in the degrees of freedom that aggravates the problem of non-uniqueness of solutions.

2.6   Performance under Extreme Conditions

In an eclipse mapping reconstruction, the brightness of a given surface element is derived from the information given by the changes in flux caused by its occultation (at ingress) and reappearance (at egress) from behind the secondary star. In the case of an eclipse light curve with incomplete phase coverage, there are regions in the disc for which only one of these pieces of information is available. Moreover, for a system with low inclination, there are regions in the back side of the disc which are never covered by the shadow of the secondary star and, therefore, there is no information about the brightness distribution of these regions on the shape of the light curve. This section presents simulations aiming to assess the reliability of eclipse mapping reconstructions obtained under the combined extreme conditions of incomplete eclipse coverage, low binary inclination and relatively low signal-to-noise data.

Four artificial brightness distributions with asymmetric polar Gaussian spots on the back, front, leading and trailing sides of the disc were constructed (Fig. 4). A low-inclination geometry ($q = 1$ and $i = 71°$) was adopted to simulate the eclipses and light curves with signal-to-noise $S/N \simeq 5 - 15$ and an incomplete set of orbital phases were produced. The artificial light curves were analyzed with the eclipse mapping method and the results are shown in Fig. 4.

For the adopted set of orbital phases, the leading side of the disc (the lower hemisphere of the eclipse maps in Fig. 4) is mapped by the moving shadow of the secondary star both during ingress and egress, whereas much of the trailing side of the disc is only mapped by the secondary star at ingress phases. Dashed lines in the eclipse maps mark the locus of the far edge of the shadow of the secondary star along the eclipse. Regions to the left of this line are never covered by the secondary star.

Despite the incomplete eclipse coverage, good quality reconstructions are obtained for the front, trailing and leading maps. The spots appear spread in radius due to the low signal-to-noise of the light curves, and are elongated in azimuth because of the intrinsic azimuthal smearing effect of the eclipse mapping method. The results are equally good for the leading and the trailing maps, despite the fact that the spot in the latter case is located in the disc region for which there is limited information in the shape of the light curve. For the back map, much of the asymmetric brightness distribution completely escapes eclipse. Not surprisingly, the eclipse map does not correctly reproduce the brightness distribution in the disc regions beyond those covered by the secondary star. The missing flux appears in the uneclipsed component.

These simulations show that eclipse mapping obviously fails to recover the brightness distribution of disc regions for which there is no information in the shape of the eclipse, but performs reasonably well in the case of incomplete phase coverage even with relatively low signal-to-noise data.

The brightness distribution of the back map approximately simulates the intrinsic front-back asymmetry of a flared disc as seen at a high inclination angle ($i > 80°$). It is fortunately that, in these cases, the shadow of the secondary star maps most (if not all) of the primary Roche lobe for any reasonable mass ratio.

Simulations of reconstructions from light curves of more limited phase coverage are presented in Baptista & Catalán [2]. Tests on the reliability of eclipse

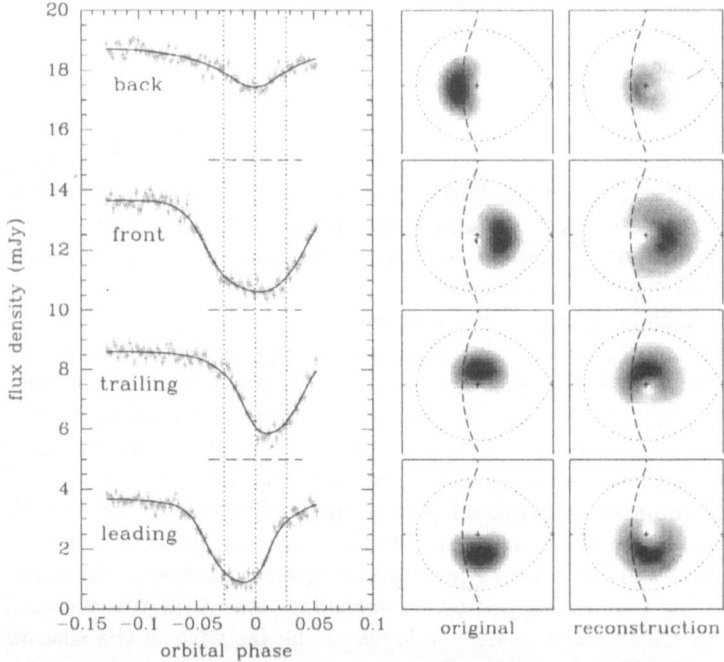

Fig. 4. Reconstructing asymmetric brightness distributions with light curves of low signal-to-noise and incomplete phase coverage. The left-hand panel shows the artificial light curves (dots with error bars) and corresponding eclipse mapping models (solid lines). Horizontal dashed lines indicate the true zero level in each case. Vertical dotted lines mark ingress/egress phases of the white dwarf and mid-eclipse. The middle and right-hand panels show, respectively, the original maps and the reconstructions in a logarithmic greyscale. Bright regions are dark; faint regions are white. A cross marks the center of the disc; dotted lines show the Roche lobe and dashed lines depict the locus of the far edge of the shadow of the secondary star along the eclipse. The secondary is to the right of each map and the stars rotate counter-clockwise

mapping reconstructions under a variety of other conditions can be found in the literature [3,9,11,14,19,39,47].

3   A Summary of Results

When eclipse mapping appeared in the mid-1980's, the standard picture of an accretion disc in a CV was that of a flat, nearly axi-symmetric disc with a bright spot on its edge. This section reviews some of the eclipse mapping results that helped to improve this picture either by allowing key tests of theoretical expectations or by revealing new and unexpected aspects of the physics of accretion discs.

3.1   Classical Results

Early applications of the technique were useful to show that accretion discs in outbursting dwarf novae [22] and in long-period nova-like variables [24,43] closely follow the expected radial dependence of temperature with radius for a steady-state disc, $T \propto R^{-3/4}$, and to reveal that the radial temperature profile is essentially flat in the short period quiescent dwarf novae [56–58] (Fig. 5). This suggests that the viscosity in these short period systems is much lower in quiescence than in outburst, lending support to the disc instability model, and that their quiescent discs are far from being in a steady-state. Eclipse mapping studies also contributed to the puzzle about the SW Sex stars – a group of mostly eclipsing nova-like variables with periods in the range 3-4 hr that display a number of unexplained phenomena – by showing that the radial temperature profile in these systems is noticeably flatter than the $T \propto R^{-3/4}$ law [9,43]. A flat radial temperature distribution is also suggested for the old novae V Per, that lies in the middle of the CV period gap [55]. Unfortunately, this result is rather uncertain because it is based on uncalibrated white-light data and there is a large uncertainty on the eclipse geometry. A comprehensive account of these pioneering results can be found in [21].

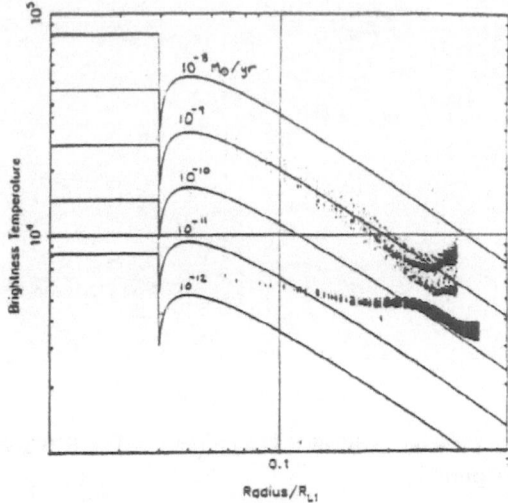

Fig. 5. The radial temperature profile of the dwarf nova Z Cha in outburst and in quiescence. The solid lines show steady-state blackbody disc models for mass accretion rates from 10^{-8} to 10^{-12} $M_\odot$ yr^{-1}. From [21]

It has been a usual practice to convert the intensities in the eclipse maps to blackbody brightness temperatures and then compare them to the radial run of the effective temperature predicted by steady state, optically thick disc models. A criticism about this procedure is that a monochromatic blackbody brightness

temperature may not always be a proper estimate of the disc effective temperature. As pointed out by [5], a relation between these two quantities is non-trivial,
and can only be properly obtained by constructing self-consistent models of the
vertical structure of the disc. Nevertheless, the brightness temperature should
be close to the effective temperature for the optically thick disc regions.

From the $T(R)$ diagram it is possible to obtain an independent estimate of
the disc mass accretion rate. Horne [21] compiled the inferred mass accretion
rates from a dozen of eclipse mapping experiments to construct an $\dot{M} \times P_{orb}$
diagram. An updated version of this diagram is shown in Fig. 6. It seems a bit
disappointing that the diagram is still loosely populated. In particular, there is
yet no eclipse mapping estimate of $\dot{M}$ for a system inside the 2-3 hr CV period
gap. There is a significant scatter in the $\dot{M}$ derived from different experiments
for a given object (e.g. in UX UMa, from $10^{-8.1}$ to $10^{-8.7}$ $M_\odot \, yr^{-1}$ at 0.1 R_{L1}).
Whereas part of this scatter is possibly a real effect due to long-term changes
in the mass transfer rates, it stands as a warning that one should be careful
in interpreting mass accretion rates derived from the brightness temperature
distributions, as discussed above.

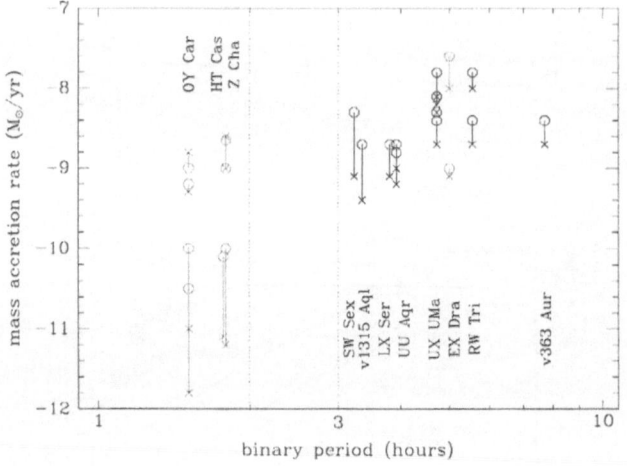

Fig. 6. Mass transfer rates at radii of 0.1 R_{L1} (crosses) and 0.3 R_{L1} (circles) as a
function of the binary period

According to current evolutionary scenarios, CVs should evolve towards shorter orbital periods with decreasing mass transfer rates as a consequence of orbital
angular momentum losses due mainly to magnetic   braking (for systems above
the period gap) or gravitational radiation (for systems below the gap) [33,34]. In
Fig. 6 it appears that there is a tendency among the steady-state discs of nova-
like variables to show larger $\dot{M}$ for longer binary period – in agreement with the
above expectation – and that the discs of nova-like variables and outbursting
dwarf novae have comparable $\dot{M}$. The mass accretion rates in the eclipse maps

of nova-like variables increase with disc radius. The departures from the steady-state disc model are more pronounced for the SW Sex stars (period range 3-4 hs). Illumination of the outer disc regions by the inner disc or mass ejection in a wind from the inner disc are possible explanations for this effect.

Multi-colour eclipse mapping is useful to probe the spectrum emitted by the different parts of the disc surface. Two-colour diagrams show that the inner disc regions of outbursting dwarf novae [14,22] and of nova-like variables [4,9,24] are optically thick with a vertical temperature gradient less steep than that of a stellar atmosphere, and that optically thin, chromospheric emission appears to be important in the outer disc regions (Fig. 7). The fact that the emission from the inner disc regions is optically thick thermal radiation opens the possibility to use a colour-magnitude diagram to obtain independent estimates of the distance to the binary with a procedure similar to cluster main-sequence fitting. Distance estimates with this method were obtained for Z Cha [22], OY Car [14], UU Aqr [9], RW Tri [24] and UX UMa [4].

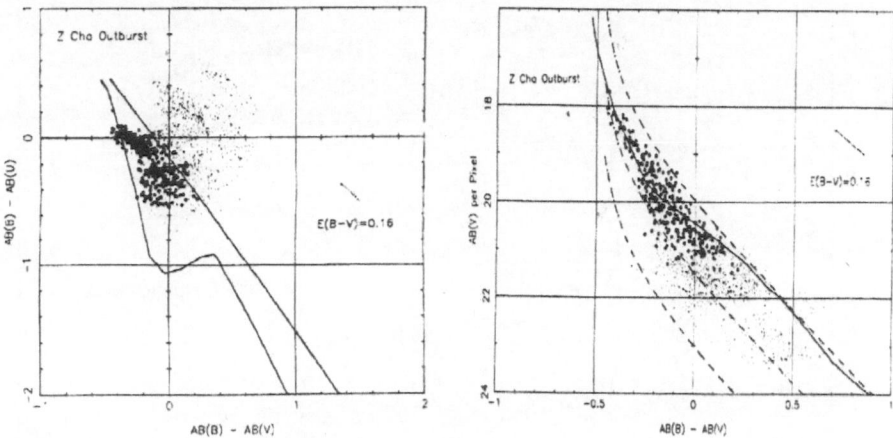

Fig. 7. Inferring the disc emission properties and the distance to Z Cha from the two-colour and colour-magnitude diagrams. Surface elements at the inner disc ($R < 0.3\ R_{L1}$) are represented by large dots, while elements in the outer disc regions are indicated by small dots. The solid and dashed curves in the right-hand panel show, respectively, the main sequence relationship for the best-fit distance and blackbody relationships for three different assumed distances. From [22]

3.2   Spectral Studies

The eclipse mapping method advanced to the stage of delivering spatially-resolved spectra of accretion discs with its application to time-resolved eclipse spectrophotometry [40]. The time-series of spectra is divided up into numerous spectral bins and light curves are extracted for each bin. The light curves are then analyzed to produce a series of monochromatic eclipse maps covering the whole spectrum.

Finally, the maps are combined to obtain the spectrum for any region of interest on the disc surface.

The spectral mapping analysis of the nova-like variables UX UMa [4,5,40,41] and UU Aqr [6] shows that the inner accretion disc is characterized by a blue continuum filled with absorption bands and lines which cross over to emission with increasing disc radius (Fig. 8). The continuum emission becomes progressively fainter and redder as one moves outwards, reflecting the radial temperature gradient. Similar results were found for SW Sex [16] and RW Tri [17]. Not surprisingly, these high-$\dot{M}$ discs seem hot and optically thick in their inner regions and cool and optically thin in their outer parts.

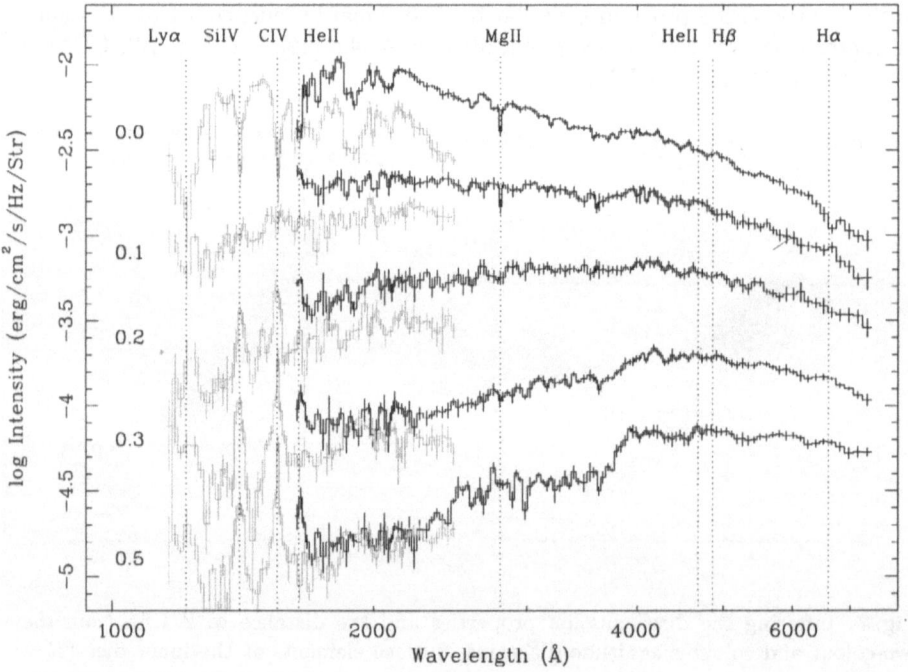

Fig. 8. Spatially resolved spectra of the UX UMa accretion disc on August 1994 (gray) and November 1994 (black). The spectra were computed for a set of concentric annular sections (mean radius indicated on the left, in units of R_{L1}). The most prominent line transitions are indicated by vertical dotted lines. From [5]

However, the unprecedent combination of spatial and spectral resolution obtained with spectral mapping started to reveal a multitude of unexpected details. In UU Aqr, the lines show clear P Cygni profiles at intermediate and large disc radii in an evidence of gas outflow [6]. In UX UMa, the comparison of spatially resolved spectra at different azimuths reveals a significant asymmetry in the disc emission at ultraviolet wavelengths, with the disc side closest to the secondary star showing pronounced absorption bands and a Balmer jump in absorption.

This effect is reminiscent of that observed previously in OY Car, where the white dwarf emission seems veiled by an "iron curtain" [23], and was attributed to absorption by cool circumstellar material [5]. The spectrum of the infalling gas stream in UX UMa and UU Aqr is noticeably different from the disc spectrum at the same radius suggesting the existence of gas stream "disk-skimming" overflow that can be seen down to $R \simeq (0.1 - 0.2)\ R_{L1}$. Spectra at the site of the bright spot suggest optically thick gas, with the Balmer jump and the Balmer lines in absorption.

The spectrum of the uneclipsed component in these nova-like systems shows strong emission lines and the Balmer jump in emission indicating that the uneclipsed light has an important contribution from optically thin gas (Fig. 9). The lines and optically thin continuum emission are most probably emitted in a vertically extended disc chromosphere + wind [5,6]. The uneclipsed spectrum of UX UMa at long wavelengths is dominated by a late-type spectrum that matches the expected contribution from the secondary star [41]. Thus, the uneclipsed component seems to provide an unexpected but interesting way of assessing the spectrum of the secondary star in eclipsing CVs.

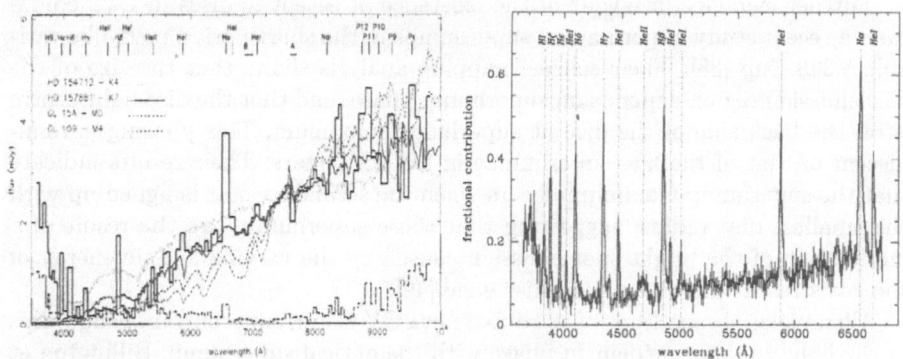

Fig. 9. The spectrum of the uneclipsed component in UX UMa (left) and UU Aqr (right, expressed as the fractional contribution to the total light). From [6,41]

Based on their spatially-resolved results, Baptista et al. [5] suggested that the reason for the long standing discrepancies between the prediction of the standard disc model and observations of accretion discs in nova-like variables (e.g. [26,30,50]) is not an inadequate treatment of radiative transfer in the disc atmosphere, but rather the presence of additional important sources of light in the system besides the accretion disc (e.g. optically thin continuum emission from a disc wind and possible absorption by circumstellar cool gas).

3.3 Spatial Studies

Eclipse mapping has also been a valuable tool to reveal that real discs have more complex structures than in the simple axi-symmetric model.

Besides the normal outbursts, short-period dwarf novae (SU UMa stars) exhibit superoutbursts in which superhumps develop with a period a few per cent longer than the binary orbital period. Normal superhumps appear early in the superoutburst and fade away by the end of the plateau phase. Late superhumps, displaced in phase by roughly 180° with respect to the normal superhumps, appear during decline and persist into quiescence [51]. Eclipse mapping experiments have been fundamental in testing superhump models.

O'Donoghue [32] analyzed light curves of Z Cha during superoutburst with a modified eclipse mapping technique. Assuming that the superhump profile is fairly stable over a timescale of a dozen of binary orbits, he separated the eclipse of the superhump source by subtracting the light curve when the superhump maximum occurs far from eclipse from that in which the superhump is centered on the eclipse. The eclipse mapping of the resulting light curve show that the superhump light arises from the outer disc, and appears to be concentrated in the disc region closest to the secondary star. This result helped to establish the superhump model of Whitehurst [53], in which the normal superhumps are the result of an increased tidal heating effect caused by the alignment of the secondary star and a slowly precessing eccentric disc.

Further evidence in favour of the existence of eccentric discs in CVs comes from a recent study of permanent superhumps in the short-period nova-like variable V348 Pup [38]. Their eclipse mapping analysis shows that the size of the disc emission region depends on superhump phase, and that the disc light centre is on the back side of the disc at superhump maximum. This phasing is reminiscent of that of the late superhumps in SU UMa stars. Their results indicate that the superhump maximum occurs when the secondary star is ligned up with the smallest disc radius, suggesting that these superhumps are the result of a modulation of the bright spot emission caused by the varying kinetic energy of the gas stream when it hits the disc edge [49].

Ultraviolet observations of the dwarf nova OY Car in superoutburst show dips in the light curve coincident in phase with the optical superhump. Billington et al. [10] analyzed this data set with a modified version of the eclipse mapping method which simultaneously maps the brightness distribution on the surface of the accretion disc and the vertical and azimuthal extent of a flaring at the edge of the disc. Their analysis indicates the presence of an opaque disc rim, the thickness of which depends on the disc azimuth and is large enough for the rim to obscure the centre of the disc at the dip phase (Fig. 10). These results are consistent with a model of normal superhumps as the consequence of time-dependent changes in the thickness of the edge of the disc, resulting in obscuration of the ultraviolet flux from the central regions and reprocessing of it into the optical part of the spectrum. Another evidence of discs with thick rims comes from the work of Robinson, Wood & Wade [37], who found that a relatively large disc opening angle is required in order to explain the ultraviolet eclipse light curves of Z Cha in outburst. It seems that the discs of dwarf novae become flared during normal outbursts and that the thickening during superoutbursts may be sufficient for the disc rim to obscure the inner disc regions.

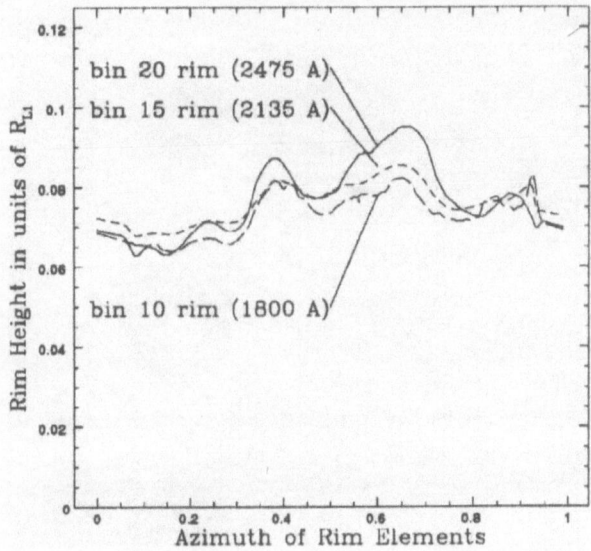

Fig. 10. The dependency of the disc thickness with azimuth at three different wavelengths for OY Car in superoutburst. From [10]

Tidally induced spiral shocks are expected to appear in dwarf novae discs during outburst as the disc expands and its outer parts feel more effectively the gravitational attraction of the secondary star (see the respective contributions of Boffin, of Makita et al. and of Steeghs in this volume). Eclipse mapping of IP Peg during outburst [3] helped to constrain the location and to investigate the spatial structure of the spiral shocks found in Doppler tomograms [18,54]. The spiral shocks are seen in the continuum and C III+N III $\lambda4650$ emission line maps as two asymmetric arcs of $\sim$ 90 degrees in azimuth extending from intermediate to the outer disc regions (Fig. 11). The He II $\lambda4686$ eclipse map also shows two asymmetric arcs diluted by a central brightness source. The central source probably corresponds to the low-velocity component seen in the Doppler tomogram and is possibly related to gas outflow in a wind emanating from the inner parts of the disc. The comparison between the Doppler and eclipse maps reveal that the Keplerian velocities derived from the radial position of the shocks are systematically larger than those inferred from the Doppler tomography indicating that the gas in the spiral shocks has sub-Keplerian velocities. This experiment illustrates the power of combining the spatial information obtained from eclipse mapping with the information on the disc dynamics derived from Doppler tomography.

3.4 Time-Resolved Studies

Eclipse maps give snapshots of the accretion disc at a given time. Time-resolved eclipse mapping may be used to track changes in the disc structure, e.g. to

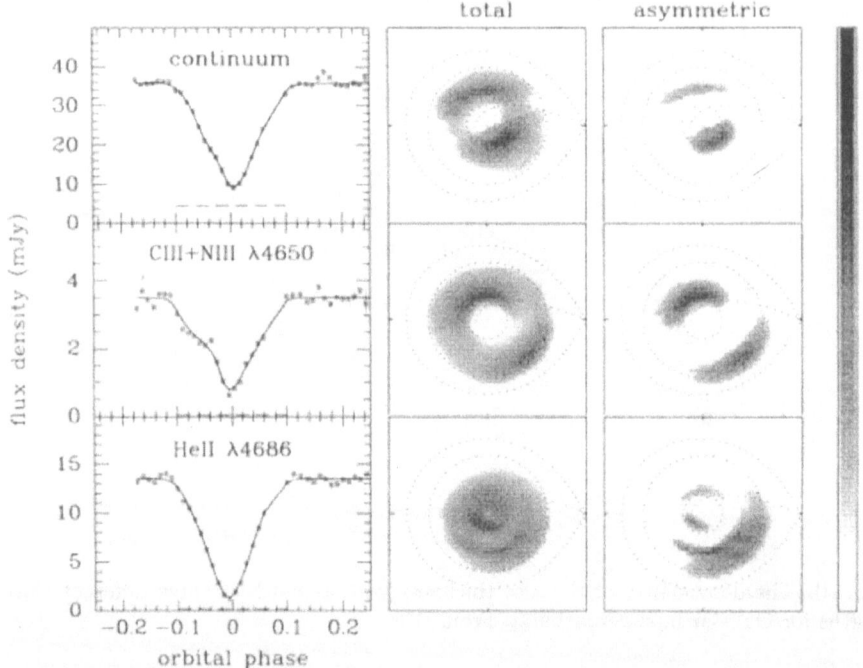

Fig. 11. Eclipse mapping of spiral shocks in IP Peg. The light curves are shown in the left-hand panels and the eclipse maps are displayed in the middle and right-hand panels in a logarithmic greyscale. The notation is the same as in Fig. 4. From [3]

assess variations in mass accretion rate or to follow the evolution of the surface brightness distribution through a dwarf nova outburst cycle.

The observed changes in the radial temperature distribution (and mass accretion rate) of eclipse maps obtained at different epochs in the high viscosity, steady-state discs of the nova-like variables UX UMa and UU Aqr are evidence that the mass transfer rate in these system is variable [4,9].

Eclipse maps of Z Cha during superoutburst show a bright rim in the outer disc regions which decreases in brightness relative to the inner regions as the superoutburst proceeds and the superhumps fade away [52]. This underscores the indications that the superhumps are sited at the outer disc rim (Sect. 3.3).

Rutten et al. [42] obtained eclipse maps of the dwarf nova OY Car along the rise to a normal outburst. Their maps show that the outburst starts in the outer disc regions with the development of a bright ring, while the inner disc regions remain at constant brightness during the rise. The flat radial temperature profile of quiescence and early rise changes, within one day, into a steep distribution that matches a steady-state disc model for $\dot{M} = 10^{-9}\ M_\odot\,\mathrm{yr}^{-1}$ at outburst maximum. Their results suggest that an uneclipsed component develops during the rise and contributes up to $\simeq 15$ per cent of the total light at outburst maximum. This

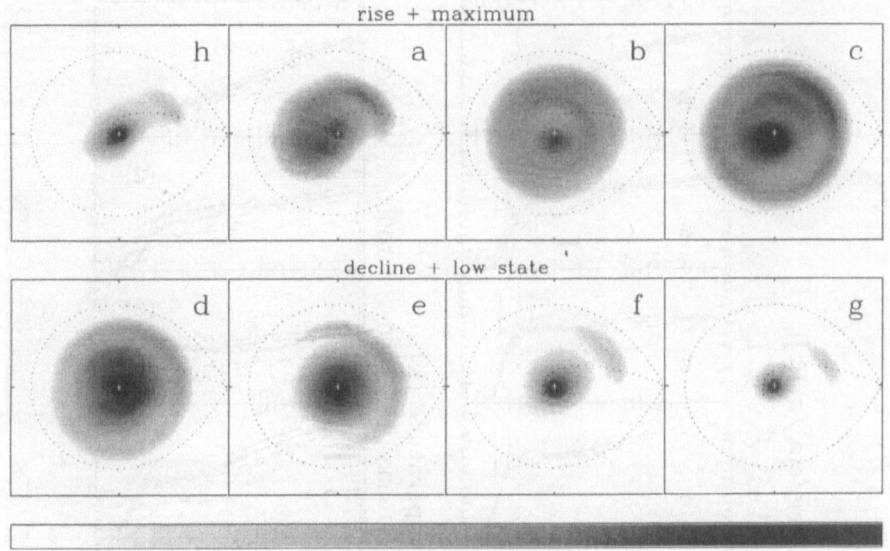

Fig. 12. Sequence of eclipse maps of the dwarf nova EX Dra. The eclipse maps capture 'snapshots' of the disc brightness distribution in quiescence (h), on the rise to maximum (a-b), during maximum light (c), through the decline phase (d-f), and at the end of the eruption, when the system goes through a low brightness state before recovering its quiescent brightness level. The notation is the same as in Fig. 4. From [2]

may indicate the development of a vertically-extended (and largely uneclipsed) disc wind, or that the disc is flared during outburst (see Sect. 3.3).

Time-resolved eclipse mapping covering the decline of an outburst and of a superoutburst were obtained, respectively, for IP Peg [11] and OY Car [14]. In both cases the radial temperature distribution evolves with the inward traveling of a transition front that leaves behind a cool disc ($T_b \simeq 5000 - 6000$ K) while the temperatures at the inner disc remain almost constant at a higher value. The derived speed of this cooling front is $\simeq 0.14\,\mathrm{km\,s^{-1}}$ for OY Car and $\simeq 0.8\,\mathrm{km\,s^{-1}}$ for IP Peg.

Eclipse maps covering the full outburst cycle of the long-period dwarf nova EX Dra [2] show the formation of a one-armed spiral structure in the disc at the early stages of the outburst [1] and reveal how the disc expands during the rise until it fills most of the primary Roche lobe at maximum light (Fig. 12). During the decline phase, the disc becomes progressively fainter until only a small bright region around the white dwarf is left at minimum light. The evolution of the radial brightness distribution suggests the presence of an inward and an outward-moving heating front during the rise and an inward-moving cooling front in the decline (Fig. 13). The inferred speed of the outward-moving heating front is of the order of 1 $\mathrm{km\,s^{-1}}$, while the speed of the cooling front is a fraction of that – in agreement with the results from IP Peg and OY Car. Their results also suggest a systematic deceleration of both the heating and the cooling fronts as

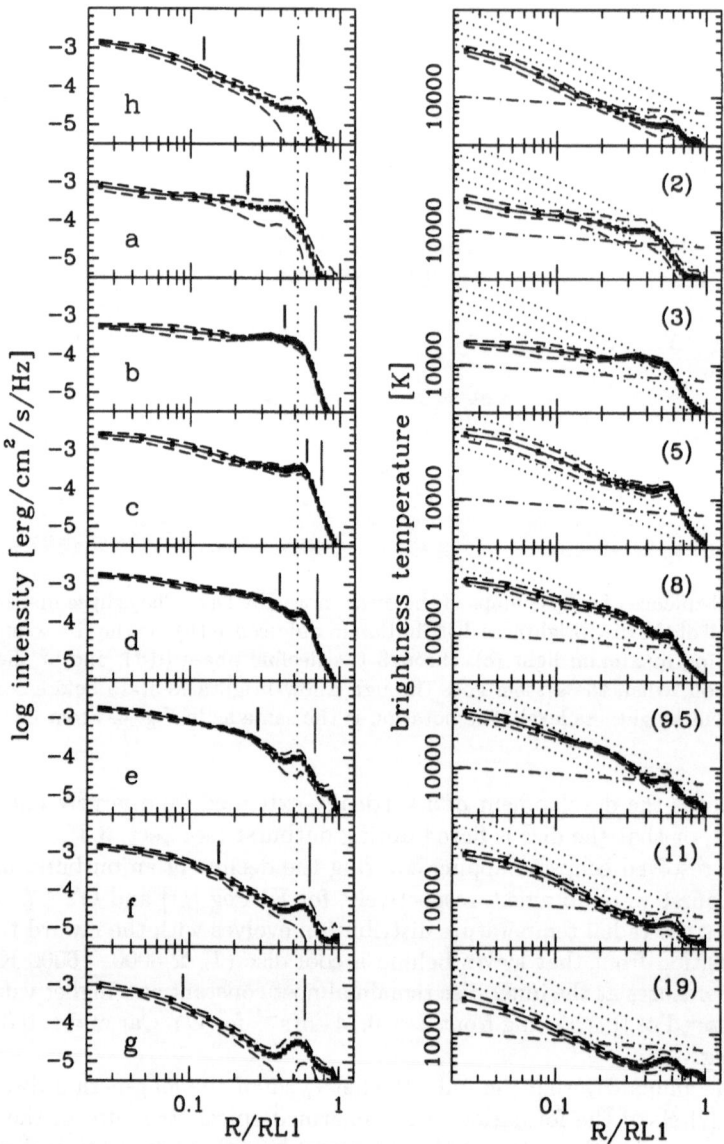

Fig. 13. Left: The radial intensity distributions of EX Dra through the outburst. Labels are the same as in Fig. 12. Dashed lines show the 1-σ limit on the average intensity for a given radius. A dotted vertical line indicates the radial position of the bright spot in quiescence. Large vertical ticks mark the position of the outer edge of the disc and short vertical ticks indicate the radial position of a reference intensity level. Right: The radial brightness temperature distributions. Steady-state disc models for mass accretion rates of log $\dot{M}$= $-7.5, -8.0, -8.5$, and -9.0 $M_\odot$ yr^{-1} are plotted as dotted lines for comparison. The dot-dashed line marks the critical temperature above which the gas should remain in a steady, high mass accretion regime. The numbers in parenthesis indicate the time (in days) from the onset of the outburst. From [2]

they travel across the disc, in agreement with predictions of the disc instability model [29]. A similar effect was seen in OY Car [14]. The radial temperature distributions shows that, as a general trend, the mass accretion rate in the outer regions is larger than in the inner disc on the rising branch, while the opposite holds during the decline branch. Most of the disc appears to be in steady-state at outburst maximum and, interestingly, also during quiescence. It may be that the mass transfer rate in EX Dra is sufficiently high to keep the inner disc regions in a permanent high viscosity, steady-state. An uneclipsed source of light was found in all maps, with a steady component associated to the secondary star and a variable component that is proportional to the out of eclipse brightness. Although disc flaring is likely in EX Dra during outburst, it seems it is not enough to account for the amplitude of the variation of the uneclipsed source. The variable component was therefore interpreted as emission arising from a disc wind, the strength of which depends on the disc mass accretion rate.

4   Future Prospects

Eclipse mapping is a powerful probe of the radial and the vertical disc structures, as well as of the physical conditions in accretion discs. Partly due to the many experiments performed over the last 15 years, we have enriched our picture of accretion discs with an impressive set of new details such as gas outflow in disc winds, gas stream overflow, flared discs with azimuthal structure at their edge, ellipsoidal precessing discs, sub-Keplerian spiral shocks, and moving transition fronts during disc outbursts.

This is however far from being the end of the road. There are still many key eclipse mapping experiments remaining to be done. The spectral mapping of dwarf novae in outburst offers a unique opportunity to probe the physical conditions of disc spiral shocks and to critically test the disc instability model by comparing the spectra of disc regions ahead and behind the transition fronts. Eclipse mapping estimates of $\dot{M}$ for CVs inside the period gap may be instrumental in testing the current theories about the origin of the period gap. A still untouched area is the mapping of the flickering sources. Our understanding of this fundamental signature of accretion processes can certainly be considerably improved with flickering mapping experiments. The unprecedent combination of high spatial and high spectral resolution on the disc surface that can be achieved from time-resolved spectroscopic data yields an unbeatable amount of information to test and improve the current disc models. Fitting disc atmosphere models to the spatially-resolved spectra is the obvious next step to the spectral mapping experiments and will certainly be rewarding.

Eclipse mapping have been continuously expanding into new domains. Of promising prospects are the direct mapping of physical parameters in the accretion disc (see the contribution of S. Vrielmann in this volume), the mapping of the brightness distribution along the accretion stream and accretion column in magnetic CVs [27], the 3-D mapping of flared discs and the surface of the secondary star ([39], see also the contribution of V. Dhillon in this volume), and the combination of eclipse mapping and Doppler tomography [12].

The leap in knowledge about accretion physics that have been (and will probably continue to be) obtained by the study of close accreting binaries with tomographic techniques can lead to a better understanding of many other astronomical scenarios in which accretion discs may play an important role, such as AGNs and quasars – the environment of which are apparently more complicated and comparatively less well known than that of these compact binaries.

Acknowledgments

Thanks to Keith Horne and René Rutten for stimulating discussions and valuable advice on the art of eclipse mapping. CNPq and PRONEX are gratefully acknowledged for financial support through grants no. 300 354/96-7 and FAU-RGS/FINEP 7697.1003.00.

References

1. Baptista R., Catalán M.S., 2000, ApJ, 539, L55
2. Baptista R., Catalán M.S., 2000, MNRAS, submitted
3. Baptista R., Harlaftis E.T., Steeghs D., 2000, MNRAS, 314, 727
4. Baptista R., Horne K., Hilditch R., Mason K.O., Drew J.E., 1995, ApJ, 448, 395
5. Baptista R., Horne K., Wade R.A., Hubeny I., Long K., Rutten R.G.M., 1998, MNRAS, 298, 1079
6. Baptista R., Silveira C., Steiner J.E., K. Horne, 2000, MNRAS, 314, 713
7. Baptista R., Steiner J.E., 1991, A&A, 249, 284
8. Baptista R., Steiner J.E., 1993, A&A, 277, 331
9. Baptista R., Steiner J.E., Horne K., 1996, MNRAS, 282, 99
10. Billington I., Marsh T.R., Horne K., Cheng F.H., Thomas G., Bruch A., O'Donoghue D., Eracleous M., 1996, MNRAS, 279, 1274
11. Bobinger A., Horne K., Mantel K.-H., Wolf S., 1997, A&A, 327, 1023
12. Bobinger A., Barwig H., Fiedler H., Mantel K.-H., Simić D., Wolf S., 1999, A&A, 348, 145
13. Bobinger A., 2000, A&A, 357, 1170
14. Bruch A., Beele D., Baptista R., 1996, A&A, 306, 151
15. Frank J., King A.R., Raine D.J., 1992, In: *Accretion Power in Astrophysics - 2nd edition* (Cambridge University Press, Cambridge)
16. Groot P.J., 1999, In: Optical Variability in Compact Source, PhD Thesis, University of Amsterdam (1999)
17. Groot P.J., 2000, presentation at the Astro-Tomography workshop held in Brussels.
18. Harlaftis E.T., Steeghs D., Horne K., Martín E., Magazzú A., 1999, MNRAS, 306, 348
19. Horne K., 1985, MNRAS, 213, 129
20. Horne K., Marsh T.R., 1986, In: *The Physics of Accretion onto Compact Objects*, eds. K. Mason, P. Charles (Springer-Verlag, Berlin), p. 1
21. Horne K., 1993, In: *Accretion Disks in Compact Stellar Systems*, ed. J. C. Wheeler (World Scientific Publ. Co., Singapore), p. 117

22. Horne K., Cook M.C., 1985, MNRAS, 214, 307
23. Horne K., Marsh T.R., Cheng F.-H., Hubeny I., Lanz T., 1994, ApJ, 426, 294
24. Horne K., Stiening R.F., 1985, MNRAS, 216, 933 933 (1985)
25. Knigge C., Long K.S., Hoard D.W., Szkody P., Dhillon V.S., 2000, ApJ, 539, L49
26. Knigge C., Long K.S., Blair W.P., Wade R.A., 1997, ApJ, 476, 291
27. Kube J., Gänsicke B.T., Beuermann K., 2000, A&A, 356, 490
28. Marsh T.R., Horne K., 1988, MNRAS, 235, 269
29. Menou K., Hameury J.-M., Stehle R., 1999, MNRAS, 305, 79
30. La Dous C., 1989, A&A, 211, 131
31. Meyer F., Meyer-Hofmeister E., 1982, A&A, 106, 34
32. O'Donoghue D., 1990, MNRAS, 246, 29
33. Paczyński B., Sienkiewicz R., 1981, ApJ, 248, 27
34. Patterson J., 1984, ApJS, 54, 443
35. Pringle J.E., Verbunt F., Wade R.A., 1986, MNRAS, 221, 169
36. Robinson E.L., Wood J.H., Bless R.C., Clemens J.C., Dolan J.F., Elliot J.L., Nelson M.J., Percival J. W., Taylor M.J., van Citters G.W., Zhang E., 1995, ApJ, 443, 295
37. Robinson E.L., Wood J.E., Wade R.A., 1999, ApJ, 514, 952
38. Rolfe D., Haswell C.A., Patterson J., 2000, MNRAS, in press
39. Rutten R.G.M., 1998, A&AS, 127, 581
40. Rutten R.G.M., Dhillon V.S., Horne K., Kuulkers E., van Paradijs J., 1993, Nature, 362, 518
41. Rutten R.G.M., Dhillon V.S., Horne K., Kuulkers E., 1994, A&A, 283, 441
42. Rutten R.G.M., Kuulkers E., Vogt N., van Paradijs J., 1992, A&A, 254, 159
43. Rutten R.G.M., van Paradijs J., Tinbergen J., 1992, A&A, 260, 213
44. Skilling J., Bryan R.K., 1984, MNRAS, 211, 111
45. Skilling J., 1987, In: *Maximum Entropy and Bayesian Methods in Applied Statistics*, ed. J. H. Justice (Cambridge University Press, Cambridge), p. 156
46. Smak J., 1992, Acta Astr., 42, 323
47. Spruit H.C., 1994, A&A, 289, 441
48. Steeghs D., Harlaftis E.T., Horne K., 1997, MNRAS, 290, L28
49. Vogt N., 1981, ApJ, 252, 653
50. Wade R.A., 1984, MNRAS, 208, 381
51. Warner B., 1995, In: *Cataclysmic Variable Stars* (Cambridge University Press, Cambridge)
52. Warner B., O'Donoghue D., 1988, MNRAS, 233, 705
53. Whitehurst R.: 1988, MNRAS, 213, 129
54. Wood J.E., 1994, In: In *Interacting Binary Stars*, ASP Conference Series, Vol. 56, ed. A. W. Shafter (ASP, USA), p. 48
55. Wood J.E., Abbott T.M.C., Shafter A.W., 1992, ApJ, 393, 729
56. Wood J.E., Horne K., Berriman G., Wade R.A., O'Donoghue D., Warner B., 1986, MNRAS, 219, 629
57. Wood J.E., Horne K., Berriman G., Wade R.A., 1989, ApJ, 341, 974
58. Wood J.E., Horne K., Vennes S., 1992, ApJ, 385, 294

Physical Parameter Eclipse Mapping

S. Vrielmann

Dept. of Astronomy, University of Cape Town, Rondebosch, 7700, South Africa

Abstract. The tomographic method *Physical Parameter Eclipse Mapping* is a tool to reconstruct spatial distributions of physical parameters (like temperatures and surface densities) in accretion discs of cataclysmic variables. After summarizing the method, we apply it to multi-colour eclipse light curves of various dwarf novae and nova-likes like VZ Scl, IP Peg in outburst, UU Aqr, V2051 Oph and HT Cas in order to derive the temperatures (and surface densities) in the disc, the white dwarf temperature, the disc size, the effective temperatures and the viscosities. The results allows us to establish or refine a physical model for the accretion disc. Our maps of HT Cas and V 2051Oph, for example, indicate that the (quiescent) disc must be structured into a cool, optically thick inner disc sandwiched by hot, optically thin chromospheres. In addition, the disc of HT Cas must be patchy with a covering factor of about 40% caused by magnetic activity in the disc.

1 Introduction

Cataclysmic variables (CVs) are close interacting binary system consisting of a main sequence star (the secondary) and a white dwarf. The Roche-lobe filling secondary loses matter via the inner Lagrangian point to the primary. If the magnetic field of the white dwarf is negligible, then angular momentum conservation drives the matter from the secondary into an accretion disc around the primary component. This disc matter is maintained by a steady mass stream from the secondary, which hits the disc in the so-called bright spot.

Some of these systems undergo frequently an outburst, in which the system brightens for a short period of time (a few days) in comparison to the interval between such eruptions (few weeks to months). It is believed that the accretion disc follows a disc instability cycle [27] in which the hydrogen in the disc switches its ionization status. Other systems, the so-called nova-like variables, appear to be in a permanent outburst state and are believed to exhibit optically thick, steady state discs. Quiescent disc, in contrast, show line emission, mainly of hydrogen and are therefore at least partially optically thin. An extensive overview over the theoretical and observational aspects of CVs is given in Warner [47].

The ultimate wish of any astronomers is probably to be able to see the stars from close by, to get a direct image of the stellar systems. Tomographic methods give us a unique opportunity to get such a picture, even if indirect, without getting close to the object. By making use of the eclipse light curve, the classical

Eclipse Mapping technique (EM, [18]) provides us with images of the accretion disc, while Physical Parameter Eclipse Mapping (PPEM, [45]) gives us some insight into the accretion physics. This paper gives an introduction to PPEM and summaries applications of this method to multi-colour light curves of five different cataclysmic variables.

2   The PPEM Method

Figure 1 illustrates the varying viewing angle a high inclination CV undergoes during an eclipse. One can see that at any one time during eclipse, only part of the accretion disc is occulted by the secondary. The flux at the given orbital phase is the total of all emission of any material not eclipsed at this phase. Any local intensity maximum in the accretion disc, e.g. the bright spot, will be eclipsed and reappears at a characteristic orbital phase and causes thereby a pair of steep gradients or steps in the observed eclipse light curve, one in ingress and the other in egress. Depending on the spatial location of this spot, the pair of steps will be shifted in relation to phase 0 which is defined as the conjunction of the white dwarf.

In EM one takes advantage of this spatial information of the intensity distribution of the accretion disc hidden in the eclipse profile. We have to make three basic assumptions: (a) the geometry of the secondary is known (it is usually a good approximation to assume it fills its Roche lobe); (b) the geometry of the disc is known (in the simplest approximation we use a geometrically infinitisimally thin disc); (c) the disc emission does not vary with time (this can be achieved by using averaged eclipse profiles in order to reduce the amount of flickering and flares from the disc). By fitting the eclipse light curve, we reconstruct the intensity distribution using a maximum entropy method (MEM, [34]). The MEM algorithm allows one to choose the simplest solution still compatible with the data in this otherwise ill-conditioned back projection problem. Further details of the EM method can be found in Horne [18] or Baptista & Steiner [1]. The usefulness of this method is extensively described by R. Baptista in these proceedings.

The PPEM approach goes a step further in that we map physical parameters, like temperature T and surface density Σ, instead of intensities. Figure 2 explains the method by means of a flowchart diagram. In addition to the three above mentioned basic assumptions (indicated by the right upper panel) we have to presuppose a spectral model for the disc emission (indicated as "Model" in the chart), relating the parameters to be mapped (e.g. T, Σ) to the radiated intensity in a given filter (e.g. $I_\nu = f(T, \Sigma)$). The model can be as simple as a pure black body spectrum, the PPEM is then called *Temperature Mapping* or as complicated as non-linear LTE disc model atmospheres as calculated by Hubeny [21]. Two simple model spectra are described in Sect. 2.2. The predicted multi-colour light curves are then compared to the observed ones. As long as the fit is not satisfactory, the parameter maps are varied according to the MEM algorithm (i.e. using gradients $\delta I/\delta T$ (and $\delta I/\delta \Sigma$)). As soon as the χ^2 that was aimed for

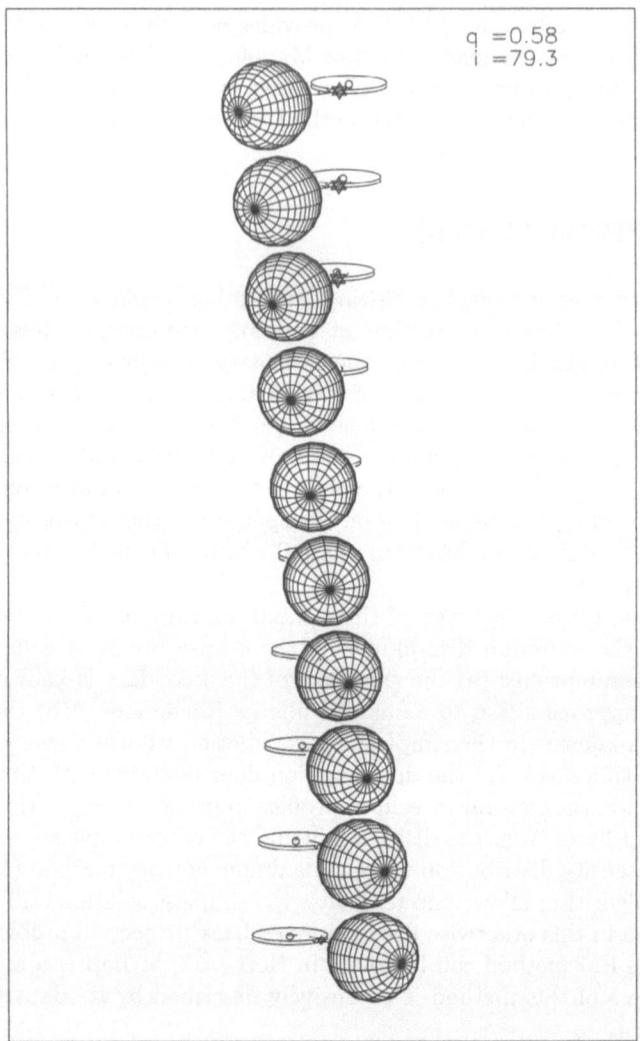

Fig. 1. Illustration of a cataclysmic variable going through an eclipse of the accretion disc. The parameters are set for IP Peg. Courtesy to K. Horne for the program *cvmovie*

is reached (and the maps have converged, see [18]) the maps T (and Σ) can be further analysed and compared to current disc models (see Sect. 4).

2.1   The Light Curves

The PPEM method requires input in the form of light curves. The number of light curves in various wavelength regions (filters/passbands) necessary to calculate a reliable parameter map depends on the used spectral model (see following

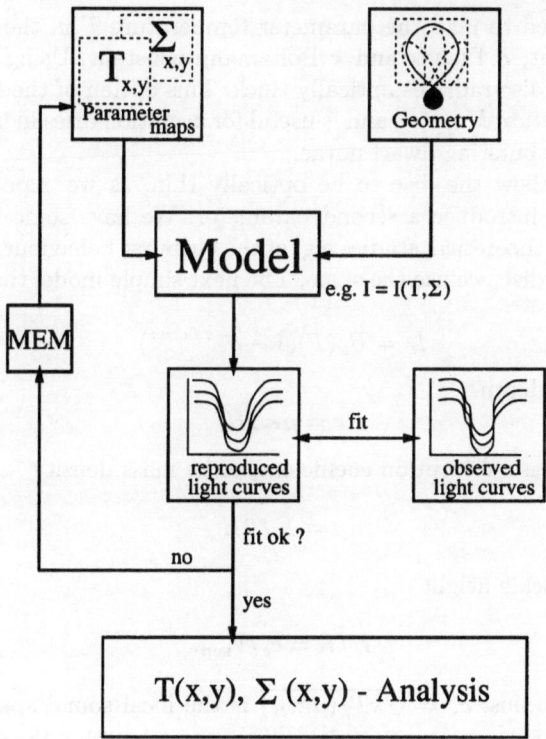

Fig. 2. A flowchart diagram to illustrate the Physical Parameter Eclipse Mapping algorithm

Sect. 2.2) and must be equal to or greater than the number of parameters to be mapped. For example, if one uses a black body spectrum to map the temperature (see Sect. 5.2 or 5.1), one light curve would be sufficient for a unique PPEM analysis. However, the more light curves (in various filters) one uses the better: the long wavelength data (e.g. IR light curves) will be ideal to determine low temperature (outer) regimes in the disc, while the short wavelength data (e.g. UV light curves) will be most appropriate to map the hot (inner) disc regions.

If one uses an appropriate spectral model, such as calculated by Hubeny [21], spectrophotometry can be analysed. PPEM ideally allows to extract a maximum of the information content of the data.

2.2   The Model Spectra

The spectral model provides a relation between the mapped parameter(s) P and the intensity I_ν at the frequency ν. If one would set $P = I_\nu$, this simply reduces PPEM to EM. In the simplest true PPEM version we use a black body spectrum for the spectral model, i.e.

$$I_\nu = B_\nu(T) = \frac{2h\nu^3}{c^2}\left(e^{\frac{h\nu}{kT}} - 1\right) \tag{1}$$

which can be used to map the parameter temperature T in the disc. Here c is the speed of light, h Planck and k Boltzmann constant. Using this model we assume that the disc radiates optically thick. This option of the PPEM method is called *Temperature Mapping* and is useful for accretion discs in novae, nova-like variables and outbursting dwarf novae.

In order to allow the disc to be optically thin, as we expect in quiescent dwarf novae, we introduce a second parameter. We have some freedom in the choice, but since theoretical studies e.g. of the outburst behaviour use the surface density Σ in the disc, we use the same. The next simple model thus is calculated as:

$$I_\nu = B_\nu(T)(1 - e^{-\tau/\cos i}) \qquad (2)$$

with the optical depth

$$\tau = \rho \kappa 2H, \qquad (3)$$

where κ is the mass absorption coefficient, ρ the mass density

$$\rho = \Sigma / 2H \qquad (4)$$

and H the disc scale height

$$H/R = c_s/V_{\text{kep}}. \qquad (5)$$

Here, R is the radius, $c_s = \sqrt{kT/(\mu m_H)}$ is the local sound speed and $V_{\text{kep}} = \sqrt{G\mathcal{M}_1/R}$ the Keplerian velocity of the disc material, with μ the mean molecular weight, m_H the mass of a hydrogen atom, G the gravitational constant and $\mathcal{M}_1$ the mass of the white dwarf.

In our studies we used a pure hydrogen slab in local thermodynamic equilibrium (LTE) including only bound-free and free-free H and H$^-$ emission. Even though this model is very simple it still is very useful in that it allows us to distinguish between optically thin and thick regions of the disc. The eclipse mapping of dwarf nova discs and the presence of emision lines suggest that at least part of the quiescent discs are optically thin.

For the usual range of temperatures and surface densities in CV accretion discs, this LTE approximation is still relatively good. In more realistic models (e.g. models by [21]) one would include metals which significantly increase the opacity at temperatures lower than about $T_m \sim 6300$K leading to lower intensities than determined by our $I(T, \Sigma)$. Therefore, using our simple model, our algorithm overestimates the temperatures (and if optically thin also the surface density) in the cooler parts of the disc, in order to meet the required intensities. This will only be important in the outer regions of the disc with temperatures below T_m.

2.3   The White Dwarf

For the white dwarf emission we used white dwarf spectra for the temperature range 10 000 K to 30 000 K. Outside this range black body spectra are used.

The white dwarf is assumed as a spherical object in terms of occultation of the accretion disc and eclipse by the secondary, but it is attributed only a single temperature. Since the absorption lines partly coincide with the mean wavelengths of the filters, pass band response functions are used. Usually, the lower hemisphere of the white dwarf is assumed to be obscured by the inner disc.

2.4 The Spatial Grid for the Accretion Disc

In the present studies the accretion disc is assumed to be infinitesimally thin. Onto this disc we constructed a two-dimensional grid of pixel. This grid consists of rings cut into a number of pixel that provides equal areas for all pixel [45]. For reference of spatial structures in the disc, we use radius and azimuth. The radius is used in units of the distance between the white dwarf and the inner Lagrangian point L_1, the white dwarf being at the origin. The second coordinate is the azimuth, the angle as seen from the white dwarf and counting from $-180°$ to $180°$. Azimuth 0 points towards the secondary, the leading lune of the disc has positive and the following lune negative azimuth angles.

2.5 The Uneclipsed Component

Apart from the disc emission we allow for another emission component that is never eclipsed, e.g. the secondary. However, we do not need to specify a geometrical location or a spectral model for this uneclipsed component, but we can reconstruct it for each wavelength/pass band independently as a constant flux contribution. It is usually given in the light curve plots as a solid horizontal line.

The uneclipsed component can then be used to give limits on the secondary or, if its contribution is known, the remaining uneclipsed flux can be analysed. It can originate in disc regions never eclipsed due to a relatively low inclination angle, or more likely regions at heights z above the disc that are never eclipsed by the secondary.

2.6 The Distance to the Systems

In order to reconstruct sensible physical parameter distributions within the accretion disc, the distance to the system has to be known. Only in some cases, relatively good estimates could be made using e.g. the secondary absorption line spectrum or the white dwarf flux.

PPEM provides also the option to determine an independent estimate of the distance, using the combined flux of the accretion disc and the white dwarf. This estimate can either be compared to present estimates (e.g. HT Cas, UU Aqr) or give the first opportunity to establish a distance (e.g. V2051 Oph).

The PPEM distance estimate is based on the fact that a spectrum with at least 3 wavelength points is not only determined by the temperature and the surface density, but also by the distance. If the assumed distance is smaller or larger than the true distance, the fit to the data will be poor. In a multi-pixel analysis like PPEM this might be compensated by neighbouring pixel leading to an overall "good" fit to the observed light curve (expressed as a small χ^2), but the resulting reconstruction will show artificial structures and the predicted light curve will have kinks not justified by the data. The amount of structure is expressed as the entropy S of the map: the smoother the map the higher the entropy. For each trial distance d we converge the maps to a specific χ^2 and then plot the parameter S against d. Where this function peaks, the map shows the least amount of structures and we will ideally find the true distance.

Only maps corresponding to fits of equal goodness can be compared for this distance estimate, since the entropy not only depends on the distance but also on the final χ^2. The lower the χ^2, i.e. the better the fit, the more structures will appear in the map, reducing thereby the entropy. At which χ^2 to stop has to be determined individually, it depends on the goodness of the spectral model or the amount of flickering in the light curve and is therefore a somewhat subjective measure.

Using this distance estimate one presupposes that the spectral model describes the true emissivity of the disc reasonably well. It is difficult to determine the error introduced by a wrong model. A generally good fit to the observed multi-colour eclipse light curves, however, seems to justify the model used.

3   The Reliability of the PPEM Method

Tests of the PPEM method were shown in Vrielmann et al.[45]. In general, the method allows to reconstruct the parameter distributions very well. However, it depends on the spectral model and the parameter values, how reliable the resulting maps are.

The black body spectrum is a non-linear function of the temperature, but since it gives an unambiguous function $I(T)$ it can be used to map the temperature uniquely. Therefore, the *Temperature Mapping* gives very reliable results. Since PPEM is using multi-colour light curves we see a huge improvement compared to classical eclipse mapping: steep gradients (e.g. at the disc edge) are much better reproduced in PPEM [45].

For a $T - \Sigma$ model as described in Sect. 2.2 one has to check where in the parameter space the solutions $I(T, \Sigma)$ are unique before the maps can be reliably analysed. For example, in the optically thick limit (large Σ), the surface density can assume any value above a certain limit without a change in the spectrum. However, the temperature in this case is very well defined. In other regions of the parameter space only the surface density is well defined (large T, small Σ), in again other regions both (intermediate values of T and Σ) or none (see the *banana* shaped contour lines for small T, small Σ in Fig.3) of the parameters can be determined uniquely. This pattern changes slightly with the disc radius. Figure 3 shows this pattern for a radius of $R = 0.32R_{L1} = 1 \cdot 10^{10}$ cm.

Parameters are most reliable in disc regions that emit the largest intensities. In disc regions with very low intensities the reconstructed parameters may be

χ²–Contour lines

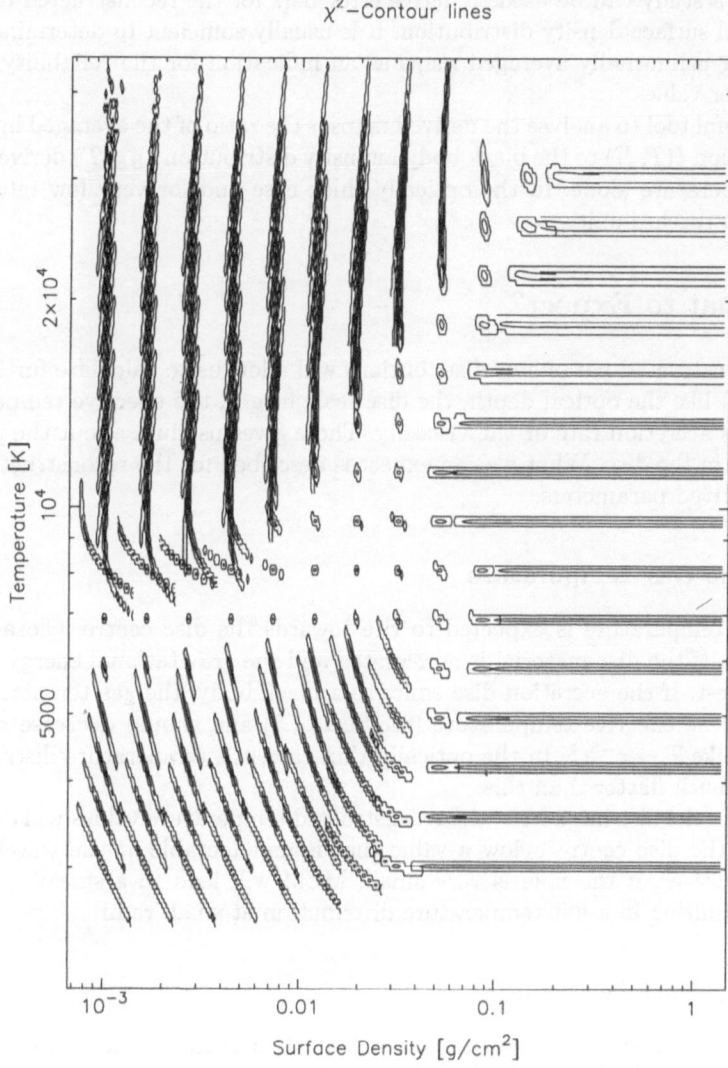

Fig. 3. χ^2 contours for different parameter combinations of temperature T and surface density Σ for a disc radius $R = 0.32R_{L1} = 1 \cdot 10^{10}$ cm. For each parameter pair T, Σ the spectrum $I(T, \Sigma)$ was calculated, then the spectra for $I(T \pm \Delta T, \Sigma \pm \Delta \Sigma)$ with small values for ΔT and $\Delta \Sigma$ and the difference in the spectra expressed as the χ^2. Contours are drawn for $\chi^2 = 3$ and 1 assuming an error in the spectra of 1% of the maximum value. The narrower the contour line, the better determined the parameter.

influenced by the MEM algorithm leading to a smoothing of the spatial gradients. This is typically the case in the outer disc regions.

Such a study can be used to derive error bars for the reconstructed temperature and surface density distribution. It is usually sufficient to determine these errors for azimuthally averaged maps as an indication for the reliability of the parameter values.

A useful tool to analyse the derived maps is the ratio of the averaged intensity distribution $I(T, \Sigma)$ to the black body intensity distribution $I_{BB}(T)$ derived from the temperature alone. In the optically thick case and for very low intensities this ratio reaches unity.

4 What to Expect?

The reconstructed parameter distributions will allow us to calculate further parameters, like the optical depth, the disc scale height, the effective temperature and mass accretion rate or the viscosity. These gives us clues about the physics going on in the disc. What we can expect is described for the reconstructed and a few derived parameters.

4.1 The Gas Temperature

The gas temperature is expected to rise towards the disc centre. Close to the white dwarf the disc material is accelerated and the gravitational energy release is strongest. If the accretion disc emits as a black body, the gas temperature is equal to the effective temperature (see Sect. 4.3) and should decrease radially roughly like $T \sim r^{-3/4}$. In the optically thin case, the temperature distribution may be much flatter than this.

In case the disc has a hole, the reconstructed temperature values will decrease towards the disc centre below a value that is undetectable at the wavelengths used. However, if the hole is very small, MEM will lead to a smearing of the values resulting in a flat temperature distribution at small radii.

4.2 The Surface Density

The literature is divided about the radial behaviour of the surface density. While Meyer & Meyer-Hofmeister's [25] calculations result in a radially decreasing surface density distribution, Ludwig, Meyer-Hofmeister & Ritter [23] and Cannizzo et al. [9] derive slightly increasing surface density distributions.

Since the critical surface density distributions within the disc instability model show an increase of Σ_{crit} with radius [8,14] we would rather expect that the actual $\Sigma(r)$ distribution also increases with radius. This would make it much easier for a disc to undergo an outburst.

Note, that if the disc is optically thick, we have no means to determine the surface density, only a lower limit that may be far from the true value.

4.3   The Effective Temperature

Considering the gravitational energy release and taking into account the slowing down of the disc material at the white dwarf surface, a steady state accretion disc should have the following radial effective temperature distribution [12]:

$$T_{\text{eff}}(r) = T_* \left(\frac{r}{R_1} \right)^{-3/4} \left[1 - \left(\frac{R_1}{r} \right)^{1/2} \right]^{1/4} \tag{6}$$

with

$$T_* = \left[\frac{3 G \mathcal{M}_1 \dot{\mathcal{M}}}{8 \pi \sigma R_1^3} \right]^{1/4} \tag{7}$$

where r is the radius in the disc, R_1 and $\mathcal{M}_1$ are the radius and mass of the white dwarf, G the Gravitational constant, $\dot{\mathcal{M}}$ the mass accretion rate of the disc and σ the Stephan-Boltzmann constant. For large radii Eq. (6) reduces to

$$T_{\text{eff}}(r) = T_* \left(\frac{r}{R_1} \right)^{-3/4} \quad \text{for } r \gg R_1. \tag{8}$$

Deviations from this distribution in real disc show that it is not in steady state. This can be expected for a quiescent accretion disc of a dwarf nova awaiting the next outburst. Discs in nova-like variables and in outbursting dwarf novae are expected to follow the steady state $T_{\text{eff}}(r)$-profile.

4.4   The Critical Temperature

The effective temperature distribution can then also be compared to a critical temperature, as e.g. calculated by Ludwig et al. [23]. If the effective temperature distribution falls below this critical value the accretion disc is in the lower branch of the $T_{\text{eff}} - \Sigma$ hysteresis curve which is used to explain the disc instability cycle. Such a disc should undergo outbursts. If the temperatures are above this critical value, the accretion disc is on the hot branch of this hysteresis curve and should therefore be currently in outburst or within a nova-like or nova system.

4.5   The Viscosity

A further parameter that can be derived is the viscosity ν, usually parametrized as $\nu = \alpha c_s H$ (where c_s is the local sound speed and H again the disc scale height) using Shakura & Sunyaev's [31] α-parameter. Theoretical studies involving the disc instability cycle predict values of α around 0.01 in quiescence [38].

We calculate the viscosity using the standard relation between the viscously dissipated and total radiated flux F_ν,

$$2 H \alpha P \frac{3}{4} \Omega_K = \int F_\nu d\nu = \sigma T_{\text{eff}}^4 \tag{9}$$

where P is the (gas) pressure and Ω_K the Keplerian angular velocity of the disc material.

5   Application to Real Data

This section gives some highlights of the application of the PPEM method to different objects. Any details about the analyses can be found in the corresponding, mentioned articles.

5.1   VZ Sculptoris

VZ Scl is a little studied eclipsing nova-like with an extreme difference between high (normal) and low states of about 4.5 mag [26]. Since in the low state the secondary dominates the spectrum at all wavelengths, Sherington et al. [32] could determine a reliable distance of 530 pc. The other system parameters, in particular the inclination angle i and the mass ratio q are somewhat uncertain. We adopted the same values O'Donoghue et al. [26] used for their Eclipse Mapping and the ephemeris determined by Warner & Thackeray [51].

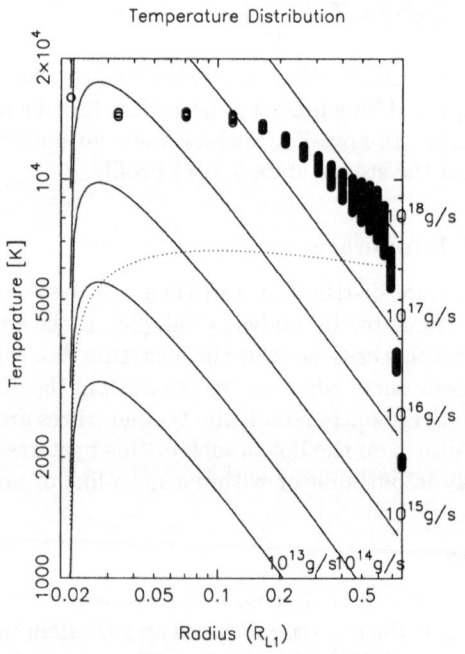

Fig. 4. The temperature distribution of **VZ Scl.**

Our observations were according to O'Donoghue et al. [26] taken when the object was in its normal state and we therefore used the *Temperature Mapping* version of PPEM [42]. As expected, the black body assumption turned out to be a fairly good approximation. We reached a χ^2 of 5. The disc appears to be in steady state between 0.15 R_{L1} to the disc edge at 0.7 R_{L1} with a mass accretion rate of about $3 \times 10^{17} \mathrm{gs}^{-1}$ (see Fig. 4). In the inner part of the disc ($r < 0.15\ R_{L1}$)

the temperature profile is very flat. This could be caused by a small hole in the disc, that is smeared out due to the MEM algorithm. Other explanations are discussed by e.g. Rutten et al. [29], but the true cause is still not clear.

Note in particular, that we simply used the published distance and still achieved a relatively good fit to the data. In a future more extensive study we will test if the disc has optically thin region and if we can reach a better fit to the data using the $T - \Sigma$ version of PPEM. We will then also determine an independent distance estimate. However, we do not expect a great deviation from the literature value.

5.2 IP Pegasi

IP Peg is probably the best studied dwarf nova, and still is not completely demystified. It is most famous for the spiral waves found by Steeghs, Harlaftis, Horne ([36], see also these proceedings) during rise to outburst.

We analysed UBVRI data taken during four nights on decline from outburst [5] by feeding them into our PPEM algorithm [41] using the *Temperature Mapping* option. We achieved relatively good fits to the data, with χ^2's ranging between 1.1 and 4.

All four temperature maps (Fig. 5) show a prominent azimuthally smeared out bright spot at a radius of 0.5 to $0.55 R_{L1}$. The temperature in the central part, up to a radius of $0.1 R_{L1}$ drops dramatically during the three nights after maximum light from about 9 000 K to 4 000 K, possibly indicating an emptying out of the inner disc. Only in the last of the four nights, when the system reached the quiescent brightness, does the temperature reach again about 10 000 K at small radii. Instead, a hot ring at radius $0.2 R_{L1}$ present in the first three nights has disappeared in the last one. This mysterious behaviour seems to indicate that the outburst takes place only in the inner parts of the disc, without evidence of a cooling front.

These studies of VZ Scl and IP Peg during decline from outburst show that even the simple *Temperature Mapping* is a very useful tool.

5.3 UU Aqarii

The eclipsing nova-like UU Aqr belongs to the SW Sex stars, i.e. in spite of the high inclination it shows single peaked emission lines with phase dependent absorption features. Doppler tomography reveals a disc, with the asymmetric part of the emission lines caused by a prominent bright spot [16]. This disc shows long term photometric variations in the form of high and low states with an amplitude of about 0.3 mag.

Baptista et al. [2] repeatedly observed UU Aqr over a period of about 6 years catching the system in a high and two low states. We applied the PPEM method to their averaged light curves separated in a high state and a low state light curve. Here, we present only the application of PPEM to the averaged high state data. An analysis of the averaged low state data (and of the individual light curves) will be presented by Vrielmann [43].

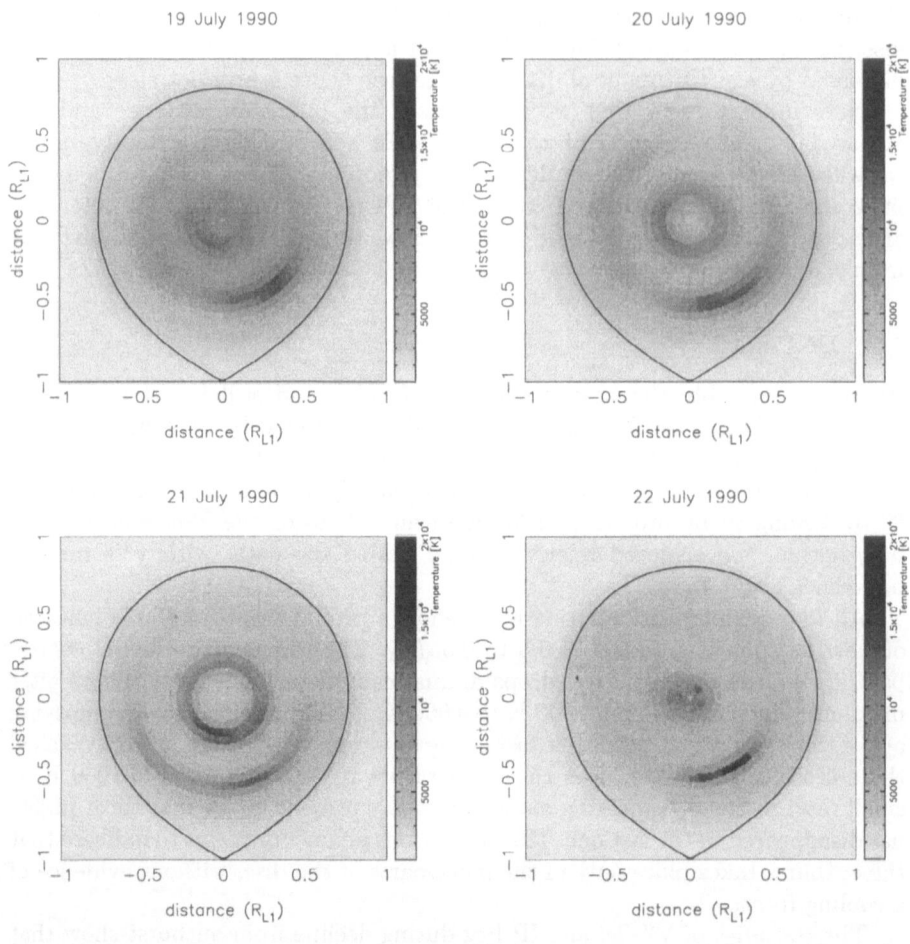

Fig. 5. The temperature maps of **IP Peg** on decline from outburst. The secondary is always at the bottom, just outside the plotted area.

Figure 6 (left) shows the averaged high state UBVRI light curves together with the PPEM fits. Baptista et al. [2] estimated a distance to UU Aqr by fitting the white dwarf fluxes. If the inner disc is opaque and obscures the lower hemisphere of the white dwarf, they derive a distance of 270±50 pc. Assuming an optically thick inner disc, Baptista et al. [3] performed a cluster main sequence fitting similar procedure to derive a distance of 200 ± 30 pc.

We independently estimated the distance to UU Aqr using the PPEM method as described in Sect. 2.6. Figure 6 (right) shows the entropy as a function of the

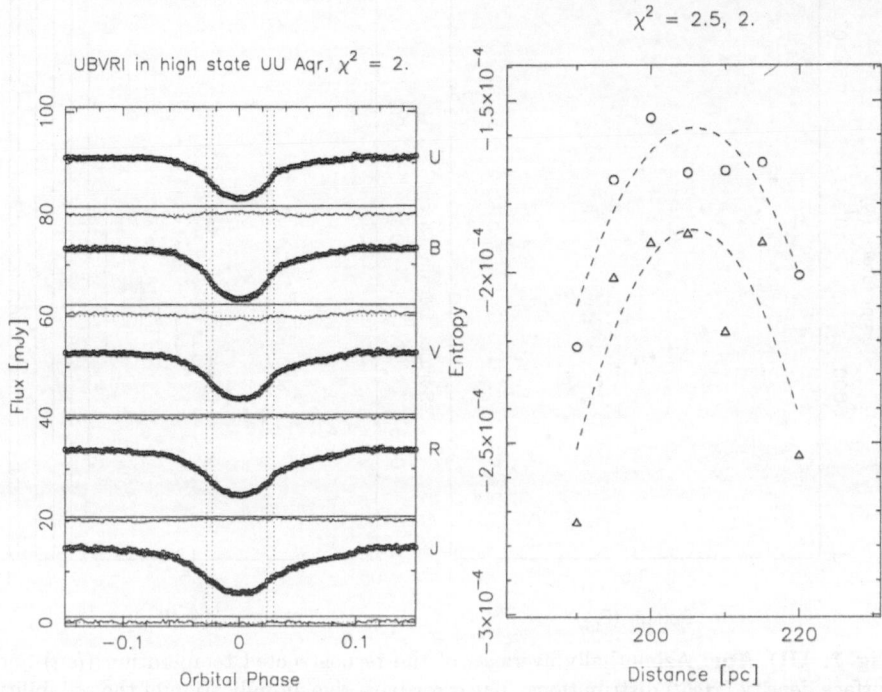

Fig. 6. *Left:* Averaged UBVRI light curves of **UU Aqr** in high state with the PPEM fits. The light curves are shifted upwards with the dotted line giving the zero line for each light curve. The residuals and the uneclipsed component are plotted in relation to the zero lines. Vertical dashed lines give the phases of the white dwarf ingress and egress. *Right:* Entropy vs. trial distance for $\chi^2 = 2.5$ (circles) and $\chi^2 = 2$ (triangles) and parabolic fits to the points peaking at 206 pc.

trial distance. The parabolic fit to the data for $\chi^2 = 2$ peaks at a distance of 206 pc, consistent with Baptista et al.'s [3] estimate.

Fig. 7 shows the averages of the reconstructed temperature T and surface density distributions Σ with error bars indicating the reliablility of the reconstructed values. The derived intensity distribution $I(T, \Sigma)$ for the filter I helps us to define a disc radius. $I(T, \Sigma)$ has values at $r = 0.6 R_{L1}$ of less then 10% of the white dwarf value. In the outer regions the intensity vanishes and therefore the Balmer Jump disappears and the error bars of Σ again become very large. For large radii, the temperature drops below T_m, i.e. the true temperatures are even lower and reach undetectable limits. Finally, the original T, Σ maps show a bright spot at a radius of $r_{\text{spot}} = 0.6 R_{L1}$ (see also Fig. 8, left) which lets us set the disc size as $0.6 R_{L1}$.

While the temperatures are everywhere well defined and decrease with radius as expected, the surface density values are only in the outer parts of the disc $(0.3 R_{L1} < r < 0.6 R_{L1})$ trustworthy. They clearly increase with radius even within the allowance of the error bars. In the inner part the disc is optically

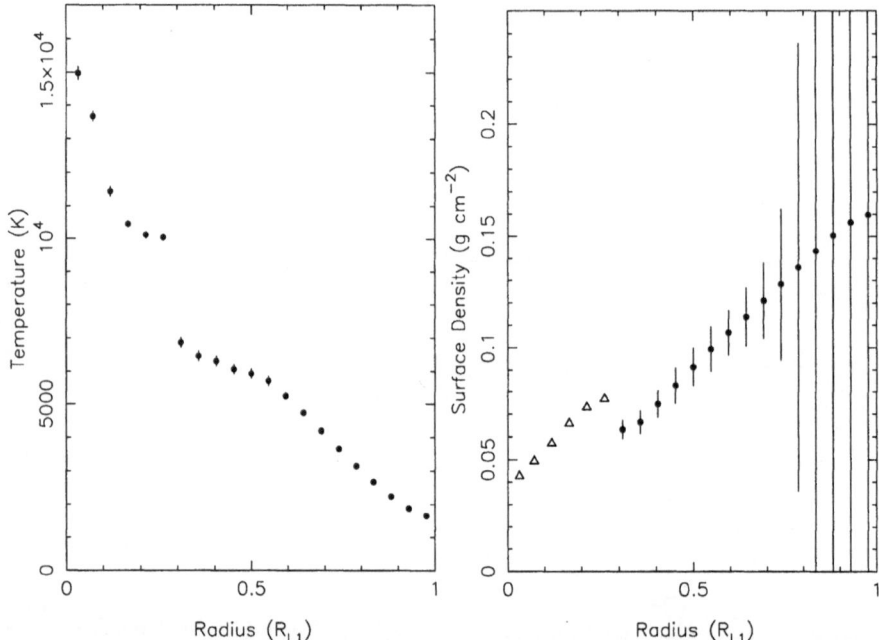

Fig. 7. UU Aqr: Azimuthally averages of the reconstructed temperature (*left*) and surface density (*right*) distributions. The error bars give an indication of the reliability of the reconstructed values according to a study of the spectral model. In the inner part is the disc optically thick, therefore the surface density there gives only lower limits (indicated by triangles). In the outer regions, the intensity $I(T, \Sigma)$ becomes very small, leading to large error bars in Σ.

thick (see Fig. 8, left) and we can only derive lower limits for Σ. The true values, however, may be much larger than those limits.

The white dwarf temperature was reconstructed to 26 100 K using white dwarf spectra as described in Sect. 2.3. This value very well lies in the usual range of white dwarf temperature estimates of between 9 000 K and 60 000 K [39] and quite close to the average effective temperatures of white dwarfs in non-magnetic CVs, $< T_{\text{eff}} >= 24\,100$ K [33].

As Fig. 8 (left) shows, only the central region up to about $0.3R_{L1}$ is optically thick, the remaining disc is optically thin, including the bright spot. This is unexpected, nova-likes are expected to have overall optically thick discs. However, this indicates a probable location for the line emission, i.e. in the outer parts of the disc.

The gray-scale plot shows that the bright spot is relatively weak compared to Hoard et al.'s [16] Doppler maps and delayed in the disc with respect to the expected position, where the ballistic stream hits the accretion disc. The stream matter apparently enters the disc in such a way that it first travels with the rotating material in the disc before it releases and radiates its energy away. It

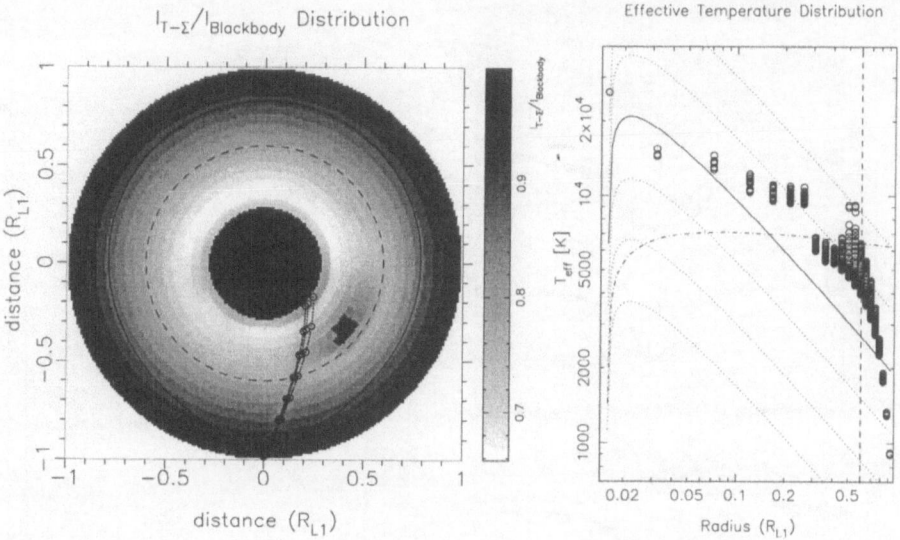

Fig. 8. UU Aqr: *Left:* The azimuthally and spectrally averaged intensity ratio $I(T, \Sigma)/I_{BB}(T)$ as explained in Sect. 3 as a gray-scale plot. The dashed circle is the disc radius, the pear shaped line is the Roche-lobe of the primary and the lines from the inner Lagrangian point are ballistic stream lines for mass ratios $q \pm 50\%$. The secondary is at the bottom. *Right:* The radial effective temperature distribution. Underlying dotted lines are effective temperature distributions for steady state discs with mass accretion rates $\log \dot{M} = 13$ to 18, the one for $\dot{M} = 10^{16} \mathrm{gs}^{-1}$ is drawn solid for reference. The dashed vertical line gives the disc radius, the dash-dotted line a critical temperature according to Ludwig et al. [23]

coincides possibly with Hoard et al.'s absorbing wall, where the line emission diminishes because the material is nearly optically thick.

Baptista et al. [2] estimate of the mass and radius of the secondary fits to a main sequence star in the range M3.5 to M4. Subtracting the flux of such a star from the uneclipsed component requires the object to be later than about M3.7. Such a star would leave us with no significant flux in I and a spectrum that rises towards shorter wavelenghts and peaks in the B filter. It is not possible to fit a black body spectrum to this distribution. The uneclipsed flux must therefore originate in an extended, possibly optically thick region with varying temperature. Contrary to HT Cas (Sect. 5.5) and V2051 Oph (Sect. 5.4), there is no cool component of the uneclipsed flux.

Further details concerning this study including an analysis of all of Baptista's individual light curves in high and low states will appear in Vrielmann [43]. This presented study shows in particular, that our PPEM analysis gives results in agreement with other methods, especially concerning the distance to the system.

$\chi^2 = 3.$

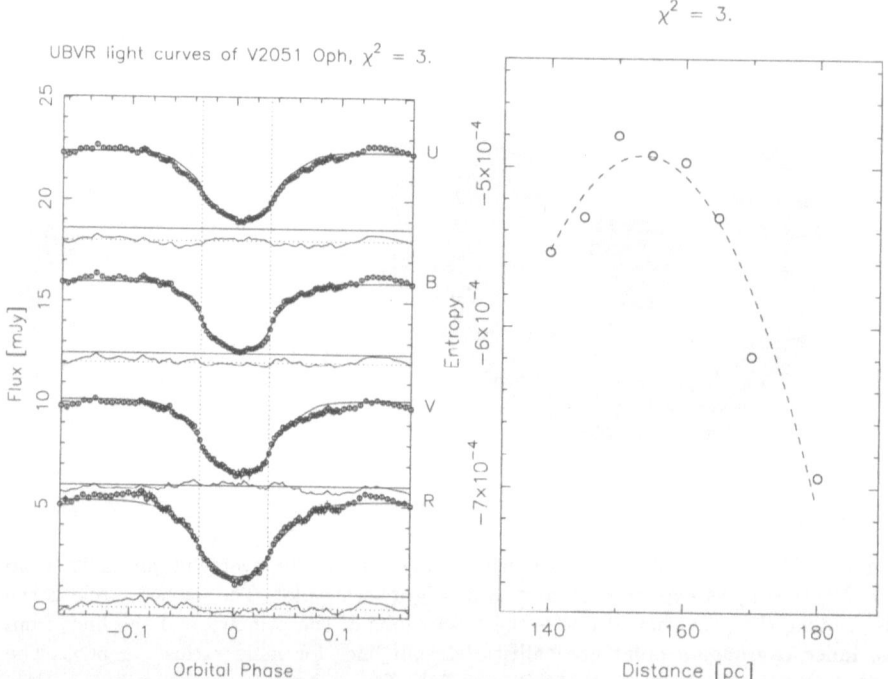

Fig. 9. *Left:* Averaged UBVRI light curves of **V2051 Oph** with the PPEM fits. For the explanation of the dashed and dotted lines see Fig. 6 (left). *Right:* Entropy vs. trial distance for $\chi^2 = 3$ and a parabolic fit to the points peaking at 154 pc

5.4   V2051 Ophiuchi

The nature of the cataclysmic variable V2051 Oph was mysterious since the discovery by Sanduleak [30]. It has been classified as a (low-field) polar [6] where the orbital hump is explained as a flaring accretion column [50] and as a dwarf nova showing outbursts, double peaked emission lines and an eclipse light curve that can be explained as caused by an accretion disc [4,49,52].

It is certain that it is a high inclination system, displaying eclipses and double peaked emission lines and after the superoutbursts in May 1998 and July 1999 it seems clear that the system must have an accretion disc. Warner [48] suggested that it might be the first system of a new class he terms polaroid. Polaroids are similar to intermediate polars (IPs), i.e. dwarf novae in which the white dwarf has an intermediately strong magnetic field that disrupts the inner part of the disc. However, while IPs have a fast spinning white dwarf, the primary in a polaroid is synchronized.

Figure 9 (left) shows UBVRI light curves of V2051 Oph averaged over eight eclipses and as used in the PPEM analysis of Vrielmann [44]. Most flares and the flickering are averaged out, except for the originally strong flare at phase 0.12.

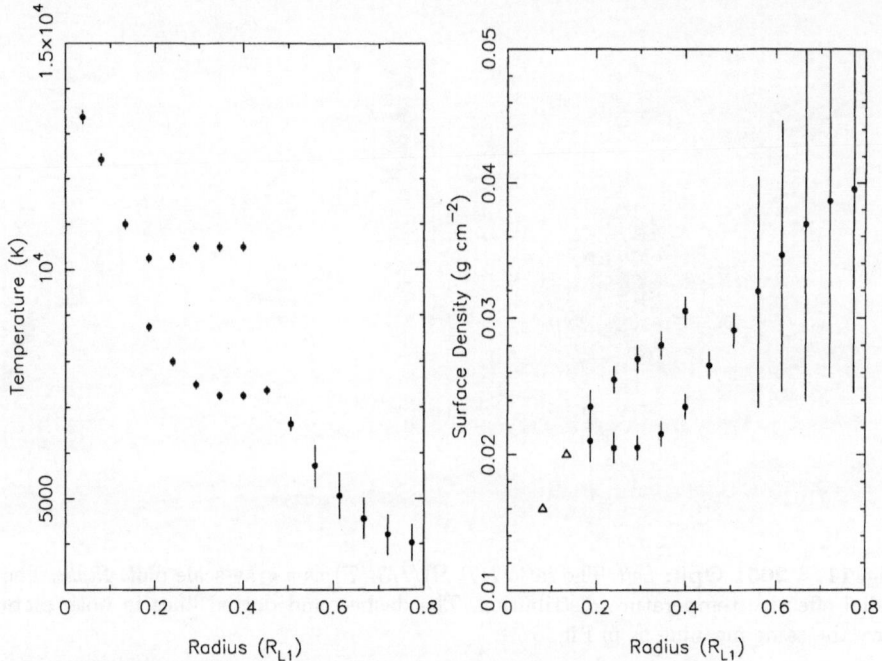

Fig. 10. V2051 Oph: The averaged reconstructed temperature (*left*) and surface density (*right*) distributions with error bars according to a study of the model. The error bars therefore indicate the reliability of the reconstructed parameter values, not the variation in azimuth. The values at intermediate radii ($0.2R_{L1}$ to $0.4R_{L1}$) have been split up to illustrate the range of parameter values. At a radius of $0.04R_{L1}$ the disc is completely optically thick, i.e. we omitted its reconstructed Σ value. For the next two annulli the disc is nearly optically thick, therefore we only give a lower limits for the surface density, indicated by a triangles.

We used the PPEM method to establish the first reliable distance estimate to V2051 Oph as described in Sect. 2.6. For each trial distance the data have been fitted to a χ^2 of 3. Figure 9 (right) shows the resulting entropy as a function of trial distance. A parabolic fit to the data points peaks at 154 pc.

Figure 10 shows averages of the reconstructed temperature and surface density distribution with error bars according to a study of the model. The values in the radial region $0.2R_{L1}$ to $0.4R_{L1}$ have been split up in an upper and lower branch because of a true spatial separation of regions and to illustrate the range of reconstructed parameters as well as that of the error bars. The upper branch values in both parameters correspond to a relatively confined region in the disc that we call *hot region*. It is located between the stars and is nearly optically thick. The lower branch values correspond to the remaining disc (cf. Fig. 11, left). In the outer regions of the disc the error bars in both parameters increase, because the parameter values enter the region with *banana* shaped χ^2 contour lines (see Fig. 3).

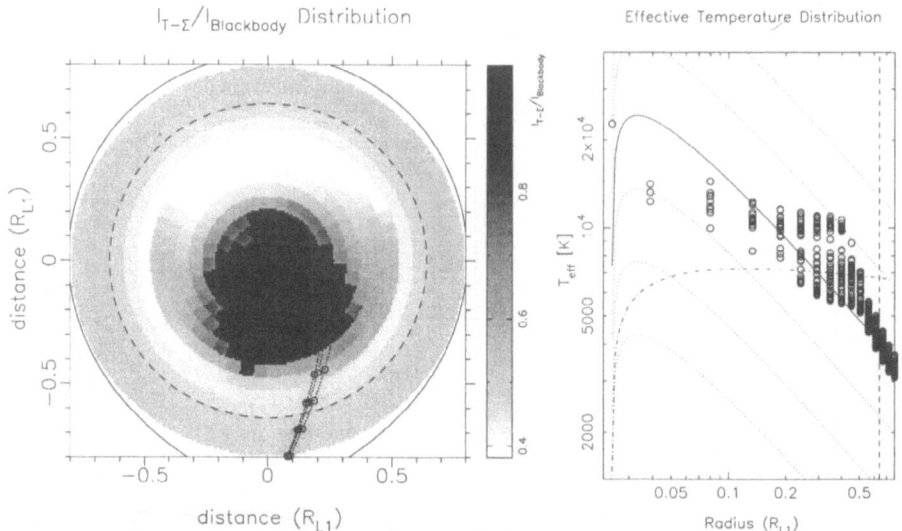

Fig. 11. V2051 Oph: *Left:* The ratio $I(T, \Sigma)/I_{BB}(T)$ as a gray-scale plot. *Right:* The radial effective temperature distribution. The dashed and dotted lines in both plots have the same meaning as in Fig. 8

The temperature T decreases radially, as expected. In contrast, the surface density Σ rises towards the disc edge, but according to the large error bars in the outer disc regions, Σ could also be almost constant throughout the disc. However, a decrease of Σ with radius appears out of the question.

Using white dwarf spectra, the white dwarf temperature was reconstructed to 22 700 K. This value is close to typical values of white dwarf temperatures, but somewhat higher than in the similar objects HT Cas, OY Car and Z Cha of around 15 000 K [13].

Figure 11 (left) gives a gray-scale display of the spatial distribution of the ratio $I(T, \Sigma)/I_{BB}(T)$. The central parts up to a radius of $0.15R_{L1}$ and the *hot region* between the stars are optically thick. In the outer region, $I(T, \Sigma)/I_{BB}(T)$ is close to unity, because both $I(T, \Sigma)$ and $I_{BB}(T)$ vanish there. Part of the optically thick *hot region* can be attributed to the bright spot, expected at the point where the ballistic gas stream hits the disc. But the *hot region* is too large and spreads out too far towards negative azimuths (into the following lune).

The right panel of Fig. 11 shows the radial distribution of the effective temperature in V2051 Oph's disc. A clear separation can be seen for the *hot region* and the remainder of the disc. While the mass accretion rate $\dot{\mathcal{M}}$ of the *hot region* lies above the critical value, the remaining disc's $\dot{\mathcal{M}}$ is approximately equal to the critical value or below. Since V2051 Oph is a dwarf nova showing occational outbursts, this mass accretion rate is far too high. It should lie well below the critial value.

A grayscale plot of the T_{eff} distribution (not shown here) shows a distinct ridge within the *hot region*. It is almost parallel to the binary axis, but tilted by about 15° towards the following lune of the disc.

In order to explain the *hot region*, the presence of an outbursting disc and the variable humps as seen by Warner & O'Donoghue [50] we propose the model for V2051 Oph as described in Sect. 6.1.

Calculating the viscosity parameter α we derive values far too large. While theory predicts values in the range of $\alpha \sim 0.01$ (see Sect. 4.5), our values lie between 30 and 1000. In Sect. 6.2 we explain how we can solve this discrepancy with our proposed model of accretion discs.

The uneclipsed component is reconstructed to 0.56 mJy, 0.43 mJy, 0.00 mJy and 0.66 mJy for the filters UBVR, respectively. Baptista et al.'s [4] mass and radius for the secondary, fits to a M4.5 main sequence star [22], however, this leads to fluxes of 0.07 mJy in V and 0.22 mJy in R. The secondary must therefore be of slightly later type or the error in the reconstructed values is of order a few hundredth mJy.

The separation of the UV and IR component means the uneclipsed component must originate in two separate sources. The UV flux probably comes from the hot chromosphere, extending to at least a few white dwarf radii above the disc plane. The IR source may be a cool disc wind. These estimates, though, are only valid as long as the spectral model is a good representation of the true emissivity of the disc.

Any further details about this analysis are described in Vrielmann [44].

5.5   HT Cassiopeiae

HT Cas is an unusual dwarf nova in that it does not show much of a bright spot and has occationally very long quiescent times of up to 9 years. On the other hand, Patterson [28] suggested it might serve as a *Rosetta stone* in explaining what drives the accretion discs. Since it is one of the few eclipsing dwarf novae it gives us a special opportunity for the analysis of a quiescent disc with PPEM.

We used UBVR light curves previously analysed and published by Horne et al. [20]. Figure 12 (left) shows the averaged UBVR eclipse light curves as prepared for our PPEM analysis [46]. Most of the flickering has disappeared and the most prominent remaining feature is the white dwarf eclipse.

We used these data to determine an independent distance estimate as described in Sect. 2.6. Previous reliable estimates involved the secondary [24] and the white dwarf [53] and resulted in values of 140 pc and 165 pc, respectively. Our PPEM estimate is significantly larger with a value of 207 pc (Fig. 12, right), however, the fits to the light curves are very good with a χ^2 of 1.75. Sect. 6.3 describes our suggested solution for this distance problem. It allows us to use the parameter values as reconstructed.

Figure 13 gives azimuthally averaged temperature and surface density distributions as reconstructed for the trial distance 205 pc. Like in the disc of V2051 Oph, the temperature decreases with radius. However, the surface den-

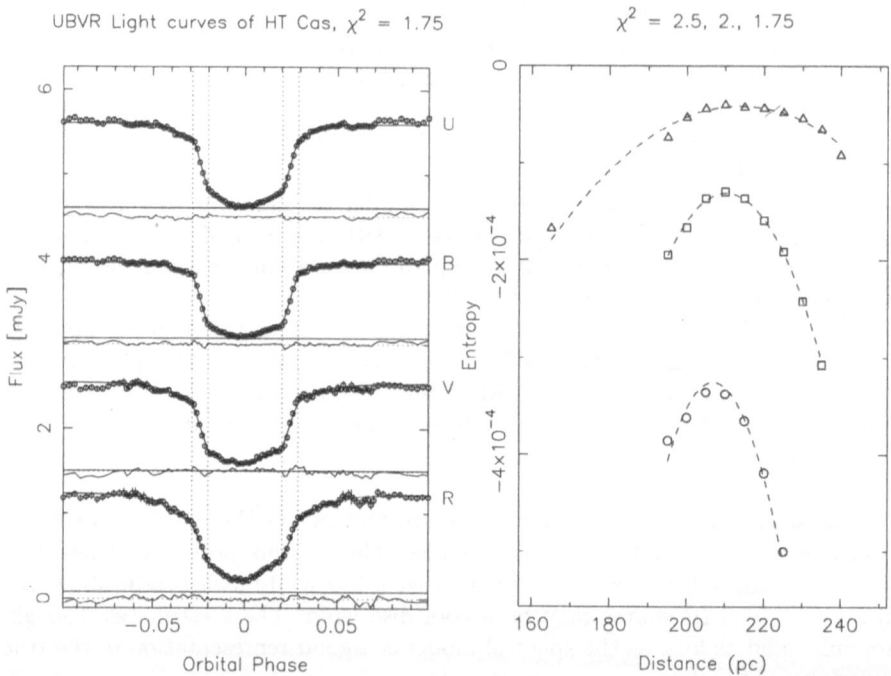

Fig. 12. *Left:* Averaged UBVRI light curves of **HT Cas** with the PPEM fits. For the explanation of the dashed and dotted lines see Fig. 6 (left). *Right:* Entropy vs. trial distance for $\chi^2 = 1.75$ and a parabolic fit to the points peaking at 207 pc

sity clearly increases radially even within the large error bars. The corresponding intensity distribution allow us to set a radius of the disc at $0.4R_{L1}$.

The reconstructed temperature of the white dwarf is 22 600 K. Because of our larger distance, this value is larger than Wood et al.'s [53] value of 18 700±1 800 K determined from the same data.

As shown by Fig. 14 (left), only the very central parts of the accretion disc are optically thick and the $I(T,\Sigma)/I_{BB}(T)$ ratio shows some asymmetry, also present in the original $T - \Sigma$ maps. We would expect the emission lines to originate only in regions of the disc with radii $r > 0.05R_{L1}$.

Assuming a distance of 205 pc, the effective temperature is significantly larger than the critical values (see Fig. 14, right). It would be difficult to explain such high effective temperature and therefore mass accretion rates in a system that shows only rare outbursts, i.e. should have an exceptionally low mass accretion rate. If we agree with the cited literature values for the distance and assume a distance of 133 pc (see Sect. 6.3), the effective temperature would drop to values below or close to the critical temperatures also shown in Fig. 14 (right). In the inner part of the disc ($r < 0.2R_{L1}$), the T_{eff} is flat, while outside this range the disc is steady state like with a mass accretion rate of about $2 \times 10^{15}\text{gs}^{-1}$.

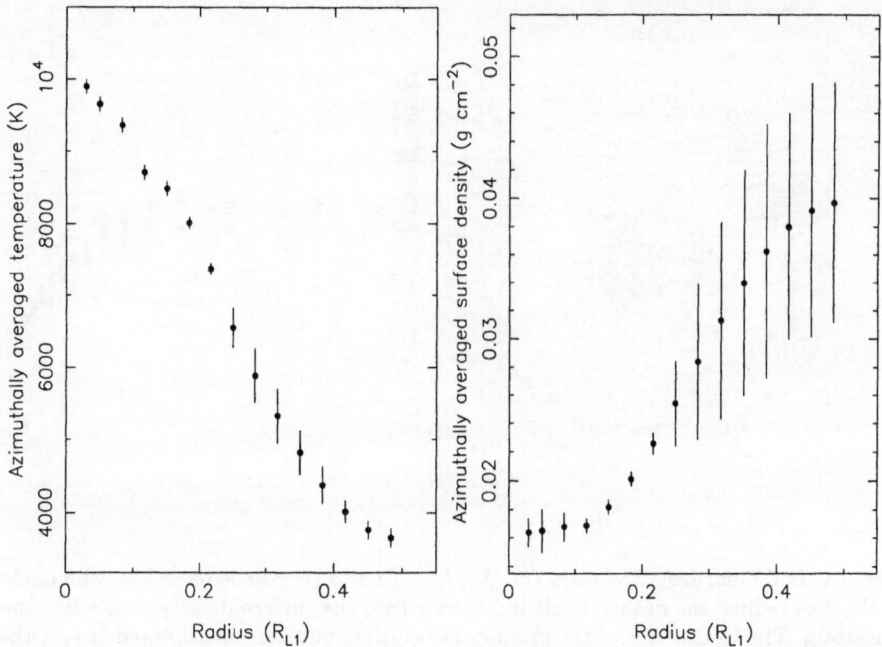

Fig. 13. HT Cas: The averaged reconstructed temperature (*left*) and surface density (*right*) distributions with error bars according to a study of the model

Similarly as in the case of V2051 Oph, the viscosity parameter α we determine from the maps using Eq. (9) assumes values too large by three to four orders of magnitude. Sect. 6.2 describes our model for the disc in both systems which qualitatively explains our solution of the discrepancy to theory.

The reconstructed uneclipsed component can partly be explained by a M5.4 secondary [24]. Subtracting such a main sequence star using a distance of 133 pc, basically no extra uneclipsed flux in the V band remains, like in V2051 Oph. However, in U, B, and R we find significant additional flux of 0.1 mJy, 0.08 mJy, and 0.05 mJy, respectively. As in V2051 Oph this must originate from two distinctly different sources, presumeably a hot chromosphere (providing the U and B flux) and a cool disc wind, giving rise to the R flux.

6   Discussion

6.1   About the Nature of V2051 Oph

V2051 Oph has often been compared to HT Cas, OY Car and Z Cha because of the similar orbital periods and all being SU UMa stars. Especially the similarly long and similarly variable outburst intervals of HT Cas and V2051 Oph is striking. One would hope that if one could explain one of these objects, all were understood.

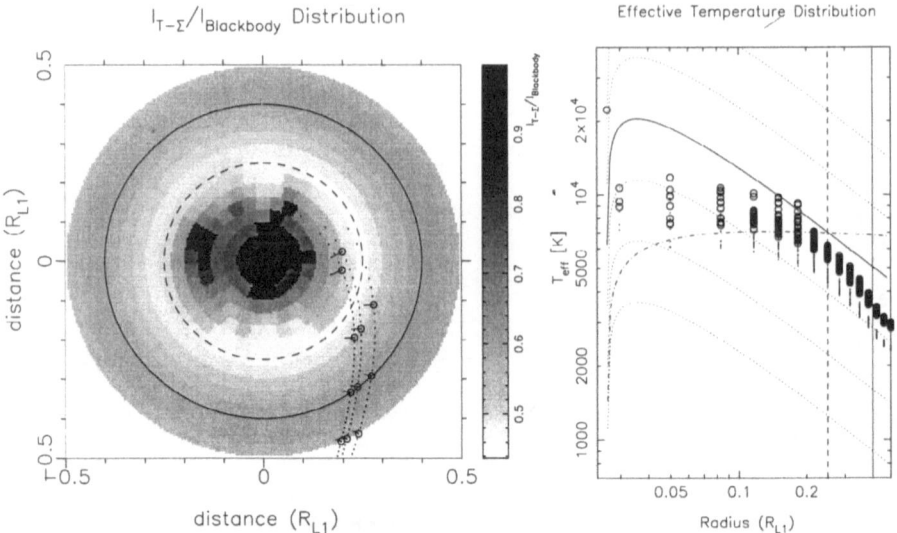

$I_{T-\Sigma}/I_{\text{Blackbody}}$ Distribution   Effective Temperature Distribution

Fig. 14. HT Cas: *Left:* The ratio $I(T,\Sigma)/I_{BB}(T)$ as a gray-scale plot. The solid circle is the disc radius, the dashed circle indicates where the surface density values become uncertain. The Roche-lobe of the primary lies entirely outside the displayed area. Otherwise same as Fig. 8 (left). *Right:* The radial effective temperature distribution for a distance of 205 pc (circles) and 133 pc (dots). The dashed vertical line gives the radius where the surface density values become uncertain and the solid line gives the disc radius. Otherwise same as Fig. 8 (right)

However, V2051 Oph has some special features, like the variable hump as seen in the light curves of Warner & Cropper [49] and Warner & O'Donoghue [50] that can only be explained by a hump source very close to the white dwarf. Objects with an (disrupted) accretion disc and an (flaring) accretion column would most easily be explained as an intermediate polar (IP). However, IPs usually show characteristic X-ray emission. V2051 Oph has only been detected as a very faint X-ray source, but on the other hand no high inclination system shows high X-ray count rates. Instead there seems to exist an anti-correlation between the observable X-ray emission measure and the inclination angle. The X-ray source therefore is very close to the white dwarf and must be obscured or absorbed in the high inclination systems [17,40].

The absence of circular polarisation in the wavelength range 3500-9200 Å [10] could be explained by a very low magnetic field of 1 MG or less [50], as expected for IPs.

If V2051 Oph is indeed an intermediate polar, the inner disc radius must be very small, of order less than 0.1 R_{L1}. Otherwise, it would have been reconstructed by PPEM. On the other hand, the accretion curtains would partially fill out the space between the white dwarf and the accretion disc.

Warner [48] suggests that V2051 Oph could be one of the systems he terms polaroid. A polaroid is an intermediate polar with a synchronized white dwarf.

With our analysis we cannot distinguish between an IP and a polaroid, but oscillations found in V2051 Oph indicate a non-synchronous white dwarf [35] and therefore we favour the IP model.

It remains to be explained what causes the *hot region* in the disc. One possibility is limb brightening of the accretion disc that we did not take into account. If the disc was flared this effect would even be enhanced. However, it would not explain the slightly asymmetric ridge we see in the effective temperature distribution. Another possibility is an illumination from a bulge where the ballistic stream hits the accretion disc as described in Buckley & Tuohy [7]. However, this would also make it difficult to explain the ridge in the effective temperature map. A third explanation is a warp in the disc in such a way that the disc region between the stars is illuminated by the white dwarf at azimuths $-15°$ and the region on the opposite side of the disc ($+165°$) has a bulge so that the outer parts of the disc cannot be illuminated by the central object (for a more detailed explanation see [44]). However, it would be difficult to explain if it is stable in the rotating frame of the binary.

A PPEM analysis of multi-colour light curves from other time periods would be helpful in deciding what causes the *hot region*, if its location in the disc is stable, rotates in the disc or is variable in its presence.

6.2   Accretion Disc Model for HT Cas and V2051 Oph

Instead of dismissing Eq. (9) or our PPEM analysis altogether, we were seeking for an explanation for the high values of α of a few hundreds that we derive for the disc in both HT Cas and V2051 Oph.

If we compare our derived surface density values with the values used in the disc instability model [11,23,37] we find that our values are much lower than those necessary for the disc entering the hysteresis curve. Theory predicts values between 10 and 100 g cm^{-2} for α's between 0.1 and 1 and even at larger Σ's for smaller α's. Since the disc cannot change the Σ from quiescence to outburst that dramatically, this means the outburst can only occur if the disc is very well optically thick already before its onset. On the other hand line emission is seen during rise and decline of an outburst [15].

To solve this problem we suggest that the disc consists of a cool, optically thick layer carrying most of the matter and which is sandwiched by a hot, optically thin chromosphere. The surface densities in the underlying disc must be a factor 10^4 (Σ a few hundreds) larger than the reconstructed values to bring the α's down to reasonable values of between 0.01 and 0.1. However, we only "see" the chromosphere, therefore we derive the large values for α.

The presence of a hot chromosphere is supported by the excess of blue emission in the uneclipsed component in HT Cas, V2051 Oph and UU Aqr. This emission must come from regions above the disc plane that are never eclipsed, i.e. a few white dwarf radii. This means, the chromosphere is an extended layer on top of the disc surface. Furthermore, this chromosphere must be the location where the line emission originates.

6.3   HT Cas' Patchy Disc

In a similar way as above, the distance problem with HT Cas does not mean we have to dismiss the PPEM algorithm right away. As we have shown, the PPEM estimate gives results in agreement with literature values for UU Aqr (see Sect. 5.3). Furthermore, our use of literature values for IP Peg and VZ Scl does not indicate any disagreement.

Instead, we can explain our large discrepancy between our PPEM distance estimate and literature values for HT Cas by allowing the disc to be *patchy*, i.e. the emitting surface is smaller then the geometrical surface. Using Marsh's [24] distance estimate with a recalibrated Barnes-Evans relation (for details see [46]), the covering factor must be of the order $C = (133/207)^2 = 41\%$.

The patchy disc can be caused by magnetic activity in localized regions on the surface or in the upper layers of the accretion disc. Magnetic flux created by dynamo action and/or the Balbous-Hawley instabilities driving the viscosity rises out of the cool midplane regions and dissipates most of the viscously generated energy via magnetic reconnection or similar coronal processes. This model also explains why the energy dissipation rate is proportional to the local orbital frequency [19].

Our PPEM analysis of HT Cas gives an indirect evidence of the hydromagnetic nature of the anomalous viscosity in accretion disks.

7   Summary

We have shown that Physical Parameter Eclipse Mapping is a powerful tool that helps us to learn about the physics of accretion discs in cataclysmic variables. This review can only highlight some interesting results. But, whether we derive "just" an independent distance estimate or get insight into the disc structure, there is always something to learn from the application of this method.

Acknowledgments

This work was funded by the South African NRF and CHL Foundation. I wish to thank Raymundo Baptista and Keith Horne for communicating me their data to apply my PPEM method. Many thanks go as well to Keith, Rick Hessman, Stephen Potter and Brian Warner for fruitful discussions. Futhermore, I am very grateful to Hilde Langenaken for babysitting my daughter Nina in Brussels and thank Nina for giving me so much joy and being so cooperative during the workshop.

References

1. Baptista R., Steiner J.E., 1993, A&A 277, 331
2. Baptista R., Steiner J.E., Cieslinski, D., 1994, ApJ, 433, 332
3. Baptista R., Steiner J.E., Horne K., 1996, MNRAS, 282, 99

4. Baptista R., Catalán M.S., Horne K., Zilli D., 1998, MNRAS, 300, 233
5. Bobinger A., Horne K., Mantel K.-H., Wolf S., 1997, A&A 327, 1023
6. Bond H., Wagner R.L., 1977, IAU Circ. 304
7. Buckley D.A.H., Tuohy I.R., 1989, ApJ, 344, 376
8. Cannizzo, J.K., Wheeler J.C., 1984, ApJS, 55, 367
9. Cannizzo J.K., Gosh P., Wheeler J.C., 1982, ApJ, 260, L83
10. Cropper M.S., 1986, MNRAS, 222, 225
11. Faulkner J., Lin D.N.C., Papaloizou J., 1983, MNRAS, 205, 359
12. Frank J., King A., Raine D., 1992, In: *Accretion Power in Astrophysics*, Cambridge Astrophysics Series, Cambridge University Press, p. 81
13. Gänsicke, B.T., Koester, D. 1999, A&A, 346, 151
14. Hameury J.-M., Lasota J.-P., Dubus G., 1999, MNRAS, 303, 39
15. Hessman F.V., Robinson E.L., Nather R.E., Zhang E.-H., 1984, ApJ, 286, 747
16. Hoard D.W., Still M.D., Szkody P., Smith R.C., Buckley D.A.H., 1998, MNRAS, 294, 689
17. Holcomb S., Caillault J.-P., Patterson J., 1994, A&AS, 185, 2117
18. Horne K., 1985, MNRAS, 213, 129
19. Horne K., Saar S.H., 1991, ApJ, 374, L55
20. Horne K., Wood J.H., Stienning R.F., 1991, ApJ, 378, 271
21. Hubeny I., 1991, In: IAU Colloq. 129, "Structure and Emission Properties of Accetion Disks", eds. C. Bertout, S. Collin, J.-P. Lasota, J. Tran Thanh Van (Singapure: Fong & Sons), p. 227
22. Kirkpatrick, J.D., McCarthy, D.W. 1994, AJ, 107, 333
23. Ludwig K., Meyer-Hofmeister E., Ritter H., 1994, A&A 290, 473
24. Marsh T.R., 1990, ApJ, 357, 621
25. Meyer F., Meyer-Hofmeister E., 1982, A&A, 106, 34
26. O'Donoghue, D., Fairall, A.P., Warner, B. 1987, MNRAS, 225, 43
27. Osaki Y., 1996, PASP, 108, 39
28. Patterson J., 1981, ApJS 45, 517
29. Rutten R.G.M., van Paradijs J., Tinbergen J., 1992 A&A, 260, 213
30. Sanduleak N. 1972, IBVS, No. 663
31. Shakura N.I., Sunyaev R.A., 1973, A&A, 24, 337
32. Sherington M.R., Bailey J., Jameson R.F., 1984, MNRAS, 206, 859
33. Sion E.M., 1999, PASP, 111, 532
34. Skilling J., Bryan R.K., 1984, MNRAS 211, 111
35. Steeghs D., 2000,private communication
36. Steeghs D., Harlaftis E.T., Horne K., 1997, MNRAS, 290L, 28
37. Smak J., 1982, AcA 32, 199
38. Smak J., 1984, PASP, 96, 5
39. Szkody P., 1998, AAS, 192, 6302
40. van Teeseling A., Beuermann K., Verbunt F., 1996, A&A, 315, 467
41. Vrielmann S., 1997, Ph.D. thesis, University of Göttingen
42. Vrielmann S., 1999, In: "Disk Instabilities in Close Binary Systems", Kyoto, October 27-30, 1998, ed. S. Mineshige and J.C. Wheeler, Universal Academy Press, p. 115
43. Vrielmann S., 2000, in preparation
44. Vrielmann S., 2000, submitted to MNRAS
45. Vrielmann S., Horne K., Hessman F.V. 1999, MNRAS, 306, 766
46. Vrielmann S., Hessman F.V., Horne K., 2000, submitted to MNRAS
47. Warner B., 1995, In: "Cataclysmic variable stars", Cambridge Astrophysics Series, Cambridge University Press, p. 207

48. Warner B., 1996, Ap&SS, 241, 263
49. Warner B., Cropper M. 1983, MNRAS, 203, 909
50. Warner B., O'Donoghue D., 1987, MNRAS, 224, 733
51. Warner B., Thackeray A.D., 1975, MNRAS, 172, 433
52. Watts D.J., Bailey J., Hill P.W., Greenhill J.G., McCowage C., Carty T., 1986, A&A, 154, 197
53. Wood J.H., Naylor T., Hassall B.J.M., Ramseyer T.F., 1995, MNRAS, 273, 772

Applications of Indirect Imaging Techniques in X-ray Binaries

E.T. Harlaftis

Institute of Astronomy and Astrophysics, National Observatory of Athens,
P.O.Box 20048, Thession, Athens – 11810, Greece

Abstract. A review is given on aspects of indirect imaging techniques in X-ray binaries which are used as diagnostics tools for probing the X-ray dominated accretion disc physics. These techniques utilize observed properties such as the emission line profile variability, the time delays between simultaneous optical/X-ray light curves, the light curves of eclipsing systems and the pulsed emission from the compact object in order to reconstruct the accretion disc's line emissivity (Doppler tomography), the irradiated disc and heated secondary (echo mapping), the outer disc structure (modified eclipse mapping) and the accreting regions onto the compact object, respectively.

1 Introduction

Low-mass X-ray binaries (LMXBs) involve mass transfer from a main-sequence companion star and accretion onto a neutron star or a black hole via a disc. Their optical and UV emission is dominated by reprocessing of X-rays, mainly in the disc but also on the companion star. The fundamental difference from cataclysmic variable (CV) discs is the amount of X-rays produced close to the accreting region and their irradiating effect on the binary components. The X-ray illumination emanating from the vicinity of the compact object heats up the surrounding disc and the surface of the nearby companion star. The X-ray heating is so intense that it controls the radial and vertical structure of the LMXB discs, thus overtaking viscous heating that controls accretion in CV discs. The study of LMXBs has been difficult since they are faint objects in the optical, and the binary parameters are not well established because there are very few eclipsing systems and irradiation is blurring out any variabilities. The reader who is interested in more details in X-ray binaries can refer to [33] or [59]. Here, we will review a few examples of image reconstruction where an observed property is used in order to gain insight into the state of the companion star, the accretion disc and the compact object of the X-ray binary.

2 The Heated Companion Star

The companion star in X-ray binaries is heated by the hard X-ray radiation, which penetrates the photosphere, and is re-emitted as blackbody flux [15]. As a consequence the light re-emitted by the companion star depends on the X-ray illumination pattern. The occultation by the irradiated disc will become apparent as a shadow on the companion star. The hard X-ray emission that does not encounter the accretion disc and hits the star directly will heat up its photosphere and cause continuum emission and absorption line production. This orbital heating effect is pronounced in systems like Her X-1 (where the companion star is varying between spectral types ∼O9 at maximum to A7/F0 at minimum [39]) and Cyg X-2 [3]) and can in certain circumstances totally dominate the star's evolution, as in e.g. the Black Widow pulsar PSR1957+20 [44,50].

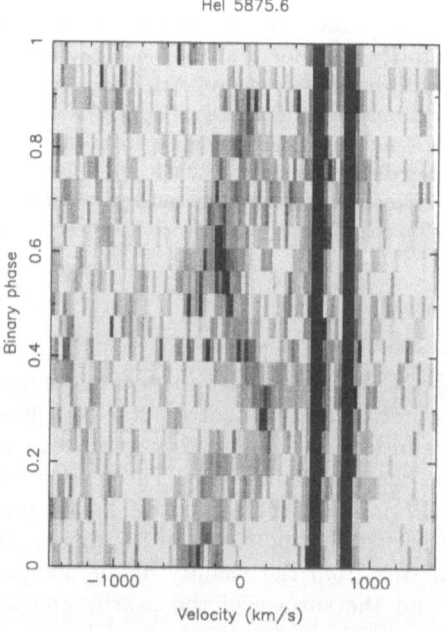

Fig. 1. The trailed spectra of HeI λ5876 from the binary X 1822-371. The line is in absorption and moves from red-to-blue at phase 0.5 which is the signature of the companion's star motion. There is also an unidentified component at phase 0.9. Reproduction from [20].

That short period LMXBs might exhibit such pronounced heating (of an otherwise cool star) is supported by the discovery of phase-dependent HeI absorption (λ5876) in the secondary of the 5.6 hour eclipsing LMXB X 1822-371 (see Fig. 1), which indicates that the inner face of the star appears to be hotter than its back face by 10000-15000 K [20]. The heating of the photosphere has never been treated correctly and any theoretical progress is expected through observational constraints [8]. Podsiadlowski [45] has shown that X-ray irradiation can drastically change the secondary's structure (expanding the atmosphere by a factor of 2-3), and thereby its evolution, provided that significant amounts of energy can be transported to the back side [16]. When applying this to outburst radial velocity data of the galactic *microquasar* Nova Sco 1994 Shahbaz et al. [53] found significant changes to the binary mass solutions. Orosz & Bailyn [6] used a conventional sinusoidal analysis and derived a black hole mass of 7.01 ± 0.22 M$_\odot$ from a K-velocity of 228 km s^{-1}, whereas Shahbaz et al. [53] obtained a much better irradiated fit to the radial velocity curve of K=215 km s^{-1} and a mass of 5.4 ± 1.3 M$_\odot$. The asymmetric distribution of the absorption line strength around the inner face of the companion star will indicate the X-ray illumination pattern through the disc and Roche tomography may be used to map the enhanced absorption line region onto the companion star Roche-lobe surface (Dhillon and Watson, this Volume).

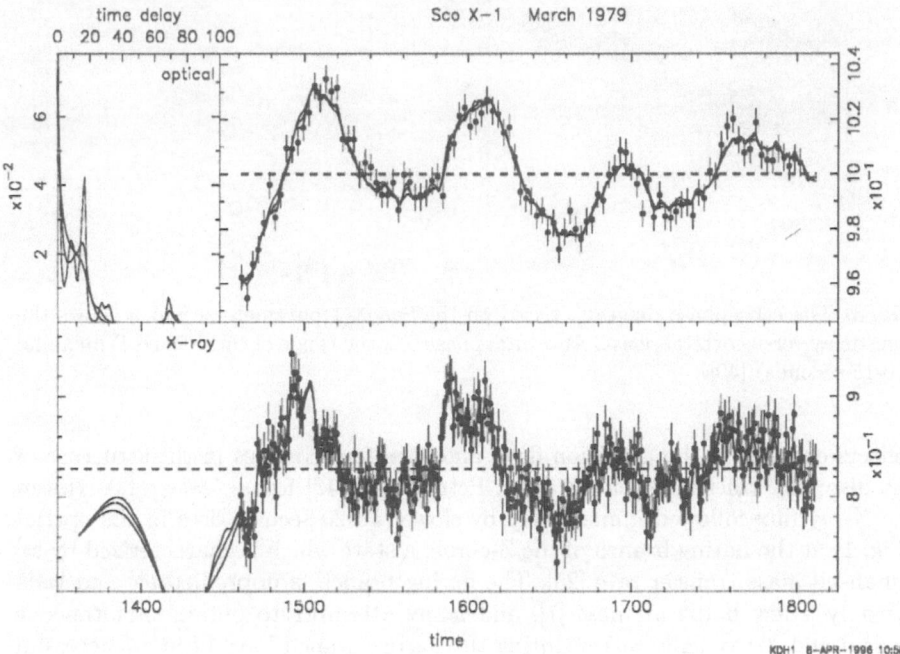

Fig. 2. Simultaneous optical and X-ray light curves of Sco X-1 [43]. A re-analysis of the data, using a maximum-entropy technique (echo mapping) gives a transfer function (top-left panel) which shows a time delay of ~13 seconds, consistent with reprocessing off the inner face of the donor star [40].

A new technique which has started producing results, is "Echo" mapping. This method utilizes the time delays between optical and X-ray photons in order to map the irradiated regions in the binary system, assuming that the X-rays are produced at the centre of the disc (O'Brien, this Volume). Here, we present an application of the method on the brightest X-ray object, the 19-hour binary system, Sco X-1, which moves along a Z-shaped curve in the X-ray colour-colour diagrams on a ~1 day timescale [27]. It is believed that the above behaviour

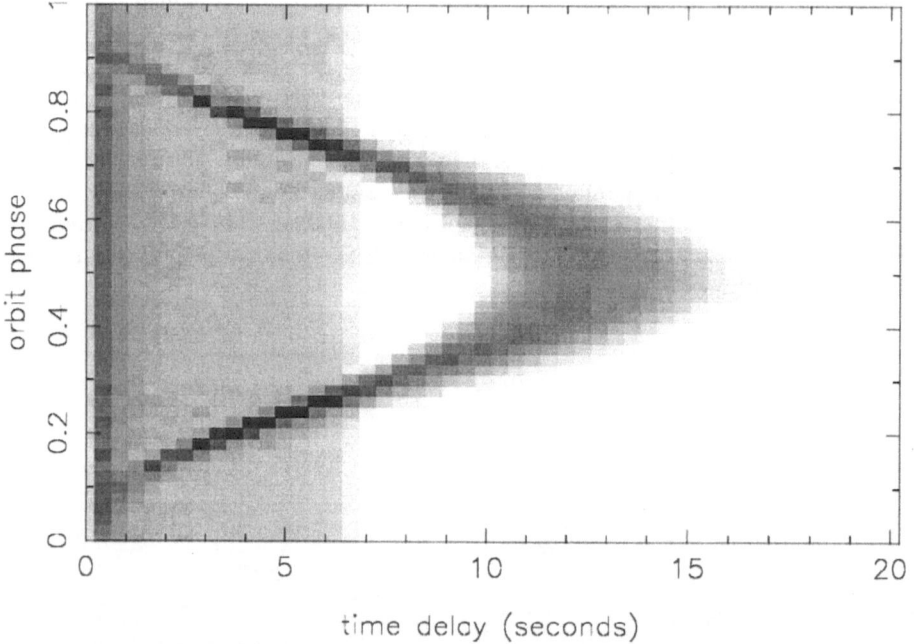

Fig. 3. The echo-phase diagram, based on the Sco X-1 parameters, which shows the time delay versus orbital phase. At orbital phase 0.5, the range of time delays is maximal (10-15 seconds) [40].

reflects changes in the accretion flow, and therefore changes in the structure of the disc (e.g. thickening of the disc). Petro et al. [43] found fast (<1s) rises in the X-ray flux followed immediately by slower 10-20 second rises in the optical (Fig. 1) at the flaring branch of the Z-curve, a state which is characterized by an enhanced mass transfer rate [25]. The flaring branch is unpredictable and lasts for only a few hours at most [7], and many attempts to obtain simultaneous optical and X-ray light curves during the flaring branch have been unsuccessful so far.

Therefore, the only data of Sco X-1 where a correlation between optical and X-ray photons is seen during the flaring branch is the data from [43]. Fig. 2 shows the simultaneous X-ray and optical light curves of Sco X-1. The transfer function between the light curves (small upper panel in Fig. 2) shows a distribution of time-delays which peak at ~13 seconds which corresponds to the light-travel distance of the neutron star to the inner face of the donor star [40]. A model diagram showing the time-delay versus binary phase is presented in Fig. 3. The constant time-delay with phase is due to irradiation in the disc and the cut-off at 6.5 seconds signifies the outer edge of the disc. The sinusoidal-like time delays with phase are due to the irradiated donor star and is minimum at the back face but maximum at the inner face of the star. The spread of time-delays at binary phase 0.5 should eventually constrain the total area of the heated surface

Fig. 4. Model of the irradiated system of Sco X-1. The reprocessing regions irradiated by a point-like X-ray source are shown using the time-delay distribution extracted from the transfer function of the optical and X-ray light curves [40].

with regard to the Roche lobe. The model of the irradiated regions in the binary system of Sco X-1 is given in Fig. 4.

3   The Accretion Disc

3.1   Doppler Tomography

Doppler tomography has shown its great diagnostic value with the discovery of spiral shocks in accretion discs (Steeghs, this Volume). Application of the technique in X-ray binaries has been proved much more difficult, mainly due to the fact that X-ray binaries are optically fainter than cataclysmic variables.

For example, in persistent X-ray sources, such as X 1822-371, the Balmer emission profiles are not clearly double-peaked due to effects which are most likely related to the irradiated, extended discs observed edge-on. The trailed spectra look more like a blurred version of typical trailed spectra in dwarf novae (e.g. OY Car [18]). The projected outer rim of the disc is quite likely the source

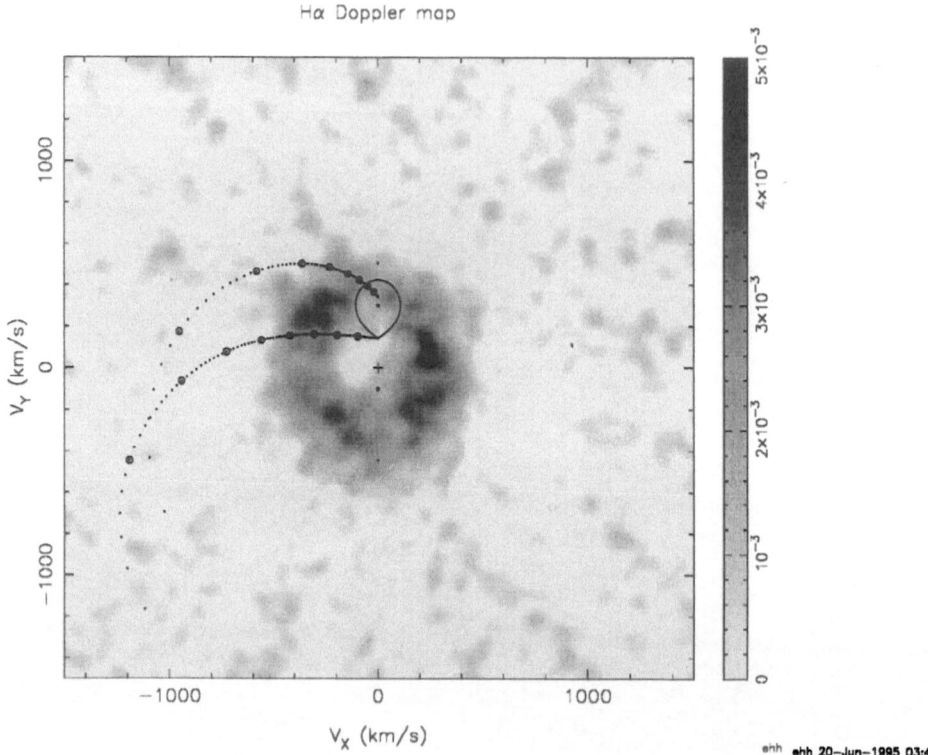

Hα Doppler map

Fig. 5. The Hα Doppler map of X1822-371 shows the emission-line distribution as the typical ring-like structure with possibly enhanced emissivity at phases 0.2 and 0.8. Reproduction from [20].

of the "blurred" trailed spectra. Thereafter, reconstruction of the emission line distribution reveals a blurred ring-like structure with no other clearly defined structure. Fig. 5 presents such a Doppler map of the X-ray persistent source X 1822-371, one of the *accretion disc corona* (ADC) sources in which X-rays from the compact object are not viewed directly, but are scattered into our line-of-sight by an extended corona above the disc (and which explains its unusually low L_X/L_{opt} ratio). Furthermore, the discovery of spiral shocks in the accretion disc of the cataclysmic variable IP Peg [54], which occurred at a time of a high mass transfer rate similar to that in LMXBs, raises the question of detecting them in X1822-371. Clearly, any hint of disc structure will be easier to reveal with observations of the high-ionisation line He II λ4686 which should provide maps with better clarity than Hα. However, He II Doppler maps of other neutron star X-ray binaries show complex line distributions, such as that of XTE J2123-058 (Hynes et al., this Volume; see also there a list of Doppler maps of neutron star LMXBs). The latter Doppler map shows a low-velocity emission at the back-side of the disc which is difficult to interpret. A magnetic propeller scenario is

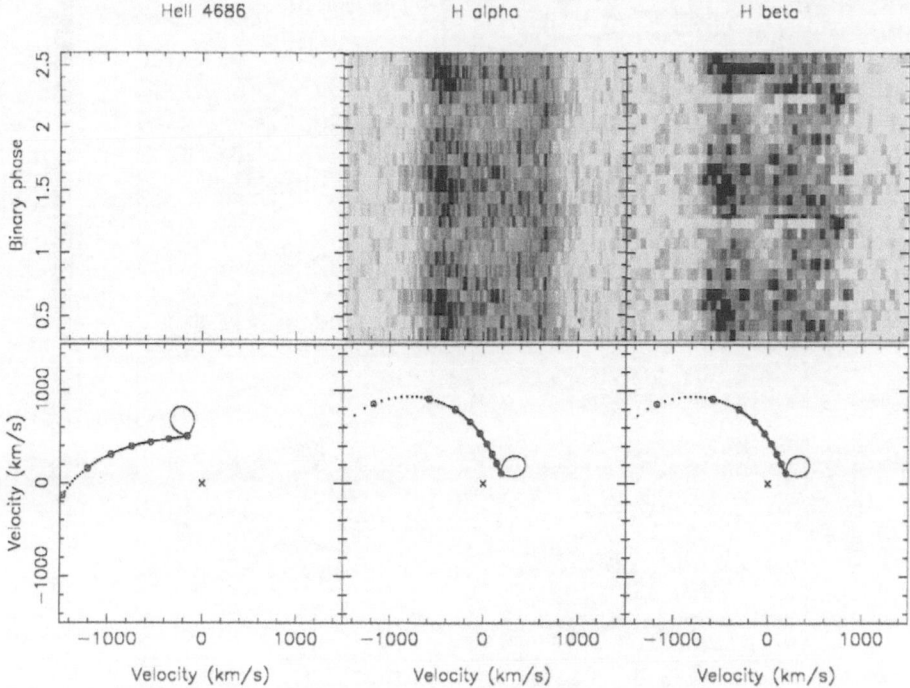

Fig. 6. Doppler images of He II, Hα and Hβ of the X-ray transient GRO J0422+32 during a mini-outburst. Two models were used to fit the He II bright-spot with most likely the one that shows the gas stream passing through the bright-spot (the latter one has a phase offset and arbitrary parameters of $K_c = 375$ km s^{-1} and $K_x = 75$ km s^{-1}; Harlaftis et al. [22] estimate $K_c = 372$ km s^{-1} and $K_x = 43$ km s^{-1}). Reproduction from [2].

favoured by Hynes et al. (this Volume) as the origin of the low-velocity emission. Alternatively, it may be that this emission is produced by the gas stream overflow crashing back on the disc, thus the gas velocities would be shocked from around 1200 km s^{-1} to 300 km s^{-1} [22].

In X-ray transient sources it has been difficult to derive Doppler maps, since they are very faint at quiescence, and when they are at outburst the binary period is not accurately known in order to tailor phase-resolved observations. The disc outshines the star and it is very difficult to derive a spectroscopic period, and thus a reliable ephemeris. This becomes apparent with the Doppler tomogram of GRO J0422+32 [2], presented in Fig. 6, where two solutions were possible at the time given the uncertainty in the definition of absolute phase zero (inferior conjunction of the companion star). However, the He II emission spot is most likely caused by the impact of the gas stream onto the disc in GRO J0422+32. When at quiescence, the object faintness prohibits any phase-resolved studies. However, the advent of a wealth of X-ray satellites in the 90's and the advent

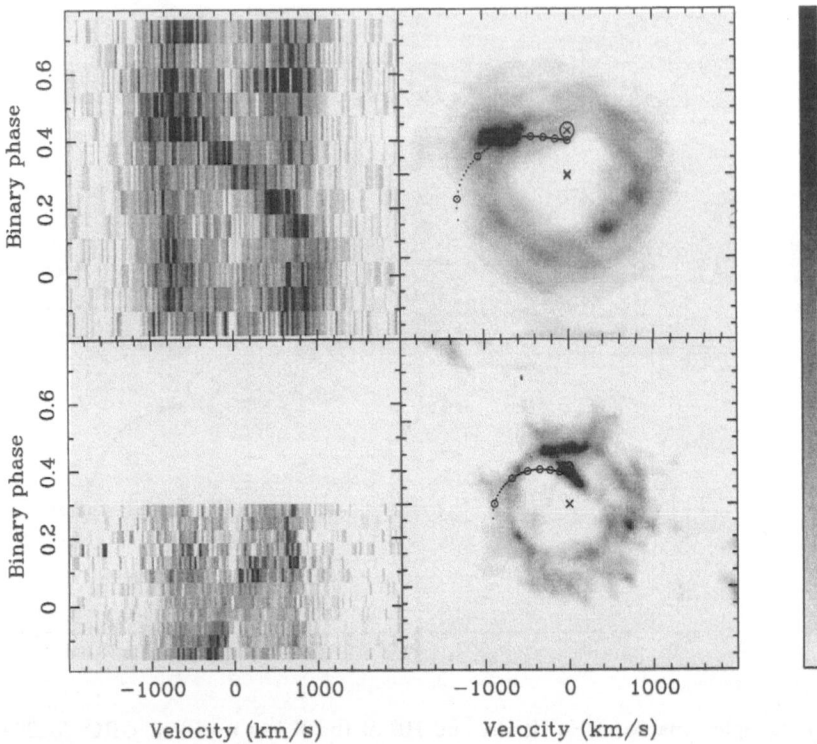

Fig. 7. The Hα Doppler image (*top-right panel*) of the accretion disc surrounding the black hole GS 2000+25 (*bottom-right panel* for Nova Oph 1977), as reconstructed from 13 Keck-I/LRIS spectra which are also presented (*top-left panel*; *bottom-left panel* for the 12 spectra of Nova Oph 1977). By projecting the image in a particular direction, one obtains the Hα emission-line profile as a function of velocity; for example, projecting toward the top results in the profile at orbital phase 0.0, which has a blueshifted peak. The path in velocity coordinates of gas streaming from the dwarf K5 secondary star is illustrated. The GS 2000+25 Doppler map shows a bright spot, at the upper left quadrant, which results from collision of the gas stream with the accretion disc around the black hole. The Nova Oph 1977 map also shows a trace of an "S"-wave component which, however, is not resolved with clarity. The image was reconstructed by applying Doppler tomography, a maximum entropy technique, to the phase-resolved spectra, as described in [19,20].

of the new generation of telescopes has enabled the first Doppler tomograms of the accretion discs around black holes. The line emissivity of such discs follows a R^{-b} law with $b = 1.5 - 2.2$ [23]. The Doppler map of GS2000+25 in quiescence clearly shows that there is on-going mass transfer onto the disc from the presence of the bright spot along the ballistic trajectory of the gas stream in Fig. 7 [19]. However, there is no detectable emission in the X-rays or UV suggesting that the inner disc may be empty or frozen (Mukai, private communication). The same behaviour has also been observed in A0620-00, i.e. a bright spot in the outer disc but no activity from the inner disc [34,36], a behaviour consistent with advection dominated accretion flow models [9]. A Doppler map of Nova Oph 1977 in quiescence - with a hint of some secondary star emission (for further details see [21]) - is also shown in order to demonstrate the difficulty in building Doppler maps of X-ray transients in quiescence even with 10m class telescopes.

3.2 The Vertical Structure of the Accretion Disc

Although many measurements of the radial disc structure exist, mainly from eclipse mapping studies of the temperature-radius relationship (e.g. [51]), very little is known about the vertical stratification of accretion discs. The vertical structure of the disc would require a temperature inversion to explain, for example, the emission lines from discs. Indeed, Hubeny [29] finds that the emergent spectrum depends on the vertical structure model and constraints on the value of the disc viscosity could be imposed from measurements of the optical thickness of the disc lines. In irradiated discs, the vertical height goes like $H/R \sim (R^{-1/8} - R^{-2/7})$ which results in a concave disc at large distances from the compact object rather than a dependence like $H/R \sim R^{-1/8}$ in a viscously-heated disc. Analysis of existing X-ray and UV light curves of X 1822-371 (Fig. 8) requires a H/R ratio for the disc that is rather large ($\sim$0.2; [46]). The question that arises is how one can utilize the vertically-extended accretion discs in order to derive constraints of the accretion disc properties. One (new) way is to observe simultaneous optical and X-ray light curves and analyse them using echo mapping. The analysis should reveal X-ray reprocessing regions in the disc, as in the case of GRO J1655-40 ([30] and O'Brien in this Volume). In principle, these regions should eventually constrain the vertical structure with azimuth. The other way is to infer the outer disc thickness from the X-ray shadow cast on the companion star. The 1.24 seconds pulsar Her X-1 irradiates the accretion disc around the neutron star and gives rise to a precessing and warped accretion disc. The shadow of the accretion disc onto the inner face of the companion star provides a diagnostic of the vertical structure of the disc (see Fig. 9; also see [24]).

Another way is to extend the traditional eclipse mapping technique (see Baptista, this Volume) in the steps of Rutten [52] in order to fit the full orbital light curve of a prototype vertically-extended disc. Indeed, Billington et al. [1] explained the UV dips seen in the light curves of OY Car during superoutburst as outer rim structure where UV light is reprocessed into optical, using the model shown in Fig. 10. This modified eclipse mapping technique fitted the light curves by varying both the disc flux distribution and the outer rim structure. The reconstructed rim structure dependence with binary phase is presented in the following figure (Fig. 11). The rim arises at the outer disc, and in particular from radii larger than 0.55 R_{L_1}. This is reflected in Fig. 12 where the rim at 0.5 R_{L_1} has a different structure than the other outer rims. This is because the rim structure is too far inside the disc and this forces the re-distribution of the flux to an artificially asymmetric flux map. A map which is not consistent with the observed axisymmetric discs in the UV. The variation of the mean rim height

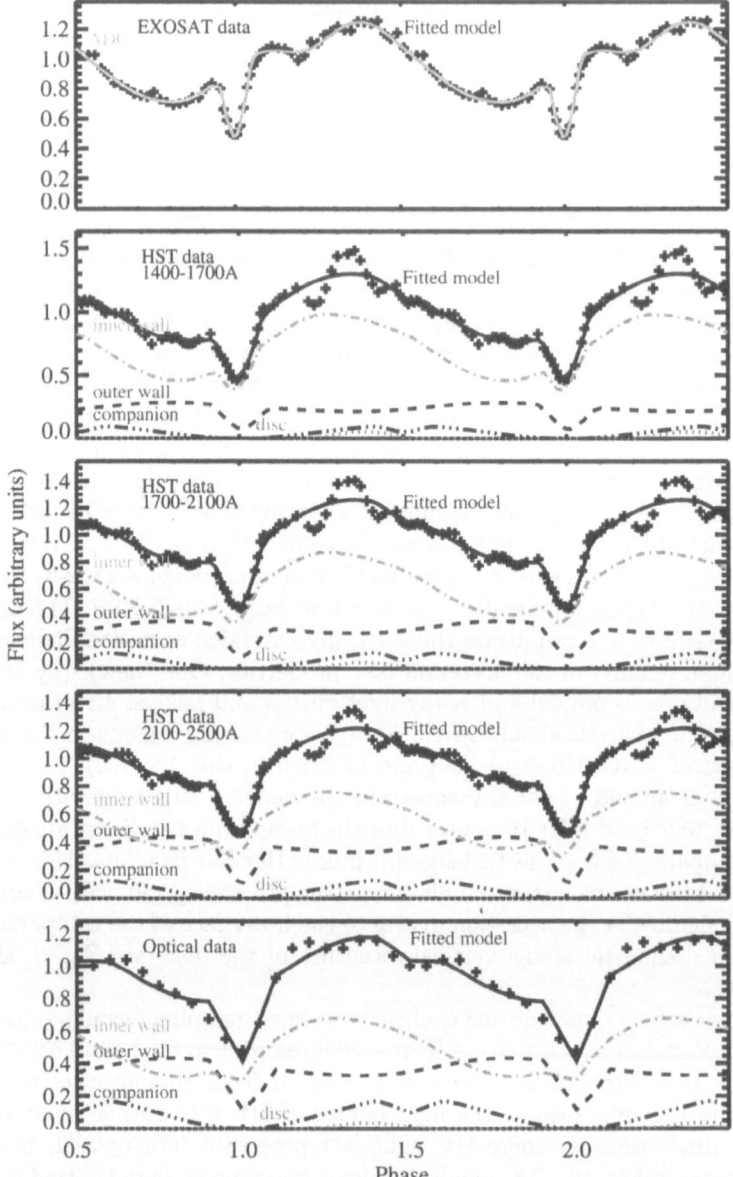

Fig. 8. The "dipping" X-ray source X 1822-371 is the prototype X-ray binary for large vertical structure, as implicated from the X-ray light curve (X-ray "dips" at binary phase 0.2 and 0.8 and eclipse). Simultaneous fits of a disc model to X-ray, UV and optical light curves. The UV-HST data have been separated into three bands.The contribution of each model component (companion star, disc, outer and inner disc wall) to each curve is shown separately [46].

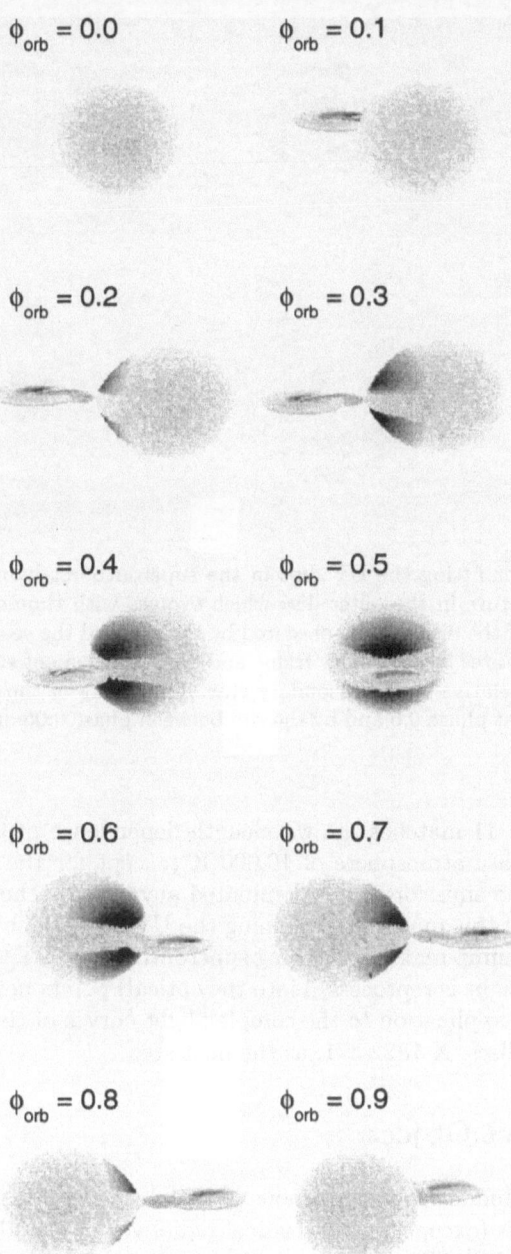

Fig. 9. Sky projections of the emitting surfaces of Her X-1 over an orbital cycle. Due to the large twist gradient in the accretion disc, the disc shadow does not display great variability over the orbital cycle [48,56].

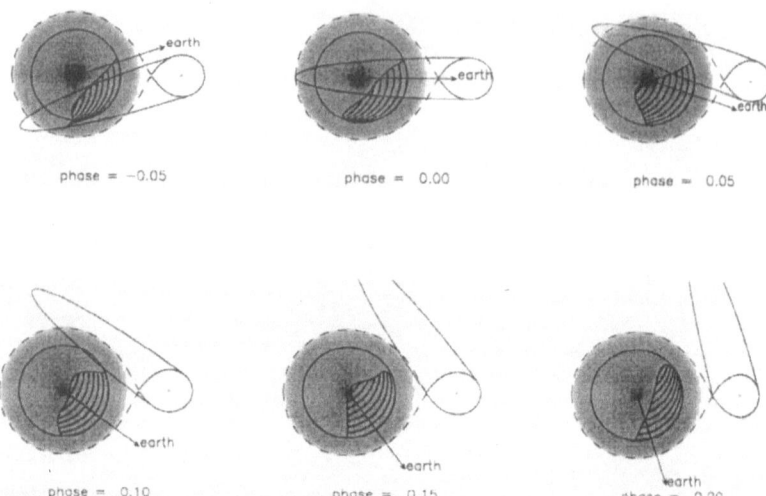

Fig. 10. The model for fitting the UV dips in the superoutburst light curves of OY Car is triggering structure in the outer disc which evolves with time on a dynamical timescale. The areas of the disc surface obscured by the rim and the secondary star are illustrated. The rim rotates in the binary frame and each rim element starts to flare up at the same position relative to the secondary star. The centre of the disc is eclipsed by the secondary star at phase 0.0 and by the rim between phase 0.05 and 0.15 causing the UV dip [1].

with azimuth in Fig. 11 matches the wavelength dependence of the absorption coefficient of a hot disc atmosphere of 10,000 K (except for the shorter wavelengths where the line emission is not dominated anymore by the disc but by a wind). The success of this model in explaining the UV dips which appear simultaneous to the superhump maximum during superoutburst of OY Car (outer disc structure where UV light is reprocessed into the optical) points now to a revision of the technique and application to the complex light curves of the prototype of vertically-extended discs, X 1822-371, as the next step.

4   The Compact Object

The compact object hidden in the luminous X-ray binary can be inferred by using various techniques (except for the classical radial velocity study of the wings of the emission lines [14]). The most interesting are based on the observed coherent pulses which must arise either from the surface of the compact object or from the coupling region between the accretion disc and the magnetosphere. Indeed, the UV continuum double pulse profile observed at the 33 seconds spinning period of the compact object in the cataclysmic variable AE Aqr was successfully modeled as the accreting spots on the surface of the compact object (Fig. 13 [10]; see also the work by Cropper and Horne [6] who mapped the accreting regions

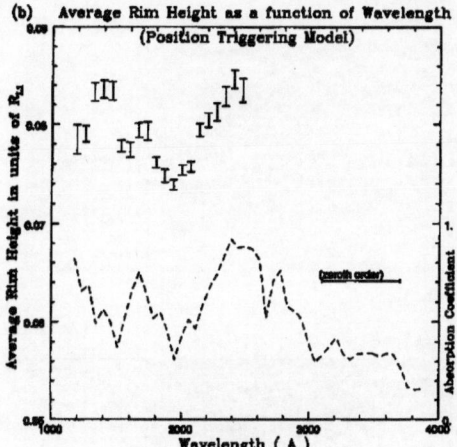

Fig. 11. The average rim height as a function of wavelength. The lower dashed line and the right-hand scale show the theoretical opacity of an accretion disc atmosphere at 10,000 K for the same wavelength band.

on the magnetic white dwarf of ST LMi). This is also the system where the magnetic propeller model has found substantial support from emission line observations (see Wynn, this Volume). According to the magnetic propeller model, the compact object rotates so fast that the gas cannot accrete on it but rather is propelled away. This concept, first proposed for neutron star X-ray binaries in the 70's [32], has recently returned as a potential model for neutron star X-ray

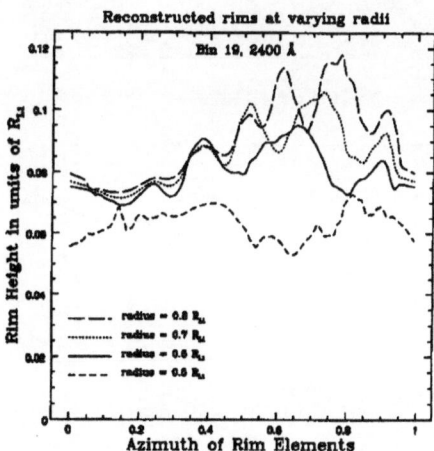

Fig. 12. The reconstructed rims at radii 0.6 R_{L_1}, 0.7 R_{L_1} and 0.8 R_{L_1}. The rim at 0.5 R_{L_1} is not well reconstructed as is evidenced by the different shape of the rim at this radius. The discrepancy on the left side of the curves is due to non-disc line emission.

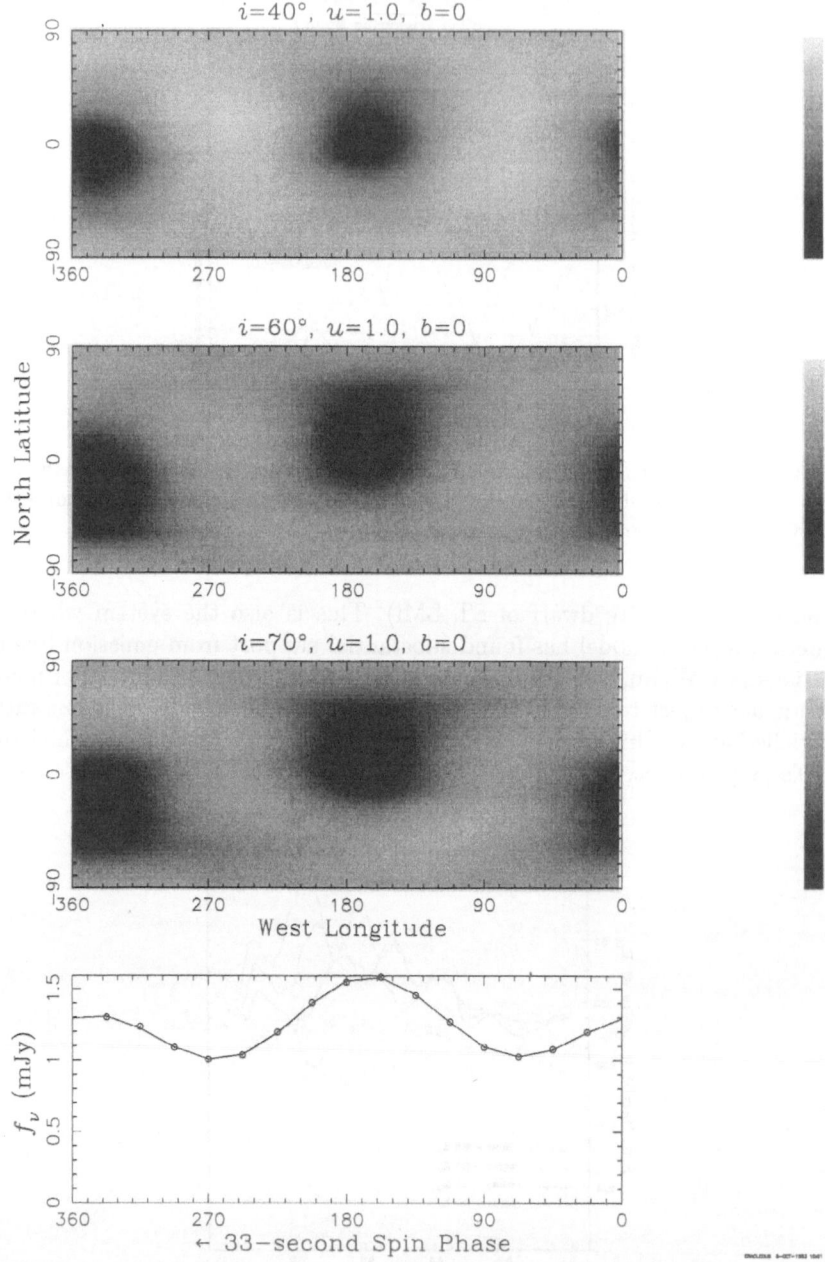

Fig. 13. Maximum entropy maps of the white dwarf surface brightness distribution reconstructed from the observed UV pulse profile showing the emission spots on the surface of the rapidly rotating white dwarf at an inclination of 60°, limb darkening coefficient of one and background light of zero [10].

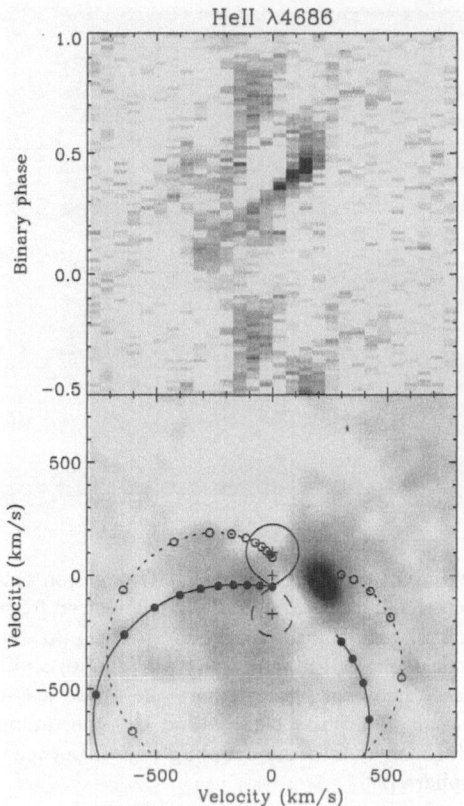

Fig. 14. The He II Doppler map of Her X-1 showing a spot of emission close to the neutron star at velocities associated to the gas stream. The parameters used were q=1.67 (mass of donor/accretor) and M=1.3 for the neutron star [47].

binaries (Hynes et al., this Volume), after it found sound support in the AE Aqr case.

Perhaps, the magnetic propeller model is the likely interpretation of the He II Doppler map of Her X-1 where a low-velocity emission spot close to the neutron star is revealed (Fig. 14). Alternatively, the 1.24 second searchlight beam from the neutron star illuminates the truncated inner disc. Gas from the inner edge of the disc is then funneled along the magnetic field lines onto the poles of the neutron star (for a review see [42] and references therein). In this case, there is current consensus that the disc feeds gas to the compact object through an 'accretion curtain' model [13,49] which produces a dipole pattern in both emission and absorption as the searchlight beam from each pole passes through the accretion curtain to the line of sight [22]. For example, the double-pulse He II emission profile coming from the 545 seconds spinning compact object in the intermediate polar RX J0558+53 is mapped as such a dipole pattern in both emission and absorption centred on the white dwarf (Fig. 15).

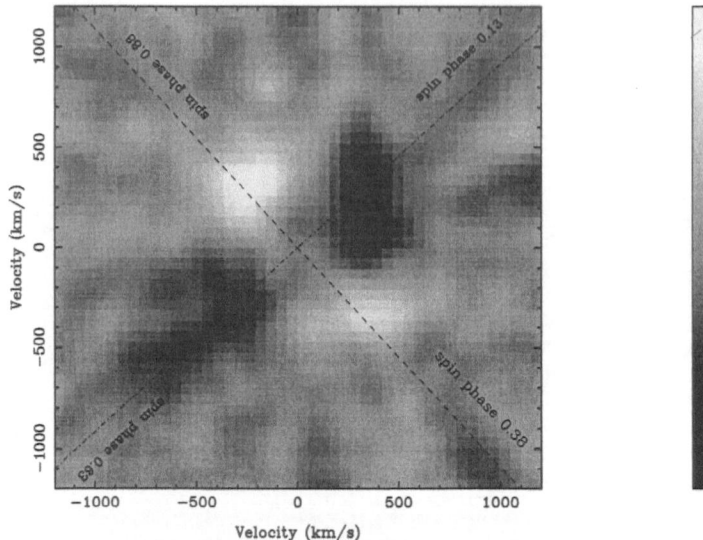

Fig. 15. The Doppler map of the double-pulse He II emission profile with spin period (545 seconds) of the intermediate polar RX J0558, as viewed from the compact object (0,0). Back projection of the He II pulsed emission profiles produces a quadrapole-like velocity distribution, consisting of the minima ('dark' shade) and the maxima ('bright' shade) of the two pulses. The spin phases where the above are more pronounced are also marked. The emission line pulse lags behind the continuum pulse by 0.12 spin cycles giving a powerful insight into the coupling region between the Kepler-orbiting gas and the magnetosphere [22].

The illuminating effects of the compact object's beams on the surrounding supersonic gas can provide insight in the inner disc of X-ray binaries through periodogram analysis of the line profiles. For example, harmonics of the beat frequency between the ω spin frequency and the Ω orbital frequency as well as different combinations of these frequencies are then suggestive of specific illuminating patterns. For example, a simple, disc-fed emission has most power in the 2ω frequency. Such a periodogram analysis of power spectra of line profiles against frequency and velocity is shown in Fig. 16 where prevalence of the $2(\omega - \Omega)$ and 2ω frequencies indicate both disc- and stream-fed emission from two diametrically-opposed poles with similar emission properties a truncated disc is implied in RX J0558+53 [22]. This analysis provides a powerful probing tool in the coupling region between the supersonic gas in the inner disc and the magnetosphere of the compact object and perhaps this will be undertaken soon for the He II line of Her X-1 .

5 Future Prospects

Indirect imaging techniques, utilizing optical spectroscopy, can now probe X-ray binary physics with sufficient signal to noise ratio and start distinguishing

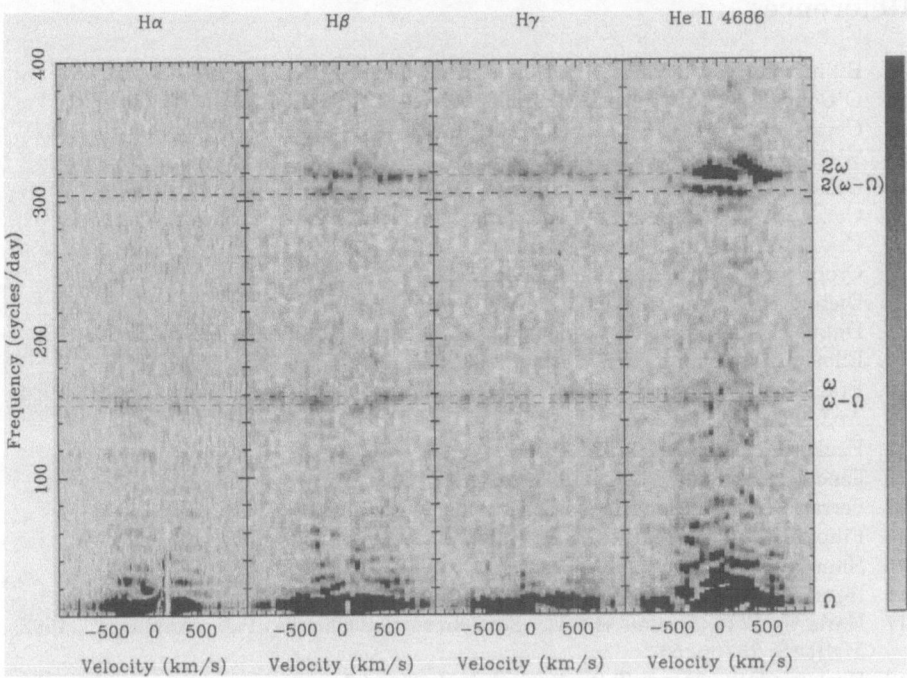

Fig. 16. The Fourier periodogram per velocity bin in the continuum and the emission lines. See text.The spin frequency is only evident in Hβ whereas the first harmonic in dominant in all power spectra except that of Hα. An orbital side-band at 2 $(\omega - \Omega)$ is also clearly present.

accretion details comparable to those observed in the brighter cataclysmic variables. Moreover, the quality of X-ray observations is such now that applications in this domain wiil be the next step forward. The behaviour of the hard X-rays with respect to the soft X-rays (time lags and spectra) can probe the size of the Compton scattering region and infer radial density profiles [28]. Image reconstruction of the hot electron plasma may result in defining better properties of the hot accretion disc and clarify its relation to advection dominated accretion flows. The spectral resolution of iron profiles has considerably increased with the advent of the ASCA satellite and has revealed in many interactive binary systems three peaks in the iron profile, namely thermal emission at 6.7 and 7.0 KeV, and fluorescent emission at 6.4 KeV [11]. Doppler tomography and echo tomography using X-ray iron line profiles and continuum, and even eclipse maps of the accretion disc from X-ray light curves [37], may become possible with the new X-ray satellites.

References

1. Billington I., Marsh T.R., Horne K., Cheng F.H., Thomas G., Bruch A., O'Donoghue D., Eracleous M., 1996, MNRAS, 279, 1274
2. Casares J., Charles P., Marsh T.R., Martin A.C., Harlaftis E.T., Pavlenko E.P., Wagner R.M., 1995, MNRAS, 274, 565
3. Casares J., Charles P.A., Kuulkers E. 1997, ApJ, 493, L39
4. Cowley A.P., Crampton D., Hutchings J.B. 1982 ApJ, 255, 596
5. Crampton D., Hutchings J.B., 1974, ApJ, 191, 483
6. Cropper M., Horne K., 1994, MNRAS, 267, 481
7. Dieters S.W., van der Klis M., 2000, MNRAS, 311, 201
8. Dubus G., Lasota J.-P., Hameury J.-M., Charles P.A., 1999, MNRAS, 303, 139
9. Eliot Q., Narayan R., 1999, ApJ, 520, 298
10. Eracleous M., Horne K., Robinson E.L., Zhang E., Marsh T.R., Wood J.H., 1994, ApJ, 433, 313
11. Ezuka H., Ishida M., 1999, ApJS, 120, 277
12. Fender R.P., Pooley G.G., 1998, MNRAS, 300, 573
13. Ferrario L., Wickramasinghe D.T., King A.R., 1993, MNRAS, 260, 149
14. Filippenko A. V., Matheson T., Barth A. J., 1995, ApJ, 455, L139
15. Hameury J.M., King A.R., Lasota J.P., 1986, A&A, 162, 79
16. Hameury J.-M. et al., 1993 A&A, 277, 81
17. Harlaftis E.T., Hassall B.J.M., Sonneborn G., Charles, P.A., Naylor T., 1992, MNRAS, 257, p. 607
18. Harlaftis E.T., Marsh T.R., 1996, A & A, 308, 97
19. Harlaftis E.T., Horne K., Filippenko A. 1996, PASP, 108, 762
20. Harlaftis E.T., Charles P.A., Horne K. 1997, MNRAS, 285, 673
21. Harlaftis E.T., Steeghs D., Horne K., Filippenko A.V., AJ, 1997, Vol. 114, No. 3, p. 1170-1175
22. Harlaftis E.T., Horne K., 1999, MNRAS, 305, 437
23. Harlaftis E.T., Collier S.J., Horne K., Filippenko A.V., 1999, A&A, 341, 491
24. Harlaftis E.T., 1999, A & A, 346, L73
25. Hasinger, G, van der Klis M., Ebisawa K., Donati T., Mitsuda K., 1990, A&A, 235, 131
26. Hellier C., Mason K.O., 1990, in *Accretion-Powered Compact Binaries*, ed. C.Mauche, Cambridge University Press, p. 185.
27. Hertz P., Vaughan B., Wood K. S., Norris J. P., Mitsuda K., Michelson P. F., Dotani T., 1992, ApJ, 396, 201
28. Hua X., Kazanas D., 1999, ApJ, 512, 793
29. Hubeny I., 1994, in *Interacting Binary stars*, A. W. Shafter (ed.), ASP Conference series, Vol. 56, ed. A. W. Shafter, p.3
30. Hynes R.I., Haswell C.A., Shrader C.R., Chen W., Horne K., Harlaftis E.T., O'Brien K., Hellier C., Fender R.P., 1998, MNRAS, 300, 64
31. Hynes R.I., O'Brien K., Horne K., Chen W., Haswell C.A. 1998, MNRAS, 299, 37
32. Illarionov A.F., Sunyaev R.A., 1975, A&A, 39, 185
33. Lewin W.H.G., van Paradijs J., van den Heuvel E.P.J. (eds.), 1995, *X-ray Binaries*, Cambridge Astrophysics Series 26
34. Marsh T.R., Robinson E.L., Wood J.H., 1994, MNRAS, 266, 137
35. Mason K.O., et al., 1982 MNRAS, 200, 793
36. McClintock J.E., Horne K., Remillard R.A., 1995, ApJ, 442, 358
37. Mukai K., Wood J.H., Naylor T., Schlegel E.M., Swank J.H., 1997, ApJ, 475, 812

38. Norton A.J., 1996, MNRAS, 280, 937
39. Oke J.B., 1976, ApJ, 209, 547
40. O'Brien K., 2000, Ph.D. Thesis, University of St. Andrews
41. Orosz J.A., Bailyn C.D. 1997 ApJ, 477, 876
42. Patterson J., 1994, PASP, 106, 209
43. Petro L.D., Bradt H.V., Kelley R.L., Horne K., Gomer R., 1981, ApJ, 251, L7-L11
44. Phinney E.S. et al., 1988 Nat, 333, 832
45. Podsiadlowski P., 1991 Nat., 350, 136
46. Puchnarewicz E.M., Mason K.O., Cordova F.A., 1995 Adv Sp Res., 16, 3, 65
47. Quaintrell H., Still M.D., Roche P.D., 2000, MNRAS, submitted.
48. Quaintrell H., 1998, Ph. D. thesis, University of Sussex
49. Rosen S.R., Mason K.O., 1988, 231, 549
50. Ruderman M., Shaham J., Tavani M., 1989, ApJ, 343, 292
51. Rutten R.J.M., Dhillon V. S., Horne K., Kuulkers E., 1993, Nature, 362, 518
52. Rutten R.J.M., 1998, A&AS, 127, 581
53. Shahbaz T., Groot P., Phillips S.N., Casares J.C., Charles P.A., van Paradijs J., 2000, MNRAS, 314, 747
54. Steeghs D., Harlaftis E.T., Horne K., MNRAS, 1997, 290, L28
55. Stehle, 1999, MNRAS, 304, 687
56. Still M.D., Quaintrell H., Roche P.D., Reynolds A.P., 1997, MNRAS, 292, 52
57. Wood J.H., Marsh T.R., Robinson E.L., Stiening R.F., Horne K., Stover R.J., Schoembs R., Allen S.L., Bond H.E., Jones D.H.P., Grauer A.D., Ciardullo R., 1989, MNRAS, 239, 809
58. Wynn G.A., King A.R., 1992, MNRAS, 255, 83
59. Wheeler J.G., 1993, *Accretion Discs in Compact Stellar Systems*, Advanced Series in Astrophysics and Cosmology, Vol. 9, World Scientific

Doppler Tomography of XTE J2123-058 and Other Neutron Star LMXBs

R.I. Hynes[1], P.A. Charles[1], C.A. Haswell[2], J. Casares[3], and C. Zurita[3]

[1] Department of Physics and Astronomy, University of Southampton, Southampton, SO17 1BJ, UK
[2] Department of Physics and Astronomy, The Open University, Walton Hall, Milton Keynes, MK7 6AA, UK
[3] Instituto de Astrofísica de Canarias, 38200 La Laguna, Tenerife, Spain

Abstract. We describe Doppler tomography obtained in the 1998 outburst of the neutron star low mass X-ray binary (LMXB) XTE J2123–058. This analysis, and other aspects of phase-resolved spectroscopy, indicate similarities to SW Sex systems, except that anomalous emission kinematics are seen in He II, whilst phase 0.5 absorption is confined to Hα. This separation of these effects may provide tighter constraints on models in the LMXB case than is possible for SW Sex systems. We will compare results for other LMXBs which appear to show similar kinematics and discuss how models for the SW Sex phenomenon can be adapted to these systems. Finally we will summarise the limited Doppler tomography performed on the class of neutron star LMXBs as a whole, and discuss whether any common patterns can yet be identified.

1 Introduction

Low mass X-ray binaries (LMXBs) contain a late-type 'normal' star of $\leq 1\,M_\odot$ accreting via Roche lobe overflow onto a black hole or neutron star. Black hole systems are mostly only active during transient outbursts, making them difficult targets for techniques requiring phase-resolved spectroscopy such as Doppler tomography. Many neutron star systems are persistently active, but with a few exceptions are sufficiently faint that they are still difficult targets. Amongst the bright exceptions, only Her X-1 has seen extensive application of Doppler tomography [13,15]. This is a very unusual system with a companion star more massive than the neutron star, and not strictly an LMXB at all. We will focus here on the more limited data available on 'typical' LMXBs, i.e. those in which the companion star is a late-type dwarf. We begin by summarising our own work on a transient neutron star system, XTE J2123–058 and then compare this with other LMXBs.

2 XTE J2123–058

The transient LMXB XTE J2123–058 was discovered by *RXTE* on 1998 June 27 [12] and promptly identified with a 17th magnitude blue star with an optical spectrum typical of transients in outburst [19]. Type-I X-ray bursts [16] indicated the compact object to be a neutron star. A pronounced photometric modulation

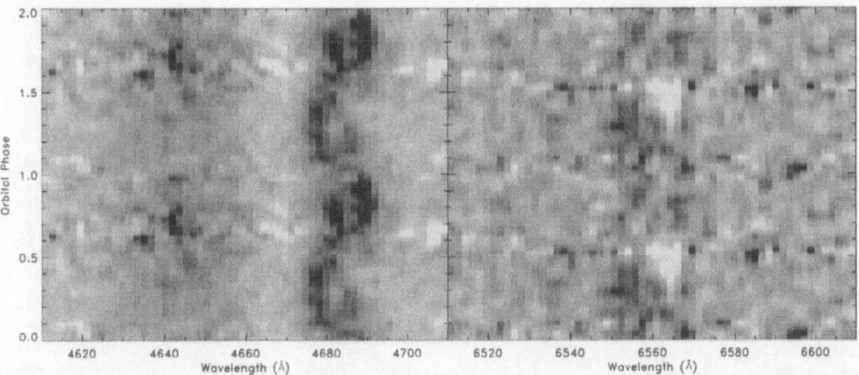

Fig. 1. Trailed spectrograms for XTE J2123–058. The left hand panel shows C III/N III and He II ; the right Hα. In the latter, the transient absorption can be seen as the white region.

[2] on the orbital period of 6.0 hr [9,11,20] is due to the changing aspect of the X-ray heated companion indicating a high inclination system. During the outburst we obtained extensive photometry [23] and phase-resolved spectroscopy [10]; this section will focus on the latter.

2.1 The Available Data

We observed XTE J2123–058 using the ISIS dual-beam spectrograph on the 4.2m William Herschel Telescope (WHT) on the nights of 1998 July 18–20. Full details and discussion of the data analysis are given in [10]. Phase resolved data suitable for Doppler tomography were obtained on July 19–20 (approximately one binary orbit each night) with an unvignetted wavelength range of ∼4000–6500 Å and spectral resolution 2.9–4.1 Å.

2.2 Doppler Tomography

The emission lines show significant changes over an orbital cycle (Fig. 1). He II 4686 Å shows complex changes in line position and structure, with two S-waves apparently interweaving. The light curve reveals a strong peak near phase 0.75 and a weaker one near 0.25.

We have used Doppler tomography (Marsh, this volume) to identify He II 4686 Å emission sites in velocity space (Fig. 2a). One of the fundamental assumptions of Doppler tomography, that we always see all of the line flux at *some* velocity, is clearly violated, however, as the integrated line flux is not constant. We attempt to account for this by normalising the line profiles. Without normalisation the structure of the derived tomogram is very similar, so our results do not appear to be sensitive to this difficulty. A better solution would be to use the modulation mapping method (Steeghs, this volume).

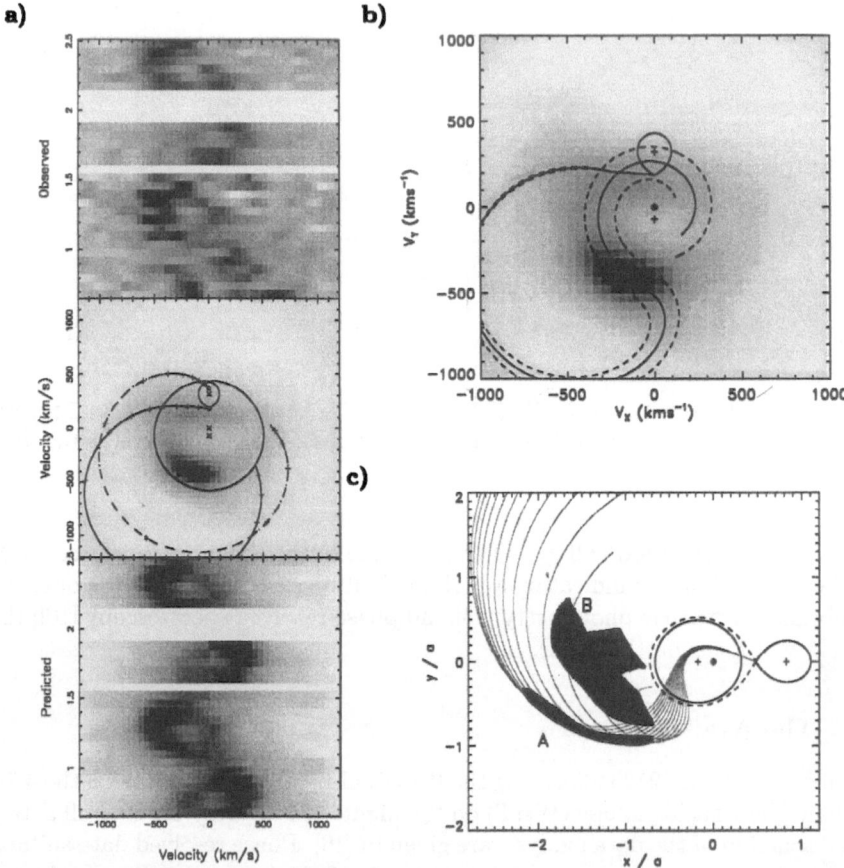

Fig. 2. Doppler tomography of XTE J2123–058. a) From top to bottom, the actual data, the tomogram and the reconstructed data. In the centre panel, the solid line is the ballistic stream trajectory, the dashed line the Keplerian velocity at the stream position and the large circle is the Keplerian velocity at the disc edge. b) Comparison between magnetic propeller trajectories and the tomogram. c) Spatial plot corresponding to (b). Points in region A produce emission with kinematics corresponding to the white hatched region in the tomogram. Points in region B will produce absorption with the phasing and velocities observed in Hα.

In order to interpret the tomogram quantitatively, we require estimates of system parameters. These were derived from results of fits to outburst light curves [23]. The parameters we assume are $P_{orb} = 0.24821$ day, $i = 73°$, $M_1 = 1.4\,M_\odot$, $M_2 = 0.3\,M_\odot$, $R_{disc} = 0.75\,R_{L_1}$; see [10] for discussion .

The dominant emission site (corresponding to the main S-wave) appears on the opposite side of the neutron star from the companion. For any system parameters, it is inconsistent with the heated face of the companion and the stream/disc impact point. There is a fainter spot near the L_1 point, corresponding to the fainter S-wave, which is possibly associated with the accretion stream. These spots in the tomogram may be joined, although this may be a smoothing artifact of the reconstruction. It is notable that nearly all the emission has a lower velocity than that of Keplerian material at the outer disc edge, and so if associated with the disc requires sub-Keplerian velocities.

2.3   A Neutron Star SW Sex System?

The phenomenon of emission concentrated in the lower-left hand quadrant of a Doppler tomogram at low velocities is not new; it is one of the distinguishing characteristics of the SW Sex class of cataclysmic variables [7,8,18]. These are all novalike variables, typically seen at a relatively high inclination.

The tomograms are not the only similarity. SW Sex systems exhibit transient absorption lines which are strongest near phase 0.5 or slightly earlier and are moderately blue shifted. Transient absorption near phase 0.5 is seen in $H\alpha$ in XTE J2123–058 (Fig. 1). It begins slightly redshifted and moves to the blue, and can be clearly seen in a trailed spectrogram. Examination of the average spectrum for phases 0.35–0.55 confirms that this is real absorption below the continuum level.

3   Other Candidate SW Sex-Like LMXBs

The only other short-period neutron star LMXB with published Doppler tomography is 2A 1822–371 ([6] and the contribution of E. Harlaftis in this volume). This star was observed in $H\alpha$ and exhibited disc emission enhanced towards the stream impact point. Similar behaviour is suggested by a radial velocity analysis of He II [3]. This is different to what we see in XTE J2123–058. Non-tomographic analysis of other short-period neutron star LMXBs in an active state, however, suggests a similar behaviour to that described for XTE J2123–058 as shown in Fig. 3 where we overplot the He II radial velocity information for other systems on our He II Doppler tomogram. 2A 1822–371 is clearly the odd one out, and the other systems (4U 2129+47 [17], EXO 0748–676 [4], 4U 1636–536 [1] and U 1735–444 [1]) all show emission centred in the lower or lower-left part of the tomogram.

4   SW Sex Models Applied to Neutron Star LMXBs

Various models have been proposed for the SW Sex phenomenon including bipolar winds, magnetic accretion and variations on a stream overflow theme. Current opinion favours the latter interpretation, with the stream either being accelerated out of the Roche lobe by a magnetic propeller, or re-impacting the disc. Both of these models are discussed in more detail below.

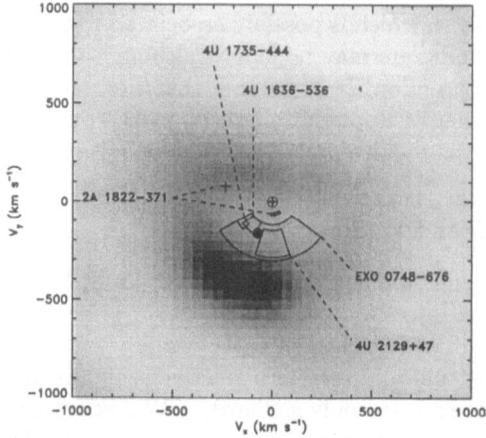

Fig. 3. Doppler tomography of XTE J2123–058 with radial velocity information on other systems overlaid. Boxes represent uncertainties in velocity semi-amplitude and phasing. The large point is the best fit semi-amplitude and phase for the XTE J2123–058 data.

4.1 The Magnetic Propeller Interpretation

It is possible to explain the behaviour of the He II 4686 Å line in terms of the magnetic propeller model. The essence of this model is that in the presence of a rapidly rotating magnetic field, accreting diamagnetic material may be accelerated tangentially, leading to it being ejected from the system. The field might be anchored on the compact object, as in the prototype propeller AE Aqr [5,22] or in the disc itself [8].

We adopt the parameterisation used by Wynn et al. [22] and construct a simple model of a propeller in XTE J2123–058. Full details are given in [10]. We can readily find a trajectory which passes through the central emission on the tomogram. This is shown in Fig. 2b, together with two bracketing trajectories corresponding to more and less acceleration. For a plausible model, we also require an explanation for why one particular place on the trajectory is bright. Such an explanation is offered by Horne [8]: there is a point outside the binary at which trajectories intersect. At this point, faster moving blobs cross the path of slower blobs and enhanced emission might be expected. This can be seen in Fig. 2c. To facilitate a quantitative comparison of our data with the model, the region A in Fig. 2c encloses points with velocities consistent with the bright emission spot. These points are indeed located where the trajectories cross. Our data are therefore consistent with the emission mechanism suggested by [8]. If the region in which accelerated blobs collide is optically thick in the line then emission is expected predominantly from the inner edge of this region where fast moving blobs impact slow moving blobs. For the geometry we have considered, as can be seen in Fig. 2c, this will result in emission predominantly towards the top of the figure. This corresponds to seeing maximum emission near phase 0.75, exactly as is observed in our data. Can we also explain the phase-dependent

Balmer absorption? The region B in Fig. 2c encloses points which would absorb disc emission with the phasing and velocities observed in Hα. This region can be attributed to lower-velocity trajectories which fall back towards the disc; thus the transient absorption can be accommodated in this model.

In summary, a magnetic propeller model can account for both He II emission kinematics and lightcurve and Hα transient absorption. This requires blobs to be ejected with a range of velocities; absorption is caused by low velocity blobs, perhaps because these account for most of the ejected mass. Emission is only seen from higher velocity blobs, which might be expected to exhibit more energetic collisions. The model has some difficulties, however, both for a field anchored on the compact object and for a disc-anchored field.

If the field is anchored on the compact object then it must rotate very fast, as spin periods of neutron stars are typically of order milliseconds. For such rapid rotation we must consider the light-cylinder, defined by the radius at which the magnetic field must rotate at the speed of light to remain synchronised with the compact object spin [22]. Outside of this radius, $\sim 6 \times 10^{-5}\, R_{\mathrm{disc}}$ for XTE J2123–058, the magnetic field will be unable to keep up and hence becomes wound up. This will make the propeller less effective, although it may still produce some acceleration (Wynn priv. comm., 2000). The disc-anchored propeller [8] also has problems, in that it should very efficiently remove angular momentum from the disc (Wynn & King priv. comm., 1999), but this could perhaps be overcome if only a small fraction of stream material passes through the propeller. Finally, when a propeller coexists with a disc, there is the added difficulty in explaining how the accelerated material clears the disc rim. In the case of the disc-anchored propeller one can argue that field loops emerging from the disc could accelerate material upwards as well as out; for a propeller anchored on the neutron star some other explanation would be needed.

4.2   The Stream Overflow and Disc Re-impact Interpretation

The most popular alternative explanation for the SW Sex phenomenon is the accretion stream overflow and re-impact model. This was originally suggested by Shafter et al [14]; for a recent exposition see [7]. This model has the advantage of being physically very plausible. Some stream overflow is implied by the observations of X-ray dips in some LMXBs, and overflowing material should re-impact in the inner disc if not supported in some way. To explain the observations, however, requires a number of additional elements.

In this model the overflowing stream can be thought of in two regions. The initial part of the stream produces the transient absorption when seen against the brighter background of the disc. The latter part, where the overflow re-impacts the disc, is seen in emission giving rise to the high velocity component. In the simplest form of the model, the overflow stream should produce absorption at all phases; it always obscures some of the disc. This problem is overcome by invoking a strongly flared disc so that the overflow stream is only visible near phase 0.5. In XTE J2123–058, however, the neutron star is directly visible, so the disc area that can be obscured by a flare is limited and it is harder to reproduce the depth

Table 1. Doppler tomography of neutron star LMXBs. Bowen refers to the C III/N III blend near 4640 Å. Cen X-4 was observed in quiescence; the others in an X-ray active state. Sources: (a) Harlaftis (this volume) (b) Torres (2000, private communication) (c) Steeghs & Casares (2000, private communication)

	P_{orb} (hr)	Companion	Disc	Stream/ Hotspot	Other	Sources
2A 1822–371	5.6		H I	H I		[6] & a
XTE J2123–058	6.0			He II ?	He II	[10]
Cen X-4	15	H I	H I	H I?		b
AC211	17			He II ?	He II ?	b
Sco X-1	19	Bowen	He II , Bowen			c
GX 349+2	22		H I			[21]
Her X-1	41	H I, He I, Bowen, N V			He II	[15,13]

and transience of the absorption. A further difficulty is that to produce emission at the low velocities observed requires disc emission which is sub-Keplerian by a significant amount. This problem is not peculiar to XTE J2123–058 but common to the SW Sex class in general.

5   Conclusions

We have demonstrated that XTE J2123–058 at least, and likely some other short-period neutron star LMXBs, show observational similarities to SW Sex cataclysmic variables. These include a Doppler tomogram with emission concentrated at low velocities in the lower left quadrant and transient absorption around phase 0.5. We have discussed the application of two current SW Sex models to LMXBs. The stream overflow and re-impact model is theoretically plausible, but does not readily account for observations. The magnetic propeller model can easily explain the main observations, at least in XTE J2123–058, but has yet to overcome several theoretical objections. Whatever the correct interpretation, it is clear that the LMXB–SW Sex connection is an intriguing and fruitful one that may enhance our understanding of both types of systems.

So far we have concentrated on short-period 'normal' LMXBs; systems similar to XTE J2123–058. In Table 1 we summarise all the Doppler tomography of neutron star LMXBs known to us. The sample is clearly small and we include this as much to highlight the lack of data compared to cataclysmic variables as to draw strong conclusions. One can perhaps suggest that the companion star mainly tends to be prominent only in the longer period systems, and that 'unusual' tomograms (the other category) are most often seen in He II , but both of these conclusions require a larger sample for confirmation. It can be hoped that increasing availability of large telescope time will make such observations possible.

Acknowledgements

RIH, PAC and CAH would like to acknowledge support from grant F/00-180/A from the Leverhulme Trust. The William Herschel Telescope is operated on the island of La Palma by the Isaac Newton Group in the Spanish Observatorio del Roque de los Muchachos of the Instituto de Astrofísica de Canarias.

References

1. Augusteijn T., van der Hooft F., de Jong J. A., van Kerkwijk M. H., van Paradijs J., 1998, A&A, 332, 561
2. Casares J., Serra-Ricart M., Zurita C., Gomez A., Alcalde D., Charles P., 1998, IAU Circ. 6971
3. Cowley A. P., Crampton D., Hutchings J. B., 1982, ApJ, 255, 596
4. Crampton D., Cowley A. P., Stauffer J., Ianna P., Hutchings J. B., 1986, ApJ, 306, 599
5. Eracleous M., Horne K., 1996, ApJ, 471, 427
6. Harlaftis E. T., Charles P. A., Horne K., 1997, MNRAS, 285, 673
7. Hellier C., 2000, NewAR, 44, 131
8. Horne K., 1999, in, *Annapolis Workshop on Magnetic Cataclysmic Variables, ASP Conference Series, Volume 157*, ed. by C. Hellier, K. Mukai, ASP Conf. Ser. 157, p. 349
9. Hynes R. I., Charles P. A., Haswell C. A., Casares J., Serra-Ricart M., Zurita C., 1998, IAU Circ. 6976
10. Hynes R. I., Charles P. A., Haswell C. A., Casares J., Zurita C., Serra-Ricart M., 2000, MNRAS, submitted
11. Ilovaisky S. A., Chevalier C., 1998, IAU Circ. 6975
12. Levine A., Swank J., Smith E., 1998, IAU Circ. 6955
13. Quaintrell H., Still M. D., Roche P. D., 2000, MNRAS, submitted
14. Shafter A. W., Hessman F. V., Zhang E. H., 1988, ApJ, 327, 248
15. Still M. D., Quaintrell H., Roche P. D., Reynolds A. P.,1997, MNRAS, 292, 52
16. Takeshima T., Strohmayer T. E., 1998, IAU Circ. 6958
17. Thorstensen J. R., Charles P. A., 1982, ApJ, 253, 756
18. Thorstensen J. R., Ringwald F. A., Wade R. A., Schmidt G. D., Norsworthy J. E., 1991, ApJ, 102, 272
19. Tomsick J. A., Halpern J. P., Leighly K. M., Perlman E., 1998, IAU Circ. 6957
20. Tomsick J. A., Halpern J. P., Kemp J., Kaaret P., 1998, ApJ, 521, 341
21. Wachter S., 1997, BAAS, 29, 1279, No. 44.11
22. Wynn G. A., King A. R., Horne K., 1997, MNRAS, 286, 436
23. Zurita C., Casares J., Shahbaz T., Charles P. A., Hynes R. I., Shugarov S., Goransky V., Pavlenko E. P., Kuznetsova Y., 2000, MNRAS, 316, 137

Echo Mapping of Active Galactic Nuclei

K. Horne

[1] Physics & Astronomy, Univ. of St.Andrews, St.Andrews, KY16 9SS, Scotland, UK
[2] Dept. of Astronomy, Univ. of Texas, Austin, TX 78712, USA

Abstract. Echo mapping exploits light travel time delays, revealed by multi-wavelength variability studies, to map the geometry, kinematics, and physical conditions of reprocessing sites in photo-ionized gas flows. In active galactic nuclei (AGN), the ultraviolet to near infrared light arises in part from reprocessing of EUV and X-ray light from a compact and erratically variable source in the nucleus. The observed time delays, 0.1-2 days for the continuum and 1-100 days for the broad emission lines, probe regions only micro-arcseconds from the nucleus. Emission-line delays reveal radially stratified ionization zones, identify the nature of the gas motions, and estimate the masses of the central black holes. Continuum time delays map the temperature-radius structure of AGN accretion discs, and provide distances that may be accurate enough to realize the potential of AGNs as cosmological probes.

1 Introduction

Geologists routinely use sound waves to search for oil and map the interior structure of the Earth. Geological tomographers employ earthquake sources or set off explosive charges at the Earth's surface, and place networks of receivers to record time of arrival of the echoes returning from below. The measured time delays reveal the depths of submerged rock layers, and measure integrals of the wave speed along various paths through the Earth's interior.

In a somewhat similar fashion, astronomers use light echoes to probe the interior structures of objects too small to be directly imaged. Astronomical tomographers generally cannot wait long enough for a return signal. They rely instead on compact variable sources of ionizing radiation embedded within the region being probed. The two main applications of echo mapping in astronomy are to X-ray binaries, where a compact variable source of X-rays is associated with an accreting neutron star or black hole in a binary system, and Active Galactic Nuclei (AGN), where a supermassive black hole accretes gas in the centre of a galaxy.

This review attempts to summarize a decade of research into echo mapping of AGN, including both methods and results, and illustrates some capabilities to be realized in the next decade as improved datasets become available from proposed facilities like KRONOS – a space-borne multi-wavelength monitoring mission – and RoboNet – a ground-based network of robotic telescopes. The remainder of section 1, is a brief introduction to AGN. Section 2 then reviews a range of techniques used for echo mapping. Sections 3, 4, and 5 follow with discussions of echo mapping experiments that probe the geometry, kinematics, and physical conditions in the emission line regions of Seyfert 1 galaxies. Section 6 discusses the use of time delays in the continuum emission to map temperature-radius profiles of AGN accretion disks, leading also to a new method of estimating their distances and hence cosmological parameters. Conclusions are summarized in section 7.

1.1   Black Holes and Accretion Discs in AGNs

According to the most widely accepted model of AGN, a supermassive black hole ($M \sim 10^{6-9} M_\odot$) residing in the core of a host galaxy feeds on gas supplied by an accretion disc ($\dot{M} \sim 1 M_\odot$ yr^{-1}). Gas in the disc, heated while spiraling into the gravitational potential well, develops a steep inward temperature gradient $T \propto (M\dot{M})^{1/4} R^{-3/4}$ and thus emits continuum radiation over a wide span of wavelengths. Outside a critical radius ($R_d \sim$ pc), temperatures remain below the dust sublimation threshold $T_d \sim 1500$ K. Here a geometrically thick torus of cool dust and molecular gas, encircles the disc. The disc and torus collect dust and gas from the galaxy's interstellar medium, and may also receive material stripped from passing stars by various processes (tides, bow shocks, winds, irradiation).

In addition to the continuum radiation, AGN spectra exhibit two types of emission lines that are thought to arise in distinct regions. The Broad Line Region (BLR) produces broad permitted emission lines ($v \gtrsim 10,000$ km s^{-1}). The Narrow Line Region (NLR) produces narrow forbidden and permitted emission lines ($v \lesssim 1000$ km s^{-1}). The narrow lines are generally constant, while the continuum and the broad emission lines vary erratically.

Seyfert nuclei divide into two types, thought to represent different viewing angles. The spectra of Type 1 Seyferts exhibit both broad and narrow emission lines, while in Type 2 Seyferts the broad lines are absent or visible only in a faint linearly polarized component of the spectrum. The current consensus is that the thick dusty torus blocks our view of the BLR when viewed at $i \gtrsim 60°$, but some BLR light scatters toward us after rising far enough above the plane to clear the torus.

The NLR in nearby Seyfert galaxies is resolved by *HST* imaging studies with resolutions of ~ 0.1 arcseconds. The NLR typically has a clumpy bi-conical morphology, aligned with the axis of radio jets, and suggesting broadly-collimated ionization cones emerging from the nucleus, perhaps collimated by the dusty torus.

The BLR and continuum production regions are unresolved by *HST*. The BLR represents warm gas ($T \sim 10^4$K) in a hostile environment, with highly supersonic velocities and exposure to powerful ionizing radiation. What is the source of the gas? How long can it persist? Is it confined, or continuously regenerated? What forces drive the motions? These are questions of current interest that we would like to answer. Many different models have been proposed, and some eliminated, but ambiguity remains. Echo mapping experiments seek answers by using light travel time delays, revealed by variability on 1-100 day timescales, to probe these regions on $\sim 10^{-6}$ arcsecond scales.

1.2 Photo-ionization Models

Just how large is the BLR? If the emission lines are powered by photo-ionization, then we can estimate the size of the photo-ionized region surrounding the AGN. Photo-ionization models such as CLOUDY [11] consider the energy and ionization balance inside a 1-dimensional gas cloud parameterized by hydrogen number density n_H, column density N_H, and distance R from the source of ionizing radiation L_λ. The calculated emission-line spectrum emerging from such a gas cloud depends primarily on the ionization parameter,

$$U = \frac{\text{ionizing photons}}{\text{target atoms}} = \frac{Q}{4\pi R^2 c\, n_H},$$ (1)

where the luminosity of hydrogen-ionizing photons is

$$Q = \int_0^{\lambda_0} \frac{\lambda L_\lambda d\lambda}{h\, c},$$ (2)

with L_λ the luminosity spectrum emitted by the nucleus and $\lambda_0 = 912\text{Å}$ the wavelength of photons just capable of ionizing hydrogen. Re-arranging this equation yields

$$R = \left(\frac{Q}{4\pi U c\, n_H}\right)^{1/2}.$$ (3)

By comparing the flux ratios of broad emission lines in observed spectra with predictions of single-cloud photo-ionization models, typical parameters in the photo-ionized gas are found to be $U \sim 10^{-2}$ and $n_H \sim 10^{8-10}\text{cm}^{-3}$. With these conditions, the light travel time across the radius of the ionized zone is

$$\frac{R}{c} \sim 200\text{d} \left(\frac{Q}{10^{54}\text{s}^{-1}}\right)^{1/2} \left(\frac{U}{10^{-2}}\right)^{-1/2} \left(\frac{n_H}{10^9\text{cm}^{-3}}\right)^{-1/2},$$ (4)

where $Q \sim 10^{54}\text{s}^{-1}(H_0/100)^{-2}$ is appropriate for the best-studied Seyfert 1 nucleus of NGC 5548. On the basis of such predictions, astronomers searched for reverberation effects on timescales of months before eventually realizing that reverberation was occurring on much shorter timescales.

2 Echo Mapping Methods

2.1 Reverberation

A compact variable source of ionizing radiation launches spherical waves of heating and cooling that travel at the speed of light, triggering changes in the radiation emitted by surrounding gas clouds. The light travel time from nucleus to reprocessing site to observer is longer than that for the direct path from nucleus to observer. A distant observer sees the reprocessed radiation arrive with a time

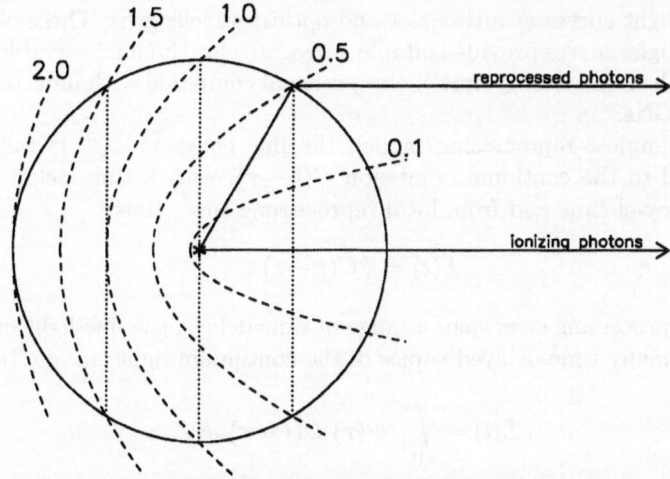

Fig. 1. Ionizing photons from a compact source are reprocessed by gas clouds in a thin spherical shell. A distant observer sees the reprocessed photons arrive with a time delay ranging from 0, for the near edge of the shell, to $2R/c$, for the far edge. The iso-delay paraboloids slice the shell into zones with areas proportional to the range of delays.

delay reflecting its position (R,θ) within the source. The time delay for a gas cloud located at a distance R from the nucleus is

$$\tau = \frac{R}{c}(1 + \cos\theta) , \qquad (5)$$

where the angle θ is measured from 0 for a cloud on the far side of the nucleus to 180° on the line of sight between the nucleus and the observer. The delay is 0 for gas on the line of sight between us and the nucleus, and $2R/c$ for gas directly behind the nucleus (Fig. 1). The iso-delay contours are concentric paraboloids wrapped around the line of sight. Such time delays are the basis of echo mapping techniques.

The validity of this basic picture is well supported by the results of intensive campaigns designed to monitor the variable optical, ultraviolet, and X-ray spectra of Seyfert 1 galaxies. Notable among these are the AGN Watch campaigns (http://www.astronomy.ohio-state.edu/~agnwatch/). In Seyfert 1 galaxies, the variations observed in continuum light are practically simultaneous at all wavelengths throughout the ultraviolet and optical, but the corresponding emission line variations exhibit time delays of 1 to 100 days, probing the photo-ionized gas 1 to 100 light days from the nucleus. This corresponds to micro-arcsecond scales in nearby Seyfert galaxies.

2.2 Linear and Non-linear Reprocessing Models

The ionizing radiation that drives the line emission includes X-ray and EUV light that cannot be directly observed. However, quite similar variations are seen in continuum light curves at ultraviolet and optical wavelengths. These observable continuum light curves provide suitable surrogates for the unobservable ionizing radiation. The reprocessing time (hours) is small compared with light travel time (days) in AGNs.

In the simplest reprocessing model, the line emission $L(t)$ is taken to be proportional to the continuum emission $C(t - \tau)$, with a time delay τ arising from light travel time and from local reprocessing time, thus

$$L(t) = \Psi C(t - \tau) . \tag{6}$$

Since the reprocessing sites span a range of time delays, the line light curve $L(t)$ is a sum of many time-delayed copies of the continuum light curve $C(t)$,

$$L(t) = \int_0^\infty \Psi(\tau) \, C(t - \tau) \, d\tau. \tag{7}$$

This convolution integral introduces $\Psi(\tau)$, the "transfer function" or "convolution kernel" or "delay map", to describe the strength of the reprocessed light that arrives with various time delays.

Since light travels at a constant speed, the surfaces of constant time delay are ellipsoids with one focus at the nucleus and the other at the observer. Near the nucleus, these ellipsoids are effectively paraboloids (Fig. 1). $\Psi(\tau)$ is in effect a 1-dimensional map that slices up the emission-line gas, revealing how much gas is present between each pair of iso-delay paraboloids.

The aim of echo mapping is to recover $\Psi(\tau)$ from measurements of $L(t)$ and $C(t)$ made at specific times t_i. To fit such observations, the linear model above is too simple in at least two respects. The first problem is additional sources of light contributing to the observed continuum and emission-line fluxes. Examples are background starlight, and narrow emission lines. When these sources do not vary on human timescales, they simply add constants to $L(t)$ and $C(t)$,

$$L(t) = \bar{L} + \int_0^\infty \Psi(\tau) \, \left[C(t - \tau) - \bar{C} \right] d\tau. \tag{8}$$

A second problem with the linear reprocessing model is that the reprocessed emission is more generally a non-linear function of the ionizing radiation. For example, when a cloud is only partially ionized (ionization bounded), its line emission generally increases with the ionizing radiation. At higher ionizing fluxes, the cloud becomes fully ionized (matter bounded), and its line emission may then saturate or even decrease with further increases in ionizing radiation. For single clouds we therefore expect the response function $L(C)$ to be highly non-linear (Fig. 2). The marginal response $\partial L/\partial C$ is generally less than the mean response L/C, and may become negative when an increase in ionizing radiation reduces the line emission.

The total response is of course a sum of responses from many different gas clouds, those closer to the nucleus being more fully ionized than those farther away. If clouds are present at many radii, so that zones of constant ionization

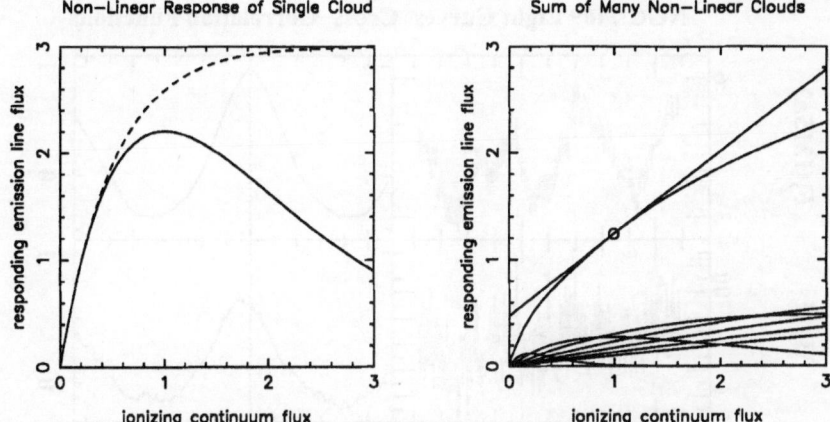

Fig. 2. The emission-line response from gas clouds exposed to different levels of photo-ionizing continuum radiation. Left: The initially linear response saturates and may even decrease as the cloud becomes more fully ionized. Right: When gas clouds are present at many radii, the ionization zone simply expands outward, resulting in an ensemble response that increases monotonically and is less non-linear than that from a single cloud. The tangent line is then a useful approximation.

parameter can simply expand, then the effect of averaging over many clouds is to produce a total response that is monotonically increasing and less strongly non-linear. Such a response may be adequately approximated by a tangent line,

$$L(C) = \bar{L} + \frac{\partial L}{\partial C} \left(C - \bar{C} \right) \;, \tag{9}$$

for ionizing radiation changes in some range above and below a mean level (Fig. 2). Observations support the use of this approximation – plots of observed line fluxes against continuum fluxes show roughly linear relationships, with $L/C > \partial L/\partial C$ so that extrapolation to zero continuum flux leaves a positive residual line flux.

For these reasons, it is usually appropriate in echo mapping to adopt the linearized reprocessing model, equation (8), with the "background" fluxes, $\bar{L}$ for the line and $\bar{C}$ for the continuum, set somewhere in the range of values spanned by the observations. In this model, the delay map $\Psi(\tau)$ weights gas clouds in proportion to their marginal response $\partial L/\partial C$. The roughly linear responses from partially-ionized clouds are fully registered, while the saturated responses of more fully ionized clouds have a reduced effect.

2.3 Cross-Correlation Analyses

With reasonably complete light curves, it is usually obvious that the highs and lows in the line light curves occur later than those in the continuum (Fig. 3). To quantify this time delay, or lag, a common practice is to cross-correlate the line

NGC 7469 Light Curves   Cross–Correlation Functions

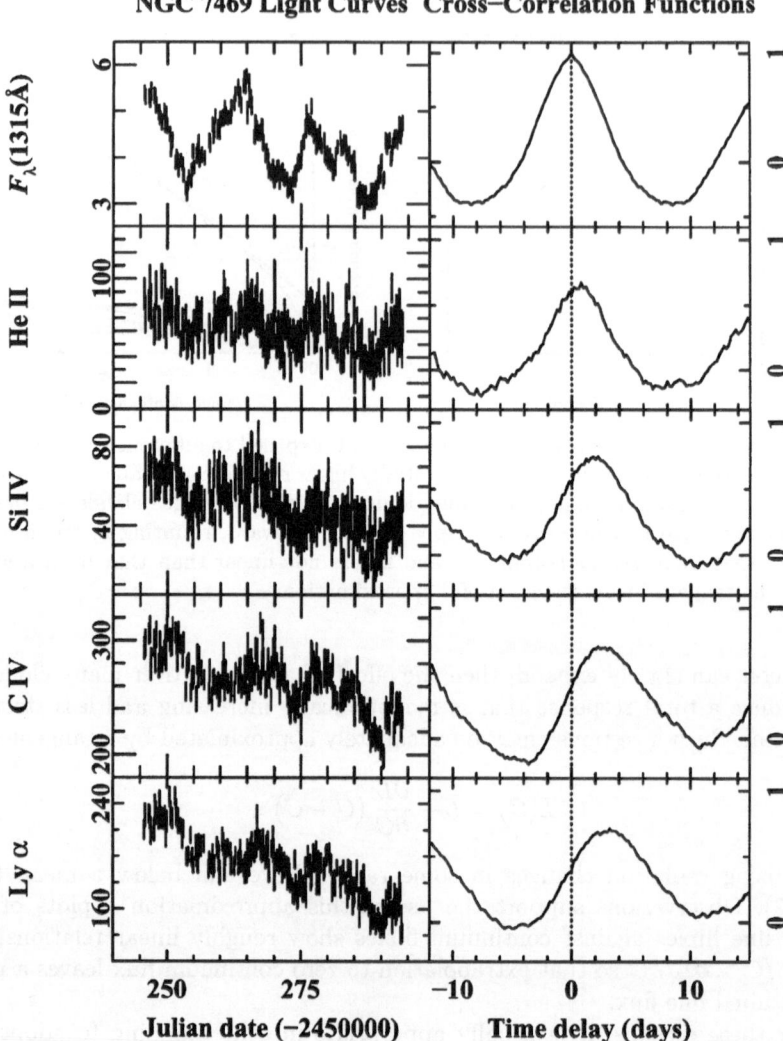

Fig. 3. The left-hand columns show light curves of NGC 7469 obtained with *IUE* during an intensive AGN Watch monitoring campaign during the summer of 1996. The right-hand column shows the result of cross-correlating the light curve immediately to the left with the 1315 Å light curve at the top of the left column; the panel at the top of the right column thus shows the 1315 Å continuum auto-correlation function. Data from [42].

and continuum light curves. The cross-correlation function (CCF) is

$$L \star C(\tau) = \int L(t)\, C(t - \tau)\, dt. \tag{10}$$

Several methods have been developed and refined to compute CCFs from noisy measurements available only at discrete unevenly-spaced times, either by interpolating the data [14,45], or binning the CCF [10,1]. The resulting CCFs generally have a peak shifted away from zero in a direction indicating that changes in the emission lines lag behind those in the continuum (Fig. 3). The CCF lag – $\tau_{\rm CCF}$ – at which $L \star C$ has its maximum value, serves to quantify roughly the size of the emission-line region.

Since a range of time delays is present, described by $\Psi(\tau)$, what delay is measured by the cross-correlation peak? Since $L(t)$ is itself a convolution between $C(t)$ and $\Psi(\tau)$, and since convolution is a linear operation, we have

$$L \star C = (\Psi \star C) \star C = \Psi \star (C \star C). \tag{11}$$

The CCF is therefore a convolution of the delay map with the continuum autocorrelation function (ACF),

$$C \star C(\tau) = \int C(t)\, C(t - \tau)\, dt. \tag{12}$$

The ACF is symmetric in τ, and thus always has a peak at $\tau = 0$. With rapid continuum variations, the ACF is sharp, and the CCF peak should be close to the strongest peak of $\Psi(\tau)$. This tends to favour short delays from the inner regions of the BLR [26]. When continuum variations are slow, however, the ACF is broad, smearing out sharp peaks in $\Psi(\tau)$, and shifting $\tau_{\rm CCF}$ toward the centroid of $\Psi(\tau)$. Thus the cross-correlation peak depends not only on the delay structure in $\Psi(\tau)$, but also on the character of the continuum variations. Different observing campaigns may therefore yield different lags even when the underlying delay map is the same.

2.4   Three Echo Mapping Methods

More refined echo mapping analyses aim to recover the delay map $\Psi(\tau)$ rather than just a characteristic time lag. These echo mapping methods generally require more complete data than the cross-correlation analyses. Three practical methods have been developed.

The **Regularized Linear Inversion (RLI)** method [38,22] notes that the convolution integral,

$$L(t) = \int C(t - \tau)\, \Psi(\tau)\, d\tau \,, \tag{13}$$

becomes a matrix equation,

$$L(t_i) = \sum_j C(t_i - \tau_j)\, \Psi(\tau_j)\, d\tau \,, \tag{14}$$

when the times t_i are evenly spaced. If the times are not evenly spaced, one may try to remedy that by interpolation. If the matrix $C_{ij} = C(t_i - \tau_j)$ can be inverted, solving the matrix equation yields

$$\Psi(\tau_k) = \sum_i C_{ki}^{-1} L(t_i)/d\tau. \tag{15}$$

If the continuum variations are unsuitable, the matrix has small eigenvalues and the inversion is unstable, strongly amplifying noise in the measurements $L(t_i)$ and $C(t_i)$. This problem is treated by altering the matrix to reduce the influence of small eigenvalues, thereby reducing noise but blurring the delay map.

A related method, **Subtractive Optimally-Localized Averages (SOLA)** [32], aims to estimate the delay map as weighted averages of the emission line measurements,

$$\hat{\Psi}(\tau) = \int K(\tau, t)\, L(t)\, dt. \tag{16}$$

Since $L = C \star \Psi$, we may write this as

$$\hat{\Psi}(\tau) = \int K(\tau, t) \int C(t - s)\, \Psi(s)\, ds\, dt \,. \tag{17}$$

The estimate $\hat{\Psi}$ is therefore a blurred version of the true delay map Ψ. With suitable continuum variations, the weights $K(\tau, t)$ can be chosen to make the blur kernel

$$b(\tau, s) = \int K(\tau, t)\, C(t - s)\, dt \tag{18}$$

resemble a narrow Gaussian of width Δ centred at $s = \tau$. The parameter Δ then controls the trade-off between noise and resolution in reconstructing $\Psi(\tau)$. These direct inversion methods work best when the observed light curves detect continuum variations on a wide range of timescales.

The **Maximum Entropy Method (MEM)** [16,15] is a more general fitting method that allows the use of any linear or non-linear reverberation model. As an extension of maximum likelihood techniques, MEM employs a "badness of fit" statistic, χ^2, to judge whether the echo model being considered achieves a satisfactory fit to the data. The requirement $\chi^2/N \sim 1 \pm \sqrt{2/N}$, where N is the number of continuum and line measurements, ensures that the model fits as well as is warranted by the error bars on the data points, without over-fitting to noise. MEM fits are required also to maximize the 'entropy', which is designed to measure the 'simplicity' of the map. For positive maps $p_i > 0$, the entropy

$$S = \sum_i p_i - q_i - p_i \ln p_i / q_i \tag{19}$$

is maximized when $p_i = q_i$. MEM thus steers the map p_i toward default values q_i. With the default map set to

$$q_i = (p_{i-1} p_{i+1})^{1/2}, \tag{20}$$

the entropy steers each pixel toward its neighbors, and MEM then finds the 'smoothest' positive map that fits the data. For further technical details see [15].

Fig. 4 illustrates recovery of a delay map from a MEM fit to fake light curves. The lower panel shows an erratically varying continuum light curve. The line light curve in the upper panel is smoother and has time-delayed peaks. A smooth continuum light curve $C(t)$ threads through the data points, and extrapolates

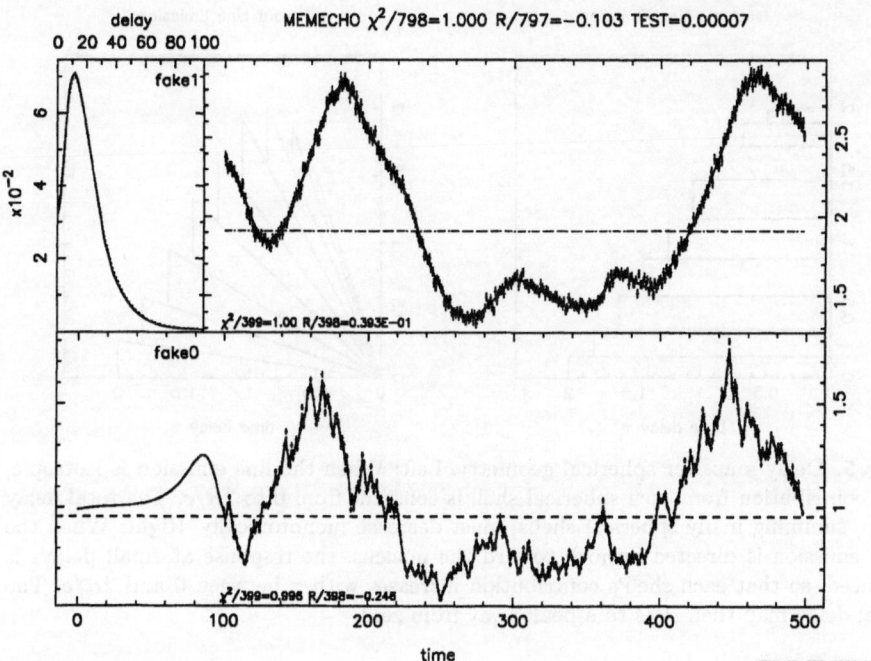

Fig. 4. Simulation test of the maximum entropy method showing the recovery of a delay map (top left) from data points sampling an erratically varying continuum light curve (bottom) and the corresponding delay-smeared emission-line light curve (top right). The reconstructed delay map closely resembles the true map $\Psi(\tau) \propto \tau e^{-\tau}$.

to earlier times. Convolving this light curve with the delay map $\Psi(\tau)$, shown in the upper left panel, gives the line light curve $L(t)$, fitting the data points in the upper right panel. Dashed lines give the backgrounds $\bar{L}$ and $\bar{C}$. The MEM fit with $\chi^2/N = 1$ adjusts $C(t)$, $\Psi(\tau)$, and $\bar{L}$ to fit the data points while keeping $C(t)$ and $\Psi(\tau)$ as smooth as possible.

In this simulation test, the true transfer function, $\Psi(\tau) \propto \tau e^{-\tau}$, is accurately recovered from the data. The fake dataset represents the type of data that could be obtained with daily sampling over a baseline of 1 year. While this is rather better than has been achieved so far, future experiments (Kronos, RoboNet) specifically designed for long-term monitoring will make this simulation more relevant.

3   Mapping the Geometry of Emission-Line Regions

3.1   Spherical Shells

Having shown that we can recover the delay map $\Psi(\tau)$, given suitable data, how may we interpret it? This is relatively straightforward if the geometry is spherically symmetric.

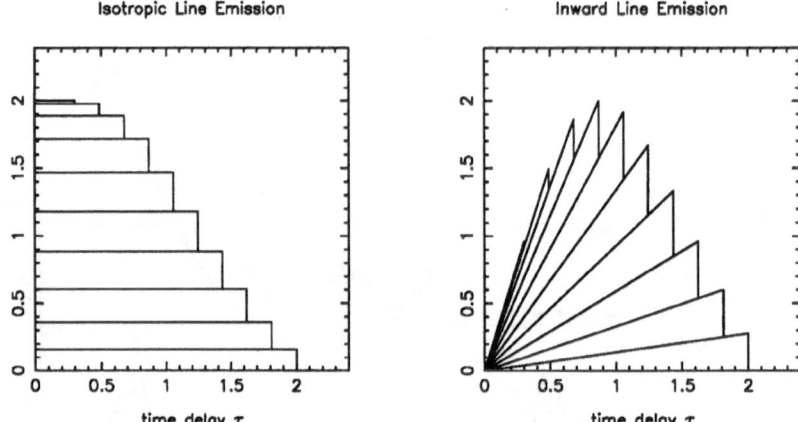

Fig. 5. Delay maps for spherical geometry. Left: When the line emission is isotropic, the contribution from each spherical shell is constant from 0 to $2R/c$. The total delay map, summing many spherical shells, must decrease monotonically. Right: When the line emission is directed inward, toward the nucleus, the response at small delays is reduced, so that each shell's contribution increases with τ between 0 and $2R/c$. The total delay map then rises to a peak away from zero.

Consider a thin spherical shell that is irradiated by a brief flash of ionizing radiation from a source located at the shell's centre. The flash reaches every point on the shell after a time R/c. Since recombination times are short, each point responds by emitting a brief flash of recombination radiation. A distant observer sees first the flash of ionizing radiation, and then the response of reprocessed light from the shell arriving with a range of time delays. The time delay is 0 for the near edge of the shell, and $2R/c$ on the far edge of the shell. The iso-delay paraboloids slice up the spherical shell into zones with areas proportional to the range of time delay (Fig. 1). The response from the spherical shell is therefore a flat-topped "boxcar" function, constant between the delay limits 0 and $2R/c$ (Fig. 5).

Any spherically symmetric geometry is just a nested set of concentric spherical shells. The delay map for a spherical geometry is therefore a sum of boxcar functions (Fig. 5). Since all the boxcars begin at delay 0, the delay map must peak at delay 0, and decrease monotonically thereafter. The contribution from the shell of radius R may be identified from the slope $\partial\Psi/\partial\tau$ evaluated at the appropriate time delay $\tau = 2R/c$. (The case of anisotropic emission is discussed in Section 3.3 below.)

3.2   The BLR Is Smaller than Expected

Fig. 6 shows a maximum entropy fit of the linearized echo model of Eqn. (8) to Hβ and optical continuum light curves of NGC 5548 [16]. Subtracting the continuum background $\bar{C}$, convolving with the delay map $\Psi(\tau)$, and adding the

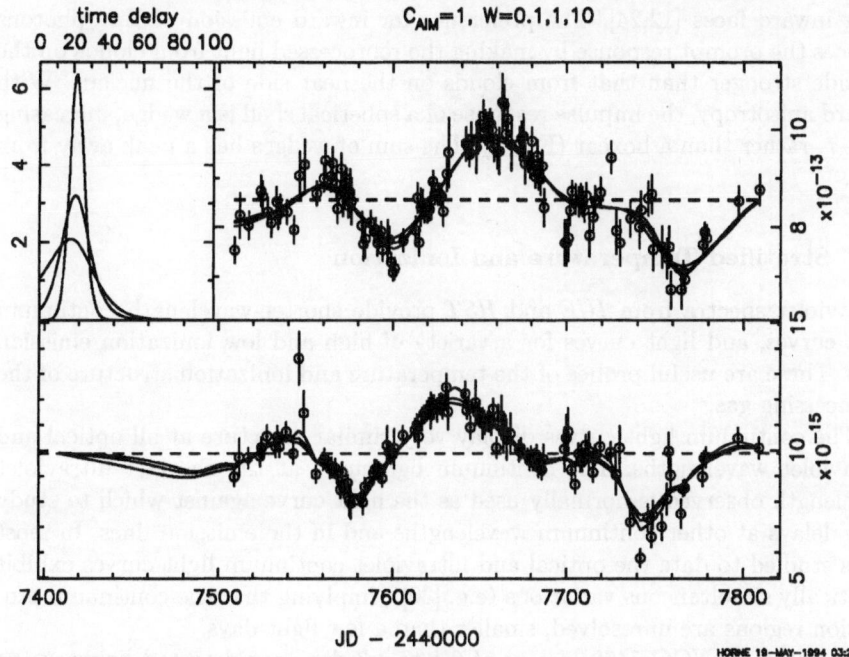

Fig. 6. Echo maps of Hβ emission in NGC 5548 found by a maximum entropy fit to data points from a 9-month AGN Watch monitoring campaign during 1989. The optical continuum light curve (lower panel) is convolved with the delay map (top left) to produce the Hβ emission line light curve (top right). Horizontal dashed lines give the mean line and continuum fluxes. Three fits are shown to indicate likely uncertainties. Data from [16].

line background $\bar{L}$, gives the Hβ light curve. The three fits shown, all with $\chi^2/N = 1$, indicate the likely range of uncertainty due to the trade-off between the 'stiffness' of $\Psi(\tau)$ and $C(t)$.

The Hβ map has a single peak at a delay of 20 days, and declines to low values by 40 days. This suggests that the size of the Hβ emission-line region is 10-20 light days. This is 10 to 20 times smaller than the 200 light day size estimated in Eqn. (4) on the basis of single-cloud photo-ionization models. With clouds this close to the nucleus, the ionization parameter U would be higher unless gas densities are increased by a factor of 100, to $n_H \sim 10^{11} \text{cm}^{-3}$.

3.3 Anisotropic Emission

The Hβ response in NGC 5548 is smaller at delay 0 than at 20 days. This lack of a prompt response conflicts with the monotonically decreasing delay map we expect for a spherical geometry. One interpretation is that there is a deficit of gas on the line of sight to the nucleus.

However, more likely this is a signature of anisotropic emission of the Hβ photons arising from optically thick gas clouds that are photo-ionized only on their inward faces [12,24]. The preference for inward emission of Hβ photons reduces the prompt response by making the reprocessed light from clouds on the far side stronger than that from clouds on the near side of the nucleus. With inward anisotropy, the impulse response of a spherical shell is a wedge, increasing with τ, rather than a boxcar (Fig. 5). The sum of wedges has a peak away from zero.

3.4   Stratified Temperature and Ionization

Ultraviolet spectra from *IUE* and *HST* provide shorter-wavelength continuum light curves, and light curves for a variety of high and low ionization emission lines. These are useful probes of the temperature and ionization structure of the reprocessing gas.

The continuum light curves display very similar structure at all optical and ultraviolet wavelengths. The continuum lightcurve at the shortest ultraviolet wavelength observed is normally used as the light curve against which to study time delays at other continuum wavelengths and in the emission lines. In most cases studied to date the optical and ultraviolet continuum light curves exhibit practically simultaneous variations (e.g. [30]), implying that the continuum production regions are unresolved, smaller than a few light days.

In one case, NGC 7469, delays of 0.1 to 1.7 days are detected between the 1350Å and 7500Å continuum light curves based on 40 days of continuous *IUE* and optical monitoring [42,6]. The delay increases with wavelength as $\tau \propto \lambda^{4/3}$, suggesting that the temperature decreases as $T \propto R^{-3/4}$. We discuss this further in section 6.

A systematic pattern is generally seen in the emission-line time delays. High-ionization lines (N v , He II) exhibit the shortest delays, while lower-ionization lines (C IV ,Lyα ,Hβ) have longer delays. This was established in the first AGN Watch campaign, by cross-correlation analysis [4] and by maximum entropy fits [23], using *IUE* and optical light curves of NGC 5548 sampled at 4 day intervals for 240 days. The effect is now seen in many objects (Table 1). This pattern is consistent with photo-ionization models, in which higher ionization zones occur closer to the nucleus.

Table 1. AGN Watch Monitoring Campaigns

target	year	τ_{CCF} (days)				references
		He II	Lyα	C IV	Hβ	
NGC 5548	1989	7	12	12	20	[4,28,7]
NGC 3783	1992	0	4	4	8	[?,36]
NGC 5548	1993	2	8	8	14	[19]
Fairall 9	1994	4	17	–	23	[34,35]
3C 390.3	1995	10	50	50	20	[42,6]
NGC 7469	1996	1	3	3	6	[25,8]

Another interesting result is that the time delay for C IV emission is generally smaller than that for C III] emission. In early work, this line ratio was the basis for estimating the gas density in the BLR to be $n_H \sim 10^{8-10} \mathrm{cm}^{-3}$. However, this result is now considered suspect because the C IV and C III] lines have different time delays and thus arise in different regions. The C IV line arises from gas clouds closer to the nucleus where a higher density, $n_H \sim 10^{11} \mathrm{cm}^{-3}$, is needed to maintain the ionization parameter.

3.5 Ionized Zones Expand with Luminosity

The ionization zones should be larger in higher luminosity objects. This prediction has been tested using echo mapping studies of a dozen active galaxies, including two low-luminosity quasars, spanning three decades in luminosity (Fig. 7, [18]). The lag for each object is determined by cross-correlating Hβ and optical continuum light curves. A log-log plot of the time delay vs luminosity

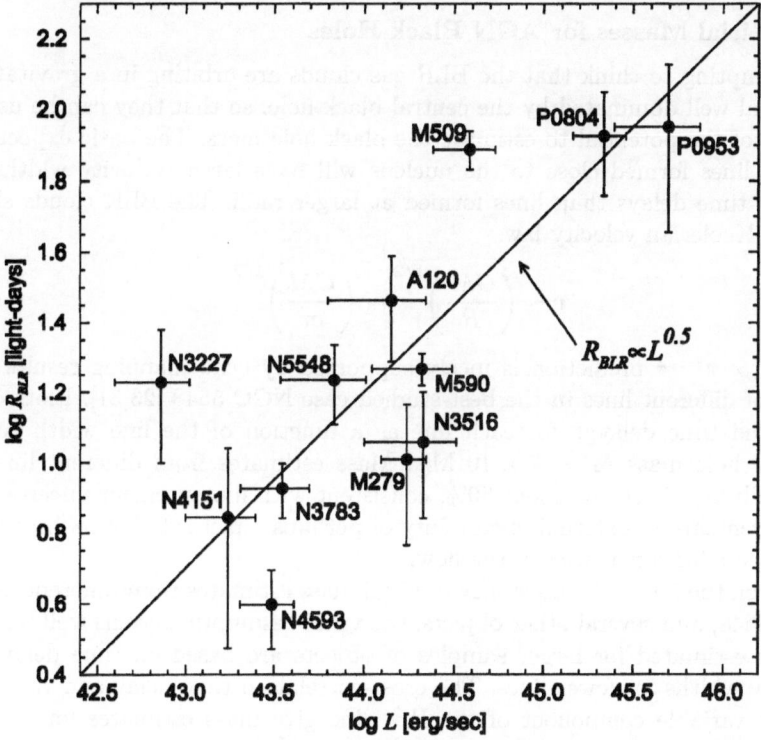

Fig. 7. The radii of ionized zones emitting broad Hβ emission lines for AGNs with different ionizing continuum luminosities. The radii are estimated from time delays that are found by cross-correlating Hβ and optical continuum light curves. The diagonal line with $R \propto L^{1/2}$ corresponds to a constant ionization parameter. Figure from [18].

shows a standard deviation of about 0.3 dex about a correlation of the form

$$\tau_{H\beta} \sim 17d \, L_{44}^{1/2}, \tag{21}$$

where L_{44} is the hydrogen-ionizing luminosity in units of 10^{44}erg s^{-1}. For comparison, if gas clouds with similar densities are present over a wide range of radius, then zones of constant ionization parameter should increase with $R \propto L^{1/2}$.

In the above correlation, derived by comparing different objects, the observed scatter may be due in part to differences in inclination, black hole mass, and accretion rate. The ionized zone in each object is expected expected to change size if the ionizing luminosity changes by a substantial factor. To map this effect, the echo model may be generalized to allow the delay map to change with the driving continuum light curve,

$$L(t) = \int_0^\infty C(t - \tau) \, \Psi(\tau, C(t - \tau)) \, d\tau \,. \tag{22}$$

4   Mapping the Kinematics of Emission-Line Regions

4.1   Virial Masses for AGN Black Holes

It is tempting to think that the BLR gas clouds are orbiting in a gravitational potential well dominated by the central black hole, so that they can be used as tracers of the potential to estimate the black hole mass. The basic expectation is that lines formed close to the nucleus will have larger velocity widths and smaller time delays than lines formed at larger radii. The BLR clouds should obey a Keplerian velocity law

$$v \sim \left(\frac{GM}{R}\right)^{1/2} \sim \left(\frac{GM}{c\tau}\right)^{1/2} \,. \tag{23}$$

This $v \propto \tau^{-1/2}$ prediction is nicely supported by echo mapping results for a range of different lines in the best-studied case NGC 5548 [23,31]. Plots of the measured time delay τ for each line as a function of the line width suggest a black hole mass $M \sim 7 \times 10^7 M_\odot$. Mass estimates from different lines are internally consistent to about 30%, consistent with measurement uncertainties. There remains an external uncertainty of perhaps a factor 3 due to uncertainty in the detailed kinematics of the flow.

Given the internal consistency of virial mass estimates from different lines in NGC 5548, and several other objects, the virial assumption seems well justified. Masses estimated for larger samples of objects are based on time delays and velocity widths for fewer lines. The cross-correlation time delay and rms width for the variable component of the Hβ line give mass estimates for 17 AGN (Fig. 8, [39]), which lie in the range $10^{6.5}$ to $10^{8.5} M_\odot$.

The Hβ reverberation masses serve also as calibrators for masses in other AGN where detailed reverberation studies have not yet been carried out but where the emission line region size may be estimated from photo-ionization physics [39]. This effectively enables mass estimates for all AGN with measured Hβ line widths.

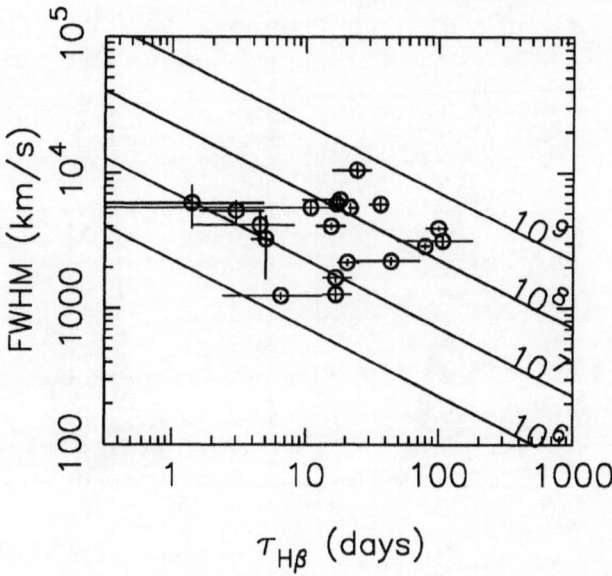

Fig. 8. Black hole masses estimated from the Hβ emission line based on cross-correlation time delay and FWHM of the line in the variable component of the light. Data from [39].

4.2 Velocity-Delay Maps

To go beyond the simple mass estimates above, and to decrease the factor of 3 uncertainty in the virial masses, we need to establish in more detail the nature of the gas flow in individual systems. Are the BLR clouds flowing inward, outward, circulating around in a plane, or orbiting in randomly-inclined orbits?

We may attempt to extract such details by using emission line profile variability to derive a 2-dimensional velocity-delay map $\Psi(v, \tau)$. Fig. 9 compares velocity-delay maps for two different flows: spherical free-fall into a point-mass potential, and a Keplerian disc [43,27]. The dramatically different appearance of the two flows highlights the power of velocity-delay maps for kinematic diagnosis.

A radial flow, inward or outward, produces a strong red–blue asymmetry in the velocity-delay map, while an azimuthal flow does not. Line emission from gas flowing in toward the nucleus is redshifted ($v > 0$) on the near side (τ small) and blueshifted ($v < 0$) on the far side (τ large). The signature of inflowing gas is therefore small time delays on the red side and large time delays on the blue side. Outflowing gas produces just the opposite red–blue asymmetry. Gas circulating around the nucleus has the same time delay on the red and blue side, producing a symmetric velocity-delay map.

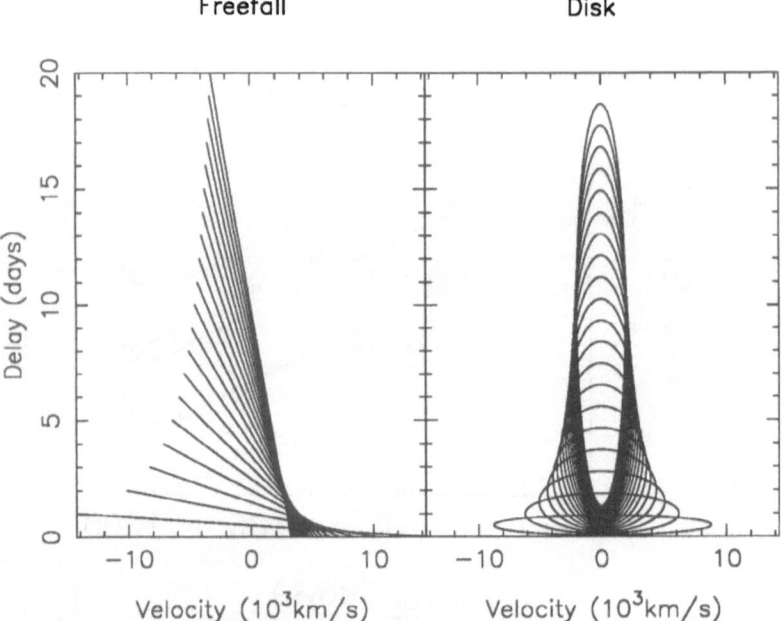

Fig. 9. Velocity–delay maps for spherical freefall and Keplerian disc kinematics. Spherical shells map to diagonal lines, and disc annuli map to ellipses. The central mass is $10^7 M_\odot$ in both cases. The disc inclination angle is 60°.

Any spherically-symmetric flow is a nested set of thin spherical shells. The time delay τ and Doppler shift v are

$$\tau = \frac{R}{c}\left(1 + \cos\theta\right), \qquad v = v_R \cos\theta, \qquad (24)$$

with R the shell radius, θ the angle from the back edge, and v_R the outflow velocity. The linear dependence of both τ and v on $\cos\theta$ implies that each shell maps into a diagonal line in the velocity-delay plane, as shown in Fig. 9.

The circulating disc flow is a set of concentric co-planar cylindrical annuli. The time delay and Doppler shift are

$$\tau = \frac{R}{c}\left(1 + \sin i \sin\phi\right), \qquad v = v_\phi \sin i \cos\phi, \qquad (25)$$

with R and i the radius and inclination of the annulus, ϕ the azimuth, and v_ϕ the azimuthal velocity. Each annulus maps to an ellipse on the velocity-delay plane, as in Fig. 9. Inner annuli map to "squashed" ellipses (large $\pm v$, small τ), while outer annuli map to "stretched" ellipses (small $\pm v$, large τ).

While $\Psi(v, \tau)$ is a distorted 2-dimensional projection of the 6-dimensional phase space, it can reveal important aspects of the flow, particularly if the velocity field is ordered and has some degree of symmetry. The envelope of the velocity-delay map may reveal the presence of virial motions $v^2 \propto GM/c\tau$, with M the mass of the central object. A red/blue asymmetry, or lack thereof, gauges the relative importance of radial and azimuthal motions. Disordered velocity fields with a range of velocities at each position smear the map in the v direction. The far side (large τ) may be enhanced by anisotropic emission from optically thick clouds radiating their lines inward toward the nucleus [12,24].

4.3   Velocity-Delay Maps from Observations

To derive Doppler-delay maps from the data, we simply slice the observed line profile into wavelength bins, and recover the delay map at each wavelength. This is a simple extension of the echo model used to fit continuum and emission-line light curves. At each wavelength λ and time t the emission-line flux

$$L(\lambda, t) = \bar{L}(\lambda) + \int_0^\infty \left[C(t - \tau) - \bar{C} \right] \Psi(\lambda, \tau) \, d\tau \tag{26}$$

is obtained by summing time-delayed responses to the continuum light curve and adding a time-independent background spectrum $\bar{L}(\lambda)$.

We observe $C(t)$ and $L(\lambda, t)$ at some set of times t_i. We fix $\bar{C}$, e.g. close to the mean of the observed values. We then adjust the continuum light curve $C(t)$, the background spectrum $\bar{L}(\lambda)$, and the Doppler-delay map $\Psi(\lambda, \tau)$, in order to fit the observations. The fitting procedure maximizes the entropy to find the "simplest" map(s) that fit the observations with $\chi^2/N = 1$.

Fig. 10 exhibits the velocity-delay map for C IV 1550 and He II 1640 emission as reconstructed from 44 *IUE* spectra of NGC 4151 covering 22 epochs during 1991 Nov 9 – Dec 15 ([37]. While spanning only 36 days, this campaign recorded favorable continuum variations, including a bumpy exponential decline followed by a rapid rise, which were sufficient to support echo mapping on delays from 0 to 20 days.

Disregard the strong C IV absorption feature, which obliterates the delay structure at small velocities. The wide range of velocities at small delays and smaller range at larger delays suggests virial motions. The dashed curves in Fig. 10 give escape velocity envelopes $v = \sqrt{2GM/c\tau}$ for masses 0.5, 1.0, and 2.0×10^7 M$_\odot$. A mass of order 10^7 M$_\odot$ may is indicated within 1 light-day of the nucleus.

The approximate red–blue symmetry of the C IV map rules out purely radial inflow or outflow kinematics. The somewhat stronger C IV response on the red side at small delays and on the blue side at larger delays suggests a gas component with freefall kinematics. However, if the C IV emission arises from the irradiated faces of optically-thick clouds, those on the far side of the nucleus will be brighter, and the redward asymmetry can then be interpreted as an outflow combined with the inward C IV anisotropy.

Velocity-delay maps have so far been constructed only for C IV emission in NGC 5548 [41,9] and NGC 4151 [37]. Both systems show the same trend of velocity dispersion decreasing with delay, and red response stronger at small delays. As more maps are constructed, we will learn whether these are general characteristics of Seyfert broad-line regions.

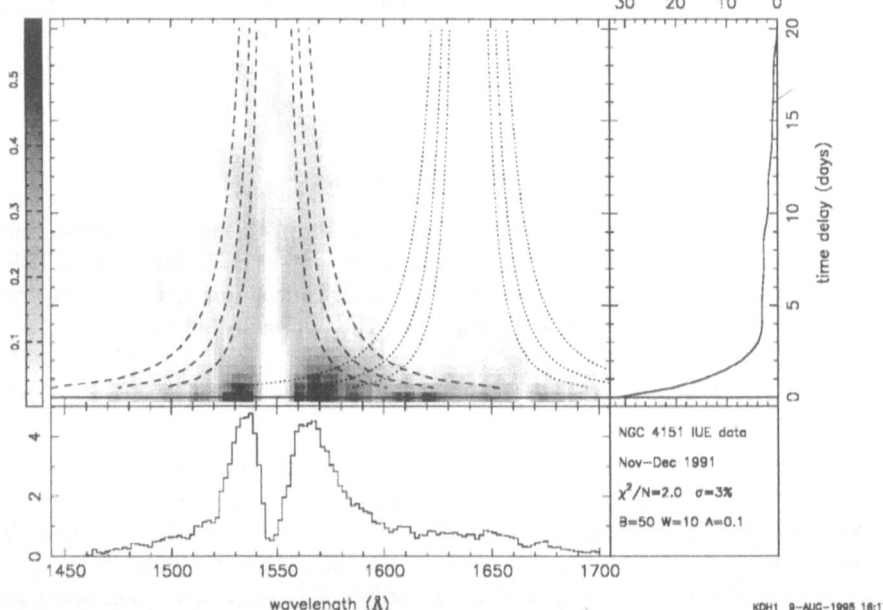

Fig. 10. MEMECHO velocity-delay map of C IV 1550 emission (and superimposed He II 1640 emission) in NGC 4151. Dashed curves give escape velocity for masses 0.5, 1.0, and $2.0 \times 10^7 M_\odot$.

5   Mapping Physical Conditions in Emission-Line Regions

The echo mapping results reveal that the BLR structure is stratified in density, ionization, and velocity. To match the quality of current and especially future datasets, the single-cloud models used in the 1980s to gain rough estimates of physical conditions are no longer adequate.

A simple upgrade in sophistication distributes clouds with some spatial distribution $\Psi(R, \theta)$, and allows the cloud properties to vary with position, $v(R, \theta)$, $n_H(R, \theta)$, $N_H(R, \theta)$. This succeeds in matching the fluxes and delays of a number of the emission lines in NGC 5548 [17]. This level of description may not be adequate, however. The observed line ratios seem to require a nearly universal ionization parameter $U \sim 10^{-2}$, which requires a finely tuned density structure $n \propto Q/R^2$.

Progress on this long-standing "fine-tuning problem" emerged with the introduction of the LOC (Locally Optimally-emitting Clouds) model [2], in which a wide range of clouds types is postulated to exist at each radius. When many cloud types are present, the line radiation at each position is dominated by those that are maximally efficient in reprocessing the ionizing radiation. The characteristic emission-line spectra of AGN then emerges naturally as a consequence of photo-ionization physics, rather than requiring specific fine-tuned density structure.

If we augment the LOC model to include also a range of velocities at each position, then our description of the BLR requires a 5-dimensional cloud map $\Psi(R, \theta, n_H, N_H, v)$. Power-laws in R, n_H, N_H and v prove to be adequate parameterizations for rough fits to NGC 5548, accounting for the fluxes, widths, and time lags of many lines simultaneously [21].

5.1 5-Dimensional Emission-Line Tomography

Rich information is coded in the fine details of a reverberating emission-line spectrum. To access that information, observations need to be fitted in far greater detail than has previously been attempted. Rather than extracting line and continuum lightcurves, we aim to fit the complete spectrum as it changes in time. This avoids the problem of de-blending lines by fitting to the blended profiles of hundreds of emission lines. The predictions of a photo-ionization model such as CLOUDY are incorporated explicitly to account for the fully non-linear and anisotropic responses of the emission-line clouds as they respond to the changing ionizing radiation.

A single gas cloud is characterized by its density n_H, column density N_H, distance from the nucleus R, azimuth θ, and Doppler shift v. To characterize the entire population of gas clouds, we employ a 5-dimensional map $\Psi(R, \theta, n_H, N_H, v)$. We can omit the cloud's position angle ϕ around the line of sight, and the two perpendicular velocity components, because the data provide no useful constraint on these.

We assume that the shape of the ionizing spectrum is known. The time-dependent ionizing photon luminosity is $Q(t)$. The flux of ionizing photons incident at time t on clouds at radius R and azimuth θ is then

$$F(t, R, \theta) = \frac{Q(t - \tau)}{4\pi R^2} \, , \tag{27}$$

where the time delay is

$$\tau = \frac{R}{c} \left(1 + \cos\theta \right) \, . \tag{28}$$

The clouds at this position respond to the ionizing radiation by emitting emission line L with a certain efficiency ϵ_L. The efficiency depends on the ionizing flux F, the cloud density n_H and column N_H, the viewing angle θ, and of course element abundances. These reprocessing efficiencies $\epsilon_L(F, n_H, N_H, \theta)$ are evaluated with a photo-ionization code, e.g. CLOUDY (Fig. 11). To save computer time, these are pre-calculated on a suitable grid of cloud parameters, e.g. equally spaced in $\log F$, $\log n_H$, and $\log N_H$ [20]. Results required for any cloud parameters are then found rapidly by interpolation in the grid.

Because clouds are irradiated on their inward faces, the reprocessed line emission is generally anisotropic (Fig. 11). We allow for this by letting ϵ_L depend on θ. CLOUDY computes the emission emerging inward, $\theta = 0$, and outward, $\theta = 90°$. For intermediate angles we interpolate linearly in $\cos\theta$.

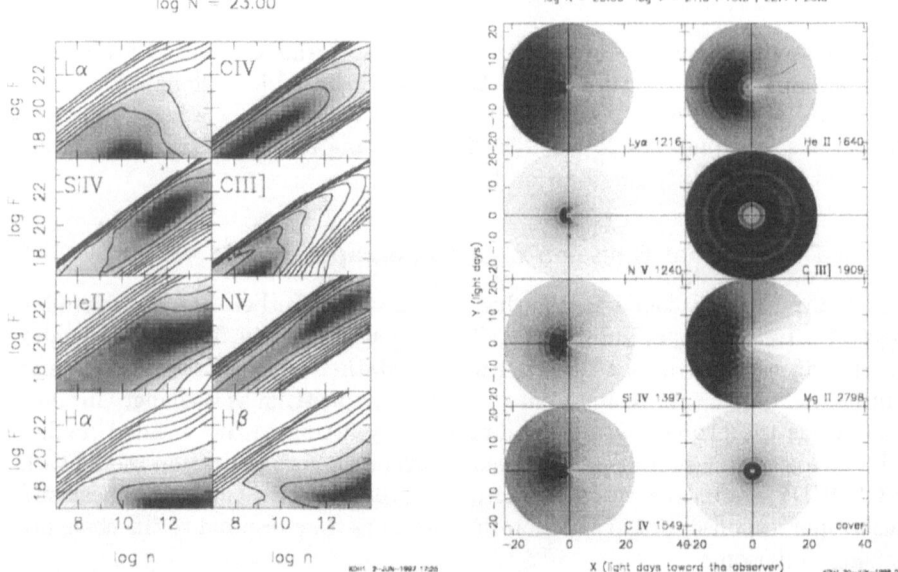

Fig. 11. Reprocessing efficiencies (left) for various emission lines are shown for clouds with a column density $N_H = 10^{23} \text{cm}^{-2}$ as functions of cloud density n_H and ionizing flux $F = Q/(4\pi R^2)$, where Q is the ionizing photon luminosity and R is distance from the active nucleus of NGC 5548. These results were calculated with the photoionization code CLOUDY. Low-density clouds close to the nucleus are fully ionized, while high density clouds far from the nucleus are under-ionized. Conditions along the diagonal ridges are just right for efficient reprocessing. Anisotropic emission (right) for 7 ultraviolet lines is predicted to be as shown by the line emissivity maps as functions of position (x, y), where the nucleus of the AGN is at the origin and the observer is at $x = +\infty$. Thus while C III] emission is isotropic, other lines are emitted more strongly toward the nucleus, making clouds more visible on the far side of the nucleus where we view their irradiated faces.

The observed spectrum is a sum of three components: direct light from the nucleus, reprocessed light from the surrounding gas clouds, and background light, – e.g. from stars,

$$f_\nu(\lambda, t) = f_\nu^D(\lambda, t) + f_\nu^R(\lambda, t) + f_\nu^B(\lambda) . \tag{29}$$

The direct light is

$$f_\nu^D(\lambda, t) = \frac{Q(t) S_\nu(\lambda)}{4\pi D^2} \tag{30}$$

where D is the distance and the spectral shape is

$$S_\nu(\lambda) = \frac{L_\nu(\lambda, t)}{Q(t)}. \tag{31}$$

The reprocessed light is a sum over numerous emission lines, where λ_L and ϵ_L are the rest wavelength and the reprocessing efficiency for line L. The gas clouds are distributed over a 3-dimensional volume. At each location the cloud population has a distribution over density n_H, column density N_H, and Doppler shift v. The reprocessed light arising from such a configuration is

$$f_\nu^R(\lambda, t) = \int_0^\infty \frac{Q(t-\tau)}{4\pi D^2} \, \Psi_\nu(\lambda, \tau, Q(t-\tau)) \, d\tau \, , \qquad (32)$$

where the (anisotropic and non-linear) transfer function is

$$\begin{aligned}
\Psi_\nu(\lambda, \tau, Q) = \sum_L \int & 2\pi R \, dR \, \sin\theta \, d\theta \, dn_H \, dN_H \, dv \\
& \Psi(R, \theta, n_H, N_H, v) \\
& \epsilon_L(F, n_H, N_H, \theta) \\
& g_\nu(\lambda - (1+v/c)\lambda_L) \\
& \delta\left(R - \left(\tfrac{Q}{4\pi F}\right)^{1/2}\right) \\
& \delta\left(\tau - \tfrac{R}{c}(1+\cos\theta)\right) \, .
\end{aligned} \qquad (33)$$

Here g_ν is a Gaussian profile to apply the appropriate Doppler shift, and the Dirac δ functions insert the appropriate time delay and ionizing flux at each reprocessing site.

To fit the above model to observations of $f_\nu(\lambda, t)$, we adjust the distance D, the ionizing radiation lightcurve $Q(t)$, the background spectrum $f_\nu^B(\lambda)$, and the 5-D cloud map $\Psi(R, \theta, n_H, N_H, v)$. The 5-D cloud map Ψ can have loads of pixels, $\sim 10^{5-6}$. Constraints available from reverberating emission-line spectra will be insufficient to fully determine the cloud map. However, once again MEM may be employed to locate the "smoothest" and "most symmetric" cloud maps that fits the data. Computers are now fast enough to support this type of detailed modelling and mapping of AGN emission regions.

5.2   Mapping a Spherical Shell from Simulated Data

Fig. 12 shows a simulation test designed to determine how well the geometry and physical conditions may be recoverable from emission-line reverberations. The adopted BLR model places clouds with density $n_H = 10^{11} \mathrm{cm}^{-3}$ and column density $N_H = 10^{23} \mathrm{cm}^{-2}$ on a thin spherical shell of radius $R/c = 12\mathrm{d}$. Synthetic light curves are computed showing reverberations in 7 ultraviolet emission lines, using CLOUDY to calculate the appropriate anisotropic and non-linear emissivities. The synthetic light curves are sampled at 121 epochs spaced by 2 days, and noise is added to simulate observational errors.

The MEM fit does not assume a spherical shell geometry, but rather it considers every possible cloud map $\Psi(R, \theta, n_H)$, and tries to find the simplest such map that fits the emission line light curves. The MEM fit adjusts 4147 pixels in the cloud map $\Psi(R, \theta, n_H)$, 143 points in the continuum light curve $C(t)$, 7 emission-line background fluxes $\bar{L}(\lambda)$, and 1 continuum flux $\bar{C}$. Note that the fit assumes the correct column density and distance. The fit to $N = 968$ data points achieves $\chi^2/N = 1$. The entropy steers each pixel in the map toward its

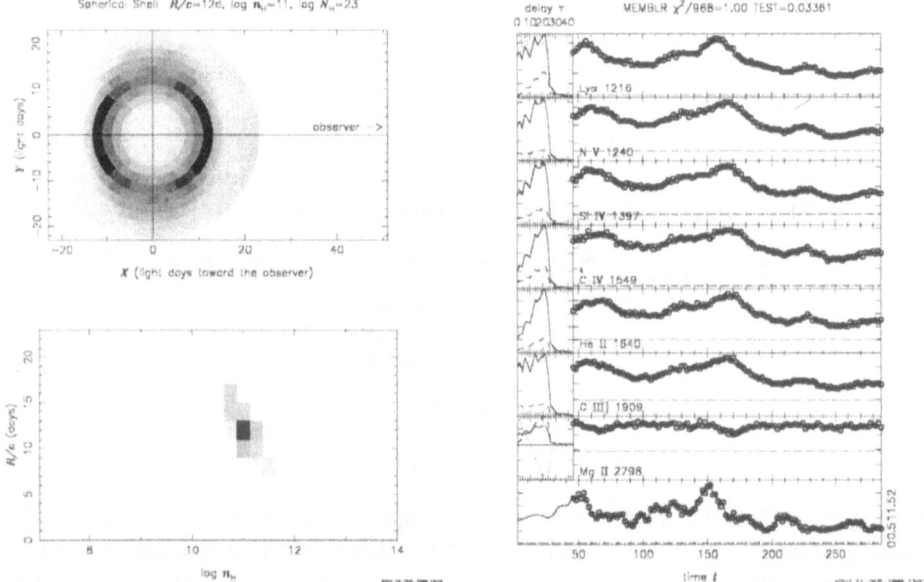

Fig. 12. Reconstructed map $\Psi(R, \theta, n_H)$ of a thin spherical shell (left) from a maximum entropy fit to 7 ultraviolet emission-line light curves (right). The light curve fits achieve $\chi^2/N = 1$. The shell radius $R/c = 12$d and density $\log n_H = 11$ are correctly recovered. For further details, see text.

nearest neighbors, thus encouraging smooth maps, and toward the pixel with the opposite sign of $\cos\theta$, to encourage front-back symmetry.

On the left-hand side of Fig. 12, we see that the recovered geometry displays the appearance of a hollow shell with the correct radius, and that the density-radius projection of the map has a peak at the correct density. The shell spreads in radius by a few light days, with lower densities at larger radii to maintain the same ionization parameter.

On the right-hand side of Fig. 12, we see that the 8 light curves are well reproduced by the fit. The highs and lows are a bit more extreme in the data – a common characteristic of regularized fits. To the left of each emission-line light curve, three delay maps are shown corresponding to the maximum brightness (solid curve), minimum brightness (dashed), and the difference (dotted). All the lines exhibit an inward anisotropy except C III] . All the lines have positive linear responses except Mg II , which has a positive response on the near side of the shell and a negative response on the far side.

This simulation test has used reverberation effects in emission-line light curves to map the geometry and density structure in the photo-ionized emission line zone. The 2-dimensional map clearly reveals the correct hollow shell geometry, and moreover the correct density is also recovered. These results suggest

that there are good prospects for probing the geometry and physical conditions
in real AGNs.

6 Temperature Maps of AGN Accretion Discs

So far we have concentrated on mapping emission line regions associated with
AGNs. For this purpose we consider the continuum variations to be essentially
simultaneous at all wavelengths. However, if the continuum radiation also arises
in part from reprocessing, we would expect small time delays, of order a day,
increasing with wavelength if the temperature decreases with distance from the
centre.

6.1 Steady State Blackbody Accretion Disks

The effective temperature on the surface of a steady-state accretion disc decreases with radius as

$$T = \left(\frac{3GM\dot{M}}{8\pi\sigma R^3} \right)^{1/4}, \tag{34}$$

where M is the black hole mass and $\dot{M}$ is the accretion rate, and we neglect
corrections important at small R.

Assuming that the disc surface radiates as a blackbody, we may calculate its
spectrum by summing blackbody spectra weighted by the projected areas of the
annuli,

$$f_\nu = \int B_\nu(T) \frac{2\pi R dR \cos i}{D^2} = \left(\frac{1200 G^2 h}{\pi^9 c^2} \right)^{1/3} I \left(\frac{\cos i}{D^2} \right) \left(M\dot{M} \right)^{2/3} \nu^{1/3}. \tag{35}$$

Here D is the distance, i is the inclination of the disc axis relative to the line of
sight, and $I = \int_0^\infty x^{5/3}/(e^x - 1) \sim 1.932$.

The optical/ultraviolet spectra of AGNs have "Big Blue Bump" components
that are attributed to thermal emission from accretion discs. Observed spectra
are generally redder than the characteristic $f_\nu \propto \nu^{1/3}$ spectrum predicted by
disc theory (e.g. [13]). The spectral signature of an accretion disc is therefore
not very convincingly demonstrated.

However, observed spectra are contaminated e.g. by starlight from the host
galaxy (e.g. [44]). Taking difference spectra cancels out the contamination and
measures the variable component of the light. The difference spectra are usually
bluer and can be in satisfactory agreement with the predicted spectrum for the
change in brightness of an irradiated accretion disc [5].

6.2 Echo Mapping Accretion Disc Temperature Profiles

In order to map the temperature profiles of accretion discs, we need some way
to measure the temperature, and some way to measure the radius at which

that temperature applies. In the simplest possible terms, we use a time delay to measure the radius, and a wavelength to measure the temperature.

The disc surface is irradiated by the erratically variable source located near its centre, launching heating and cooling waves that propogate at the speed of light outward from the centre of the disc. This effectively reprocesses the hard X-ray and EUV photons that irradiate the disc into softer ultraviolet, optical, and infrared photons. A distant observer will note that the reprocessed light arrives with a time delay

$$\tau = \frac{R}{c} \left(1 + \sin i \cos \theta \right) \ . \tag{36}$$

The heating wave requires a time R/c to reach radius R, and the reprocessing site at radius R and azimuth θ is at a distance from the observer that is larger by $R \sin i \cos \theta$ relative to the centre of the disc.

Note that for the annulus at radius R, the mean time delay, averaged around the annulus, is always R/c regardless of the inclination, but the range of time delays depends on the inclination. If the disc is face on ($i = 0$) then all azimuths have the same time delay. If the disc is edge on ($i = 90°$) then the far edge of the disc at $\theta = 0$ has the maximum time delay $\tau = 2R/c$, and the near edge at $\theta = 180°$ has a time delay of zero.

We use the blackbody spectrum to associate each wavelength with a temperature. The Planck spectrum for a blackbody temperature T,

$$B_\nu(T) = \frac{2hc}{\lambda^3 \left(e^X - 1\right)} \ , \tag{37}$$

peaks at the dimensionless frequency $X = hc/\lambda kT \sim 2.8$. When irradiation increases T slightly, the change in the spectrum, proportional to $\partial B_\nu/\partial T$, peaks near $X \sim 3.8$.

To map the temperature profile of an irradiated accretion disc, we measure the time delay at different wavelengths. When we observe a change in the spectrum at wavelength λ, we are measuring the reprocessed light from disc annuli where temperatures are

$$T \sim \frac{hc}{\lambda k X} \tag{38}$$

with $X \sim 4$. Ultraviolet, optical, and near infrared light curves give time delays for hot gas with $T \sim 10^5 \mathrm{K}$, warm gas with $T \sim 10^4 \mathrm{K}$, and cold gas with $T \sim 10^3 \mathrm{K}$, respectively. If the temperature decreases with radius, the time delay should increase with wavelength.

For the temperature profile of the steady-state disc, the time delay is

$$\tau = \left(\frac{45G}{16\pi^6 c^5 h}\right)^{1/3} \left(M\dot{M}\right)^{1/3} (X\lambda)^{4/3} \ . \tag{39}$$

This $\tau \propto \lambda^{4/3}$ prediction has been verified in the case of the Seyfert 1 galaxy NGC 7469 [6]. The observed $\tau(\lambda)$ provides a measurement of $M\dot{M}$,

$$M\dot{M} = \left(\frac{16\pi^6 hc^5}{45G}\right) (X\lambda)^{-4} \ \tau^3 \ . \tag{40}$$

Substituting for $M\dot{M}$ in the expression for f_ν, we find the distance,

$$D = \left(\frac{16\pi hc^3}{3}\right)^{1/2} I^{1/3} \left(\frac{\cos i}{f_\nu}\right)^{1/2} (X\lambda)^{-4/3} \, \tau \, . \tag{41}$$

Note that this distance is determined independently of the redshift of the AGN. Finally, for redshift $z \ll 1$, the Hubble constant is

$$H_0 = \frac{cz}{D} = \left(\frac{3}{16\pi hc}\right)^{1/2} I^{-1/3} \left(\frac{f_\nu}{\cos i}\right)^{1/2} (X\lambda)^{4/3} \left(\frac{z}{\tau}\right) \, . \tag{42}$$

The above is of course only an outline description of the method. In practice one needs to fit a model of reverberations in an irradiated disc to observed light curves. This new method is based on fairly straightforward physics – light travel time delays to measure radii, and blackbody spectra to associate a temperature with each wavelength.

6.3 Evidence of a Steady-State Blackbody Disc in NGC 7469

Wavelength-dependent time delays in the continuum light have been found in the analysis of a 40-night campaign of ultraviolet and optical spectra of NGC 7469 [42,5]. Individual spectra (top panel of Fig. 13) are quite red due at least in part to the contribution of starlight from the host galaxy. However, the variable component of the light, isolated for example by taking the difference between bright and faint spectra, has a much bluer continuum that is in fact consistent with $f_\nu \propto \nu^{1/3}$. Thus while the individual spectra are inconsistent with a steady-state blackbody accretion disk, the variable component of the light may well arise from such a source.

Further indications of an accretion disk origin for the variable component of the continuum light can be seen in the time delay spectrum (lower panel of Fig. 13). The delay at each wavelength is measured by cross-correlating a monochromatic lightcurve extracted from the spectra against the lightcurve in the bluest continuum band, in this case near 1315Å. The time delay increases near each of the emission lines, where the light is a mix of continuum and line emission. The lower envelope of the delay spectrum shows that the continuum light variations have a delay that increases with wavelength. The optical variations lag behind those at 1315Å by ~ 1 d at 5500 Å up to ~ 1.8 d near 8000 Å. The observed continuum delays are in reasonable agreement with the $\tau \propto \lambda^{4/3}$ prediction (curved line on Fig. 13) based on a steady-state blackbody accretion disk with $T \propto R^{-3/4}$ and $M\dot{M} \sim 6 \times 10^5 M_\odot^2 \mathrm{yr}^{-1}$.

6.4 AGN Discs as Cosmological Probes

AGN are promising candidates for use as cosmological probes because they are bright enough to be visible at large redshifts. They are continuously bright, unlike supernovae, so that it is not necessary to expend large quantities of telescope time

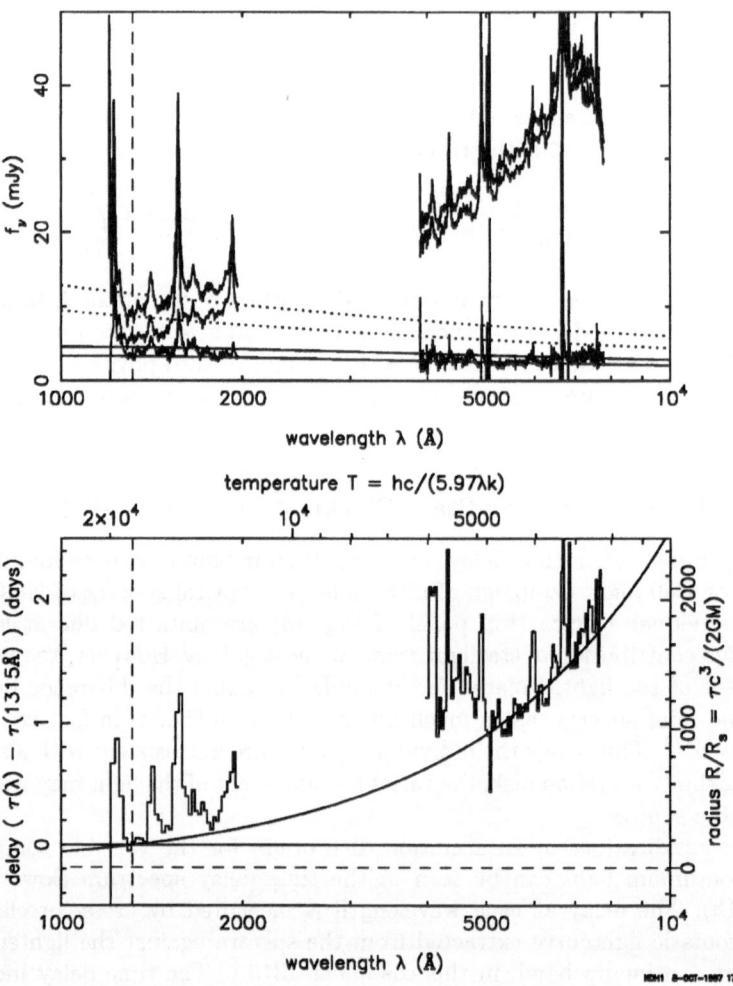

7469 z=0.0164 E(B–V)=0.14 H$_0$=60,70 M=10^{7}M$_\odot$ Ṁ=0.07M$_\odot$/yr i=60° R=(3,10

Fig. 13. Continuum time delays in NGC 7469. The top panel shows that the brightest and faintest spectra from a 40-night observing campaign in 1996, are considerably redder than the difference spectrum, which characterises the variable component of the light. After de-reddening ($E(B-V) = 0.14$), the difference spectrum has a power-law form with $f_\nu \propto \nu^{1/3}$, consistent with the predicted spectrum of a steady-state blackbody accretion disc. The time delay spectrum $\tau(\lambda)$ in the lower panel was obtained from the same dataset by cross-correlating monochromatic lightcurves derived from the spectra against the lightcurve at ~ 1315Å. The delay increases near each of the emission lines, and the delay in the continuum increases with wavelength as $\tau \propto \lambda^{4/3}$, also consistent with the prediction for a steady-state disc with $T \propto R^{-3/4}$.

just to find them. However, while Type Ia supernovae may well be "standard bombs", whose individual luminosities can be calibrated to accuracies of 10-20%, AGNs cover a wide range of luminosities and there has so far been no reliable method of measuring their individual luminosities and distances with comparable accuracy.

By using continuum echoes to map the $T(R)$ profiles of AGN accretion disks, we may have found a way to measure their luminosities and distances. By monitoring variability, we measure $\tau(\lambda)$. Assuming blackbody spectra, we use $\tau(\lambda)$ to infer $T(R)$, and hence we calculate the absolute flux of the disc. Comparing the observed and calculated fluxes, we find the distance, or rather $D/\sqrt{\cos i}$ because foreshortening scales the disc flux by $\cos i$. As this distance is obtained without reference to the redshift, the method estimates the Hubble constant. The results for NGC 7469 give a Hubble constant $H_0\sqrt{\cos i/0.7} = 42 \pm 9$ km s^{-1}Mpc^{-1} [5]. This result is not completely silly, suggesting that the underlying assumptions may be largely correct.

The unknown disc inclination appears to introduce a large uncertainty in the method. However, this may not be so serious a problem as it at first appears. The reason is that if the inclination is high, $i \gtrsim 60°$, then the dusty torus will obscure the broad emission line region. Thus all Seyfert 1 galaxies are expected to be roughly face-on, with $0.7 \lesssim \sqrt{\cos i} < 1$. This plausibly justifies adopting $\sqrt{\cos i} = 0.85 \pm 0.15$. The assumption can be checked by applying the method to a larger sample of Seyfert 1 galaxies, which should give independent estimates of H_0 with an rms scatter less than 18%. The $\sqrt{\cos i}$ uncertainty could then be reduced further by averaging over the sample of presumably random inclinations in the above range.

The use of blackbody spectra is also a questionable assumption. However, it could be a good assumption if the irradiated disk surfaces have relatively weak vertical temperature gradients. The internal consistency of the blackbody assumption may be checked by seeing if the same distance is found at different wavelengths. This test holds up fairly well for NGC 7469 [5].

If the method can be shown to work for a larger sample of objects, it may serve to calibrate AGNs as standard candles for cosmology. The AGN disk reverberation method should be at least as accurate as current estimates based on the Sunyaev-Zeldovich effect. Additional work is needed to determine whether it can approach the accuracy of supernova distances.

7   Conclusion

To summarize briefly, echo mapping is being used to resolve structures in active galactic nuclei on micro-arcsecond scales. Achievements in the first decade of this new field are based largely on analysis of data from IUE and ground-based monitoring campaigns. The main results are direct measurements of the sizes of broad emission-line regions, providing several critical tests of photo-ionization models (radial ionization structure, luminosity-radius correlation), and virial mass estimates for dozens of super-massive black holes.

Selected results are also presented of simulations evaluating key capabilities of echo mapping that may be realized with future facilities (KRONOS, RoboNet) designed to acquire high-quality time-lapse monitoring datasets. The simulations demonstrate that if we can acquire accurate well-sampled records of the evolving spectra of AGNs, and if the basic physical assumptions (photo-ionization models) are correct, then we can map the emission line regions in 5 dimensions including the geometry (R, θ), kinematics (v), and physical conditions (n_H, N_H).

From wavelength-dependent continuum time delays, we are able also to map the temperature-radius profile $T(R)$ in the continuum production region, which is expected to be the surface of the accretion disc. The resulting distance estimates, could allow AGNs to realize their potential as cosmological probes.

Acknowledgements

For financial support during the preparation of this review I am grateful to PPARC for a Sr. Research Fellowship at University of St.Andrews and to the Astronomy Department at the University of Texas at Austin for a temporary appointment as Beatrice Tinsley Visiting Professor.

References

1. Alexander, T. 1997, in Astronomical Time Series, ed. D.Maoz, A.Sternberg, E.M.Leibowitz (Dordrecht:Kluwer), 163
2. Baldwin, J., Ferland, G., Korista, K., Verner, D. 1995, ApJL, 455, L119
3. Blandford, R.D., McKee, C.F. 1982, ApJ, 255, 419
4. Clavel J., et al. 1991, ApJ, 366, 64
5. Collier, S.J., Horne, K., Wanders, I., Peterson, B.M. 1999, MNRAS, 302, 24
6. Collier, S.J., et al. 1998, ApJ, 500, 162
7. Dietrich M., et al. 1993, ApJ, 408, 416
8. Dietrich, M., et al. 1998, ApJS, 115, 185
9. Done, C., Krolik, J.H. 1996, ApJ, 463, 144
10. Edelson, R.A., Krolik, J.H. 1988, ApJ, 333, 646
11. Ferland, G.J., Korista, K.T., Verner, D.A., Ferguson, J.W., Kingdon, J.B., Verner, E.M. 1998, PASP, 110, 761
12. Ferland, G.J., Peterson, B.M., Horne, K., Welsh, W.F., Nahar, S.N. 1992, ApJ, 387, 95
13. Francis, P.J., Hewett, P.C., Foltz, C.B., Chaffee, F.H., Weymann, R.J., Morris, S.L. 1991, ApJ, 373, 465
14. Gaskell, C.M., Sparke, L.S. 1986, ApJ, 305, 175
15. Horne, K. 1994, in Reverberation Mapping of the Broad Line Region in Active Galactic Nuclei, ed. P.M.Gondhalekar, K.Horne, B.M.Peterson (Astron.Soc.Pac: San Francisco), 23
16. Horne, K., Welsh, W.F., Peterson, B.M. 1991, ApJL, 367, L5
17. Kaspi, S., Netzer, H. 1999 ApJ, 524, 71
18. Kaspi, S., Smith, P.S., Maoz, D., Netzer, H., Jannuzi, B.T. 1996, ApJ, 471, 75
19. Korista, K.T., et al. 1995, ApJS, 97, 285
20. Korista, K.T., Baldwin, J., Ferland, G., Verner, D. 1997, ApJS, 108, 401

21. Korista, K.T., Goad, M.R. 2000, ApJ, 536, 284
22. Krolik, J.H., Done, C. 1995, ApJ, 440, 166
23. Krolik, J.H., Horne, K., Kallman, T.R., Malkan, M.A., Edelson, R.A., Kriss, G.A. 1991, ApJ, 371, 541
24. O'Brien, P.T., Goad, M.R., Gondhalekar, P.M. 1994, MNRAS, 268, 845
25. O'Brien, P.T., et al. 1998, ApJ, 509, 163
26. Perez, E., Robinson, A., de la Fuente, L. 1992a, MNRAS, 255, 502
27. Perez, E., Robinson, A., de la Fuente, L. 1992b, MNRAS, 256, 103
28. Peterson B.M., et al. 1991, ApJ, 368, 119
29. Peterson, B.M., Ali, B., Horne, K., Bertram, R., Lame, N.J., Poggee, R.W., Wagner, M., R. 1993, ApJ, 402, 469
30. Peterson, B.M., Wanders, I., Horne, K., Collier, S., Alexander, T., Kaspi, S., Maoz, D. 1998, PASP, 110, 660
31. Peterson, B.M., Wandel, A., 1999, ApJL, 521, L95
32. Pijpers, F.P., Wanders, I. 1994, MNRAS, 271, 183
33. Reichert, G.A, et al., 1994, ApJ, 425, 582
34. Rodriguez-Pascual, P.M., et al. 1997, ApJS, 110, 9
35. Santos-Lleo, M., et al. 1997, ApJS, 112, 271
36. Stirpe G.M., et al. 1994, ApJ, 425, 609
37. Ulrich, M.H., Horne, K. 1996, MNRAS, 283, 748
38. Vio, R., Horne, K., Wamsteker, W. 1994, PASP, 106, 1091
39. Wandel, A., Peterson, B.M., Malkan, M.A., 1999, ApJ, 526, 579
40. Wanders, I., Horne, K. 1994, A&A, 289, 76
41. Wanders, I., Goad, M.R., Korista, K.T., Peterson, B.M., Horne, K., Ferland, G., Koratkar, A.P., Pogge, R.W., Shields, J.C. 1995, ApJL, 453, L87
42. Wanders, I., et al. 1997, ApJS, 113, 69
43. Welsh, W.F., Horne, K. 1991, ApJ, 379, 586
44. Welsh, W.F., Peterson, B.M., Koratkar, A.P., Korista, K.T. 1999, ApJ, in press.
45. White, R.J., Peterson, B.M. 1994, PASP, 106, 879

Echoes in X-ray Binaries: Mapping the Accretion Flow

K. O'Brien and K. Horne

University of St Andrews, St Andrews, UK KY16 9SS

Abstract. In X-ray binaries much of the optical/UV emission arises from X-rays reprocessed by material in the accretion disc, stream and the companion star. The resulting reprocessed variability will be delayed in time with respect to the X-ray variability by an amount depending on the position of the reprocessing regions. From this, we can determine a range of time-delays present in the system. This time-delay transfer function can be used to 'echo-map' the geometry of the reprocessing regions in the binary system.

We present our modeling of this transfer function and show results from our echo-mapping campaign using X-ray lightcurves from RXTE, simultaneous with HST. In the X-ray transient, GRO J1655-40, shortly after the 1996 outburst, we find evidence for reprocessing in the outer regions of a thick accretion disc.

1 Introduction

X-ray binaries (XRBs) are close binaries that contain a relatively un-evolved donor star and a neutron star or black hole that is thought to be accreting material through Roche-lobe overflow. Material passing through the inner Lagrangian point moves along a ballistic trajectory until impacting onto the outer regions of an accretion disc. This material spirals through the disc, losing angular momentum, until it accretes onto the central compact object, where X-rays are emitted from the inner disc regions.

Much of the optical emission in XRBs arises from reprocessing of these X-rays by material in regions around the central compact object. Light travel times within the system are of order 10s of seconds. Optical variability may thus be delayed in time relative to the X-ray driving variability by an amount characteristic of the position of the reprocessing region in the binary, which depends in turn on the geometry of the binary. The optical variability may be modelled as a convolution of the X-ray variability with a time-delay transfer function.

This time delay is the basis of an indirect imaging technique, known as echo tomography, which probes the structure of accretion flows on scales that cannot be imaged directly. Echo mapping has already been developed to interpret lightcurves of Active Galactic Nuclei (AGN), where time delays are used to resolve photoionized emission-line regions near the compact variable source of ionizing radiation in the nucleus (Horne, this volume). In AGN, the timescale of detectable variations is days to weeks, giving a resolution in the transfer functions of 1-10 light days [4,1]. In XRBs the binary separation is light seconds rather than light days, requiring high-speed optical/UV and X-ray lightcurves to probe the structure of the components of the binary in detail. The detectable X-ray and optical variations in the lightcurves of such systems are also suitably fast.

We present a simple geometric model for the time-delay transfer functions of XRBs, using a synthetic binary code. We analyze correlated X-ray and UV variability in GRO J1655-40, using our computed transfer functions, to constrain the size, thickness and geometric shape of the accretion disc in the system.

2   Reprocessing of X-rays

In the standard model of reprocessing, X-rays are emitted by material in the deep potential well of the compact object. These photoionize and heat the surrounding regions of gas, which later recombine and cool, producing lower energy photons. The optical emission seen by a distant observer is delayed in time of arrival relative to the X-rays by two mechanisms. The first is a finite reprocessing time for the X-ray photons, which for this work is assumed to be negligible [5], and the second is the light travel times between the X-ray source and the reprocessing sites within the binary system.

2.1   Light Travel Times

The light travel times arise from the time of flight differences for photons that are observed directly and those that are reprocessed and re-emitted before travelling to the observer. These delays can be up to twice the binary separation, obtained from Kepler's third law,

$$\frac{a}{c} = 9.76\text{s} \left(\frac{M_{\mathrm{x}} + M_d}{M_\odot} \right)^{\frac{1}{3}} \left(\frac{P}{\text{days}} \right)^{\frac{2}{3}}, \tag{1}$$

where a is the binary separation, M_{x} and M_d are the masses of the compact object and donor star, P is the orbital period. In LMXBs the binary separation is of the order of several light seconds.

The time delay τ at binary phase ϕ for a reprocessing site with cylindrical coordinates (R, θ, Z) is

$$\tau(\boldsymbol{x}, \phi) = \frac{\sqrt{R^2 + Z^2}}{c} (1 + \sin i \cos(\phi - \theta)) - \frac{Z}{c} \cos i \tag{2}$$

where i is the inclination of the system and c is the speed of light.

The dynamic response function is found by considering how a change in X-ray flux, $\Delta f_{\mathrm{x}}(t)$, drives a change in the reprocessed flux. We can define the dynamic time delay transfer function to be

$$\Psi_\nu (\lambda, \tau, \phi) = \int \left[\frac{\delta I_\nu (\lambda, \boldsymbol{x}, \Delta f_{\mathrm{x}}(t - \tau))}{\delta f_{\mathrm{x}}(t - \tau)} \right] \delta(\tau - \tau(\boldsymbol{x}, \phi)) \, d\boldsymbol{x}, \tag{3}$$

where $\tau(\boldsymbol{x}, \phi)$ is the geometric time delay of a reprocessing site at position $\boldsymbol{x}$, see (2).

3 Model X-ray Binary Code

We have developed a code to model time delay transfer functions based on determining the contributions from different regions in the binary. In this section we describe the models used to construct the individual regions of the binary; the donor star, the accretion stream and the accretion disc. The code uses distances scaled to the binary separation in a right-handed cartesian coordinate system corotating with the binary. Each surface panel is a triangle, characterized by its area dA, orientation n, position x and temperature T.

3.1 Donor Star

The donor star is modeled assuming it fills its critical Roche potential, so that mass transfer occurs via Roche lobe overflow through the inner Lagrangian point. Optically thick panels are placed over the surface of the Roche potential. The panels are triangular so that the curved surfaces of the binary are mapped more accurately than is possible using 4-sided shapes [7]. The panels, when unirradiated, are assigned an effective temperature T_{star} given by the spectral type of the donor star.

3.2 Accretion Stream

The accretion stream is modeled by following the ballistic trajectories of 4 test particles. The 'width' and the 'height' of the stream (its extent in the y- and z-directions respectively) is determined by the initial positions of the test particles. The stream is symmetric about the x-y plane. The unirradiated accretion stream is assumed to have a constant temperature T_s along it's length and the effects of irradiation are considered in the same way as those of the donor star.

3.3 Accretion Disc

The disc thickness is assumed to increase with radius from 0 at R = R_{in} to H_{out} at R = R_{out}, with the form,

$$H = R_{out} \left(\frac{H}{R}\right)_{out} \left(\frac{R - R_{in}}{R_{out} - R_{in}}\right)^{\beta},$$ (4)

where the parameters are the inner and outer disc radii, R_{in} and R_{out} in units of R(L1), the half thickness of the outer disc $(H/R)_{out}$ and the exponent β which describes the overall shape of the disc. The temperature structure of the un-irradiated disc is that of a steady state disc, in the absence of irradiation,

$$T_{disc}(R) = T_{out} \left(\frac{R}{R_{out}}\right)^{-\frac{3}{4}},$$ (5)

where T_{out} and T_{in} are the temperatures of the outer and inner disc respectively.

3.4   Irradiation Model

The effective temperature of a region at a distance R from the X-ray source, assumed in our model to be a point source located at the centre of the accretion disc, is found from the accretion luminosity for a typical LMXB,

$$T_x^4 = \frac{L_x(1-A)}{4\pi\sigma R^2} \tag{6}$$

and

$$L_x = \eta\frac{GM_x\dot{M}}{R_{ns}}, \tag{7}$$

where T_x is the temperature, A the albedo, η the efficiency, M_x, the mass of the compact object, $\dot{M}$ the accretion rate onto the compact object, R_{ns} is the size of the compact object and R is the distance between the compact object and the irradiated panel. This is normalised using the binary separation, a, the distance between the centres of mass of the stars, as is the coordinate system for the binary.

The irradiation of the binary takes place in three stages. The first stage is to calculate the temperature structure of the binary in the absence of any irradiation. This is done with characteristic temperatures for the donor star and the accretion stream and disc. The temperature structure of the disc is assumed to be that of an unirradiated disc as given in (5). The surface panels of the binary exposed to X-rays are determined by projecting the binary surfaces onto a spherical polar representation of the sky, as it appears from the X-ray source. Each triangular panel is mapped to the sky starting with the one furthest from the source and ending with the panel closest. Those panels remaining visible and unocculted on the sky map are irradiated. The change in effective temperature of a panel is scaled by the projected area with respect to the X-ray source at a distance R from the source. Hence the temperature after irradiation is given by,

$$T^4 = T_x^4\cos\theta_x\left(\frac{a}{R}\right)^2 + T_{eff}^4 \tag{8}$$

where T is the temperature of the panel, θ_x the angle between the line of sight from the central source and the normal to the surface of the panel and T_{eff} is the unirradiated effective temperature of the panel.

The second stage is to irradiate the binary with the constant component of the X-ray flux. This component of the X-ray flux is equated to the mean effective temperature of the X-ray source, as given in (8), where $T_x \equiv \overline{T_x}$. The third and final stage is to repeat stage two with $T_x \equiv T(t)_x$, which represents irradiating the binary with a time varying component. The difference between stages two and three represents the temperature change of the panels due to the time varying component of the X-ray flux alone, $\Delta f_x(t)$.

The irradiated regions of the binary can be clearly seen in Fig. 1, where the left hand panels show a typical X-ray binary, using binary parameters based on those of Scorpius X-1, viewed from an inclination of $60°$.

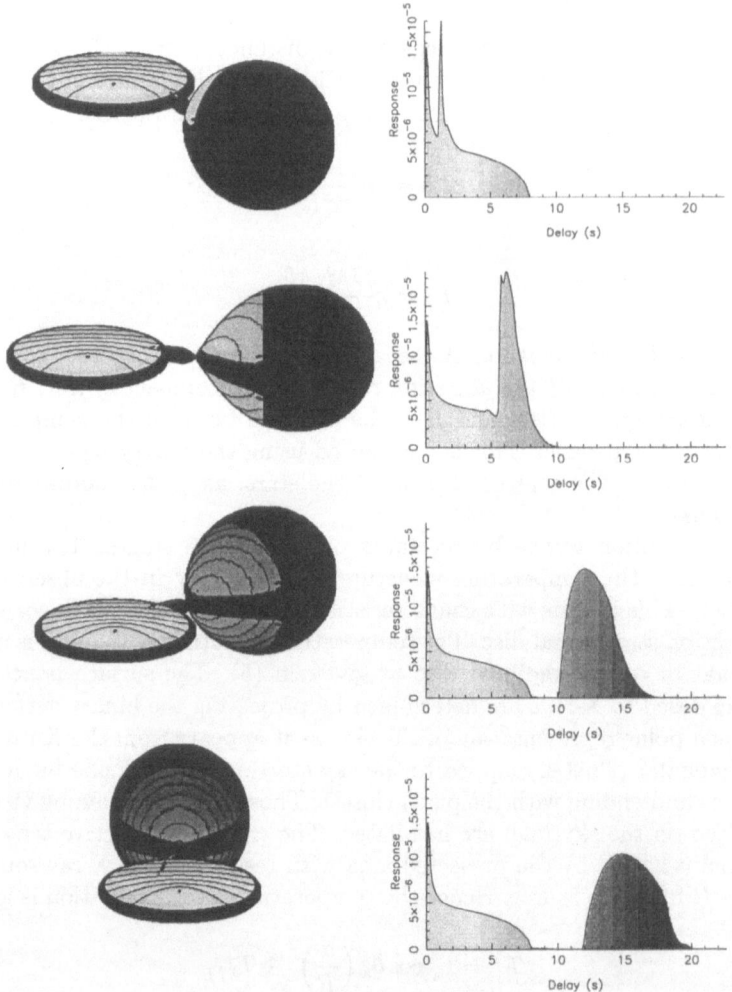

Fig. 1. Left, model X-ray binaries, based on the Scorpius X-1 binary parameters, showing iso-delay surfaces projected onto the irradiated surfaces of the binary. Right, the associated time delay transfer functions, showing the relative contributions from the regions highlighted in the model X-ray binaries.

The response of a panel to the variable component of the irradiating X-ray flux is given by,

$$I_\nu(\lambda, \boldsymbol{x}, \Delta f_{\mathrm{x}}(t)) = \int \left[B_\nu(\lambda, T_{\mathrm{x}}(t)) - B_\nu(\lambda, \overline{T_{\mathrm{x}}}) \right] P(\lambda) \, I(u, \alpha) \, d\Omega(\boldsymbol{x}, \phi) \, d\lambda, \quad (9)$$

where $P(\lambda)$ is the passband of the detector, $I(u, \alpha)$ is the amount of limb darkening for the panel, where u is the limb darkening coefficient and α is the angle between a vector normal to the panel and one pointing towards the observer and $d\Omega(\boldsymbol{x}, \phi)$ is the observability of a reprocessing region at $\boldsymbol{x}$, observed at binary phase, ϕ.

This response is substituted into the expression for the dynamic response given in (3). The resulting time delays are mapped onto a time delay grid to produce our time-delay transfer function. Examples of this transfer function can be seen in the right-hand panels of Fig. 1. The time delay of the donor star can be seen to change with binary phase in the individual images, whereas the accretion disc remains constant throughout the binary orbit. This effect can be seen more clearly in Fig. 2, where the individual transfer functions have been trailed in orbital phase to create an echo-phase diagram, showing clearly how the time delays of the individual components change with binary phase.

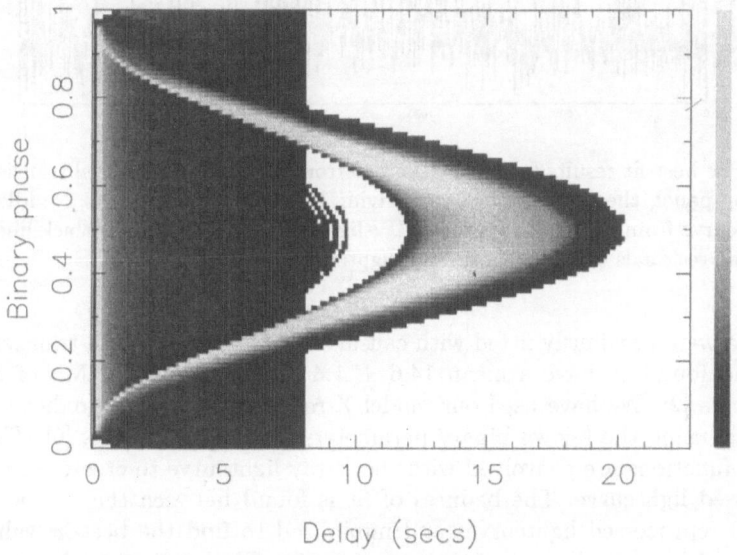

Fig. 2. A plot of time-delay transfer functions as a function of binary phase, based on the binary parameters of Scorpius X-1 [8,3]. The accretion disc has constant time delays in the region 0-8 seconds, whereas the time delays from the companion star are seen to vary sinusoidally with binary phase between 0-20 seconds. The greyscale represents the relative strength of the reprocessing.

4   Results for GRO J1655-40

We have used these model transfer functions to constrain the geometric parameters for the soft X-ray transient GRO J1655-40. The X-ray data was taken with the PCA onboard *RXTE* on June 8 1996, simultaneous with the *HST* data. These lightcurves are shown in the top and middle panels of Fig.3 respectively.

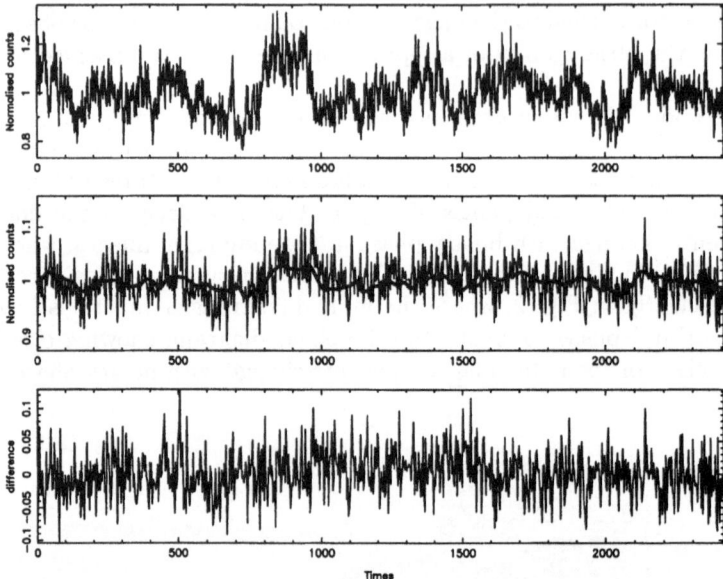

Fig. 3. The best-fit results for GRO J1655-40 from a grid-search of trial transfer functions. Top panel, the normalised X-ray driving lightcurve from RXTE. Middle panel, UV lightcurve from HST with synthetic UV lightcurve superimposed (thick line). Bottom panel, residuals of the fit to the UV lightcurve.

The data were previously fitted with causal and acausal Gaussian transfer functions and found to have a mean 14.6 ± 1.4 seconds, with a RMS of 10.5 ± 1.9 seconds [2]. We have used our model X-ray binary code to predict transfer functions, using the known binary parameters for GRO J1655-40, [6]. The trial transfer functions are convolved with the X-ray lightcurve to create a synthetic reprocessed lightcurve. The badness of fit is found between the synthetic and observed reprocessed lightcurves and minimized to find the best-fit values for the size, thickness and shape of the accretion disc. The synthetic lightcurve from the best-fit transfer function is shown by the thick line in the middle panel of Fig. 3, while the bottom panel shows the residuals of the fit. The best-fit model transfer function is shown in Fig. 4, together with the best-fit Gaussian transfer function from [2].

We find that the disc extends to 67% of the way to the inner Lagrangian point, is geometrically thick, with an opening angle ∼ 14°and is somewhat flared. This has the effect of shadowing the donor star from much of the irradiation.

5 Discussion

We have used the time delays observed between the X-ray and optical/UV variability in X-ray binaries to echo-map the irradiated regions. We have developed a code to simulate the time-delay transfer functions for such systems and find

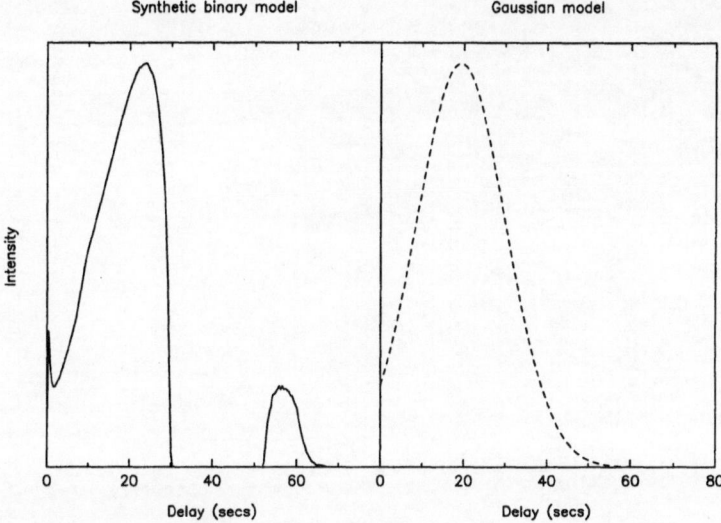

Synthetic binary model

Gaussian model

Fig. 4. A comparison of the best fit transfer functions for GRO J1655-40 from our two modeling methods. On the left is the synthetic X-ray binary model transfer function and on the right is the acausal Gaussian transfer function.

that, in the case of the SXT, GRO J1655-40, there is evidence for a geometrically thick outer accretion disc that shields the inner face of the donor star from irradiation.

While the method of echo-tomography of X-ray binaries is still in its infancy, we have shown that with just a small amount of data, from co-ordinated observing campaigns using ground-based and satellite observatories, this technique can reveal interesting insights into the geometry of X-ray binaries. Furthermore this technique has the promise of probing the structure and geometry of such systems on scales unobtainable with any other current technique.

References

1. Keith Horne, W. F. Welsh and B. M. Peterson, 1991, ApJL, 367, L5
2. R. I. Hynes, K. O'Brien, Keith Horne, W. Chen, and C. Haswell, 1998, MNRAS, 299, L37
3. T. Kallman, J. Raymond and S. Vrtilek, 1991, ApJ, 370, 717
4. J. H. Krolik, Keith Horne, T. R. Kallman, M. A. Malkan, R. A. Edelson and G. A. Kriss, 1991, ApJ, 371, 541
5. K. O'Brien, Keith Horne, R. I. Hynes, W. Chen, C. Haswell, M. Still, 2000, MNRAS, submitted
6. J. A. Orosz and C. D. Bailyn, 1997, ApJ, 477, 876
7. R. Rutten and V. Dhillon, 1994, A&A, 271, 793
8. S. Vrtilek, J. Penninx, J. Raymond, F. Verbunt, P. Hertz, K. Wood, W. Lewin and K. Mitsuda, 1991, ApJ, 376, 278

Object Index

Subject Index